INSTRUCTOR RESOURCE MANUAL WITH COMPLETE SOLUTIONS

Matthew Johll

Mark Ott

Introductory CHEMISTRY

SIXTH EDITION

NIVALDO J. TRO

330 Hudson Street, NY NY 10013

Courseware Portfolio Manager: Scott Dustan
Content Producer: Chandrika Madhavan
Managing Producer: Kristen Flathman
Courseware Editorial Assistant: Shercian Kinosian
Full-Service Vendor: codeMantra
Cover Art: Quade Paul
Cover Design: Seventeenth Street Studios
Manufacturing Buyer: Stacey Weinberger

Copyright © **2018, 2015, 2012** Pearson Education, Inc. All Rights Reserved. Printed in the United States of America. This publication is protected by copyright, and permission should be obtained from the publisher prior to any prohibited reproduction, storage in a retrieval system, or transmission in any form or by any means, electronic, mechanical, photocopying, recording, or otherwise. For information regarding permissions, request forms and the appropriate contacts within the Pearson Education Global Rights & Permissions department, please visit www.pearsoned.com/permissions/.

Unless otherwise indicated herein, any third-party trademarks that may appear in this work are the property of their respective owners and any references to third-party trademarks, logos or other trade dress are for demonstrative or descriptive purposes only. Such references are not intended to imply any sponsorship, endorsement, authorization, or promotion of Pearson's products by the owners of such marks, or any relationship between the owner and Pearson Education, Inc. or its affiliates, authors, licensees or distributors.

ISBN-10: 0-134-56405-7; ISBN-13: 978-0-134-56405-0

www.pearsonhighered.com

Full Solutions Manual
Table of Contents

Chapter 1	The Chemical World	1
Chapter 2	Measurement and Problem Solving	4
Chapter 3	Matter and Energy	23
Chapter 4	Atoms and Elements	39
Chapter 5	Molecules and Compounds	56
Chapter 6	Chemical Composition	71
Chapter 7	Chemical Reactions	99
Chapter 8	Quantities in Chemical Reactions	115
Chapter 9	Electrons in Atoms and the Periodic Table	138
Chapter 10	Chemical Bonding	152
Chapter 11	Gases	171
Chapter 12	Liquids, Solids, and Intermolecular Forces	197
Chapter 13	Solutions	212
Chapter 14	Acids and Bases	232
Chapter 15	Chemical Equilibrium	249
Chapter 16	Oxidation and Reduction	262
Chapter 17	Radioactivity and Nuclear Chemistry	282
Chapter 18	Organic Chemistry	293
Chapter 19	Biochemistry	315

The *Instructor's Resource Manual* appears after the *Instructor's Solutions Manual*.

Copyright © 2018 Pearson Education, Inc.

The Chemical World

Questions

1. Soda contains carbon dioxide molecules that are forced into a mixture with water molecules because of the increased pressure. Opening the can of soda will release the pressure, which allows the carbon dioxide to escape from the mixture and form bubbles.

2. Chemicals are what make up everything around us, including ourselves! Some examples of chemicals include sugar, carbon dioxide, and water, which make up soda.

3. Chemists study the world around us and try to explain how it works. By studying the interactions of atoms and molecules, chemists hope to better understand the world around us.

4. Because matter is made up of molecules, the behavior of matter is dictated by the behavior of the molecules. For example, water expands when it freezes because the distance between water molecules gets farther apart during the freezing process.

5. Chemistry is the science that seeks to understand what matter does by studying what atoms and molecules do.

6. Chemistry is connected to everyday life because molecules are interacting with you to produce all the experiences you have, from your dreaming while asleep to watching the stars at night. Chemistry is not confined to a laboratory unless you consider the entire universe a giant laboratory.

7. The scientific method is a way of approaching a problem that emphasizes making observations, planning experiments, performing experiments, and then using logic to come to a conclusion.

8. An example of the scientific method being used from this chapter is the discovery of what happens during combustion reactions, which requires proving phlogiston theory as incorrect.

9. A scientific law is a brief statement that summarizes previous observations and can be used to predict the results of future related experiments. The scientific law does not explain why the observations occur. That is the role of a scientific theory.

10. Before a scientific theory is well established, it is referred to as a hypothesis.

11. A scientific theory is the best current explanation of a phenomenon that has been validated by years' worth of experiments.

12. Antoine Lavoisier established the law of conservation of mass, which states, "In a chemical reaction, matter is neither created nor destroyed."

13. John Dalton proposed his atomic theory, which states that all matter is composed of small, indestructible particles called atoms.

14. The three things needed to succeed in chemistry are curiosity, calculation, and commitment.

Problems

15. a) observation
    b) theory
    c) law
    d) observation

16. a) observation
    b) law
    c) theory
    d) observation

17. Examination of the data table reveals that when the mass of the gas increases, a corresponding increase in the volume of the gas occurs. Two equivalent ways of expressing this information as a scientific law would be: (1) the mass of a gas is directly proportional to its volume or (2) the mass-to-volume ratio of a gas is constant.

18. Examination of the data table reveals that when the temperature of the gas increases, a corresponding increase in the volume of the gas occurs. Two equivalent ways of expressing this information as a scientific law would be: (1) the temperature of a gas is directly proportional to its volume or (2) the temperature-to-volume ratio of a gas is constant.

19. a) The reactivity of a chemical is proportional to the molecular size.
    b) Reactivity is due to the momentum of collisions, and because the larger molecules have a greater momentum, they have a greater reactivity.
    Remember this problem is for an imaginary universe!

20. a) Water is always made up of 1 part hydrogen to 8 parts oxygen by mass.
    b) Carbon dioxide is always made of 8 parts oxygen to 3 parts carbon by mass.
    c) A compound is always made of atoms of two or more elements combined in a specific mass ratio.
    d) A compound is made up of a specific ratio of atoms of each element.

21. The company is using consumer fear of "chemicals" to imply that their product is safe. The "no chemicals" could refer to additives, preservatives, pesticides, or insecticides, which are all groups of chemicals that many consumers do not want present in their products. A scientist would not find this statement correct because all matter is composed of chemicals.

22. Answers may vary; a representative group of answers: water, carbon dioxide, oxygen, carbon monoxide, copper, gold, aluminum, iron, carbohydrates, proteins.

23. Your answers must be consistent with the following definitions:
    Observation: measuring or observing some aspect of nature.
    Law: summarizing the results of a large number of observations.
    Hypothesis: tentative interpretations of a group of observations.
    Theory: a model that explains and gives underlying causes for observations and laws.

24. Answers vary depending on each group.

Data Interpretation and Analysis

25.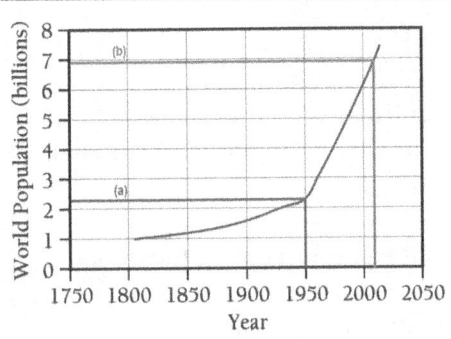
    a) 2.3 billion
    b) 6.9 billion
    c) 6.9 − 2.3 = 4.6 billion
    d) 4.6/60 = 0.077 billion/year
    e) 0.077 billion/year × 25 years = 1.9 billion
       6.9 billion + 1.9 billion = 8.8 billion

Measurement and Problem Solving 2

1. Units must be included with a number so that there is no doubt as to the meaning of the number. For example, if you reported a mass of 5, that would be unclear. However, if you reported a mass of 5 grams, there would be no confusion.

2. The number of digits reflects how well you know a value or, in other words, how precisely the value is known. The more precisely a number is known, the more digits you can use. All digits that are certain, plus one digit that contains some uncertainty, are reported.

3. Scientific notation is used to make very small numbers and very large numbers easier to write and to understand.

4. A measured quantity is written so that all but the very last digit are certain. The last digit is the only digit that contains uncertainty.

5. a) Zeros in the middle of two nonzero numbers ARE significant.
   b) Zeros at the end of a number with a decimal point ARE significant.
   c) Zeros at the beginning of a number are NOT significant.
   d) Zeros at the end of a number with no decimal point are ambiguous. Avoid using these numbers and use scientific notation instead.

6. Exact numbers have an infinite number of significant figures. That is, they contain no uncertainty. Exact numbers are:
   – measurements that can be counted, such as the number of students in class;
   – defined quantities (e.g., 1000 mL in 1 liter);
   – integers that are part of a mathematical equation (e.g., $4/3\pi r^2$).

7. The number of significant digits in multiplication and division problems is determined by whichever number contains the fewest significant digits.

8. The number of significant digits in addition and subtraction problems is determined by the number that has the fewest decimal places.

9. When calculations involve a combination of multiplication, division, addition, and subtraction, it is necessary to do the mathematical steps inside the parentheses first. Then, determine how many significant digits there should be during each step, but instead of rounding, underline the last significant digit. Keep all digits in the problem until it is completely finished; then round based on the significant figure rules.

10. Only consider the first digit that is going to be dropped; numbers beyond that are irrelevant. If that number is four or less, round down. If the number is five or higher, round up.

11. The SI unit for length is the meter (m), for mass the kilogram (kg), and for time the second (s).

12. The common units for volume are distance raised to the third power (e.g., m^3, cm^3). For liquids, the common units of volume that are used in science are the liter (L) and the milliliter (mL). (*Note*: 1 mL = 1 cm^3)

13. The Frisbee would be measured in meters with a prefix of deci (0.1). One might choose to use a prefix of centi (0.01), as it tends to be more commonly used.

14. The weight of an object is determined by the pull of gravity on that object. The mass of an object is a measure of the amount of matter that is contained in the object.

15. Correct answers may vary based on the marking division of the ruler.

16. Correct answers will vary.

17. Units can act as a guide in the calculation and are able to show if the calculation is off track. More importantly, the units provide important information about the final answer and must be included if the answer is to be correctly written and understood.

18. Units are treated exactly the same as numbers in calculations and can be multiplied, divided, and canceled.

19. A conversion factor is a fraction composed of two equivalent quantities and is used to convert information from one set of units to another.

20. The fundamental value of a quantity is not changed when it is multiplied by a conversion factor. The conversion factor consists of two values that are equivalent. For example, the length of the textbook was measured in inches but then converted to feet using the conversion factor: 1 ft = 12 in. The unit changed; however, the length of the textbook itself has not changed.

21. inches → feet   feet → inches

    $\dfrac{1 \text{ foot}}{12 \text{ inches}}$ $\dfrac{12 \text{ inches}}{1 \text{ foot}}$

Copyright © 2018 Pearson Education, Inc.

22. a) miles → kilometers

    $$\frac{1 \text{ kilometer}}{0.6214 \text{ mile}}$$

    b) kilometers → miles

    $$\frac{0.6214 \text{ mile}}{1 \text{ kilometer}}$$

    c) gallons → liters

    $$\frac{3.785 \text{ liters}}{1 \text{ gallon}}$$

    d) liters → gallons

    $$\frac{1 \text{ gallon}}{3.785 \text{ liters}}$$

23. Sort: Sort the information into the *given* information and what the problem is asking you to *find*.
    Strategize: You formulate a series of steps that form a solution map.
    Solve: Carry out the mathematical operations set up by the solution map.
    Check: Verify that the answer makes sense and has the correct units.

24. In multiple-step problems, it is easy for errors to creep into the solution. By checking your answer and making sure the value makes sense and has the correct units, you can catch most errors.

25. grams → pounds

    $$\frac{1 \text{ lb}}{453.6 \text{ g}}$$

26. milliliters → liters → quarts → gallons

    $$\frac{1 \text{ L}}{1000 \text{ mL}} \quad \frac{1.057 \text{ qt}}{1 \text{ L}} \quad \frac{1 \text{ gal}}{4 \text{ qt}}$$

27. meters → centimeters → inches → feet

    $$\frac{100 \text{ cm}}{1 \text{ m}} \quad \frac{1 \text{ in}}{2.54 \text{ cm}} \quad \frac{1 \text{ ft}}{12 \text{ in}}$$

28. ounces → pounds → grams

    $$\frac{1 \text{ lb}}{16 \text{ oz}} \quad \frac{453.6 \text{ g}}{1 \text{ lb}}$$

29. Density is the ratio of the mass of an object to the volume of the object. Objects that have a large density are perceived to be heavy. Objects that have a small density are perceived to be light. Density can work as a conversion factor because it is an equality that converts between mass and volume. For example, 1 gram of water will occupy a volume of 1 mL. Therefore, if you have a mass of 100 grams, you can use the density to calculate the volume.

30. The density of an object can be calculated by determining both the mass and the volume of an object and then dividing the mass by the volume. The solution map is:
mass, volume → density
$$d = \frac{m}{V}$$

Problems

Scientific Notation

31. a) 3.8802×10^7
    b) 1.419×10^6
    c) 1.9746×10^7
    d) 5.84×10^5

32. a) 7.376×10^9
    b) 1.404×10^9
    c) 1.1258×10^7
    d) 4.677×10^6

33. a) 7.461×10^{-11} m
    b) 1.58×10^{-5} mi
    c) 6.32×10^{-7} m
    d) 1.5×10^{-5} m

34. a) 1×10^{-9} s
    b) 1.43×10^{-1} s
    c) 1×10^{-12} s
    d) 1×10^{-6} m

35. a) 602,200,000,000,000,000,000,000 atoms
    b) 0.00000000000000000016 C
    c) 299,000,000 m/s
    d) 344 m/s

36. a) 0.000000450 m
    b) 13,700,000,000 yrs
    c) 5,000,000,000 yrs
    d) 50 yrs

37. a) 32,200,000
    b) 0.0072
    c) 118,000,000,000
    d) 0.00000943

38. a) 1,300,000
    b) 0.00011
    c) 190
    d) 0.000000000741

39. 2,000,000,000 $\underline{2 \times 10^9}$

    $\underline{1,211,000,000}$ 1.211×10^9

    0.000874 $\underline{8.74 \times 10^{-4}}$

    $\underline{320,000,000,000}$ 3.2×10^{11}

40. $\underline{0.0042}$ 4.2×10^{-3}

    315,171,000 $\underline{3.15171 \times 10^8}$

    $\underline{0.000000000018}$ 1.8×10^{-11}

    1,232,000 $\underline{1.232 \times 10^6}$

Significant Figures

41. a) 54.3 mL
    b) 48.7 °C
    c) 46.8 °C
    d) 64 mL

42. a) 5.55 mL
    b) 55.9 mL
    c) 7.18 mL
    d) 4.144 g

43. a) 0.00̶5̶0̶5̶0̶ m (with 0.00 crossed out, 5050 underlined)
    b) 0.000000000000060 s (leading zeros crossed out, 60 underlined)
    c) 220,103 kg (0 and 103 underlined)
    d) 0.00̶1̶0̶8̶ in (0.00 crossed out, 108 underlined)

44. a) 0.00010320 s (0.000 crossed out, 10320 underlined)
    b) 1322600324 kg (00 underlined)
    c) 0.0001240 in (0.000 crossed out, 1240 underlined)
    d) 0.02061 m (0.0 crossed out, 2061 underlined)

45. a) 4
    b) 4
    c) 6
    d) 5

46. a) 5
    b) 5
    c) 3
    d) 2

47. | Number | Significant Figures | |
| --- | --- | --- |
| a) 895675 | 6 | correct |
| b) 0.000869 | 6 | incorrect, 3: leading zeros are not significant |
| c) 0.5672100 | 5 | incorrect, 7: terminal zeros are significant |
| d) 6.022×10^{23} | 4 | correct |

48. | Number | Significant Figures | |
| --- | --- | --- |
| a) 24 | 2 | correct |
| b) 5.6×10^{-17} | 3 | incorrect, 2: all numbers in scientific notation are significant |
| c) 3.14 | 3 | correct |
| d) 0.00383 | 5 | incorrect, 3: leading zeros are not significant |

Rounding

49. a) 256.0
    b) 0.0004893
    c) 2.901×10^{-4}
    d) 2.231×10^{6}

50. a) 1.08×10^{4}
    b) 5.00×10^{6}
    c) 1.35
    d) 0.0000345

51. a) 2.3
    b) 2.4
    c) 2.3
    d) 2.4

52. a) 65.7
    b) 65.7
    c) 65.8
    d) 65.8

53. a) incorrect, 42.3
    b) correct
    c) correct
    d) incorrect, 0.0456

54. a) incorrect, 1.2×10^3
    b) incorrect, 4.0×10^2
    c) incorrect, 56
    d) correct

55.

Number	Rounded to 4 significant figures	Rounded to 2 significant figures	Rounded to 1 significant figure
1.45815	1.458	1.5	1
8.32466	8.325	8.3	8
84.57225	84.57	85	8×10^1
132.5512	132.6	1.3×10^2	1×10^2

56.

Number	Rounded to 4 significant figures	Rounded to 2 significant figures	Rounded to 1 significant figure
94.52118	94.52	95	9×10^1
105.4545	105.5	1.1×10^2	1×10^2
0.455981	0.4560	0.46	0.5
0.009999991	0.01000	0.010	0.01

Significant Figures in Calculations

57. a) 0.054
    b) 0.619
    c) 1.2×10^8
    d) 6.6

58. a) 6×10^2
    b) 6.4×10^2
    c) 2
    d) 223

59. a) incorrect, 4.22×10^3
    b) correct
    c) incorrect, 3.9969
    d) correct

60. a) correct
    b) correct
    c) incorrect, 0.055098
    d) incorrect, 54.61

61. a) 110.6
    b) 41.4
    c) 183.3
    d) 1.22

62. a) 1473.4
    b) 0.103
    c) 411
    d) 4.2

63. a) correct
    b) incorrect, 1.0982
    c) correct
    d) incorrect, 3.53

64. a) incorrect, 9.46×10^2
    b) incorrect, 5899
    c) correct
    d) incorrect, 0.002

65. a) 3.9×10^3
    b) 632
    c) 8.93×10^4
    d) 6.34

66. a) 13.8
    b) 64
    c) 7.7×10^{10}
    d) 5.31

67. a) incorrect, 3.15×10^3
    b) correct
    c) correct
    d) correct

68. a) incorrect, 0.0257
    b) incorrect, 1.1×10^5
    c) correct
    d) correct

Unit Conversions

69. a) $3.55 \text{ kg} \times \dfrac{1000 \text{ g}}{1 \text{ kg}} = 3.55 \times 10^3 \text{ g}$

    b) $8944 \text{ mm} \times \dfrac{1 \text{ m}}{1000 \text{ mm}} = 8.944 \text{ m}$

    c) $4598 \text{ mg} \times \dfrac{1 \text{ g}}{1000 \text{ mg}} \times \dfrac{1 \text{ kg}}{1000 \text{ g}} = 4.598 \times 10^{-3} \text{ kg}$

    d) $0.0187 \text{ L} \times \dfrac{1000 \text{ mL}}{1 \text{ L}} = 18.7 \text{ mL}$

70. a) $155.5 \text{ cm} \times \dfrac{1 \text{ m}}{100 \text{ cm}} = 1.555 \text{ m}$

    b) $2491.6 \text{ g} \times \dfrac{1 \text{ kg}}{1000 \text{ g}} = 2.4916 \text{ kg}$

    c) $248 \text{ cm} \times \dfrac{10 \text{ mm}}{1 \text{ cm}} = 2.48 \times 10^3 \text{ mm}$

    d) $6781 \text{ mL} \times \dfrac{1 \text{ L}}{1000 \text{ mL}} = 6.781 \text{ L}$

71. a) $5.88 \text{ dL} \times \dfrac{1 \text{ L}}{10 \text{ dL}} = 0.588 \text{ L}$

    b) $3.41 \times 10^{-5} \text{ g} \times \dfrac{1{,}000{,}000 \text{ µg}}{1 \text{ g}} = 34.1 \text{ µg}$

    c) $1.01 \times 10^{-8} \text{ s} \times \dfrac{1 \times 10^9 \text{ ns}}{1 \text{ s}} = 10.1 \text{ ns}$

    d) $2.19 \text{ pm} \times \dfrac{1 \text{ m}}{1 \times 10^{12} \text{ pm}} = 2.19 \times 10^{-12} \text{ m}$

72. a) $1.08 \text{ Mm} \times \dfrac{1 \times 10^6 \text{ m}}{1 \text{ Mm}} \times \dfrac{1 \text{ km}}{1000 \text{ m}} = 1.08 \times 10^3 \text{ km}$

    b) $4.88 \text{ fs} \times \dfrac{1 \times 10^{-15} \text{ s}}{1 \text{ fs}} \times \dfrac{1 \times 10^{12} \text{ ps}}{1 \text{ s}} = 4.88 \times 10^{-3} \text{ s}$

    c) $7.39 \times 10^{11} \text{ m} \times \dfrac{1 \text{ Gm}}{1 \times 10^9 \text{ m}} = 739 \text{ Gm}$

    d) $1.15 \times 10^{-10} \text{ m} \times \dfrac{1 \text{ pm}}{1 \times 10^{-12} \text{ m}} = 115 \text{ pm}$

73. a) $22.5 \text{ in} \times \dfrac{2.54 \text{ cm}}{1 \text{ in}} = 57.2 \text{ cm}$

b) $126 \text{ ft} \times \dfrac{12 \text{ in}}{1 \text{ ft}} \times \dfrac{2.54 \text{ cm}}{1 \text{ in}} \times \dfrac{1 \text{ m}}{100 \text{ cm}} = 38.4 \text{ m}$

c) $825 \text{ yd} \times \dfrac{1 \text{ m}}{1.094 \text{ yd}} \times \dfrac{1 \text{ km}}{1000 \text{ m}} = 0.754 \text{ km}$

d) $2.4 \text{ in} \times \dfrac{2.54 \text{ cm}}{1 \text{ in}} \times \dfrac{10 \text{ mm}}{1 \text{ cm}} = 61 \text{ mm}$

74. a) $78.3 \text{ in} \times \dfrac{2.54 \text{ cm}}{1 \text{ in}} = 199 \text{ cm}$

b) $445 \text{ yd} \times \dfrac{1 \text{ m}}{1.094 \text{ yd}} = 407 \text{ m}$

c) $336 \text{ ft} \times \dfrac{30.48 \text{ cm}}{1 \text{ ft}} = 1.02 \times 10^4 \text{ cm}$

d) $45.3 \text{ in} \times \dfrac{2.54 \text{ cm}}{1 \text{ in}} \times \dfrac{10 \text{ mm}}{1 \text{ cm}} = 1.15 \times 10^3 \text{ mm}$

75. a) $40.0 \text{ cm} \times \dfrac{1 \text{ in}}{2.54 \text{ cm}} = 15.7 \text{ in}$

b) $27.8 \text{ m} \times \dfrac{39.37 \text{ in}}{1 \text{ m}} \times \dfrac{1 \text{ ft}}{12 \text{ in}} = 91.2 \text{ ft}$

c) $10.0 \text{ km} \times \dfrac{0.6214 \text{ mi}}{1 \text{ km}} = 6.21 \text{ mi}$

d) $3845 \text{ kg} \times \dfrac{2.205 \text{ lb}}{1 \text{ kg}} = 8478 \text{ lb}$

76. a) $254 \text{ cm} \times \dfrac{1 \text{ in}}{2.54 \text{ cm}} = 1.00 \times 10^2 \text{ in}$

b) $89 \text{ mm} \times \dfrac{1 \text{ cm}}{10 \text{ mm}} \times \dfrac{1 \text{ in}}{2.54 \text{ cm}} = 3.5 \text{ in}$

c) $7.5 \text{ L} \times \dfrac{1.057 \text{ qt}}{1 \text{ L}} = 7.9 \text{ qt}$

d) $122 \text{ kg} \times \dfrac{2.205 \text{ lb}}{1 \text{ kg}} = 269 \text{ lb}$

77.

	m	km	Mm	Gm	Tm
	5.08×10^8 m	5.08×10^5 km	508 Mm	0.508 Gm	5.08×10^{-4} Tm
	2.7976×10^{10} m	2.7976×10^7 km	27,976 Mm	27.976 Gm	2.7976×10^{-1} Tm
	1.77×10^{12} m	1.77×10^9 km	1.77×10^6 Mm	1.77×10^3 Gm	1.77 Tm
	1.5×10^8 m	1.5×10^5 km	1.5×10^2 Mm	0.15 Gm	1.5×10^{-4} Tm
	4.23×10^{11} m	4.23×10^8 km	4.23×10^5 Mm	423 Gm	0.423 Tm

78.

	s	ms	μs	ns	ps
	1.31×10^{-4} s	0.131 ms	131 μs	1.31×10^5 ns	1.31×10^8 ps
	1.26×10^{-11} s	1.26×10^{-8} ms	1.26×10^{-5} μs	1.26×10^{-2} ns	12.6 ps
	1.55×10^{-7} s	1.55×10^{-4} ms	0.155 μs	155 ns	1.55×10^5 ps
	1.99×10^{-6} s	1.99×10^{-3} ms	1.99 μs	1.99×10^3 ns	1.99×10^6 ps
	8.66×10^{-11} s	8.66×10^{-8} ms	8.66×10^{-5} μs	8.66×10^{-2} ns	86.6 ps

79. a) 2.255×10^{10} g $\times \dfrac{1 \text{ kg}}{1000 \text{ g}} = 2.255 \times 10^7$ kg

b) 2.255×10^{10} g $\times \dfrac{1 \text{ Mg}}{1 \times 10^6 \text{ g}} = 2.255 \times 10^4$ Mg

c) 2.255×10^{10} g $\times \dfrac{1000 \text{ mg}}{1 \text{ g}} = 2.255 \times 10^{13}$ mg

d) 2.255×10^{10} g $\times \dfrac{1 \text{ kg}}{1000 \text{ g}} \times \dfrac{1 \text{ metric ton}}{1000 \text{ kg}} = 2.255 \times 10^4$ metric tons

80. a) 1.88×10^{-6} g $\times \dfrac{1 \times 10^3 \text{ mg}}{\text{g}} = 1.88 \times 10^{-3}$ mg

b) 1.88×10^{-6} g $\times \dfrac{1 \times 10^2 \text{ cg}}{\text{g}} = 1.88 \times 10^{-4}$ cg

c) 1.88×10^{-6} g $\times \dfrac{1 \times 10^9 \text{ ng}}{\text{g}} = 1.88 \times 10^3$ ng

d) 1.88×10^{-6} g $\times \dfrac{1 \times 10^6 \text{ μg}}{\text{g}} = 1.88$ μg

81. $3.3 \text{ lb} \times \dfrac{1 \text{ kg}}{2.205 \text{ lb}} \times \dfrac{1000 \text{ g}}{1 \text{ kg}} = 1.5 \times 10^3 \text{ g}$

82. $1.9 \text{ lb} \times \dfrac{1 \text{ kg}}{2.205 \text{ lb}} \times \dfrac{1000 \text{ g}}{1 \text{ kg}} = 8.6 \times 10^2 \text{ g}$

83. $10.0 \text{ km} \times \dfrac{0.6214 \text{ mi}}{1 \text{ km}} \times \dfrac{1 \text{ hr}}{7.5 \text{ mi}} \times \dfrac{60 \text{ min}}{1 \text{ hr}} = 5.0 \times 10^1 \text{ min}$

84. $195 \text{ km} \times \dfrac{0.6214 \text{ mi}}{1 \text{ km}} \times \dfrac{1 \text{ hr}}{24 \text{ mi}} = 5.0 \text{ hr}$

85. $5.0 \text{ qt} \times \dfrac{1 \text{ L}}{1.057 \text{ qt}} \times \dfrac{1000 \text{ cm}^3}{1 \text{ L}} = 4.7 \times 10^3 \text{ cm}^3$

86. $2.0 \text{ gal} \times \dfrac{3.785 \text{ L}}{1 \text{ gal}} \times \dfrac{1000 \text{ cm}^3}{1 \text{ L}} = 7.6 \times 10^3 \text{ cm}^3$

Units Raised to a Power

87. a) $1.0 \text{ km}^2 \times \dfrac{(1000 \text{ m})^2}{(1 \text{ km})^2} = 1.0 \times 10^6 \text{ m}^2$

    b) $1.0 \text{ cm}^3 \times \dfrac{(1 \text{ m})^3}{(100 \text{ cm})^3} = 1.0 \times 10^{-6} \text{ m}^3$

    c) $1.0 \text{ mm}^3 \times \dfrac{(1 \text{ m})^3}{(1000 \text{ mm})^3} = 1.0 \times 10^{-9} \text{ m}^3$

88. a) $1.0 \text{ ft}^2 \times \dfrac{(12 \text{ in})^2}{(1 \text{ ft})^2} = 1.4 \times 10^2 \text{ in}^2$

    b) $1.0 \text{ yd}^2 \times \dfrac{(3 \text{ ft})^2}{(1 \text{ yd})^2} = 9.0 \text{ ft}^2$

    c) $1.0 \text{ m}^2 \times \dfrac{(1.094 \text{ yd})^2}{(1 \text{ m})^2} = 1.2 \text{ yd}^2$

89. a) $6.2 \times 10^{-31} \text{ m}^3 \times \dfrac{(1 \text{ pm})^3}{(1 \times 10^{-12} \text{ m})^3} = 6.2 \times 10^5 \text{ pm}^3$

b) $6.2 \times 10^{-31} \text{ m}^3 \times \dfrac{(1 \text{ nm})^3}{(1 \times 10^{-9} \text{ m})^3} = 6.2 \times 10^{-4} \text{ nm}^3$

c) $6.2 \times 10^{-31} \text{ m}^3 \times \dfrac{(1 \text{ Å})^3}{(1 \times 10^{-10} \text{ m})^3} = 6.2 \times 10^{-1} \text{ Å}^3$

90. a) $197 \times 10^6 \text{ mi}^2 \times \dfrac{(1 \text{ km})^2}{(0.6214 \text{ mi})^2} = 5.10 \times 10^8 \text{ km}^2$

b) $197 \times 10^6 \text{ mi}^2 \times \dfrac{(1 \text{ km})^2}{(0.6214 \text{ mi})^2} \times \dfrac{(1000 \text{ m})^2}{(1 \text{ km})^2} \times \dfrac{(1 \text{ Mm})^2}{(1 \times 10^6 \text{ m})^2} = 5.10 \times 10^2 \text{ Mm}^2$

c) $197 \times 10^6 \text{ mi}^2 \times \dfrac{(1 \text{ km})^2}{(0.6214 \text{ mi})^2} \times \dfrac{(1000 \text{ m})^2}{(1 \text{ km})^2} \times \dfrac{(10 \text{ dm})^2}{(1 \text{ m})^2} = 5.10 \times 10^{16} \text{ dm}^2$

91. a) $215 \text{ m}^2 \times \dfrac{(1 \text{ km})^2}{(1000 \text{ m})^2} = 2.15 \times 10^{-4} \text{ km}^2$

b) $215 \text{ m}^2 \times \dfrac{(10 \text{ dm})^2}{(1 \text{ m})^2} = 2.15 \times 10^4 \text{ dm}^2$

c) $215 \text{ m}^2 \times \dfrac{(100 \text{ cm})^2}{(1 \text{ m})^2} = 2.15 \times 10^6 \text{ cm}^2$

92. a) $285 \text{ m}^3 \times \dfrac{(1 \text{ km})^3}{(1000 \text{ m})^3} = 2.85 \times 10^{-7} \text{ km}^3$

b) $285 \text{ m}^3 \times \dfrac{(10 \text{ dm})^3}{(1 \text{ m})^3} = 2.85 \times 10^5 \text{ dm}^3$

c) $285 \text{ m}^3 \times \dfrac{(100 \text{ cm})^3}{(1 \text{ m})^3} = 2.85 \times 10^8 \text{ cm}^3$

93. 954 million acres = 954,000,000 acres = 9.54×10^8 acres;

$9.54 \times 10^8 \text{ acres} \times \dfrac{43{,}560 \text{ ft}^2}{1 \text{ acre}} \times \dfrac{(1 \text{ mi})^2}{(5280 \text{ ft})^2} = 1.49 \times 10^6 \text{ mi}^2$

94. $435 \text{ acres} \times \dfrac{43{,}560 \text{ ft}^2}{1 \text{ acre}} \times \dfrac{(1 \text{ mi})^2}{(5280 \text{ ft})^2} = 0.680 \text{ mi}^2$

Unit Conversion in Both the Numerator and Denominator

95. $65 \dfrac{\cancel{mi}}{\cancel{hr}} \times \dfrac{1.609 \text{ km}}{1 \cancel{mi}} \times \dfrac{24 \cancel{hr}}{1 \text{ day}} = 2.5 \times 10^3 \dfrac{\text{km}}{\text{day}}$

96. $65 \dfrac{\cancel{mi}}{\cancel{hr}} \times 5280 \dfrac{\text{ft}}{\cancel{mi}} \times \dfrac{1 \cancel{hr}}{60 \cancel{min}} \times \dfrac{1 \cancel{min}}{60 \text{ s}} = 95 \dfrac{\text{ft}}{\text{s}}$

97. $65 \dfrac{\cancel{mi}}{\cancel{hr}} \times \dfrac{1609 \text{ m}}{1 \cancel{mi}} \times \dfrac{1 \cancel{hr}}{60 \cancel{min}} \times \dfrac{1 \cancel{min}}{60 \text{ s}} = 29 \dfrac{\text{m}}{\text{s}}$

98. $65 \dfrac{\cancel{mi}}{\cancel{hr}} \times \dfrac{5280 \cancel{ft}}{1 \cancel{mi}} \times \dfrac{1 \text{ yd}}{3 \cancel{ft}} \times \dfrac{1 \cancel{hr}}{60 \text{ min}} = 1.9 \times 10^3 \dfrac{\text{yd}}{\text{min}}$

Density

99. $d = \dfrac{35.4 \text{ g}}{3.11 \text{ cm}^3} = 11.4 \text{ g/cm}^3$; This matches the density of lead.

100. $d = \dfrac{2.49 \text{ g}}{0.349 \text{ cm}^3} = 7.13 \text{ g/cm}^3$; No, the density does not match copper.

101. $d = \dfrac{3.15 \times 10^3 \text{ g}}{2.50 \text{ L}} \times \dfrac{1 \text{ L}}{1000 \text{ mL}} \times \dfrac{1 \text{ mL}}{1 \text{ cm}^3} = 1.26 \text{ g/cm}^3$

102. $d = \dfrac{12.88 \text{ kg}}{4.77 \text{ L}} \times \dfrac{1 \text{ L}}{1000 \text{ cm}^3} \times \dfrac{1000 \text{ g}}{1 \text{ kg}} = 2.70 \text{ g/cm}^3$

103. $d = \dfrac{206 \text{ g}}{10.7 \text{ mL}} \times \dfrac{1 \text{ mL}}{1 \text{ cm}^3} = 19.3 \text{ g/cm}^3$
Yes, the density matches gold.

104. $d = \dfrac{157 \text{ g}}{18.65 \text{ mL}} \times \dfrac{1 \text{ mL}}{1 \text{ cm}^3} = 8.42 \text{ g/cm}^3$;
No, the density does not match platinum.

105. a) $387 \text{ mL} \times \dfrac{1 \text{ cm}^3}{1 \text{ mL}} \times \dfrac{1.11 \text{ g}}{1 \text{ cm}^3} = 4.30 \times 10^2 \text{ g}$

 b) $3.46 \text{ kg} \times \dfrac{1000 \text{ g}}{1 \text{ kg}} \times \dfrac{1 \text{ cm}^3}{1.11 \text{ g}} \times \dfrac{1 \text{ mL}}{1 \text{ cm}^3} \times \dfrac{1 \text{ L}}{1000 \text{ mL}} = 3.12 \text{ L}$

106. a) $17.56 \text{ mL} \times \dfrac{1 \text{ cm}^3}{1 \text{ mL}} \times \dfrac{0.7857 \text{ g}}{1 \text{ cm}^3} = 13.80 \text{ g}$

b) $7.22 \text{ g} \times \dfrac{1 \text{ cm}^3}{0.7857 \text{ g}} \times \dfrac{1 \text{ mL}}{1 \text{ cm}^3} = 9.19 \text{ mL}$

Cumulative Problems

107. a) $d = \dfrac{m}{V}$ and $m = d \times V$

$m_{gold} = 19.3 \dfrac{g}{cm^3} \times 1.75 \text{ L} \times \dfrac{1000 \text{ cm}^3}{1 \text{ L}} = 3.38 \times 10^4 \text{ g}$

$m_{sand} = 3.00 \dfrac{g}{cm^3} \times 1.75 \text{ L} \times \dfrac{1000 \text{ cm}^3}{1 \text{ L}} = 5.25 \times 10^3 \text{ g}$

b) Yes, the thief set off the alarm because the sand was much lighter than the gold vase.

108. $V = (4/3)\pi r^3 = (4/3)(3.14)(1.0 \times 10^{-13})^3 = 4.187 \times 10^{-39} \text{ cm}^3$

$d = \dfrac{m}{V} = \dfrac{1.7 \times 10^{-24} \text{ g}}{4.187 \times 10^{-39} \text{ cm}^3} = 4.1 \times 10^{14} \text{ g/cm}^3$

109. $d = \dfrac{m}{V} = \dfrac{5.14 \text{ lb}}{13.4 \text{ in}^3} \times \dfrac{453.6 \text{ g}}{1 \text{ lb.}} \times \left(\dfrac{1 \text{ in}}{2.54 \text{ cm}}\right)^3 = 10.6 \text{ g/cm}^3$

110. $d = \dfrac{19.8 \text{ lb}}{2.7 \text{ gal}} \times \dfrac{1 \text{ gal}}{3.785 \text{ L}} \times \dfrac{1 \text{ L}}{1000 \text{ cm}^3} \times \dfrac{453.6 \text{ g}}{1 \text{ lb}} = 0.88 \text{ g/cm}^3$

The log must be made of oak (d = 0.9 g/cm^3).

111. $d = 2.7 \dfrac{g}{cm^3} \times \dfrac{1 \text{ kg}}{1000 \text{ g}} \times \dfrac{(100 \text{ cm})^3}{(1 \text{ m})^3} = 2.7 \times 10^3 \text{ kg/cm}^3$

112. $d = 21.4 \dfrac{g}{cm^3} \times \dfrac{1 \text{ lb}}{453.59 \text{ g}} \times \dfrac{(2.54 \text{ cm})^3}{(1 \text{ in})^3} = 0.773 \text{ lb/in}^3$

113. $150 \text{ yd}^3 \times \dfrac{(1 \text{ m})^3}{(1.094 \text{ yd})^3} \times \dfrac{(100 \text{ cm})^3}{(1 \text{ m})^3} \times \dfrac{1.0 \text{ g}}{1 \text{ cm}^3} \times \dfrac{1 \text{ lb}}{453.6 \text{ g}} = 2.5 \times 10^5 \text{ lb}$

114. $8975 \text{ ft}^3 \times \dfrac{(30.48 \text{ cm})^3}{(1 \text{ ft})^3} \times \dfrac{0.92 \text{ g}}{1 \text{ cm}^3} \times \dfrac{1 \text{ kg}}{1000 \text{ g}} = 2.3 \times 10^5 \text{ kg}$

115. $155{,}211 \text{ L} \times \dfrac{1000 \text{ cm}^3}{1 \text{ L}} \times \dfrac{0.768 \text{ g}}{1 \text{ cm}^3} \times \dfrac{1 \text{ kg}}{1000 \text{ g}} = 1.19 \times 10^5 \text{ kg}$

116. $2.5 \text{ L} \times \dfrac{1000 \text{ cm}^3}{1 \text{ L}} \times \dfrac{0.79 \text{ g}}{1 \text{ cm}^3} \times \dfrac{1 \text{ lb}}{453.6 \text{ g}} = 4.4 \text{ lb}$

117. $\dfrac{43 \text{ mi}}{1 \text{ gal}} \times \dfrac{1 \text{ gal}}{3.785 \text{ L}} \times \dfrac{1 \text{ km}}{0.6214 \text{ mi}} = 18 \text{ km/L}$

118. $\dfrac{12.8 \text{ km}}{1 \text{ L}} \times \dfrac{3.785 \text{ L}}{1 \text{ gal}} \times \dfrac{0.6214 \text{ mi}}{1 \text{ km}} = 30.1 \text{ mi/gal}$

119. $76.5 \text{ L} \times \dfrac{1 \text{ gal}}{3.785 \text{ L}} \times \dfrac{38 \text{ mi}}{1 \text{ gal}} = 7.7 \times 10^2 \text{ mi}$

120. $22.5 \text{ gal} \times \dfrac{3.785 \text{ L}}{1 \text{ gal}} \times \dfrac{12.8 \text{ km}}{1 \text{ L}} \times \dfrac{0.6214 \text{ mi}}{1 \text{ km}} = 677 \text{ miles}$

121. Because block A has the larger mass and the lesser volume, and the density is the ratio of mass divided by volume, the density of block A will be greater than the density of block B.

122. Because block A has the larger mass and the larger volume, when the ratio of mass to volume is calculated for the density, the relative densities of the two blocks may vary depending on the mass of the blocks. If the mass of block A = 1.19 × mass block B, the density of block A and B is the same; if the mass of block A < 1.19 × mass block B, then the density of block A is less than that of block B. Finally, if the mass of block A is >1.19 × mass of block B, then the density of block A is greater than that of block B. The value of 1.19 came from the ratio of the two volumes.

123. Mass of cylinder 1 = 1.35 × Mass cylinder 2

   Volume of cylinder 1 = 0.792 × Volume cylinder 2

   Density of cylinder 1 = 3.85 g/cm^3

   Density of cylinder 2 = ?

   $D_1 = \dfrac{M_1}{V_1} = \dfrac{1.35 \times M_2}{0.792 \times V_2} = 3.85 \text{ g/cm}^3$

   $\dfrac{M_2}{V_2} = \dfrac{0.792}{1.35} \times 3.85 \text{ g/cm}^3 = 2.26 \text{ g/cm}^3$

124. $D_{ave} = 9.87$ g/cm³
$D_{Cu} = 8.96$ g/cm³
$D_{Pb} = 11.4$ g/cm³
Let x = fraction of Cu, then 1.0 – x = fraction Pb.
Set the average density equal to the sum of each element's density multiplied by its fraction.
9.87 g/cm³ = (8.96 g/cm³)(x) + (11.4 g/cm³) × (1.0 – x)
9.87 –11.4 = 8.96x –11.4x
–1.5 = –2.4x
x = –1.5/–2.4 = 0.63 = fraction of Cu (or 63%)
Fraction of Pb = 1.0 – 0.63 = 0.37 (or 37%)

Highlight Problems

125. $D = \dfrac{m}{V} \Rightarrow V = \dfrac{m}{D}$

$V = \dfrac{25.8 \text{ g}}{2.7 \text{ g/cm}^3} = 9.6 \text{ cm}^3$

$V = \dfrac{4}{3}\pi r^3 \Rightarrow r = \sqrt[3]{\dfrac{3V}{4\pi}}$

$r = \sqrt[3]{\dfrac{3(9.6 \text{ cm}^3)}{4\pi}} = \sqrt[3]{2.3} = 1.3 \text{ cm}^3$

126. $D = \dfrac{m}{V} \Rightarrow V = \dfrac{m}{D}$

$V = \dfrac{87.2 \text{ g}}{8.96 \text{ g/cm}^3} = 9.73 \text{ cm}^3$

$V = L^3 \Rightarrow L = \sqrt[3]{V}$

$L = \sqrt[3]{9.73 \text{ cm}^3} = 2.14 \text{ cm}$

127. $1.55 \times 10^5 \text{ ft} \times \dfrac{0.3048 \text{ m}}{1 \text{ ft}} \times \dfrac{1 \text{ km}}{1000 \text{ m}} = 47.2 \text{ km}$

155 km – 47.2 = 108 km difference in altitude.
The orbiter would have attempted to establish an orbit 47.2 km above Mars.

128. For a circle: $d = 2\sqrt{A/\pi}$

$2006: d = 2\sqrt{29.6 \times 10^6 \text{ km}^2/\pi} = 6.14 \times 10^3 \text{ km}$

$2009: d = 2\sqrt{24.1 \times 10^6 \text{ km}^2/\pi} = 5.54 \times 10^3 \text{ km}$

Difference $= 6.14 \times 10^3 \text{ km} - 5.54 \times 10^3 \text{ km} = 0.600 \text{ km}$

$0.600 \text{ km} \times \dfrac{1000 \text{ m}}{1 \text{ km}} = 6.00 \times 10^2 \text{ m}$

129. $m = (1 \times 10^3 \text{ suns})(2.0 \times 10^{30} \text{ kg/sun}) \dfrac{1000 \text{ g}}{1 \text{ kg}} = 2.0 \times 10^{36} \text{ g}$

$V = 4/3(3.14)\left(\dfrac{2.16 \times 10^3 \text{ mi}}{2}\right)^3 = 5.27 \times 10^9 \text{ mi}^3$

$V = 5.27 \times 10^9 \text{ mi}^3 \times \dfrac{(1 \text{ km})^3}{(0.6214 \text{ mi})^3} \times \dfrac{(1000 \text{ m})^3}{(1 \text{ km})^3} \times \dfrac{(100 \text{ cm})^3}{(1 \text{ m})^3} = 2.20 \times 10^{25} \text{ cm}^3$

$d = \dfrac{2.0 \times 10^{36} \text{ g}}{2.20 \times 10^{25} \text{ cm}^3} = 9.1 \times 10^{10} \text{ g/cm}^3$

130. Titanium density $= 4.50 \text{ g/cm}^3$, Volume of cube $= (\text{Length})^3 = (6.8 \text{ cm})^3$

mass $= d \times V = \dfrac{4.50 \text{ g}}{\text{cm}^3} \times (6.8 \text{ cm})^3 \times \dfrac{1 \text{ kg}}{1000 \text{ g}} = 1.4 \text{ kg}$

If made from iron:

mass $= d \times V = \dfrac{7.86 \text{ g}}{\text{cm}^3} \times (6.8 \text{ cm})^3 \times \dfrac{1 \text{ kg}}{1000 \text{ g}} = 2.5 \text{ kg}$

131. Answers may vary; a representative set of answers is shown here.
   For human hair
   a. 0.0020 in.
   b. 5.1×10^{-5} m
   c. 0.000051 m
   d. 51 μm

   For Earth to sun
   a. 92,960,000 miles
   b. 1.496×10^{11} m
   c. 149,600,000,000 m
   d. 149.6 Gm

132. Each statement can be true based upon the accuracy of the measurement device being used. The number of digits in the answer reflects the accuracy of the measuring device. Each answer is measured by a different method such as (a) pedometer, (b) odometer, and (c) global positioning system (GPS).

133. Answers will vary by group.

Data Interpretation and Analysis

134.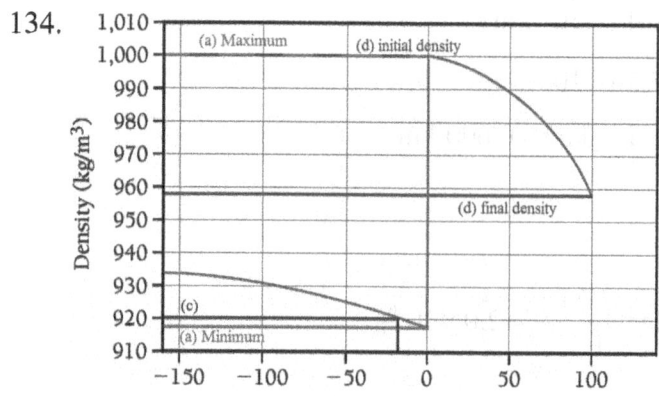

a) % Change = $\dfrac{\text{Final Value} - \text{Initial Value}}{\text{Initial Value}} \times 100$

% Change = $\dfrac{918 - 1000}{1000} \times 100 = -8.20\%$

b) Volume @ 1°C: $54 \text{ g} \times \dfrac{1 \text{ m}^3}{1000 \text{ kg}} \times \dfrac{1 \text{ kg}}{1000 \text{ g}} \times \dfrac{100^3 \text{ cm}^3}{1^3 \text{ m}^3} = 54 \text{ cm}^3$

Volume @ −1°C: $54 \text{ g} \times \dfrac{1 \text{ m}^3}{918 \text{ kg}} \times \dfrac{1 \text{ kg}}{1000 \text{ g}} \times \dfrac{100^3 \text{ cm}^3}{1^3 \text{ m}^3} = 59 \text{ cm}^3$

Change in Volume: $59 \text{ cm}^3 - 54 \text{ cm}^3 = 5 \text{ cm}^3$

c) Convert volume of ice to mass of ice:

$2.65 \times 10^6 \text{ km}^3 \times \dfrac{1000^3 \text{ m}^3}{1^3 \text{ km}^3} \times \dfrac{920 \text{ kg}}{1 \text{ m}^3} = 2.44 \times 10^{18} \text{ kg}$

Convert mass of ice to volume of water:

$2.44 \times 10^{18} \text{ kg} \times \dfrac{1 \text{ m}^3}{1000 \text{ kg}} = 2.44 \times 10^{15} \text{ m}^3$

d) Convert to mass of water at 1 °C:

$1.00 \text{ L} \times \dfrac{1000 \text{ cm}^3}{1 \text{ L}} \times \dfrac{1^3 \text{ m}^3}{100^3 \text{ cm}^3} \times \dfrac{1000 \text{ kg}}{1 \text{ m}^3} = 1.00 \text{ kg}$

Convert to volume of water at 100°C:

$1.00 \text{ kg} \times \dfrac{1 \text{ m}^3}{958 \text{ kg}} \times \dfrac{100^3 \text{ cm}^3}{1^3 \text{ m}^3} \times \dfrac{1 \text{ L}}{1000 \text{ cm}^3} = 1.04 \text{ L}$

Matter and Energy 3

Questions

1. Matter is defined as anything that occupies space and has mass. It can be thought of as the physical material that makes up the universe. Examples are steel, wood, water, and plastic.

2. Matter is ultimately composed of tiny particles called atoms. Atoms are commonly found bonded to other atoms to form larger particles called molecules.

3. Matter can be found as either a solid, a liquid, or a gas.

4. A solid has a fixed volume and a definite shape, and is not compressible.

5. A crystalline solid has atoms arranged in a repeating geometric pattern, whereas an amorphous solid has atoms that do not form repeating patterns.

6. A liquid has a fixed volume but will assume the shape of the container it fills. Like solids, liquids are not compressible.

7. A gas has no fixed volume or shape. Rather, it assumes both the shape and the volume of the container it occupies.

8. Gases are compressible because there is a large distance, which can be decreased, between gas particles. Solids and liquids cannot be compressed because the particles that make them up are touching.

9. A mixture is composed of two or more pure substances that have been mixed together in a variable proportion.

10. A heterogeneous mixture has two or more distinct regions that have different compositions. A homogeneous mixture has only one region, and the composition of the mixture does not change.

11. A pure substance is composed of only one type of atom or one type of molecule.

12. An element is a substance that cannot be broken down into two or more simpler substances. A compound is a substance that is made up of two or more elements in a fixed ratio. A compound can be broken down into simpler substances (the atoms that make up the compound).

13. A mixture is formed when two or more pure substances are mixed together, but a new substance is not formed. A compound is formed when two or more elements are bonded together in a fixed ratio to form a new substance.

14. A property of a substance that can be determined or observed without changing the chemical composition of a compound is called a physical property. A property in which the chemical composition of a substance is altered is called a chemical property.

15. A physical change is one that does not alter the chemical composition of a compound. A chemical change, however, will alter the chemical composition of a compound.

16. The law of conservation of mass states, "Matter is neither created nor destroyed in a chemical reaction."

17. Energy is the capacity to do work or generate heat.

18. The law of conservation of energy states: "Energy can neither be created nor destroyed."

19. Kinetic energy is the energy associated with motion. Potential energy is the energy associated with position or composition.

20. Chemical energy is a form of potential energy associated with the position of particles that compose the chemical system. Examples of chemical energy are the burning of gasoline and the burning of natural gas that can move objects or provide heat.

21. The common unit for energy in the laboratory is the joule (J). Other commonly used units of energy found in nutritional information are the calorie (cal) and the Calorie (Cal). Finally, another common unit of energy is the kilowatt-hour (kWh), which is how electricity is measured from utilities.

22. An exothermic reaction is a chemical reaction that releases energy. The reactants have greater energy than the products.

23. An endothermic reaction is a chemical reaction that absorbs energy. The products have greater energy than the reactants.

24. Temperature can be measured by the Fahrenheit scale (°F), the Celsius scale (°C), and the Kelvin scale (K).

25. Temperature is the measure of thermal energy of matter, whereas heat is the transfer or exchange of thermal energy caused by a temperature difference.

26. The temperature scales differ in the size of a degree and in how they define zero. The Kelvin, or absolute temperature scale, defines absolute zero as the temperature at which all atomic and molecular motion ceases. There can be no temperature lower than this. Celsius defines zero as the temperature at which water freezes and the boiling point of water is 100 °C. On the Fahrenheit scale, water freezes at 32 °F and boils at 212 °F. The size of a 1° change in the Celsius and Kelvin scales is the same, whereas a change of 1° in the Celsius scale equals a 1.8 °F change.

27. Heat capacity is a measure of how much heat is needed to raise the temperature of a given substance by 1 °C.

28. Water has a very high heat capacity; therefore, it can absorb much of the heat from the sun and not have a large increase in temperature. The heat capacity of land is much lower, which means that as it absorbs the energy from the sun, it will increase in temperature more quickly than water. Land that borders large bodies of water is therefore cooled by the lower temperature of the nearby water.

29. $°C = \dfrac{[°F - 32]}{1.8} \Rightarrow 1.8 \times °C = [°F - 32] \Rightarrow °F = (1.8 \times °C) + 32$

30. $K = °C + 273 \Rightarrow °C = K - 273$

Problems

Classifying Matter

31. a) element
    b) element
    c) compound
    d) compound

32. a) element
    b) compound
    c) element
    d) element

33. a) homogeneous
    b) heterogeneous
    c) homogeneous
    d) homogeneous

34. a) homogeneous
    b) heterogeneous
    c) heterogeneous
    d) homogeneous

35. a) pure, element
    b) mixture, homogeneous
    c) mixture, heterogeneous
    d) mixture, heterogeneous

36. a) mixture, homogeneous
    b) pure, compound
    c) mixture, heterogeneous
    d) mixture, heterogeneous

Physical and Chemical Properties and Physical and Chemical Changes

37. a) chemical
    b) physical
    c) physical
    d) chemical

38. a) physical
    b) physical
    c) chemical
    d) physical

39. Physical Properties: colorless, odorless, gas at room temperature; one liter has a mass of 1.260 g under standard conditions; mixes with acetone
    Chemical Properties: flammable, polymerizes to form polyethylene

40. Physical Properties: bluish color, pungent odor, gas at room temperature
    Chemical Properties: very reactive, decomposes on exposure to ultraviolet light

41. a) chemical
    b) physical
    c) chemical
    d) chemical

42. a) physical
    b) chemical
    c) physical
    d) chemical

43. a) physical
    b) chemical

44. a) physical
    b) chemical

The Conservation of Mass

45. Mass of reactants = Mass of products; 42 kg + 168 kg = 2.10×10^2 kg

46. Mass of reactants = Mass of products; 0.50 g + 4.0 g = 4.5 g

47. a) Yes. 67.5 g reactants = 67.5 g products
    b) No. 303.5 g reactants ≠ 294 g products

48. a) Yes. 32.4 g reactants = 32.4 g products
    b) No. 4.0×10^1 g reactants ≠ 33 g products

49. Mass of reactants = Mass of products
    9.7 g + 34.7 g = 29.3 g + g water
    g water = 44.4 − 29.3 = 15.1 g

50. Mass of reactants = Mass of products
    56 + 24 = 8.0×10^1 g of iron oxide

Conversion of Energy Units

51. a) calories → joules
    $$588 \text{ cal} \times \frac{4.184 \text{ J}}{1 \text{ cal}} = 2.46 \times 10^3 \text{ J}$$
    b) joules → calories → Calories
    $$17.4 \text{ J} \times \frac{1 \text{ cal}}{4.184 \text{ J}} \times \frac{1 \text{ Cal}}{1000 \text{ cal}} = 4.16 \times 10^{-3} \text{ Cal}$$
    c) kilojoules → joules → calories → Calories
    $$134 \text{ kJ} \times \frac{1000 \text{ J}}{1 \text{ kJ}} \times \frac{1 \text{ cal}}{4.184 \text{ J}} \times \frac{1 \text{ Cal}}{1000 \text{ cal}} = 32.0 \text{ Cal}$$
    d) Calories → calories → joules
    $$56.2 \text{ Cal} \times \frac{1000 \text{ cal}}{1 \text{ Cal}} \times \frac{4.184 \text{ J}}{1 \text{ cal}} = 2.35 \times 10^5 \text{ J}$$

52. a) joules → calories

$$45.6 \text{ J} \times \frac{1 \text{ cal}}{4.184 \text{ J}} = 10.9 \text{ cal}$$

b) calories → joules

$$355 \text{ cal} \times \frac{4.184 \text{ J}}{1 \text{ cal}} = 1.49 \times 10^3 \text{ J}$$

c) kilojoules → joules → calories

$$43.8 \text{ kJ} \times \frac{1000 \text{ J}}{1 \text{ kJ}} \times \frac{1 \text{ cal}}{4.184 \text{ J}} = 1.05 \times 10^4 \text{ cal}$$

d) calories → joules → kilojoules

$$215 \text{ cal} \times \frac{4.184 \text{ J}}{1 \text{ cal}} \times \frac{1 \text{ kJ}}{1000 \text{ J}} = 0.900 \text{ kJ}$$

53. a) $25 \text{ kWh} \times \dfrac{3.60 \times 10^6 \text{ J}}{1 \text{ kWh}} = 9.0 \times 10^7 \text{ J}$

b) $249 \text{ cal} \times \dfrac{1 \text{ Cal}}{1000 \text{ cal}} = 0.249 \text{ Cal}$

c) $113 \text{ cal} \times \dfrac{4.184 \text{ J}}{1 \text{ cal}} \times \dfrac{1 \text{ kWh}}{3.60 \times 10^6 \text{ J}} = 1.31 \times 10^{-4} \text{ kWh}$

d) $44 \text{ kJ} \times \dfrac{1000 \text{ J}}{1 \text{ kJ}} \times \dfrac{1 \text{ cal}}{4.184 \text{ J}} = 1.1 \times 10^4 \text{ cal}$

54. a) $345 \text{ Cal} \times \dfrac{1000 \text{ cal}}{1 \text{ Cal}} \times \dfrac{4.184 \text{ J}}{1 \text{ cal}} \times \dfrac{1 \text{ kWh}}{3.60 \times 10^6 \text{ J}} = 0.401 \text{ kWh}$

b) $23 \text{ J} \times \dfrac{1 \text{ cal}}{4.184 \text{ J}} = 5.5 \text{ cal}$

c) $5.7 \times 10^3 \text{ J} \times \dfrac{1 \text{ kJ}}{1000 \text{ J}} = 5.7 \text{ kJ}$

d) $326 \text{ kJ} \times \dfrac{1000 \text{ J}}{1 \text{ kJ}} = 3.26 \times 10^5 \text{ J}$

55.

J	cal	Cal	kWh
225 J	53.8 cal	5.38×10^{-2} Cal	6.25×10^{-5} kWh
3.44×10^6 J	8.21×10^5 cal	8.21×10^2 Cal	9.56×10^{-1} kWh
1.06×10^9 J	2.54×10^8 cal	2.54×10^5 Cal	295 kWh
6.49×10^5 J	1.55×10^5 cal	155 Cal	1.80×10^{-1} kWh

56.

J	cal	Cal	kWh
7.88×10^6 J	1.88×10^6 cal	1.88×10^3 Cal	2.19 kWh
4.828×10^6 J	1.154×10^6 cal	1154 Cal	1.34 kWh
3.70×10^2 J	88.4 cal	8.84×10^{-2} Cal	1.03×10^{-4} kWh
4.50×10^8 J	1.08×10^8 cal	1.08×10^5 Cal	125 kWh

57. $1027 \text{ kWh} \times \dfrac{3.60 \times 10^6 \text{ J}}{1 \text{ kWh}} = 3.697 \times 10^9 \text{ J}$

58. $32 \text{ kWh} \times \dfrac{3.60 \times 10^6 \text{ J}}{1 \text{ kWh}} = 1.2 \times 10^8 \text{ J}$

59. $2.2 \times 10^3 \text{ Cal} - 2.0 \times 10^3 \text{ Cal} = 2 \times 10^2 \text{ Cal}$

$\dfrac{2 \times 10^2 \text{ Cal}}{\text{day}} \times \dfrac{4.184 \text{ J}}{\text{cal}} \times \dfrac{1000 \text{ cal}}{1 \text{ Cal}} \times \dfrac{1 \text{ kJ}}{1000 \text{ J}} = 8 \times 10^2 \text{ kJ/day}$

$\dfrac{14.6 \times 10^3 \text{ kJ}}{1 \text{ lb.}} \times \dfrac{1 \text{ day}}{8 \times 10^2 \text{ kJ}} = 20 \text{ days}$

60. $\dfrac{245 \text{ Cal}}{1 \text{ bag}} \times \dfrac{1000 \text{ cal}}{1 \text{ Cal}} \times \dfrac{4.184 \text{ J}}{1 \text{ cal}} \times = \dfrac{1.03 \times 10^6 \text{ J}}{1 \text{ bag}}$

$\dfrac{1 \text{ bag}}{1.03 \times 10^6 \text{ J}} \times \dfrac{1000 \text{ J}}{1 \text{ kJ}} \times \dfrac{14.6 \times 10^3 \text{ kJ}}{1 \text{ lb fat}} = \dfrac{14.2 \text{ bags}}{1 \text{ lb fat}}$

Energy and Chemical and Physical Change

61. The reaction of iron with oxygen in the atmosphere releases heat; therefore it is an exothermic reaction. See Figure 3.17a.

62. The reaction within the cold pack absorbs heat; therefore, it is an endothermic reaction. See Figure 3.17b.

63. a) exothermic, $-\Delta H$
    b) endothermic, $+\Delta H$
    c) exothermic, $-\Delta H$

64. a) endothermic, $+\Delta H$
    b) exothermic, $-\Delta H$
    c) exothermic, $-\Delta H$

Converting Between Temperature Scales

65. a) $°C = \dfrac{[°F - 32]}{1.8} = \dfrac{[212 - 32]}{1.8} = \dfrac{1.80 \times 10^2}{1.8} = 1.00 \times 10^2 \, °C$

    b) $°C = K - 273 = 77 - 273 = -196 \, °C$

    $°F = (1.8 \times °C) + 32 = (1.8 \times -196) + 32 = -3.2 \times 10^2 \, °F$

    c) $K = 273 + °C = 273 + 25 = 298 \, K$

    d) $°C = \dfrac{[°F - 32]}{1.8} = \dfrac{[98.6 - 32]}{1.8} = 37 \, °C; \quad K = 273 + °C = 3.10 \times 10^2 \, K$

66. a) $°C = \dfrac{[°F - 32]}{1.8} = \dfrac{[102 - 32]}{1.8} = \dfrac{7.0 \times 10^1}{1.8} = 39 \, °C$

    b) $°C = K - 273 = -273 \, °C$

    $°F = (1.8 \times °C) + 32 = (1.8 \times -273) + 32 = -491 + 32 = -459 \, °F$

    c) $°F = (1.8 \times °C) + 32 = (1.8 \times -48) + 32 = -86.4 + 32 = -54 \, °F$

    d) $°C = K - 273 \Rightarrow °C = 273 - 273 = 0.00 \, °C$

67. $°C = \dfrac{[°F - 32]}{1.8} = \dfrac{[-80 - 32]}{1.8} = \dfrac{-112}{1.8} = -62 \, °C$

    $K = 273 + °C \Rightarrow K = 273 + -62 = 211 \, K$

68. $°C = \dfrac{[°F - 32]}{1.8} = \dfrac{[134 - 32]}{1.8} = \dfrac{102}{1.8} = 57 \, °C$

    $K = 273 + °C \Rightarrow K = 273 + 57 = 3.30 \times 10^2 \, K$

69. $°F = (1.8 \times °C) + 32 = (1.8 \times -114) + 32 = -205.2 + 32 = -173 \, °F$

    $K = °C + 273 \Rightarrow K = -114 + 273 = 159 \, K$

70. $°C = K - 273 = 4.2 - 273 = -269 \, °C$

    $°F = (1.8 \times °C) + 32 \Rightarrow °F = (1.8 \times -269) + 32 = -484.2 + 32 = -452 \, °F$

71. $°F = (1.8 \times -59.7 \, °C) + 32 = -75.5 \, °C$

72. $°C = \dfrac{(-47\,°F - 32)}{1.8} = -44\,°C$

   $K = 273 + -44\,°C = 229\,K$

73. | Kelvin | Fahrenheit | Celsius |
     |---|---|---|
     | 0.0 | −459 °F | −273.0 °C |
     | 301 K | 82.5 °F | 28.1 °C |
     | 282 K | 47 °F | 8.5 °C |

74. | Kelvin | Fahrenheit | Celsius |
     |---|---|---|
     | 273.0 | 32 °F | 0.0 °C |
     | 233 K | −40.0 °F | −40.0 °C |
     | 385 K | 233 °F | 112 °C |

Energy, Heat Capacity, and Temperature Changes

75. $q = m \times c \times \Delta T$

   $q = 65\,g \times 4.184\,\dfrac{J}{g\,°C} \times (65\,°C - 32\,°C) = 9.0 \times 10^3\,J$

76. $q = m \times c \times \Delta T$

   $q = 22\,g \times 4.184\,\dfrac{J}{g\,°C} \times (18\,°C - 7\,°C) = 1.0 \times 10^3\,J$

77. $45\,kg \times \dfrac{1000\,g}{1\,kg} \times \dfrac{2.42\,J}{g\,°C} \times (19.0\,°C - 11.0\,°C) = 8.7 \times 10^5\,J$

78. $3.5\,kg \times \dfrac{1000\,g}{1\,kg} \times \dfrac{0.128\,J}{g\,°C} \times (67\,°C - 21\,°C) = 2.1 \times 10^4\,J$

79. $89\,J = 12\,g \times \dfrac{0.128\,J}{g\,°C} \times (\Delta T) \Rightarrow \Delta T = \dfrac{89\,J}{12\,g \times \dfrac{0.128\,J}{g\,°C}} = 58\,°C$

80. $57\,J = 17.1\,g \times \dfrac{0.903\,J}{g\,°C} \times (\Delta T) \Rightarrow \Delta T = \dfrac{57\,J}{17.1\,g \times \dfrac{0.903\,J}{g\,°C}} = 3.7\,°C$

81. $15 \text{ J} = 12 \text{ g} \times \dfrac{0.449 \text{ J}}{\text{g °C}} \times (T_f - 28 \text{ °C}) \Rightarrow (T_f - 28 \text{ °C}) = \dfrac{15 \text{ J}}{12 \text{ g} \times \dfrac{0.449 \text{ J}}{\text{g °C}}} \Rightarrow$

$T_f = 2.78 + 28 \text{ °C} = 31 \text{ °C}$

82. $345 \text{ kJ} \times \dfrac{1000 \text{ J}}{1 \text{ kJ}} = 45 \text{ kg} \times \dfrac{1000 \text{ g}}{1 \text{ kg}} \times \dfrac{4.184 \text{ J}}{\text{g °C}} \times (T_f - 22.1\text{°C}) \Rightarrow$

$(T_f - 22.1\text{°C}) = \dfrac{3.45 \times 10^5 \text{ J}}{4.5 \times 10^4 \text{g} \times \dfrac{4.184 \text{ J}}{\text{g °C}}} \Rightarrow T_f = 1.83 + 22.1 \text{°C} = 23.9 \text{ °C}$

83. $248 \text{ cal} \times \dfrac{4.184 \text{ J}}{1 \text{ cal}} = 24 \text{ g} \times \dfrac{4.184 \text{ J}}{\text{g °C}} \times (\Delta T) \Rightarrow \Delta T = \dfrac{1037.6}{100.3} = 1.0 \times 10^1 \text{ °C}$

84. $146 \text{ cal} \times \dfrac{4.184 \text{ J}}{1 \text{ cal}} = 57 \text{ g} \times \dfrac{0.128 \text{ J}}{\text{g °C}} \times (T_f - 47 \text{°C}) \Rightarrow$

$(T_f - 47 \text{ °C}) = \dfrac{610.9}{7.296} \Rightarrow T_f = 83.7 + 47 = 131 \text{°C}$

85. $58 \text{ J} = 28 \text{ g} \times (\text{heat capacity}) \times (39.9 - 31.1) \text{ °C} \Rightarrow$

heat capacity $= \dfrac{58 \text{ J}}{(28 \text{ g})(8.8 \text{ °C})} = 0.24 \dfrac{\text{J}}{\text{g °C}}$

∴ It is consistent with silver metal.

86. $2.8 \text{ J} = 5.6 \text{ g} \times (\text{heat capacity}) \times (3.9) \text{°C} \Rightarrow$

heat capacity $= \dfrac{2.8 \text{ J}}{(5.6 \text{ g})(3.9 \text{°C})} = 0.13 \dfrac{\text{J}}{\text{g °C}}$

∴ Yes, it is consistent with gold.

87. $56 \text{ J} = 11 \text{ g} \times (\text{heat capacity}) \times (12.7 - 10.4) \text{ °C} \Rightarrow$

heat capacity $= \dfrac{56 \text{ J}}{(11 \text{ g})(2.3 \text{ °C})} = 2.2 \dfrac{\text{J}}{\text{g °C}}$

88. $47.5 \text{ J} = 13.2 \text{ g} \times (\text{heat capacity}) \times (1.72) \text{°C} \Rightarrow$

heat capacity $= \dfrac{47.5 \text{ J}}{(13.2 \text{ g})(1.72 \text{ °C})} = 2.09 \dfrac{\text{J}}{\text{g °C}}$

89. When warm drinks are placed into the ice, they release heat, which then melts the ice. The pre-chilled drinks, on the other hand, are already cold, so they do not release much heat and do not melt the ice.

90. Since the specific heat of water (4.184 J/(g °C)) is much higher than that of iron metal (0.449 J/(g °C)), 100 grams of water at 75 °C contains much more heat than 100 grams of iron at the same temperature. When they are placed into insulated containers and allowed to equilibrate, the final temperature of the water container will be higher, reflecting the greater amount of heat present initially.

91. $0.449 \dfrac{J}{g\ °C} \times 25.7\ g \times (22.0\ °C - 75.0\ °C) = -612\ J$

92. $0.903 \dfrac{J}{g\ °C} \times 53.2\ g \times (T_{final} - 155.0\ °C) = -2.87 \times 10^3\ J$

$48.0(T_{final} - 155.0\ °C) = -2.87 \times 10^3\ J$

$48.0 T_{final} - 7.45 \times 10^3 = -2.87 \times 10^3\ J$

$T_{final} = \dfrac{-2.87 \times 10^3\ J + 7.45 \times 10^3}{48.0} = 95.3\ °C$

Cumulative Problems

93. $17\ kJ \times \dfrac{1000\ J}{1\ kJ} = (245\ mL \times \dfrac{1.00\ g}{1\ mL}) \times \dfrac{4.184\ J}{g\ °C} \times (T_{final} - 32\ °C) \Rightarrow$

$(T_{final} - 32\ °C) = \dfrac{1.7 \times 10^4\ J}{245\ g \times \dfrac{4.184\ J}{g\ °C}} \Rightarrow T_{final} = 16.6 + 32\ °C = 49\ °C$

*Density of water is 1.00 g/mL.

94. $562\ J = 32\ mL \times \dfrac{0.789\ g}{1\ ml} \times \dfrac{2.42\ J}{g\ °C} \times (T_{final} - 11\ °C) \Rightarrow$

$(T_{final} - 11\ °C) = \dfrac{562\ J}{32\ mL \times \dfrac{0.789\ g}{1\ mL} \times \dfrac{2.42\ J}{g\ °C}} \Rightarrow T_{final} = \dfrac{562}{61.10} + 11\ °C = 2.0 \times 10^1\ °C$

95. $\text{Heat} = 1.57\ cm^3 \times \dfrac{19.3\ g}{1\ cm^3} \times \dfrac{0.128\ J}{g\ °C} \times (29.5 - 11.4\ °C) = 70.2\ J$

96. $67.4 \text{ J} = 98.5 \text{ cm}^3 \times \dfrac{2.70 \text{ g}}{1 \text{ cm}^3} \times \dfrac{0.903 \text{ J}}{\text{g °C}} \times (T_{final} - 32.5 \text{°C}) \Rightarrow$

$(T_{final} - 32.5 \text{°C}) = \dfrac{67.4 \text{ J}}{98.5 \text{ cm}^3 \times \dfrac{2.70 \text{ g}}{1 \text{ cm}^3} \times \dfrac{0.903 \text{ J}}{\text{g °C}}} \Rightarrow$

$T_{final} = \dfrac{67.4}{240.15} + 32.5 \text{°C} = 32.8 \text{°C}$

97. $T_{initial} = (85 \text{°F} - 32)/1.8 = 29 \text{°C}$

$T_{final} = (212 \text{°F} - 32)/1.8 = 1.00 \times 10^2 \text{°C}$

You must convert each temperature prior to taking the difference.

$\Delta T = 1.00 \times 10^2 \text{°C} - 29 \text{°C} = 71 \text{°C}$

$\text{Heat} = 56 \text{ L} \times \dfrac{1000 \text{ mL}}{1 \text{ L}} \times \dfrac{1 \text{ g}}{1 \text{ mL}} \times \dfrac{4.184 \text{ J}}{\text{g °C}} \times 71 \text{°C} \times \dfrac{1 \text{ kJ}}{1000 \text{ J}} = 1.7 \times 10^4 \text{ kJ}$

98. $T_{initial} = (72 \text{°F} - 32)/1.8 = 22 \text{°C}$

$T_{final} = (145 \text{°F} - 32)/1.8 = 62.8 \text{°C}$

$\Delta T = 62.8 \text{°C} - 22 \text{°C} = 41 \text{°C}$

You must convert each temperature prior to taking the difference.

$\text{Heat} = 43 \text{ g} \times \dfrac{0.903 \text{ J}}{\text{g °C}} \times 41 \text{°C} = 1.6 \times 10^3 \text{ J}$

99. $2.3 \text{ kWh} \times \dfrac{3.60 \times 10^6 \text{ J}}{1 \text{ kWh}} = 29.5 \text{ L} \times \dfrac{1000 \text{ mL}}{1 \text{ L}} \times \dfrac{1.00 \text{ g}}{1 \text{ mL}} \times \dfrac{4.184 \text{ J}}{\text{g °C}} \times \Delta T \Rightarrow$

$\Delta T = \dfrac{2.3 \text{ kWh} \times \dfrac{3.60 \times 10^6 \text{ J}}{1 \text{ kWh}}}{29.5 \text{ L} \times \dfrac{1000 \text{ mL}}{1 \text{ L}} \times \dfrac{1 \text{ g}}{1 \text{ mL}} \times \dfrac{4.184 \text{ J}}{\text{g °C}}} = \dfrac{8.28 \times 10^6}{1.045 \times 10^5} = 67 \text{ °C}$

100. $9.4 \times 10^{-2} \text{ kWh} \times \dfrac{3.60 \times 10^6 \text{ J}}{1 \text{ kWh}} = 1.45 \text{ L} \times \dfrac{1000 \text{ mL}}{1 \text{ L}} \times \dfrac{1.00 \text{ g}}{1 \text{ mL}} \times \dfrac{4.184 \text{ J}}{\text{g °C}} \times (T_{final} - 25.0)$

$= (T_{final} - 25) = \dfrac{9.4 \times 10^{-2} \text{ kWh} \times \dfrac{3.60 \times 10^6 \text{ J}}{1 \text{ kWh}}}{1.45 \text{ L} \times \dfrac{1000 \text{ mL}}{1 \text{ L}} \times \dfrac{1 \text{ g}}{1 \text{ mL}} \times \dfrac{4.184 \text{ J}}{\text{g °C}}} \Rightarrow$

$T_{final} = \dfrac{3.384 \times 10^5}{6061} + 25 = 55.8 + 25 = 81 \text{ °C}$

101. $55 \text{ gal} \times \dfrac{3.785 \text{ L}}{1 \text{ gal}} \times \dfrac{1000 \text{ mL}}{1 \text{ L}} \times \dfrac{1.00 \text{ g}}{1 \text{ mL}} \times \dfrac{4.184 \text{ J}}{\text{g °C}} \times 25 \text{ °C} \times \dfrac{1 \text{ kWh}}{3.60 \times 10^6 \text{ J}} = 6.0 \text{ kWh}$

102. $\text{kWh} = 48 \text{ kg} \times \dfrac{1000 \text{ g}}{1 \text{ kg}} \times \dfrac{1.03 \text{ J}}{\text{g °C}} \times (28 - 7) \text{ °C} \times \dfrac{1 \text{ kWh}}{3.60 \times 10^6 \text{ J}} = 0.29 \text{ kWh}$

103. $2.5 \text{ kg H}_2\text{O} \times \dfrac{1000 \text{ g}}{1 \text{ kg}} \times \dfrac{4.184 \text{ J}}{\text{g °C}} \times 75 \text{ °C} \times \dfrac{1 \text{ kJ}}{1000 \text{ J}} \times \dfrac{1 \text{ g Fuel}}{36 \text{ kJ}} = 22 \text{ g Fuel}$

104. $1.35 \text{ kg H}_2\text{O} \times \dfrac{1000 \text{ g}}{1 \text{ kg}} \times \dfrac{4.184 \text{ J}}{\text{g °C}} \times 68.0 \text{ °C} \times \dfrac{1 \text{ kJ}}{1000 \text{ J}} \times \dfrac{1 \text{ g Fuel}}{49.3 \text{ kJ}} = 7.79 \text{ g Fuel}$

105. $95 \text{ kg} \times \dfrac{1000 \text{ g}}{1 \text{ kg}} \times \dfrac{4.0 \text{ J}}{\text{g °C}} \times 0.50 \text{ °C} \times \dfrac{1 \text{ kJ}}{1000 \text{ J}} \times \dfrac{1 \text{ g H}_2\text{O}}{2.44 \text{ kJ}} = 78 \text{ g H}_2\text{O}$

106. $12 \text{ oz} \times \dfrac{1 \text{ g}}{0.034 \text{ oz}} = 3\underline{5}3 \text{ g}$

$\Delta T = 75 \text{ °F} - 35 \text{ °F} \Rightarrow \text{°C} = \dfrac{[75 - 32]}{1.8} - \dfrac{[35 - 32]}{1.8} = 2\underline{2}.2 \text{ °C}$

$3\underline{5}3 \text{ g} \times \dfrac{4.184 \text{ J}}{\text{g °C}} \times 2\underline{2}.2 \text{ °C} \times \dfrac{1 \text{ kJ}}{1000 \text{ J}} \times \dfrac{1 \text{ g Ice}}{0.33 \text{ kJ}} = 99 \text{ g ice}$

107. Heat lost by aluminum = −heat gained by water

$15.7 \text{ g Al} \times \dfrac{0.903 \text{ J}}{\text{g °C}} \times (T_{final} - 53.2 \text{ °C}) = -32.5 \text{ g H}_2\text{O} \times \dfrac{4.184 \text{ J}}{\text{g °C}} \times (T_{final} - 24.5 \text{ °C}) \Rightarrow$

$\dfrac{14.\underline{1}8 \text{ J}}{\text{°C}} \times (T_{final} - 53.2 \text{ °C}) = -\dfrac{13\underline{6}.0 \text{ J}}{\text{°C}} \times (T_{final} - 24.5 \text{ °C}) \Rightarrow$

$\dfrac{14.\underline{1}8 \text{ J}}{\text{°C}}(T_f) - 75\underline{4} \text{ J} = -\dfrac{13\underline{6}.0 \text{ J}}{\text{°C}}(T_{final}) + 33\underline{3}2 \text{ J} \Rightarrow$

$\left(\dfrac{14.\underline{1}8 \text{ J}}{\text{°C}} + \dfrac{13\underline{6}.0 \text{ J}}{\text{°C}}\right)(T_{final}) = (75\underline{4} \text{ J} + 33\underline{3}2 \text{ J})$

$\left(\dfrac{15\underline{0}.1 \text{ J}}{\text{°C}}\right)T_{final} = 40\underline{8}6 \text{ J} \Rightarrow T_{final} = \dfrac{40\underline{8}6 \text{ J}}{15\underline{0}.1 \text{ J/°C}} = 27.2 \text{ °C}$

108. Heat gained by ethanol = −heat lost by water

$$25.0 \text{ mL} \times \frac{0.789 \text{ g}}{\text{mL}} \times \frac{2.42 \text{ J}}{\text{g °C}} \times (T_{final} - 7.0 \text{ °C}) = -35.0 \text{ mL} \times \frac{1.0 \text{ g}}{\text{mL}} \times \frac{4.184 \text{ J}}{\text{g °C}} \times (T_{final} - 25.3) \Rightarrow$$

$$\frac{47.74 \text{ J}}{\text{°C}} \times (T_{final} - 7.0 \text{ °C}) = -\frac{146 \text{ J}}{\text{°C}} \times (T_{final} - 25.3) \Rightarrow$$

$$\frac{47.74 \text{ J}}{\text{°C}} (T_{final}) - 3\underline{3}4 \text{ J} = -\frac{146 \text{ J}}{\text{°C}} (T_{final}) + 3\underline{7}00 \text{ J} \Rightarrow$$

$$\left(\frac{47.74 \text{ J}}{\text{°C}} + \frac{146 \text{ J}}{\text{°C}}\right)(T_{final}) = 3\underline{7}00 \text{ J} + 3\underline{3}4 \text{ J} \Rightarrow \frac{194 \text{ J}}{\text{°C}}(T_{final}) = 40\underline{4}0 \text{ J} \Rightarrow$$

$$T_{final} = \frac{40\underline{4}0 \text{ J}}{194 \frac{\text{J}}{\text{°C}}} = 21 \text{ °C}$$

109. Watts → J/s → kJ/month; kJ/month → $/month

$$855 \text{ W} - 625 \text{ W} = 2.30 \times 10^2 \text{ W} = 2.30 \times 10^2 \text{ J/s}$$

$$\frac{2.30 \times 10^2 \text{ J}}{\text{s}} \times \frac{60 \text{ s}}{1 \text{ min}} \times \frac{60 \text{ min}}{1 \text{ hr}} \times \frac{24 \text{ hr}}{1 \text{ day}} \times \frac{30 \text{ day}}{1 \text{ month}} \times \frac{1 \text{ kJ}}{1000 \text{ J}} = 5.96 \times 10^5 \frac{\text{kJ}}{\text{month}}$$

$$5.96 \times 10^5 \frac{\text{kJ}}{\text{month}} \times \frac{1 \text{ kWh}}{3.60 \times 10^3 \text{ kJ}} \times \frac{\$0.15}{\text{kWh}} = \$25$$

110. Calculate joules → convert to time using watts

$$5.5 \text{ L} \times \frac{1000 \text{ mL}}{1 \text{ L}} \times \underbrace{\frac{1.00 \text{ g}}{1 \text{ mL}}}_{\text{density}} = 5.5 \times 10^3 \text{ g}$$

$$q = 5.5 \times 10^3 \text{ g} \times 4.184 \frac{\text{J}}{\text{g °C}} \times (42 \text{ °C} - 25 \text{ °C}) = 3.9 \times 10^5 \text{ J}$$

$$3.9 \times 10^5 \text{ J} \times \frac{1 \text{ s}}{255 \text{ J}} \times \frac{1 \text{ min}}{60 \text{ s}} = 25 \text{ min}$$

111. Set °C = °F in the equation to convert temperatures between scales; solve.
1.8(°C) + 32 = °F
1.8x + 32 = x
1.8x − x = −32
0.8x = −32
x = −32/0.8 = −40
−40 °C = −40 °F

112. Set °F = 2 × °C in the equation to convert temperatures between scales; solve.
1.8(°C) + 32 = °F
1.8(x) + 32 = 2x
32 = 0.2x
x = 32/0.2 = 160
160 °C = 320 °F

Highlight Problems

113. a) pure substance
    b) pure substance
    c) pure substance
    d) mixture

114. a) pure substance; compound
    b) mixture; homogeneous
    c) mixture; heterogeneous
    d) pure substance; element

115. physical change

116. chemical change

117. The weather patterns that develop over the ocean exchange heat with the water. If the water is several degrees colder than the air above it, an extremely large amount of heat is removed from the air to the water. This is because water has a large heat capacity and air has a low heat capacity. Conversely, if the water is several degrees warmer than the air above it, a tremendous amount of heat can be transferred to the air. This occurs all of the time and is part of our normal weather patterns. However, during the El Niño/La Niña cycle, this heat transfer to and from water and air increases dramatically, which is why it alters weather patterns on a global scale.

118. $137 \times 10^7 \text{ km}^3 \times \dfrac{(1000)^3 \text{ m}^3}{1 \text{ km}^3} \times \dfrac{(100)^3 \text{ cm}^3}{1 \text{ m}^3} \times \dfrac{1.03 \text{ g}}{1 \text{ cm}^3} \times \dfrac{4.184 \text{ J}}{\text{g °C}} \times 1 \text{ °C} = 5.9 \times 10^{24} \text{ J}$

119. a) San Francisco is nearly surrounded by the ocean. Because water can absorb large amounts of heat without an increase in temperature (i.e., water has a large heat capacity), San Francisco enjoys moderate temperatures. Sacramento, on the other hand, is a land-locked city. The earth has a relatively low heat capacity, which means that as the earth absorbs heat, the temperature quickly increases (i.e., a small heat capacity).
    b) In the winter, San Francisco is a warmer place. The ocean releases a large amount of heat back into the colder atmosphere because it can store a tremendous amount of heat. Sacramento, in contrast, is surrounded by land and is cooler because the earth's heat capacity is much lower than the ocean's.

120.   Solid   Liquid   Homogeneous   Heterogeneous

 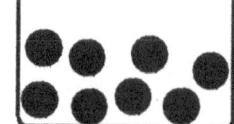  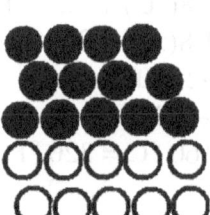

121.   Before —— After

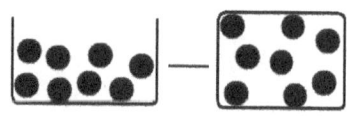

122. Where does the energy come from since it is impossible to violate the law of conservation of energy?

123. Heat and temperature are similar in that they both involve thermal energy. They differ in that temperature is a measure of thermal energy while heat is a transfer of thermal energy.

Data Interpretation and Analysis

124.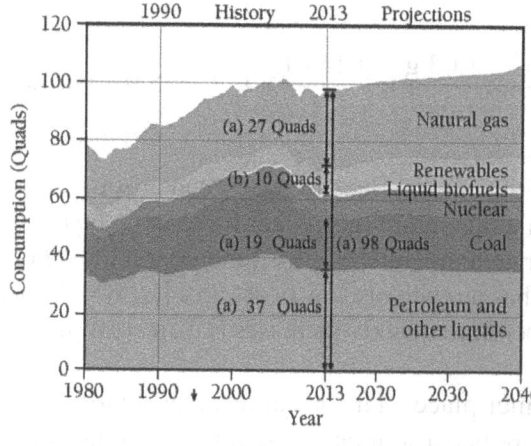

(a) Petroleum – 37 Quads
Natural Gas – 27 Quads
Coal – 19 Quads
$$\frac{27+19+37}{98} \times 100 = 85\%$$

(b) $\frac{10}{98} \times 100 = 10\%$

(c) Coal and Petroleum

(d) $6 \text{ Quads} \times \frac{1.055 \times 10^{18} \text{ J}}{1 \text{ Quad}} = 6 \times 10^{18} \text{ J}$

Atoms and Elements

4

Questions

1. Democritus reasoned that matter was made of small, indivisible, indestructible particles.

2. The three main ideas in Dalton's atomic theory are: (1) elements are composed of atoms; (2) all atoms of an element are identical, and each element has unique atoms; (3) atoms combine to form compounds in simple, whole-number ratios.

3. Rutherford had shot alpha particles at an extremely thin gold foil target. The majority of alpha particles passed directly through the foil, some particles were slightly deflected, and 1 in 20,000 bounced back toward the alpha source. If the plum pudding model had been correct, the only result would have been a slight deflection of some of the particles.

4. The main ideas of the nuclear theory of the atom are: (1) the positive charge of an atom is located in the center of the atom called the nucleus, which also contains nearly all of the mass of the atom; (2) an atom is almost entirely empty space, which is occupied by extremely small, negatively charged particles called electrons; and (3) an atom is electrically neutral, so the number of negative particles (electrons) is equal to the number of positive particles (protons).

5.

Particle	Mass (kg)	Mass (amu)	Charge	Location
Proton	1.67262×10^{-27}	1	+1	Inside Nucleus
Neutron	1.67493×10^{-27}	1	0	Inside Nucleus
Electron	0.00091×10^{-27}	0.00055	−1	Outside Nucleus

6. Electrical charge is a fundamental property of the proton (positive charge) and the electron (negative charge). The positive charge is attracted to the negative charge, while it is repelled by another positive charge, and vice versa. When the number of protons is equal to the number of electrons, the charges are effectively canceled and the molecule is neutral in charge.

7. Matter is usually charge neutral because the protons and electrons cancel each other, and, when an imbalance does exist, it usually corrects itself very quickly. If matter was not charge neutral, objects around us would be attracted and repelled by other objects near them.

8. The atomic number of an element is equal to the number of protons found within the atomic nucleus.

9. A chemical symbol is a one- or two-letter abbreviation that is unique to each element.

10. Many elements have names that describe their physical or chemical properties. For example, argon derives its name from the root Latin word *argos*, which means inactive. Some elements have been named to honor a scientist (einsteinium, curium, mendelevium), while still other elements have been named for locations (berkelium, americium, californium, europium).

11. Mendeleev was the first person to organize the elements into what we would recognize as a version of the modern periodic table. Mendeleev based his table on increasing the molecular mass of elements and grouping elements with similar properties into columns.

12. The periodic law was a recognition that while the physical and chemical properties of the elements are unique, there is a periodic repetition of similar properties. For example, lithium is a soft metal that reacts vigorously with water to produce hydrogen gas. Similarly, sodium, potassium, and rubidium are also soft metals that react vigorously with water to produce hydrogen gas.

13. The modern periodic table is organized by increasing atomic number, which is the number of protons found in each element.

14. Metals (1) conduct heat and electricity; (2) are malleable; (3) are ductile; (4) have a shiny appearance; and (5) tend to lose electrons in reactions. Metals make up the left side and the middle section of the periodic table.

15. Nonmetals (1) are poor conductors of heat and electricity; (2) tend to gain electrons in reactions; and (3) can be found as solids, liquids, or gases. Nonmetals make up the upper right side of the periodic table.

16. The metalloids make up the zigzag section that separates the metals and the nonmetals, which is found running from boron to astatine.

17. A family or group of elements is the term given to the column on a periodic table that contains elements that have similar, periodic properties.

18. a) 1 or IA   b) 2 or IIA   c) 17 or VIIA   d) 18 or VIIIA

19. An ion is an atom (or group of atoms) that becomes electrically charged due to a gain or loss of electrons.

20. A cation is a positive ion formed when an atom or a group of atoms loses one or more electrons. An anion is a negative ion formed when an atom or group of atoms gains one or more electrons.

21. a) +1   b) +2   c) +3   d) −2   e) −1

22. Isotopes are atoms of the same element that have different masses because they contain a different number of neutrons.

23. The percent natural abundance of isotopes provides the relative amounts of each isotope found in a sample of the element. These numbers are constant no matter the source of the element and are unique to each different element.

24. The mass number of an isotope is the sum of the number of protons plus the number of neutrons found in the nucleus of the particular isotope.

25. The first method for specifying isotopes is to superscript and subscript the mass number and atomic number, respectively, on the left side of the chemical symbol (e.g., ^2_1H). The second method is to identify the element followed by the mass number (e.g., H-2 or hydrogen-2).

26. Because of isotopes, there can be a small difference in the mass of atoms of the same element. Therefore, the atomic mass of an element is really the average mass of the element. The average mass is determined by accounting for the mass of each isotope and how common the isotope is found in nature.

Problems

Atomic and Nuclear Theory

27. a) consistent; all atoms of a given element have the same mass and other properties that distinguish it from the atoms of other elements.
    b) inconsistent; each element is composed of tiny indestructible particles called atoms.
    c) inconsistent; atoms combine in simple, whole-number ratios to form compounds.
    d) consistent; atoms combine in simple, whole-number ratios to form compounds.

28. a) inconsistent; all atoms of a given element have the same mass; atoms of different elements (Ca vs. Ti) should have different masses.
    b) inconsistent; all atoms of a given element have the same mass and other properties that distinguish it from the atoms of other elements.
    c) consistent; all atoms of a given element have the same mass and other properties that distinguish it from the atoms of other elements.
    d) consistent; atoms combine in simple, whole-number ratios to form compounds.

29. a) consistent; there are as many negatively charged electrons outside the nucleus as there are positively charged particles (called protons) inside the nucleus; therefore, the atom is electrically neutral.
    b) inconsistent; most of the volume of the atom is empty space occupied by tiny, negatively charged electrons.
    c) inconsistent; there are as many negatively charged electrons outside the nucleus as there are positively charged particles (called protons) inside the nucleus, therefore, the atom is electrically neutral.
    d) inconsistent; most of the atom's mass and all of its positive charge are contained in a small core called the nucleus.

30. a) consistent; most of the volume of the atom is empty space occupied by tiny, negatively charged electrons.
    b) consistent; most of the volume of the atom is empty space occupied by tiny, negatively charged electrons.
    c) inconsistent; there are as many negatively charged electrons outside the nucleus as there are positively charged particles (called protons) inside the nucleus; therefore, the atom is electrically neutral.
    d) inconsistent; Rutherford's nuclear theory does not comment on the relationship of neutrons and protons.

31. Matter appears solid because the variation in the density is on such a small scale that our eyes cannot distinguish the difference.

32. In order to explain the deflections Rutherford observed, the mass and positive charge of an atom must all be concentrated in a space much smaller than the size of the atom itself. Therefore, Rutherford concluded that matter must not be as uniform as it appears. It must contain large regions of empty space dotted with small regions of very dense matter.

Protons, Neutrons, and Electrons

33. a) False; like charges repel.
    b) False; opposite charges attract.
    c) False; all electrons have the same mass and charge.
    d) True

34. a) True
    b) True
    c) True
    d) True

35. a) False; protons and neutrons are nearly identical in mass.
    b) True
    c) False; neutral atoms have an equal number of protons and electrons.
    d) False; protons have a charge of 1+.

36. a) True
    b) False; protons have about the same mass as neutrons, but are about 2000 times heavier than electrons.
    c) True
    d) False; neutrons have no charge. The magnitude of the charge on a proton and electron are the same, but opposite in sign. Protons have about the same mass as neutrons.

37. (Mass of 1 Electron)(X) = (Mass of 1 Proton) $\Rightarrow$

$$X = \frac{\text{(Mass of 1 Proton)}}{\text{(Mass of 1 Electron)}} = \frac{1.67262 \times 10^{-27} \text{kg}}{0.00091 \times 10^{-27} \text{kg}} = 1.8 \times 10^3$$

38. (Mass of 1 Electron)(X) = 2(Mass of 1 Proton) + 2(Mass of 1 Neutron)

$$X = \frac{2(\text{Mass of 1 Proton}) + 2(\text{Mass of 1 Neutron})}{\text{(Mass of 1 Electron)}} \Rightarrow$$

$$X = \frac{2(1.67262 \times 10^{-27} \text{kg}) + 2(1.67493 \times 10^{-27} \text{kg})}{0.00091 \times 10^{-27} \text{kg}} = 7.4 \times 10^3$$

39. $1.0 \text{ g protons} \times \dfrac{1 \text{ proton}}{1.67262 \times 10^{-24} \text{g}} \times \dfrac{1 \text{ electron}}{1 \text{ proton}} \times \dfrac{9.1 \times 10^{-28} \text{g}}{1 \text{ electron}} = 5.4 \times 10^{-4} \text{g}$

40. $1.0 \text{ g electrons} \times \dfrac{1 \text{ electron}}{9.1 \times 10^{-28} \text{g}} \times \dfrac{1 \text{ proton}}{1 \text{ electron}} \times \dfrac{1.67262 \times 10^{-24} \text{g}}{1 \text{ proton}} = 1.8 \times 10^3 \text{g}$

Elements, Symbols, and Names

41. a) 87
    b) 36
    c) 91
    d) 32
    e) 13

42. a) 14
    b) 74
    c) 28
    d) 86
    e) 38

43. a) 18
    b) 50
    c) 54
    d) 8
    e) 81

44. a) 22
    b) 3
    c) 92
    d) 35
    e) 9

45. a) C, 6
    b) N, 7
    c) Na, 11
    d) K, 19
    e) Cu, 29

46. a) B, 5
    b) Ne, 10
    c) Ag, 47
    d) Hg, 80
    e) Cm, 96

47. a) Manganese, 25
    b) Silver, 47
    c) Gold, 79
    d) Lead, 82
    e) Sulfur, 16

48. a) Yttrium, 39
    b) Nitrogen, 7
    c) Neon, 10
    d) Potassium, 19
    e) Molybdenum, 42

49.

Element Name	**Element Symbol**	**Atomic Number**
Gold	Au	79
Tin	Sn	50
Arsenic	As	33
Copper	Cu	29
Iron	Fe	26
Mercury	Hg	80

50. | Element Name | Element Symbol | Atomic Number |
|---|---|---|
| Aluminum | Al | 13 |
| Iodine | I | 53 |
| Antimony | Sb | 51 |
| Sodium | Na | 11 |
| Radon | Rn | 86 |
| Lead | Pb | 82 |

The Periodic Table

51. a) metal   b) metal   c) nonmetal   d) nonmetal   e) metalloid

52. a) metal   b) metalloid   c) metalloid   d) nonmetal   e) metal

53. Metals lose electrons in reactions; therefore, a) potassium, d) barium, and e) copper.

54. Nonmetals gain electrons in reactions; therefore, a) nitrogen and b) iodine.

55. a) Te and b) K

56. c) Mo

57. c) calcium and d) barium

58. c) magnesium and e) beryllium

59. b) sodium and e) rubidium

60. c) potassium and d) lithium

61. a) halogen   b) noble gas   c) halogen   d) neither   e) noble gas

62. a) noble gas   b) halogen   c) neither   d) noble gas   e) halogen

63. a) 16 or VIA   b) 13 or IIIA   c) 14 or IVA   d) 14 or IVA   e) 15 or VA

64. a) 14 or IVA   b) 15 or VA   c) 16 or VIA   d) 14 or IVA   e) 13 or IIIA

65. The elements most like sulfur would be those elements in the same group (6A) because they all have similar physical and chemical properties. Therefore, the answer is b) oxygen.

66. The elements most like magnesium would be those elements in the same group (2A) because they all have similar physical and chemical properties. Therefore, the answer is d) calcium.

67. The elements that come from the same group would be the most similar because they have similar chemical and physical properties. Therefore, the answer is b) Cl and F.

68. The elements that come from the same group would be the most similar because they have similar chemical and physical properties. Therefore, the answer is c) Li and Na.

69. (d) S is the main-group nonmetal.

70. (b) Pd is the row 5 transition element.

71.
Chemical Symbol	Group Number	Group Name	Metal or Nonmetal
K	1A or 1	alkali metal	metal
Br	7A or 17	halogen	nonmetal
Sr	2A or 2	alkaline earth	metal
He	8A or 18	noble gas	nonmetal
Ar	8A or 18	noble gas	nonmetal

72.
Chemical Symbol	Group Number	Group Name	Metal or Nonmetal
Cl	7A or 17	halogen	nonmetal
Ca	2A or 2	alkaline earth	metal
Xe	8A or 18	noble gas	nonmetal
Na	1A or 1	alkali metal	metal
F	7A or 17	halogen	nonmetal

Ions

73. a) $Na \rightarrow Na^+ + \underline{e^-}$
    b) $O + 2e^- \rightarrow \underline{O^{2-}}$
    c) $Ca \rightarrow Ca^{2+} + \underline{2e^-}$
    d) $Cl + e^- \rightarrow \underline{Cl^-}$

74. a) $Mg \rightarrow \underline{Mg^{2+}} + 2e^-$
    b) $Ba \rightarrow Ba^{2+} + \underline{2e^-}$
    c) $I + e^- \rightarrow \underline{I^-}$
    d) $Al \rightarrow \underline{Al^{3+}} + 3e^-$

75. a) oxygen ion charge = 8 (+1) + 10 (−1) = −2
    b) aluminum ion charge = 13 (+1) + 10 (−1) = +3
    c) titanium ion charge = 22 (+1) + 18 (−1) = +4
    d) iodine ion charge = 53 (+1) + 54 (−1) = −1

76. a) tungsten ion charge = 74 (+1) + 68 (−1) = +6
    b) tellurium ion charge = 52 (+1) + 54 (−1) = −2
    c) nitrogen ion charge = 7 (+1) + 10 (−1) = −3
    d) barium ion charge = 56 (+1) + 54 (−1) = +2

77. The number of protons is determined using the atomic number of each element. The number of electrons is determined by examining the net charge on the ion.
    a) 11p + 10e⁻ = +1
    b) 56p + 54e⁻ = +2
    c) 8p + 10e⁻ = −2
    d) 27p + 24e⁻ = +3

78. The number of protons is determined using the atomic number of each element. The number of electrons is determined by examining the net charge on the ion.
    a) 13p + 10e⁻ = +3
    b) 16p + 18e⁻ = −2
    c) 53p + 54e⁻ = −1
    d) 47p + 46e⁻ = +1

79. a) False; The Ti^{2+} ion contains 22 protons and 20 electrons.
    b) True
    c) False; The Mg^{2+} ion contains 12 protons and 10 electrons.
    d) True

80. a) False; The Fe^+ ion contains 26 protons and 25 electrons.
    b) False; The Cs^+ ion contains 55 protons and 54 electrons.
    c) False; The Se^{2-} ion contains 34 protons and 36 electrons.
    d) True

81. a) Rb is in group 1A; therefore, it will form Rb^+.
    b) K is in group 1A; therefore, it will form K^+.
    c) Al is in group 3A; therefore, it will form Al^{3+}.
    d) O is in group 6A; therefore, it will form O^{2-}.

82. a) F is in group 7A; therefore, it will form F^-.
    b) N is in group 5A; therefore, it will form N^{3-}.
    c) Mg is in group 2A; therefore, it will form Mg^{2+}.
    d) Na is in group 1A; therefore, it will form Na^+.

83. a) Ga is in group 3A; therefore, it will lose 3 electrons.
    b) Li is in group 1A; therefore, it will lose 1 electron.
    c) Br is in group 7A; therefore, it will gain 1 electron.
    d) S is in group 6A; therefore, it will gain 2 electrons.

84. a) I is in group 7A; therefore, it will gain 1 electron.
    b) Ba is in group 2A; therefore, it will lose 2 electrons.
    c) Cs is in group 1A; therefore, it will lose 1 electron.
    d) Se is in group 6A; therefore, it will gain 2 electrons.

85.

Symbol	Ion Formed	# electrons in ion	# protons in ion
Te	Te^{2-}	54	52
In	In^{3+}	46	49
Sr	Sr^{2+}	36	38
Mg	Mg^{2+}	10	12
Cl	Cl^-	18	17

86.

Symbol	Ion Formed	# electrons in ion	# protons in ion
F	F^-	10	9
Be	Be^{2+}	2	4
Br	Br^-	36	35
Al	Al^{3+}	10	13
O	O^{2-}	10	8

Isotopes

87. a) Z = 1, A = 3
    b) Z = 24, A = 52
    c) Z = 20, A = 42
    d) Z = 73, A = 182

88. a) 59 = 28 + n → n = 31
    b) 235 = 92 + n → n = 143
    c) 46 = 21 + n → n = 25
    d) 42 = 18 + n → n = 24

89. a) $^{16}_{8}O$

    b) $^{19}_{9}F$

    c) $^{23}_{11}Na$

    d) $^{27}_{13}Al$

90. a) I-127
    b) P-31
    c) U-326
    d) Ar-40

91. a) $^{60}_{27}$Co

    b) $^{22}_{10}$Ne

    c) $^{131}_{53}$I

    d) $^{244}_{94}$Pu

92. a) $^{235}_{92}$U

    b) $^{52}_{23}$V

    c) $^{32}_{15}$P

    d) $^{144}_{54}$Xe

93. a) protons = Z = 11, neutrons = A − Z = 23 − 11 = 12
    b) protons = Z = 88, neutrons = A − Z = 266 − 88 = 178
    c) protons = Z = 82, neutrons = A − Z = 208 − 82 = 126
    d) protons = Z = 7, neutrons = A − Z = 14 − 7 = 7

94. a) protons = Z = 15, neutrons = A − Z = 33 − 15 = 18
    b) protons = Z = 19, neutrons = A − Z = 40 − 19 = 21
    c) protons = Z = 86, neutrons = A − Z = 222 − 86 = 136
    d) protons = Z = 43, neutrons = A − Z = 99 − 43 = 56

95. Carbon-14 has Z = 6; therefore, protons = Z = 6, and neutrons = A − Z = 14 − 6 = 8. The correct isotope symbol would be $^{14}_{6}$C.

96. Plutonium-239 has Z = 94; therefore, protons = Z = 94, and neutrons = A − Z = 239 − 94 = 145. The correct isotope symbol would be $^{239}_{94}$Pu.

Atomic Mass

97. Calculating the atomic mass of an element involves taking a weighted average in which you multiply the percent natural abundance by the atomic mass for each isotope and then adding these products together. For rubidium:
    Atomic Mass = Σ(isotopic abundance) × (isotopic mass)
    (0.7217)(84.9118 amu) + (0.2783)(86.9092 amu) = 85.47 amu

98. Calculating the atomic mass of an element involves taking a weighted average in which you multiply the percent natural abundance by the atomic mass for each isotope and then adding these products together. For silicon:
Atomic Mass = Σ(isotopic abundance) × (isotopic mass)
(0.9221)(27.9769 amu) + (0.0467)(28.9765 amu) + (0.0310)(29.9737 amu) = 28.09 amu

99. a) 100 − 50.69 = 49.31%
   b) (0.5069)(mass) + (0.4931)(80.9163) = 79.904 amu
   Mass = (79.904 − 39.90)/0.5069
   Mass = 40.00/0.5069 = 78.92 amu

100. a) 107.87 amu
   b) 100 − 51.84 = 48.16%
   c) (0.5184)(106.905 amu) + (0.4816)(Mass) = 107.87 amu
   Mass = (107.87 − 55.4196)/0.4816
   Mass = 52.4504/0.4816 = 108.9 amu

101. MW = (0.574)(120.9038) + (0.426)(122.9042)   MW = 69.4 + 52.4 = 121.8 amu
   The atomic weight is closest to that of antimony (Sb), which is 121.75 amu.

102. 100 − 69.17 = 30.83%
   (0.6917)(62.939 amu) + (0.3083)(Mass) = 63.55 amu
   Mass = (63.55 − 43.53491)/0.3083
   Mass = 20.0151/0.3083 = 64.92 amu

Cumulative Problems

103. $-125 \text{ mC} \times \dfrac{1 \text{ C}}{1000 \text{ mC}} \times \dfrac{1 \text{ e}^-}{-1.6 \times 10^{-19} \text{ C}} = 7.8 \times 10^{17} \text{ e}^-$

104. $398 \text{ mC} \times \dfrac{1 \text{ C}}{1000 \text{ mC}} \times \dfrac{1 \text{ p}}{1.6 \times 10^{-19} \text{ C}} = 2.5 \times 10^{18} \text{ p}$

105. Nucleus: $V = \dfrac{4}{3}\pi(1.0 \times 10^{-15} \text{ m})^3 = 4.2 \times 10^{-45} \text{ m}^3$

   Hydrogen: $V = \dfrac{4}{3}\pi(53 \times 10^{-12} \text{ m})^3 = 6.2 \times 10^{-31} \text{ m}^3$

   Percentage: $\dfrac{4.2 \times 10^{-45} \text{ m}^3}{6.2 \times 10^{-31} \text{ m}^3} \times 100 = 6.8 \times 10^{-13}\%$

106. Nucleus: $V = \frac{4}{3}\pi(2.7\times10^{-15}\text{m})^3 = 8.2448\times10^{-44}\text{m}^3$

   Carbon: $V = \frac{4}{3}\pi(70\times10^{-12}\text{m})^3 = 1.4368\times10^{-30}\text{m}^3$

   Percentage: $\frac{8.2448\times10^{-44}\text{m}^3}{1.4368\times10^{-30}\text{m}^3}\times100 = 5.7\times10^{-12}\%$

107.

Symbol	#p	#n	A (Mass Number)	Natural Abundance
Sr-84 or $^{84}_{38}$Sr	38	46	84	0.56%
Sr-86 or $^{86}_{38}$Sr	38	48	86	9.86%
Sr-87 or $^{87}_{38}$Sr	38	49	87	7.00%
Sr-88 or $^{88}_{38}$Sr	38	50	88	82.58%

Atomic Mass of Sr is the weighted average of each isotope:

Sr = $(0.0056\times83.9134)+(0.0986\times85.9093)+(0.0700\times86.9089)+$
   (0.8258×87.9056)

Sr = $0.47+8.47+6.08+72.59 = 87.61$ amu

108.

Symbol	#p	#n
Cr-50	24	26
Cr-52	24	28
Cr-53	24	29
Cr-54	24	30

The atomic mass of Cr is the weighted average of each isotope:

Cr = $(0.04345\times49.9460)+(0.8379\times51.9405)+(0.0950\times52.9407)+$
   (0.02365×53.9389)

Cr = $2.170+43.52+5.03+1.276 = 52.00$ amu

109.

Symbol	Z	A	#p	#e⁻	#n	Charge
Zn^{2+}	30	64	30	28	34	2+
Mn^{3+}	25	55	25	22	30	3+
P	15	31	15	15	16	0
O^{2-}	8	16	8	10	8	2−
S^{2-}	16	34	16	18	18	2−

110.

Symbol	Z	A	#p	#e⁻	#n	Charge
Mg^{2+}	12	25	12	10	13	2+
Ti^{4+}	22	48	22	18	26	4+
S^{2-}	16	32	16	18	16	2−
Ga^{3+}	31	71	31	28	40	3+
Pb^{2+}	82	207	82	80	125	2+

111. % abundance of Eu-153: $100 - 47.8 = 52.2\%$
$(0.478)(150.9198 \text{ amu}) + (0.522)(\text{Mass }^{153}\text{Eu}) = 151.97 \text{ amu}$
Mass ^{153}Eu $= (151.97 - 72.1397)/(0.522) = 153$ amu

112. This problem is solved by setting up two equations with two unknowns and solving for each variable. The first equation we can write is for determining the MW of Re, and the second equation is the sum of the two isotope weights.
    (1) MW of Re:
    $(0.3740)(\text{Re-185}) + (0.6260)(\text{Re-187}) = 186.21$ amu

    (2) Isotope Sum:
    Re-185 + Re-187 = 371.9087

    Solve Eqn (2) for one of the variables:
    Re-185 = 371.9087 − Re-187

    Substitute this value into Eqn (1):
    $(0.3740)(371.9087 - \text{Re-187}) + (0.6260)(\text{Re-187}) = 186.21$

    Solve this equation for Re-187:
    $139.0939 - 0.3740(\text{Re-187}) + 0.6260(\text{Re-187}) = 186.21$
    $139.0939 + 0.2520(\text{Re-187}) = 186.21$
    $0.2520(\text{Re-187}) = 186.21 - 139.0939$
    Re-187 = 47.1161/0.2520 = 186.969   (187.0 amu)

    Using this value of Re-187, solve for Re-185 using Eqn (2):
    Re-185 + 186.989 = 371.9087
    Re-185 = 371.9087 − 186.989 = 184.9197 (185.0 amu)

113. The atomic theory and nuclear model of the atom are both theories because they attempt to provide a broader understanding and model behavior of chemical systems.

114. Mendeleev's periodic law is an example from this chapter. It is a scientific law because it summarizes many observations but does not give an explanation for the observations.

115. The atomic mass reported on the periodic table is a weighted average of all natural stable isotopes. As fluorine only has one isotope, the atomic mass is identical to the mass of the isotope. Chlorine, however, must have more than one stable isotope that occurs naturally. The relative abundance of each isotope factors into creating the average atomic mass reported on the periodic table.

116. No, most elements have more than one stable natural isotope; therefore, the reported atomic mass is a weighted average that reflects the natural distribution of isotopes and does not represent any one single atom.

117. Set x = abundance of Cu-63, then 1 − x = abundance of Cu-65
$(62.9396)(x) + (64.9278)(1-x) = 63.55$
$62.9396x + 64.9278 - 64.9278x = 63.55$
$62.9396x - 64.9278x = 63.55 - 64.9278$
$-1.9882x = -1.38 \Rightarrow x = -1.38 / -1.9882 = 0.693$ (69.3%)
Cu-65 = 1 − 0.693 = 0.307 (30.7%)

118. Set x = abundance of Ga-69, then 1.0 − x = abundance of Ga-71
$(68.9256)(x) + (70.9247)(1-x) = 69.72$
$68.9256x + 70.9247 - 70.9247x = 69.72$
$68.9256x - 70.9247x = 69.72 - 70.9247$
$-1.9991x = -1.20 \Rightarrow x = -1.20 / -1.9991 = 0.603$ (60.3%)
Ga-71 = 1 − 0.603 = 0.397 (39.7%)

Highlight Problems

119. a) Nt-304: $\frac{36}{50} \times 100\% = 72\%$; Nt-305: $\frac{2}{50} \times 100\% = 4.0\%$;

Nt-306: $\frac{12}{50} \times 100\% = 24\%$

b) MW of Nt = (0.72)(303.956) + (0.040)(304.962) + (0.24)(305.978)
= 304.4 amu (assuming the percentages have 4 significant digits)

```
120
Nt
304.5
```

120. a) mass = 1.67493×10^{-27} kg $\times \frac{1000 \text{ g}}{1 \text{ kg}} = 1.67493 \times 10^{-24}$ g

$V = \frac{4}{3}\pi(1.0 \times 10^{-13} \text{ cm})^3 = 4.1888 \times 10^{-39} \text{ cm}^3$

$d = \frac{1.67493 \times 10^{-24}}{4.1888 \times 10^{-39}} = 4.0 \times 10^{14} \text{ g/cm}^3$

b) $V = \frac{4}{3}\pi(1.0 \times 10^{-2} \text{ cm})^3 = 4.1888 \times 10^{-6} \text{ cm}^3$

$4.1888 \times 10^{-6} \text{ cm}^3 \times 4.0 \times 10^{14} \frac{\text{g}}{\text{cm}^3} \times \frac{1 \text{ kg}}{1000 \text{ g}} = 1.7 \times 10^{6}$ kg

121.

Particle	Mass (amu)	Charge	In the nucleus? (yes/no)	# in ^{32}S atom	# in ^{79}Br$^-$ ion
Proton	1.0073	+1	yes	16	35
Neutron	1.0087	0	yes	16	44
Electron	0.00055	−1	no	16	36

122. In the style of Figure 4.6:

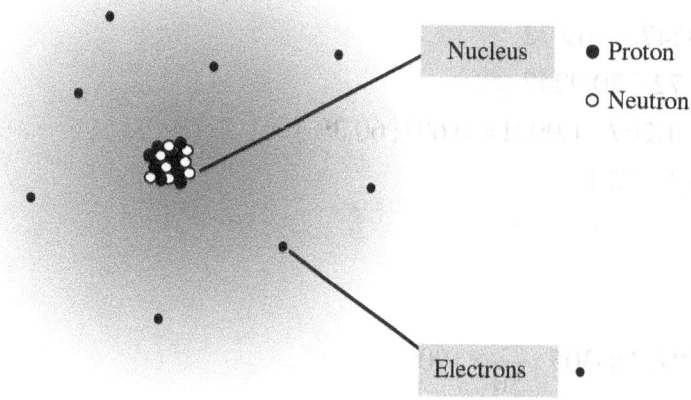

123. Mass increase is typically ~2 amu; missing element is phosphorus.

| Li 7 LiCl | Be 9 BeCl$_2$ | B 10.8 BH$_3$ | C 12 CH$_4$ | N 14 NF$_3$ | O 16 H$_2$O | F 19 F$_2$ | Na 23 NaCl | Mg 24.3 MgCl$_2$ | Al 27 AlH$_3$ | Si 28 SiH$_4$ | ? | S 32 H$_2$S | Cl 35.4 Cl$_2$ | K 39 KCl | Ca 40 CaCl$_2$ | Ga 69.7 GaH$_3$ | Ge 72.6 GeH$_4$ | As 75 AsF$_3$ | Se 79 H$_2$Se | Br 80 Br$_2$ |

124. Patterns for stable compounds are more noticeable:

Li 7 LiCl	Be 9 BeCl$_2$	B 10.8 BH$_3$	C 12 CH$_4$	N 14 NF$_3$	O 16 H$_2$O	F 19 F$_2$
Na 23 NaCl	Mg 24.3 MgCl$_2$	Al 27 AlH$_3$	Si 28 SiH$_4$	? 32 H$_2$S		S 32 H$_2$S	Cl 35.4 Cl$_2$
K 39 KCl	Ca 40 CaCl$_2$	Ga 69.7 GaH$_3$	Ge 72.6 GeH$_4$	As 75 AsF$_3$	Se 79 H$_2$Se	Br 80 Br$_2$

? = 30 XF$_3$

Data Interpretation and Analysis

125.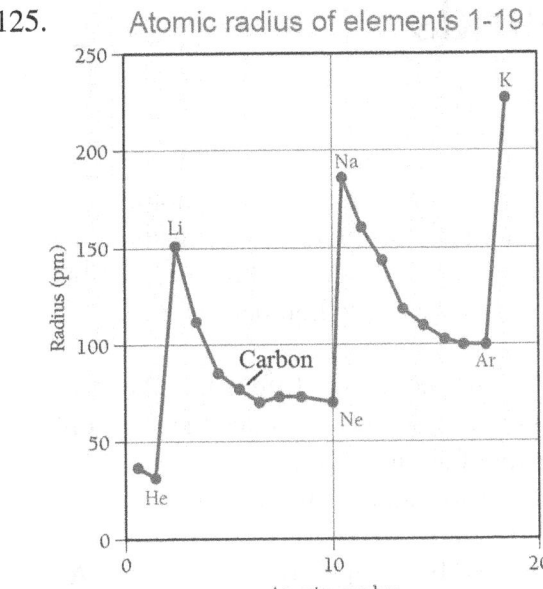

a) The atomic radius decreases as you go across each period; however, the atomic radius increases as you go down each group.
b) The elements in blue (Li, Na, K) are the largest elements of their respective periods.
c) The elements in red (He, Ne, Ar) are the smallest elements of their respective periods.
d) 75 pm

Molecules and Compounds 5

Questions

1. The properties of an element completely change when it combines with another element to form a compound. For example, water is made out of the elements hydrogen and oxygen, both of which are gases. When hydrogen and oxygen combine to form water, their properties change and they become a new liquid substance.

2. The world would be very different because life could not exist! Life is based on the formation of compounds from elements in many different ratios. Since there are only about 90 naturally occurring elements, the ability of atoms to combine to form compounds gives rise to the amazing diversity of substances in the world.

3. The law of constant composition was first expressed by Joseph Proust and states: All samples of a given compound have the same proportions of their constituent elements.

4. A chemical formula is a symbolic representation of the types and number of atoms of each element present in a compound. The elements are represented by their chemical symbol, and a subscript indicates how many atoms of that element are present. For example, the formula for water, H_2O, indicates that a molecule of water contains 2 hydrogen atoms (H) and 1 oxygen atom (O). The formula for table salt, NaCl (sodium chloride), indicates that sodium atoms and chlorine atoms are present in a 1:1 ratio.

5. The general rule for listing elements in a compound is that the most metallic element is listed first. For compounds that contain a metal, it is listed first. For compounds that do not contain a metal, the most metal-like element is listed first. For nonmetals, you list the element that is found farthest to the left and/or the element that is lowest on the periodic table.

6. To determine the number of atoms in a compound that contains parentheses, you multiply the parentheses subscript by the subscript of each element within the parentheses. Remember, if there is no subscript by an element, it is equal to 1. For example, $Mg(C_2H_3O_2)_2$ has the following number of atoms:
Mg = 1, C = 2 × 2 = 4, H = 2 × 3 = 6, O = 2 × 2 = 4.

7. An empirical formula gives the simplest whole-number ratio of atoms of each element in a compound, whereas a molecular formula gives the actual number of atoms of each element in a compound.

8. A structural formula uses lines to represent chemical bonds and shows how atoms in a molecule are connected to each other in a two-dimensional form. Molecular models are three-dimensional models that show how atoms are connected to each other and their relative positions in space.

9. Most elements can be found in nature as single atoms, which we call the **atomic elements**. However, some elements can only be found in nature in the diatomic state—that is, two atoms of the same element bonded together, which we call the **molecular elements**. There are only seven diatomic molecular elements: hydrogen (H_2), nitrogen (N_2), oxygen (O_2), fluorine (F_2), chlorine (Cl_2), bromine (Br_2), and iodine (I_2). *It is tradition that when chemists say a molecular element by name, they are referring to the diatomic state. When your instructor says oxygen reacts with hydrogen, you know to write O_2 and H_2. For a reaction that involves just a single atom of these elements, you would say monatomic oxygen (O).*

10. In ionic compounds, the atoms (or group of atoms) exist as ions and are held together by electrostatic forces called ionic bonds. In most cases, the formation of charged particles occurs as a result of a transfer of electrons between a metal and a nonmetal atom. The chemical formula for ionic compounds represents the smallest whole-numbered ratios of positive and negative ions in the neutral compound. In molecular compounds the atoms are held together by covalent bonds formed by the sharing of electrons. The chemical formula for molecular compounds represents the actual number of atoms present in the molecule.

11. The systematic name for a compound will provide the reader with enough information that the formula of the compound can be determined. The common name can be considered a "nickname" for the compound that must be memorized.

12. The metals that form Type I Ionic Compounds are as follows: From group 1A, we have lithium, sodium, potassium, rubidium, and cesium; from group 2A, we have beryllium, magnesium, calcium, strontium, and barium; from group 3A we have aluminum; from group 9B we have silver; and from group 10B we have zinc.

13. The metals that form Type II Ionic Compounds are most commonly found in the center section of the periodic table, known as the transition metals.

14. The basic form for naming Type I Binary Compounds is:
    [Cation Name][Anion Base Name + *-ide*]

15. The basic form for naming Type II Binary Compounds is the same as Type I, with the addition of the charge of the cation, written in roman numerals, inserted in parentheses between the cation name and the anion name. The pattern is:
    [Cation Name](cation charge in roman numerals) [Anion Base Name + *-ide*]

16. The charge of the cation must be indicated in Type II compounds because, in nature, the metal cation is known to have multiple charges that form different compounds; therefore, to be clear as to which compound we are naming, we have to indicate the cation charge in the name.

17. When you name compounds that contain polyatomic ions, you use the same rules from Type I and Type II; however, you insert the name of the polyatomic ions, without altering the name, into their proper place.

18. Polyatomic ions with 2− charge: sulfate (SO_4^{2-}), sulfite (SO_3^{2-}), chromate (CrO_4^{2-}), dichromate ($Cr_2O_7^{2-}$), carbonate (CO_3^{2-}), and hydrogen phosphate (HPO_4^{2-})
Polyatomic ions with 3− charge: phosphate (PO_4^{3-}) and phosphite (PO_3^{3-})

19. The form for naming molecular compounds is to name the more metallic element (i.e., the leftmost element in the periodic table) first. The less metallic element (i.e., the rightmost in the periodic table) is named second, adding the -ide suffix. Each element name is preceded by a numerical prefix to indicate the number of each atom in the compound.

20. mono = 1, di = 2, tri = 3, tetra = 4, penta = 5, hexa = 6

21. The basic form for naming binary acids is:
[hydro + base nonmetal name + ic][acid].

22. The basic form for naming oxyanions with the -ate ending is:
[base oxyanion name + ic][acid].

23. The basic form for naming oxyanions with the -ite ending is:
[base oxyanion name + ous][acid].

24. The formula mass is the average mass of the molecules that compose a compound. The formula mass is calculated by taking the sum of the atomic mass of all elements in the chemical formula.

Problems

Constant Composition of Compounds

25. Sample 1: $\dfrac{\text{mass Cl}}{\text{mass Na}} = \dfrac{7.16}{4.65} = 1.54$

    Sample 2: $\dfrac{\text{mass Cl}}{\text{mass Na}} = \dfrac{11.5}{7.45} = 1.54$, Yes

26. Sample 1: $\dfrac{\text{mass Cl}}{\text{mass C}} = \dfrac{373}{32.4} = 11.5$

　　Sample 2: $\dfrac{\text{mass Cl}}{\text{mass C}} = \dfrac{112}{12.3} = 9.11$, No

27. Sample 1: $\dfrac{\text{mass F}}{\text{mass Mg}} = \dfrac{2.57}{1.65} = 1.56$

　　Sample 2: $\dfrac{\text{mass F}}{\text{mass Mg}} = \dfrac{\text{mass F}}{1.32} = 1.56$

　　mass F $= 1.56 \times 1.32 = 2.06$ kg

28. $\dfrac{\text{mass Na}}{\text{mass F}} = 1.21$; Sample: $\dfrac{\text{mass Na}}{\text{mass F}} = \dfrac{34.5}{\text{mass F}} = 1.21 \Rightarrow$

　　mass F $= \dfrac{34.5}{1.21} = 28.5$ g

29.

	Mass N₂O	Mass N	Mass O
Sample A	2.85	1.82	1.04
Sample B	4.55	<u>2.91</u>	<u>1.66</u>
Sample C	<u>3.70</u>	<u>2.36</u>	1.35
Sample D	<u>1.74</u>	1.11	<u>0.634</u>

30.

	Mass FeCl₃	Mass Fe	Mass Cl
Sample A	3.785	1.303	2.482
Sample B	2.175	<u>0.7489</u>	<u>1.426</u>
Sample C	<u>5.844</u>	2.012	<u>3.832</u>
Sample D	<u>3.552</u>	<u>1.223</u>	2.329

Chemical Formulas

31. NI₃

32. CBr₄

33. a) Fe₃O₄
　　b) PCl₃
　　c) PCl₅
　　d) Ag₂O

34. a) CaI₂
　　b) N₂O₄
　　c) SiO₂
　　d) ZnCl₂

35. a) 4
    b) 4
    c) 2 × 3 = 6
    d) 2 × 2 = 4

36. a) 3
    b) 1
    c) 2 × 1 = 2
    d) 2 × 3 = 6

37. a) Mg = 1, Cl = 2
    b) Na = 1, N = 1, O = 3
    c) Ca = 1, N = 2 × 1 = 2, O = 2 × 2 = 4
    d) Sr = 1, O = 2 × 1 = 2, H = 2 × 1 = 2

38. a) N = 1, H = 4, Cl = 1
    b) Mg = 3, P = 2 × 1 = 2, O = 2 × 4 = 8
    c) Na = 1, C = 1, N = 1
    d) Ba = 1, H = 2 × 1 = 2, C = 2 × 1 = 2, O = 3 × 2 = 6

39. Complete the following table

Formula	No. of $C_2H_3O_2^-$ units	No. of C atoms	No. of H atoms	No. of O atoms	No. of metal atoms
$Mg(C_2H_3O_2)_2$	2	4	6	4	1
$NaC_2H_3O_2$	1	2	3	2	1
$Cr_2(C_2H_3O_2)_4$	4	8	12	8	2

40. Complete the following table

Formula	No. of SO_4^{2-} units	No. of S atoms	No. of O atoms	No. of metal atoms
$CaSO_4$	1	1	4	1
$Al_2(SO_4)_3$	3	3	12	2
K_2SO_4	1	1	4	2

41. a) CH_3
    b) NO_2
    c) C_2H_3O
    d) NH_3

42. a) CH
    b) CO_2
    c) CH_2O
    d) BH_3

Molecular View of Elements and Compounds

43. a) molecular
    b) atomic
    c) atomic
    d) molecular

44. b) oxygen and d) bromine are molecular elements

45. a) molecular
    b) ionic
    c) ionic
    d) molecular

46. a) ionic
    b) molecular
    c) molecular
    d) molecular

47. Helium—single atoms
    CCl_4—molecules
    K_2SO_4—formula units
    bromine—diatomic molecules

48. NI_3—molecules
    copper metal—single atoms
    $SrCl_2$—formula units
    nitrogen—diatomic molecules

49. a) formula units
    b) single atoms
    c) molecules
    d) molecules

50. a) formula units
    b) molecules
    c) formula units
    d) molecules

51. a) ionic, single ion type
    b) molecular
    c) molecular
    d) ionic, multiple ion types

52. a) ionic, multiple ion types
    b) molecular
    c) ionic, single ion type
    d) molecular

Writing Formulas for Ionic Compounds

53. a) Na = +1, S = −2: Na_2S
    b) Sr = +2, O = −2: SrO
    c) Al = +3, S = −2: Al_2S_3
    d) Mg = +2, Cl = −1: $MgCl_2$

54. a) Al = +3, O = −2: Al_2O_3
    b) Be = +2, I = −1: BeI_2
    c) Ca = +2, S = −2: CaS
    d) Ca = +2, I = −1: CaI_2

55. a) $KC_2H_3O_2$
    b) K_2CrO_4
    c) K_3PO_4
    d) KCN

56. a) $Ca(OH)_2$
    b) $CaCO_3$
    c) $Ca_3(PO_4)_2$
    d) $CaHPO_4$

57. N = −3, O = −2, F = −1
    a) Li = +1: Li_3N, Li_2O, LiF
    b) Ba = +2: Ba_3N_2, BaO, BaF_2
    c) Al = +3: AlN, Al_2O_3, AlF_3

58. a) Rb = +1: $RbNO_3$, Rb_2SO_4, Rb_3PO_4
    b) Sr = +2: $Sr(NO_3)_2$, $SrSO_4$, $Sr_3(PO_4)_2$
    c) In = +3: $In(NO_3)_3$, $In_2(SO_4)_3$, $InPO_4$

Naming Ionic Compounds

59. a) cesium chloride
    b) strontium bromide
    c) potassium oxide
    d) lithium fluoride

60. a) lithium iodide
    b) magnesium sulfide
    c) barium fluoride
    d) sodium fluoride

61. a) chromium(II) chloride
    b) chromium(III) chloride
    c) tin(IV) oxide
    d) lead(II) iodide

62. a) mercury(II) bromide
    b) iron(III) oxide
    c) copper(II) iodide
    d) tin(IV) chloride

63. a) multiple ion types
    b) single ion type
    c) single ion type
    d) multiple ion types

64. a) multiple ion types
    b) multiple ion types
    c) single ion type
    d) single ion type

65. a) barium nitrate
    b) lead(II) acetate
    c) ammonium iodide
    d) potassium chlorate
    e) cobalt(II) sulfate
    f) sodium perchlorate

66. a) barium hydroxide
    b) iron(III) hydroxide
    c) copper(II) nitrite
    d) lead(II) sulfate
    e) potassium hypochlorite
    f) magnesium acetate

67. a) hypobromite ion
    b) bromite ion
    c) bromate ion
    d) perbromate ion

68. a) hypoiodite ion
    b) iodite ion
    c) iodate ion
    d) periodate ion

69. a) $CuBr_2$
    b) $AgNO_3$
    c) KOH
    d) Na_2SO_4
    e) $KHSO_4$
    f) $NaHCO_3$

70. a) $CuClO_3$
    b) $KMnO_4$
    c) $PbCrO_4$
    d) CaF_2
    e) $Fe_3(PO_4)_2$
    f) $LiHSO_3$

Naming Molecular Compounds

71. a) sulfur dioxide
    b) nitrogen triiodide
    c) bromine pentafluoride
    d) nitrogen monoxide
    e) tetranitrogen tetraselenide

72. a) xenon tetrafluoride
    b) phosphorus triiodide
    c) sulfur trioxide
    d) silicon tetrachloride
    e) diiodine pentoxide

73. a) CO
    b) S_2F_4
    c) Cl_2O
    d) PF_5
    e) BBr_3
    f) P_2S_5

74. a) ClO
    b) XeO$_4$
    c) XeF$_6$
    d) CBr$_4$
    e) B$_2$Cl$_4$
    f) P$_4$Se$_3$

75. a) incorrect, phosphorus pentabromide
    b) incorrect, diphosphorus trioxide
    c) incorrect, sulfur tetrafluoride
    d) incorrect, nitrogen tetrafluoride

76. a) incorrect, nitrogen trichloride
    b) incorrect, carbon tetraiodide
    c) incorrect, carbon monoxide
    d) correct

Naming Acids

77. a) oxyacid, nitrous acid, nitrite ion
    b) binary, hydroiodic acid
    c) oxyacid, sulfuric acid, sulfate ion
    d) oxyacid, nitric acid, nitrate ion

78. a) oxyacid, carbonic acid, carbonate ion
    b) oxyacid, acetic acid, acetate ion
    c) oxyacid, phosphoric acid, phosphate ion
    d) binary, hydrochloric acid

79. a) hypochlorous acid
    b) chlorous acid
    c) chloric acid
    d) perchloric acid

80. a) bromic acid
    b) iodic acid

81. a) H$_3$PO$_4$
    b) HBr (*aq*)
    c) H$_2$SO$_3$

82. a) HF (*aq*)
    b) HCN (*aq*)
    c) HClO$_2$

Formula Mass

83. a) FM = 1.01 + 14.01 + 3(16.00) = 63.02 amu
    b) FM = 40.08 + 2(79.90) = 199.88 amu
    c) FM = 12.01 + 4(35.45) = 153.81 amu
    d) FM = 87.62 + 2(14.01) + 6(16.00) = 211.64 amu

84. a) FM = 12.01 + 2(32.07) = 76.13 amu
    b) FM = 6(12.01) + 12(1.01) + 6(16.00) = 180.18 amu
    c) FM = 55.85 + 3(14.01) + 9(16.00) = 241.88 amu
    d) FM = 7(12.01) + 16(1.01) = 100.23 amu

85. PBr_3(FM = 270.67 amu) > Ag_2O(FM = 231.74 amu) > PtO_2(FM = 227.08 amu) > $Al(NO_3)_3$(FM = 213.01 amu)

86. $Pb(C_2H_3O_2)_2$(FM = 325.3 amu) > Rb_2SO_4(FM = 267.01 amu) > WO_2(FM = 215.85 amu) > RbI(FM = 212.37 amu)

Cumulative Problems

87. a) CH_4
    b) SO_3
    c) NO_2

88. a) NH_3
    b) H_2S
    c) CO_2

89. a) 3 × 4 = 12
    b) 2 × 2 = 4
    c) 4 × 3 = 12
    d) 7 × 1 = 7

90. a) 4 × 1 = 4
    b) 2 × 3 = 6
    c) 3 × 2 = 6
    d) 5 × 4 = 20

91. a) 8
    b) 12
    c) 12

92. a) 10
    b) 6
    c) 3

93. | Formula | Type | Name |
|---|---|---|
| N₂H₄ | molecular | dinitrogen tetrahydride |
| KCl | ionic | potassium chloride |
| H₂CrO₄ (*aq*) | acid | chromic acid |
| Co(CN)₃ | ionic | cobalt(III) cyanide |

94. | Formula | Type | Name |
|---|---|---|
| K₂Cr₂O₇ | ionic | potassium dichromate |
| HBr (*aq*) | acid | hydrobromic acid |
| N₂O₅ | molecular | dinitrogen pentoxide |
| PbO₂ | ionic | lead(IV) oxide |

95. a) incorrect, calcium nitrite
    b) incorrect, potassium oxide
    c) incorrect, phosphorus trichloride
    d) correct
    e) potassium iodite

96. a) incorrect, nitric acid
    b) correct
    c) incorrect, calcium iodide
    d) incorrect, tin(II) chromate
    e) incorrect, sodium bromate

97. a) Sn(SO₄)₂, FM = 118.71 + 2(32.07) + 8(16.00) = 310.85 amu
    b) HNO₂, FM = 1.008 + 14.01 + 2(16.00) = 47.02 amu
    c) NaHCO₃, FM = 22.99 + 1.008 + 12.01 + 3(16.00) = 84.01 amu
    d) PF₅, FM = 30.97 + 5(19.00) = 125.97 amu

98. a) BaBr₂, FM = 137.33 + 2(79.90) = 297.13 amu
    b) N₂O₃, FM = 2(14.01) + 3(16.00) = 76.02 amu
    c) Cu₂SO₄, FM = 2(63.55) + 32.07 + 4(16.00) = 223.17 amu
    d) HBr (*aq*), FM = 1.008 + 79.90 = 80.91 amu

99. a) platinum(IV) oxide, FM = 195.08 + 2(16.00) = 227.08 amu
    b) dinitrogen pentoxide, FM = 2(14.01) + 5(16.00) = 108.02 amu
    c) aluminum chlorate, FM = 26.98 + 3(35.45) + 9(16.00) = 277.33 amu
    d) phosphorus pentabromide, FM = 30.97 + 5(79.90) = 430.47 amu

100. a) aluminum sulfate, FM = 2(26.98) + 3(32.07) + 12(16.00) = 342.17 amu
     b) diphosphorus trioxide, FM = 2(30.97) + 3(16.00) = 109.94 amu
     c) hypochlorous acid, FM = 1.008 + 35.45 + 16.00 = 52.46 amu
     d) chromium(III) acetate, FM = 52.00 + 6(12.01) + 9(1.008) + 6(16.00)
        = 229.13 amu

101. C = 12.01 g/mol, H = 1.01 g/mol, and the compound has a formula mass of 28.06 amu. If the compound has 1 carbon atom, the remaining mass would be equal to 15.89 hydrogen atoms. If the compound has 2 carbon atoms, the remaining mass is equal to exactly 4 hydrogen atoms; therefore, the formula of the compound is C_2H_4.

102. N = 14.01 g/mol, O = 16.00 g/mol, and the compound has a formula mass of 44.02 amu. If the compound has 1 nitrogen atom, the remaining mass would be equal to 1.88 oxygen atoms. If the compound has 2 nitrogen atoms, the remaining mass is equal to exactly 1 oxygen atom; therefore, the formula of the compound is N_2O.

103. Ten different masses of CCl_4 masses, as shown in the table.

C Isotope	No. of Cl-35	No. of Cl-37	Formula Mass
C-12	4	0	151.88
	3	1	153.88
	2	2	155.88
	1	3	157.88
	0	4	159.88
C-13	4	0	152.88
	3	1	154.88
	2	2	156.88
	1	3	158.88
	0	4	160.88

104. Eight different masses of NBr_3 masses, as shown in the table

N Isotope	No. of Br-79	No. of Br-81	Formula Mass
N-14	3	0	250.76
	2	1	252.76
	1	2	254.76
	0	3	256.76
N-15	3	0	251.76
	2	1	253.76
	1	2	255.76
	0	3	257.76

Highlight Problems

105. a) molecular element
     b) atomic element
     c) ionic compound
     d) molecular compound

106. FM = 2952(12.01) + 4664(1.01) + 832(16.00) + 812(14.01) + 8(32.07) + 4(55.85)
     $= 6.533 \times 10^4$ amu

107. Image 1: sodium hypochlorite: NaClO
     Image 2: calcium carbonate: $CaCO_3$
     Image 3: aluminum hydroxide: $Al(OH)_3$, magnesium hydroxide: $Mg(OH)_2$
     Image 4: sodium bicarbonate: $NaHCO_3$, sodium aluminum sulfate: $NaAl(SO_4)_2$

108. Carbon monoxide: CO; carbon dioxide: CO_2; carbonate ion: CO_3^{2-}
     Similarities: all contain carbon and oxygen, all contain 1 carbon atom
     Differences: number of oxygen atoms differ, covalent (CO, CO_2) and ion (CO_3^{2-})

109. Atomic elements: Does it have single atoms as its base units?
     Molecular elements: Does it have a diatomic molecule as its base units?
     Molecular compounds: Is it composed of two (or more) nonmetals?
     Ionic compounds: Is it composed of a metal cation and a nonmetal anion?
     Instructions: Determine first if the substance has more than one type of element; if it does not, then it is either an atomic element or a molecular element. Determine which one by counting the number of atoms in the base unit. If the substance has more than one type of element, determine between molecular and ionic by classifying the elements as metals and nonmetals.

110. Calcium nitrate: ionic compound, name cation first followed by name of anion.
     Dinitrogen monoxide: covalent compound, name elements using prefixes to indicate number of atoms.
     Sodium sulfide: ionic compound, name cation first followed by name of anion.
     Chromium(III) chloride: ionic compound, name of cation must indicate charge, which is determined by examining the total anion charge followed by the name of the anion.

111. Calcium nitrate: 164.09 amu
     Dinitrogen monoxide: 44.01 amu
     Sodium sulfide: 78.05 amu
     Chromium(III) chloride: 158.36 amu

Data Interpretation and Analysis

112.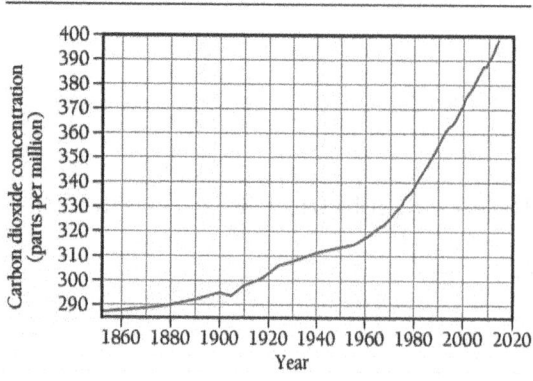

a) 1950: 285 ppm
   2000: 370 ppm
   Increase of 85 ppm

b) $\dfrac{85 \text{ ppm}}{50 \text{ yr}} = 1.7$ ppm/yr

c) $\dfrac{1.7 \text{ ppm}}{\text{yr}} \times 40 \text{ yr} = 68$ ppm
   390 ppm + 68 ppm = 458 ppm

Chemical Composition   6

Questions

1. Chemical composition is important to understand because it provides an analysis of the amount of an element found within a compound.

2. You can determine the number of atoms in a sample by first determining the mass of the sample and then converting it to the number of atoms using Avogadro's number and the molar mass of the element. Simply counting the atoms is out of the question as they are so incredibly small it makes traditional counting impossible.

3. There are 6.022×10^{23} atoms in 1 mole of atoms.

4. There are 6.022×10^{23} molecules in 1 mole of molecules.

5. One mole of atoms of an element has a mass in grams that is equal to the mass of the atom in atomic mass units (amu).

6. The mass, in grams, of 1 mole of molecules of a compound is equal to the molar mass or formula unit mass of that compound expressed in amu.

7. a) 30.97 g   b) 195.08 g   c) 12.01 g   d) 52.00 g

8. a) $12.01 + 2(16.00) = 44.01$ g
   b) $12.01 + 2(1.008) + 2(35.45) = 84.93$ g
   c) $12(12.01) + 22(1.01) + 11(16.00) = 342.34$ g
   d) $32.07 + 2(16.00) = 64.07$ g

9. The subscripts provide a ratio of atoms of one element to atoms of another element within a compound. This ratio does not apply to mass because every element has a different mass. A one-to-one atomic ratio for hydrogen to oxygen is very different than the one-to-one atomic ratio for hydrogen to sulfur because sulfur has a mass of over twice the mass of oxygen.

10. 1 mole $C_{12}H_{22}O_{11}$ / 12 moles C atoms
    1 mole $C_{12}H_{22}O_{11}$ / 22 moles H atoms
    1 mole $C_{12}H_{22}O_{11}$ / 11 moles O atoms

11. a) 11.19 g H / 100 g water
    b) 53.29 g O / 100 g fructose
    c) 84.12 g C / 100 g gasoline
    d) 52.14 g C / 100 g ethanol

12. mass % of element X = $\dfrac{\text{mass of X in 1 mole of compound}}{\text{mass of 1 mole of the compound}} \times 100\%$

13. The molecular formula is a whole-number multiple of the empirical formula.

14. It is important to be able to calculate an empirical formula from experimental data because this is how a chemist analyzes an unknown compound to determine the relative masses of the elements it contains.

15. The empirical formula mass is the sum of the masses of all atoms found in the simplest whole-number ratio of each atom type found in a compound.

16. The molar mass is a whole-number multiple of the empirical formula mass.

Problems

The Mole Concept

17. moles → atoms

$$5.8 \text{ mol Hg} \times \dfrac{6.022 \times 10^{23} \text{ Hg atoms}}{1 \text{ mol Hg}} = 3.5 \times 10^{24} \text{ Hg atoms}$$

18. atoms → moles

$$3.45 \times 10^{24} \text{ Au atoms} \times \dfrac{1 \text{ mol Au}}{6.022 \times 10^{23} \text{ Au atoms}} = 5.73 \text{ mol Au}$$

19. a) $3.4 \text{ mol Cu} \times \dfrac{6.022 \times 10^{23} \text{ Cu atoms}}{1 \text{ mole Cu}} = 2.0 \times 10^{24} \text{ Cu atoms}$

b) $9.7 \times 10^{-3} \text{ mol C} \times \dfrac{6.022 \times 10^{23} \text{ C atoms}}{1 \text{ mole C}} = 5.8 \times 10^{21} \text{ C atoms}$

c) $22.9 \text{ mol Hg} \times \dfrac{6.022 \times 10^{23} \text{ Hg atoms}}{1 \text{ mole Hg}} = 1.38 \times 10^{25} \text{ Hg atoms}$

d) $0.215 \text{ mol Na} \times \dfrac{6.022 \times 10^{23} \text{ Na atoms}}{1 \text{ mole Na}} = 1.29 \times 10^{23} \text{ Na atoms}$

20. a) $4.6 \times 10^{24} \text{ Pb atoms} \times \dfrac{1 \text{ mole Pb}}{6.022 \times 10^{23} \text{ Pb atoms}} = 7.6 \text{ moles Pb}$

b) $2.87 \times 10^{22} \text{ He atoms} \times \dfrac{1 \text{ mole He}}{6.022 \times 10^{23} \text{ He atoms}} = 4.77 \times 10^{-2} \text{ moles He}$

c) $7.91 \times 10^{23} \text{ K atoms} \times \dfrac{1 \text{ mole K}}{6.022 \times 10^{23} \text{ K atoms}} = 1.31 \text{ moles K}$

d) $4.41 \times 10^{21} \text{ Ca atoms} \times \dfrac{1 \text{ mole Ca}}{6.022 \times 10^{23} \text{ Ca atoms}} = 7.32 \times 10^{-3} \text{ moles Ca}$

21. | Element | Moles | Number of Atoms |
|---|---|---|
| Ne | 0.552 | 3.32×10^{23} |
| Ar | 5.40 | 3.25×10^{24} |
| Xe | 1.78 | 1.07×10^{24} |
| He | 1.79×10^{-4} | 1.08×10^{20} |

22. | Element | Moles | Number of Atoms |
|---|---|---|
| Cr | 1.60 | 9.61×10^{23} |
| Fe | 1.52×10^{-5} | 9.15×10^{18} |
| Ti | 0.0365 | 2.20×10^{22} |
| Hg | 0.181 | 1.09×10^{23} |

23. a) $872 \text{ sheets} \times \dfrac{1 \text{ dozen}}{12 \text{ sheets}} = 72.7 \text{ dozen}$

b) $872 \text{ sheets} \times \dfrac{1 \text{ gross}}{144 \text{ sheets}} = 6.06 \text{ gross}$

c) $872 \text{ sheets} \times \dfrac{1 \text{ ream}}{500 \text{ sheets}} = 1.74 \text{ reams}$

d) $872 \text{ sheets} \times \dfrac{1 \text{ mole}}{6.022 \times 10^{23} \text{ sheets}} = 1.45 \times 10^{-21} \text{ moles}$

24. a) $3.0 \times 10^{22} \text{ Cu atoms} \times \dfrac{1 \text{ doz Cu atoms}}{12 \text{ Cu atoms}} = 2.5 \times 10^{21} \text{ doz Cu atoms}$

b) $3.0 \times 10^{22} \text{ Cu atoms} \times \dfrac{1 \text{ gross Cu atoms}}{144 \text{ Cu atoms}} = 2.1 \times 10^{20} \text{ gross Cu atoms}$

c) $3.0 \times 10^{22} \text{ Cu atoms} \times \dfrac{1 \text{ ream Cu atoms}}{500 \text{ Cu atoms}} = 6.0 \times 10^{19} \text{ reams Cu atoms}$

d) $3.0 \times 10^{22} \text{ Cu atoms} \times \dfrac{1 \text{ mol Cu atoms}}{6.02 \times 10^{23} \text{ Cu atoms}} = 5.0 \times 10^{-2} \text{ mol Cu atoms}$

25. $38.1 \text{ g Sn} \times \dfrac{1 \text{ mole Sn}}{118.71 \text{ g Sn}} = 0.321 \text{ mole Sn}$

26. moles → grams

$0.12 \text{ mol Pb} \times \dfrac{207.2 \text{ g Pb}}{1 \text{ mol Pb}} = 25 \text{ g Pb}$

27. $0.145 \text{ mol Au} \times \dfrac{196.97 \text{ g Au}}{1 \text{ mol Au}} = 28.6 \text{ g Au}$

28. grams → moles

$0.46 \text{ g He} \times \dfrac{1 \text{ mol He}}{4.00 \text{ g He}} = 0.12 \text{ mol He}$

29. a) $1.34 \text{ g Zn} \times \dfrac{1 \text{ mol Zn}}{65.39 \text{ g Zn}} = 2.05 \times 10^{-2} \text{ mol Zn}$

b) $24.9 \text{ g Ar} \times \dfrac{1 \text{ mol Ar}}{39.95 \text{ g Ar}} = 0.623 \text{ mol Ar}$

c) $72.5 \text{ g Ta} \times \dfrac{1 \text{ mol Ta}}{180.95 \text{ g Ta}} = 0.401 \text{ mol Ta}$

d) $0.0223 \text{ g Li} \times \dfrac{1 \text{ mol Li}}{6.941 \text{ g Li}} = 3.21 \times 10^{-3} \text{ mol Li}$

30. a) $6.64 \text{ mol W} \times \dfrac{183.85 \text{ g W}}{1 \text{ mol W}} = 1.22 \times 10^{3} \text{ g W}$

b) $0.581 \text{ mol Ba} \times \dfrac{137.33 \text{ g Ba}}{1 \text{ mol Ba}} = 79.8 \text{ g Ba}$

c) $68.1 \text{ mol Xe} \times \dfrac{131.29 \text{ g Xe}}{1 \text{ mol Xe}} = 8.940 \times 10^{3} \text{ g Xe}$

d) $1.57 \text{ mol S} \times \dfrac{32.06 \text{ g S}}{1 \text{ mol S}} = 50.3 \text{ g S}$

31.
Element	Moles	Mass
Ne	1.11	22.5 g
Ar	0.117	4.67 g
Xe	7.62	1.00 kg
He	1.44×10^{-4}	5.76×10^{-4}

32.
Element	Moles	Mass
Cr	0.00442	0.230 g
Fe	0.00132	73.5 mg
Ti	1.009×10^{-3}	48.31 mg
Hg	8.87	1.78 kg

33. $0.0134 \text{ mmol Ag} \times \dfrac{1 \times 10^{-3} \text{ mole Ag}}{1 \text{ mmole Ag}} \times \dfrac{6.022 \times 10^{23} \text{ Ag atoms}}{1 \text{ mole Ag}} = 8.07 \times 10^{18} \text{ Ag atoms}$

34. $0.0102 \text{ mmol Au} \times \dfrac{1 \times 10^{-3} \text{ mole Au}}{1 \text{ mmole Au}} \times \dfrac{6.022 \times 10^{23} \text{ Au atoms}}{1 \text{ mole Au}} = 6.14 \times 10^{18} \text{ Au atoms}$

35. $3.78 \text{ g Al} \times \dfrac{1 \text{ mole Al}}{26.98 \text{ g Al}} \times \dfrac{6.022 \times 10^{23} \text{ Al atoms}}{1 \text{ mole Al}} = 8.44 \times 10^{22} \text{ Al atoms}$

36. $4.91 \times 10^{21} \text{ Pt atoms} \times \dfrac{1 \text{ mole Pt}}{6.022 \times 10^{23} \text{ Pt atoms}} \times \dfrac{195.08 \text{ g Pt}}{1 \text{ mol Pt}} = 1.59 \text{ g Pt}$

37. a) $16.9 \text{ g Sr} \times \dfrac{1 \text{ mol Sr}}{87.62 \text{ g Sr}} \times \dfrac{6.022 \times 10^{23} \text{ Sr atoms}}{1 \text{ mol Sr}} = 1.16 \times 10^{23} \text{ Sr atoms}$

b) $26.1 \text{ g Fe} \times \dfrac{1 \text{ mol Fe}}{55.85 \text{ g Fe}} \times \dfrac{6.022 \times 10^{23} \text{ Fe atoms}}{1 \text{ mol Fe}} = 2.81 \times 10^{23} \text{ Fe atoms}$

c) $8.55 \text{ g Bi} \times \dfrac{1 \text{ mol Bi}}{209.0 \text{ g Bi}} \times \dfrac{6.022 \times 10^{23} \text{ Bi atoms}}{1 \text{ mol Bi}} = 2.46 \times 10^{22} \text{ Bi atoms}$

d) $38.2 \text{ g P} \times \dfrac{1 \text{ mole P}}{30.97 \text{ g P}} \times \dfrac{6.022 \times 10^{23} \text{ P atoms}}{1 \text{ mole P}} = 7.43 \times 10^{23} \text{ P atoms}$

38. a) $1.32 \times 10^{20} \text{ U atoms} \times \dfrac{1 \text{ mol U}}{6.022 \times 10^{23} \text{ U atoms}} \times \dfrac{238 \text{ g U}}{1 \text{ mol U}} = 0.0522 \text{ g U}$

b) $2.55 \times 10^{22} \text{ Zn atoms} \times \dfrac{1 \text{ mol Zn}}{6.022 \times 10^{23} \text{ Zn atoms}} \times \dfrac{65.39 \text{ g Zn}}{1 \text{ mol Zn}} = 2.77 \text{ g Zn}$

c) $4.11 \times 10^{23} \text{ Pb atoms} \times \dfrac{1 \text{ mol Pb}}{6.022 \times 10^{23} \text{ atoms}} \times \dfrac{207.2 \text{ g Pb}}{1 \text{ mol Pb}} = 141 \text{ g Pb}$

d) $6.59 \times 10^{24} \text{ Si atoms} \times \dfrac{1 \text{ mol Si}}{6.022 \times 10^{23} \text{ Si atoms}} \times \dfrac{28.09 \text{ g Si}}{1 \text{ mol Si}} = 307 \text{ g Si}$

39. $38 \text{ mg C} \times \dfrac{1 \text{ g}}{1000 \text{ mg}} \times \dfrac{1 \text{ mol C}}{12.01 \text{ g}} \times \dfrac{6.022 \times 10^{23} \text{ atoms}}{1 \text{ mol C}} = 1.9 \times 10^{21} \text{ C atoms}$

40. $495 \text{ kg He} \times \dfrac{1000 \text{ g}}{1 \text{ kg}} \times \dfrac{1 \text{ mol He}}{4.00 \text{ g He}} \times \dfrac{6.022 \times 10^{23} \text{ atoms}}{1 \text{ mol He}} = 7.45 \times 10^{28} \text{ He atoms}$

41. $1.28 \text{ kg Ti} \times \dfrac{1 \times 10^{3} \text{ g}}{1 \text{ kg}} \times \dfrac{1 \text{ mole Ti}}{47.88 \text{ g Ti}} \times \dfrac{6.022 \times 10^{23} \text{ Ti atoms}}{1 \text{ mole Ti}} = 1.61 \times 10^{25} \text{ He atoms}$

42. $133 \text{ kg Cu} \times \dfrac{1000 \text{ g}}{1 \text{ kg}} \times \dfrac{1 \text{ mole Cu}}{63.55 \text{ g Cu}} \times \dfrac{6.022 \times 10^{23} \text{ atoms}}{1 \text{ mole Cu}} = 1.26 \times 10^{27} \text{ Cu atoms}$

43.

Element	Mass	Moles	Number of Atoms
Na	38.5 mg	1.67×10^{-3}	1.01×10^{21}
C	13.5 g	1.12	6.74×10^{23}
V	1.81×10^{-20} g	3.55×10^{-22}	214
Hg	1.44 kg	7.18	4.32×10^{24}

44.

Element	Mass	Moles	Number of Atoms
Pt	8.76 g	0.0449	2.70×10^{22}
Fe	1.06 kg	18.9	1.14×10^{25}
Ti	23.8 mg	4.97×10^{-4}	2.99×10^{20}
Hg	411 g	2.05	1.23×10^{24}

45. a) $27.2 \text{ g Cr} \times \dfrac{1 \text{ mol Cr}}{52.00 \text{ g}} \times \dfrac{6.022 \times 10^{23} \text{ atoms}}{1 \text{ mol Cr}} = 3.15 \times 10^{23}$ Cr atoms

b) $55.1 \text{ g Ti} \times \dfrac{1 \text{ mol Ti}}{47.87 \text{ g}} \times \dfrac{6.022 \times 10^{23} \text{ atoms}}{1 \text{ mol Ti}} = 6.93 \times 10^{23}$ Ti atoms

c) $205 \text{ g Pb} \times \dfrac{1 \text{ mol Pb}}{207.2 \text{ g}} \times \dfrac{6.022 \times 10^{23} \text{ atoms}}{1 \text{ mol Pb}} = 5.96 \times 10^{23}$ Pb atoms

The greatest number of atoms is found in b) 55.1 g of titanium.

46. a) $10.0 \text{ g He} \times \dfrac{1 \text{ mol He}}{4.00 \text{ g}} \times \dfrac{6.022 \times 10^{23} \text{ atoms}}{1 \text{ mol He}} = 1.51 \times 10^{24}$ He atoms

b) $25.0 \text{ g Ne} \times \dfrac{1 \text{ mol Ne}}{20.18 \text{ g}} \times \dfrac{6.022 \times 10^{23} \text{ atoms}}{1 \text{ mol Ne}} = 7.46 \times 10^{23}$ Ne atoms

c) $115 \text{ g Xe} \times \dfrac{1 \text{ mol Xe}}{131.3 \text{ g}} \times \dfrac{6.022 \times 10^{23} \text{ atoms}}{1 \text{ mol Xe}} = 5.27 \times 10^{23}$ Xe atoms

The greatest number of atoms is found in a) 10.0 g of helium.

47. a) $38.2 \text{ g NaCl} \times \dfrac{1 \text{ mol NaCl}}{58.44 \text{ g}} = 0.654 \text{ mol NaCl}$

b) $36.5 \text{ g NO} \times \dfrac{1 \text{ mol NO}}{30.01 \text{ g}} = 1.22 \text{ mol NO}$

c) $4.25 \text{ kg CO}_2 \times \dfrac{1000 \text{ g}}{1 \text{ kg}} \times \dfrac{1 \text{ mol CO}_2}{44.01 \text{ g}} = 96.6 \text{ mol CO}_2$

d) $2.71 \text{ mg CCl}_4 \times \dfrac{1 \text{ g}}{1000 \text{ mg}} \times \dfrac{1 \text{ mol CCl}_4}{153.8 \text{ g}} = 1.76 \times 10^{-5} \text{ mol CCl}_4$

48. a) $1.32 \text{ mol CF}_4 \times \dfrac{88.01 \text{ g}}{1 \text{ mol CF}_4} = 116 \text{ g CF}_4$

b) $0.555 \text{ mol MgF}_2 \times \dfrac{62.31 \text{ g}}{1 \text{ mol MgF}_2} = 34.6 \text{ g MgF}_2$

c) $1.29 \text{ mmol CS}_2 \times \dfrac{1 \text{ mol}}{1000 \text{ mmol}} \times \dfrac{76.13 \text{ g}}{1 \text{ mol CS}_2} = 0.0982 \text{ g CS}_2$

d) $1.89 \text{ kmol SO}_3 \times \dfrac{1000 \text{ mol SO}_3}{1 \text{ kmol SO}_3} \times \dfrac{80.07 \text{ g}}{1 \text{ mol SO}_3} = 1.51 \times 10^{5} \text{ g SO}_3$

49.

Compound	Mass	Moles	Number of Molecules
H_2O	112 kg	6.22×10^3	3.74×10^{27}
N_2O	6.33 g	0.144	8.67×10^{22}
SO_2	156 g	2.44	1.47×10^{24}
CH_2Cl_2	5.46 g	0.0643	3.87×10^{22}

50.

Compound	Mass	Moles	Number of Molecules
CO_2	0.673 g	0.0153	9.21×10^{21}
CO	0.420 g	0.0150	9.03×10^{21}
BrI	23.8 mg	1.15×10^{-4}	6.93×10^{19}
CF_2Cl_2	1.02 kg	8.44	5.08×10^{24}

51. $1.32 \text{ g } C_{10}H_8 \times \dfrac{1 \text{ mole } C_{10}H_8}{128.18 \text{ g } C_{10}H_8} \times \dfrac{6.022 \times 10^{23} \, C_{10}H_8}{1 \text{ mole } C_{10}H_8} = 6.20 \times 10^{21} \, C_{10}H_8 \text{ molecules}$

52. $1 \text{ H}_2\text{O molecule} \times \dfrac{1 \text{ mole H}_2\text{O}}{6.022 \times 10^{23} \text{ molecules}} \times \dfrac{18.02 \text{ g H}_2\text{O}}{1 \text{ mole H}_2\text{O}} = 2.992 \times 10^{-23} \text{g H}_2\text{O}$

53. a) $3.5 \text{ g H}_2\text{O} \times \dfrac{1 \text{ mole H}_2\text{O}}{18.02 \text{ g}} \times \dfrac{6.022 \times 10^{23} \text{ molecules}}{1 \text{ mole H}_2\text{O}} = 1.2 \times 10^{23} \text{ H}_2\text{O molecules}$

b) $56.1 \text{ g N}_2 \times \dfrac{1 \text{ mole N}_2}{28.02 \text{ g}} \times \dfrac{6.022 \times 10^{23} \text{ molecules}}{1 \text{ mole N}_2} = 1.21 \times 10^{24} \text{ N}_2 \text{ molecules}$

c) $89 \text{ g CCl}_4 \times \dfrac{1 \text{ mole CCl}_4}{153.81 \text{ g}} \times \dfrac{6.022 \times 10^{23} \text{ molecules}}{1 \text{ mole CCl}_4} = 3.5 \times 10^{23} \text{ CCl}_4 \text{ molecules}$

d) $19 \text{ g C}_6\text{H}_{12}\text{O}_6 \times \dfrac{1 \text{ mole C}_6\text{H}_{12}\text{O}_6}{180.18 \text{ g}} \times \dfrac{6.022 \times 10^{23} \text{ molecules}}{1 \text{ mole C}_6\text{H}_{12}\text{O}_6}$

$= 6.4 \times 10^{22} \text{ C}_6\text{H}_{12}\text{O}_6 \text{ molecules}$

54. a) $5.94 \times 10^{20} \text{ H}_2\text{O}_2 \text{ molecules} \times \dfrac{1 \text{ mole H}_2\text{O}_2}{6.022 \times 10^{23} \text{ molecules}} \times \dfrac{34.02 \text{ g H}_2\text{O}_2}{1 \text{ mole H}_2\text{O}_2}$

$= 3.36 \times 10^{-2} \text{ g H}_2\text{O}_2$

b) $2.8 \times 10^{22} \text{ SO}_2 \text{ molecules} \times \dfrac{1 \text{ mole SO}_2}{6.022 \times 10^{23} \text{ molecules}} \times \dfrac{64.07 \text{ g SO}_2}{1 \text{ mole SO}_2} = 3.0 \text{ g SO}_2$

c) $4.5 \times 10^{25} \text{ O}_3 \text{ molecules} \times \dfrac{1 \text{ mole O}_3}{6.022 \times 10^{23} \text{ molecules}} \times \dfrac{48.00 \text{ g O}_3}{1 \text{ mole O}_3} = 3.6 \times 10^3 \text{ g O}_3$

d) $9.85 \times 10^{19} \text{ CH}_4 \text{ molecules} \times \dfrac{1 \text{ mole CH}_4}{6.022 \times 10^{23} \text{ molecules}} \times \dfrac{16.05 \text{ g CH}_4}{1 \text{ mole CH}_4}$

$= 2.63 \times 10^{-3} \text{ g CH}_4$

55. $1.8 \times 10^{17} \text{ C}_{12}\text{H}_{22}\text{O}_{11} \text{ molecules} \times \dfrac{1 \text{ mole C}_{12}\text{H}_{22}\text{O}_{11}}{6.022 \times 10^{23} \text{ molecules}} \times \dfrac{342.34 \text{ g}}{1 \text{ mole C}_{12}\text{H}_{22}\text{O}_{11}} \times$

$\dfrac{1000 \text{ mg}}{1 \text{ g}} = 0.10 \text{ mg C}_{12}\text{H}_{22}\text{O}_{11}$

56. $0.12 \text{ mg NaCl} \times \dfrac{1 \text{ g}}{1000 \text{ mg}} \times \dfrac{1 \text{ mole NaCl}}{58.44 \text{ g}} \times \dfrac{6.022 \times 10^{23} \text{ formula units}}{1 \text{ mole NaCl}}$

$= 1.2 \times 10^{18}$ NaCl formula units

57. mole → pennies → dollars → dollars/person

$1 \text{ mol} \times \dfrac{6.022 \times 10^{23} \text{ pennies}}{1 \text{ mol}} = 6.022 \times 10^{23}$ pennies

$6.022 \times 10^{23} \text{ pennies} \times \dfrac{1 \text{ dollar}}{100 \text{ pennies}} = 6.022 \times 10^{21}$ dollars

$\dfrac{6.022 \times 10^{21} \text{ dollars}}{7.1 \times 10^{9} \text{ people}} = 8.5 \times 10^{11}$ dollars/person

$= 8.5 \times 10^{2}$ billion dollars per person, each person would be a billionaire

58. moles → dust particles → μm → m → km → # of circumferences

$6.022 \times 10^{23} \text{ particles} \times \dfrac{10 \text{ } \mu m}{1 \text{ particle}} \times \dfrac{1 \times 10^{-6} \text{ m}}{1 \text{ } \mu m} \times \dfrac{1 \text{ km}}{1000 \text{ m}} \times \dfrac{1 \text{ circumference}}{40{,}076 \text{ km}} =$

1.5×10^{11} times around the equator

Chemical Formulas as Conversion Factors

59. moles $CaCl_2$ → moles Cl

$2.7 \text{ moles } CaCl_2 \times \dfrac{2 \text{ moles Cl}}{1 \text{ mole } CaCl_2} = 5.4 \text{ mol Cl}$

60. moles $Fe(NO_3)_3$ → moles O

$12.4 \text{ moles } Fe(NO_3)_3 \times \dfrac{9 \text{ moles O}}{1 \text{ mole } Fe(NO_3)_3} = 112 \text{ mol O}$

61. a) $2.3 \text{ moles } H_2O \times \dfrac{1 \text{ mole O}}{1 \text{ mole } H_2O} = 2.3$ moles O

b) $1.2 \text{ moles } H_2O_2 \times \dfrac{2 \text{ moles O}}{1 \text{ mole } H_2O_2} = 2.4$ moles O

c) $0.9 \text{ moles } NaNO_3 \times \dfrac{3 \text{ moles O}}{1 \text{ mole } NaNO_3} = 2.7$ moles O

d) $0.5 \text{ moles } Ca(NO_3)_2 \times \dfrac{6 \text{ moles O}}{1 \text{ mole } Ca(NO_3)_2} = 3.0$ moles O

The correct answer is (d).

62. a) 3.8 moles HCl $\times \dfrac{1 \text{ mole Cl}}{1 \text{ mole HCl}} = 3.8$ moles Cl

b) 1.7 moles $CH_2Cl_2 \times \dfrac{2 \text{ moles Cl}}{1 \text{ mole } CH_2Cl_2} = 3.4$ moles Cl

c) 4.2 moles $NaClO_3 \times \dfrac{1 \text{ mole Cl}}{1 \text{ mole } NaClO_3} = 4.2$ moles Cl

d) 2.2 moles $Mg(ClO_4)_2 \times \dfrac{2 \text{ moles Cl}}{1 \text{ mole } Mg(ClO_4)_2} = 4.4$ moles Cl

The correct answer is (d).

63. a) 2.5 moles $CH_4 \times \dfrac{1 \text{ mole C}}{1 \text{ mole } CH_4} = 2.5$ moles C

b) 0.115 moles $C_2H_6 \times \dfrac{2 \text{ moles C}}{1 \text{ mole } C_2H_6} = 0.230$ moles C

c) 5.67 moles $C_4H_{10} \times \dfrac{4 \text{ moles C}}{1 \text{ mole } C_4H_{10}} = 22.7$ moles C

d) 25.1 moles $C_8H_{18} \times \dfrac{8 \text{ moles C}}{1 \text{ mole } C_8H_{18}} = 201$ moles C

64. a) 4.67 moles $H_2O \times \dfrac{2 \text{ moles H}}{1 \text{ mole } H_2O} = 9.34$ moles H

b) 8.39 moles $NH_3 \times \dfrac{3 \text{ moles H}}{1 \text{ mole } NH_3} = 25.2$ moles H

c) 0.117 moles $N_2H_4 \times \dfrac{4 \text{ moles H}}{1 \text{ mole } N_2H_4} = 0.468$ moles H

d) 35.8 moles $C_{10}H_{22} \times \dfrac{22 \text{ moles H}}{1 \text{ mole } C_{10}H_{22}} = 788$ moles H

65. a) 2 moles H per mole of molecules; 8 H atoms present
b) 4 moles H per mole of molecules; 20 H atoms present
c) 3 moles H per mole of molecules; 9 H atoms present

66. a) 3 moles O per mole of molecules; 9 O atoms present
b) 4 moles O per mole of molecules; 16 O atoms present
c) 2 moles O per mole of molecules; 6 O atoms present

67. a) $38.0 \text{ g CF}_2\text{Cl}_2 \times \dfrac{1 \text{ mol CF}_2\text{Cl}_2}{120.91 \text{ g}} \times \dfrac{2 \text{ mol Cl}}{1 \text{ mol CF}_2\text{Cl}_2} \times \dfrac{35.45 \text{ g}}{1 \text{ mol Cl}} = 22.3 \text{ g Cl}$

b) $38.0 \text{ g CFCl}_3 \times \dfrac{1 \text{ mol CFCl}_3}{137.36 \text{ g}} \times \dfrac{3 \text{ mol Cl}}{1 \text{ mol CFCl}_3} \times \dfrac{35.45 \text{ g}}{1 \text{ mol Cl}} = 29.4 \text{ g Cl}$

c) $38.0 \text{ g C}_2\text{F}_3\text{Cl}_3 \times \dfrac{1 \text{ mol C}_2\text{F}_3\text{Cl}_3}{187.37 \text{ g}} \times \dfrac{3 \text{ mol Cl}}{1 \text{ mol C}_2\text{F}_3\text{Cl}_3} \times \dfrac{35.45 \text{ g}}{1 \text{ mol Cl}} = 21.6 \text{ g Cl}$

d) $38.0 \text{ g CF}_3\text{Cl} \times \dfrac{1 \text{ mol CF}_3\text{Cl}}{104.46 \text{ g}} \times \dfrac{1 \text{ mol Cl}}{1 \text{ mol CF}_3\text{Cl}} \times \dfrac{35.45 \text{ g}}{1 \text{ mol Cl}} = 12.9 \text{ g Cl}$

68. a) $1.00 \text{ g NaCl} \times \dfrac{1 \text{ mol NaCl}}{58.44 \text{ g}} \times \dfrac{1 \text{ mol Na}}{1 \text{ mol NaCl}} \times \dfrac{22.99 \text{ g}}{1 \text{ mol Na}} = 0.393 \text{ g Na}$

b) $1.00 \text{ g Na}_3\text{PO}_4 \times \dfrac{1 \text{ mol Na}_3\text{PO}_4}{163.94 \text{ g}} \times \dfrac{3 \text{ mol Na}}{1 \text{ mol Na}_3\text{PO}_4} \times \dfrac{22.99 \text{ g}}{1 \text{ mol Na}} = 0.421 \text{ g Na}$

c) $1.00 \text{ g NaC}_7\text{H}_5\text{O}_2 \times \dfrac{1 \text{ mol NaC}_7\text{H}_5\text{O}_2}{144.10 \text{ g}} \times \dfrac{1 \text{ mol Na}}{1 \text{ mol NaC}_7\text{H}_5\text{O}_2} \times \dfrac{22.99 \text{ g}}{1 \text{ mol Na}}$

$= 0.160 \text{ g Na}$

d) $1.00 \text{ g Na}_2\text{C}_6\text{H}_6\text{O}_7 \times \dfrac{1 \text{ mol Na}_2\text{C}_6\text{H}_6\text{O}_7}{236.10 \text{ g}} \times \dfrac{2 \text{ mol Na}}{1 \text{ mol Na}_2\text{C}_6\text{H}_6\text{O}_7} \times \dfrac{22.99 \text{ g}}{1 \text{ mol Na}}$

$= 0.195 \text{ g Na}$

69. a) $1.0 \times 10^3 \text{ kg Fe} \times \dfrac{1000 \text{ g}}{1 \text{ kg}} \times \dfrac{1 \text{ mol Fe}}{55.85 \text{ g}} \times \dfrac{1 \text{ mol Fe}_2\text{O}_3}{2 \text{ mol Fe}} \times \dfrac{159.70 \text{ g}}{1 \text{ mol Fe}_2\text{O}_3} \times \dfrac{1 \text{ kg}}{1 \times 10^3 \text{ g}}$

$= 1.4 \times 10^3 \text{ kg Fe}_2\text{O}_3$

b) $1.0 \times 10^3 \text{ kg Fe} \times \dfrac{1 \times 10^3 \text{ g}}{1 \text{ kg}} \times \dfrac{1 \text{ mol Fe}}{55.85 \text{ g}} \times \dfrac{1 \text{ mol Fe}_3\text{O}_4}{3 \text{ mol Fe}} \times \dfrac{231.55 \text{ g}}{1 \text{ mol Fe}_3\text{O}_4} \times \dfrac{1 \text{ kg}}{1 \times 10^3 \text{ g}}$

$= 1.4 \times 10^3 \text{ kg Fe}_3\text{O}_4$

c) $1.0 \times 10^3 \text{ kg Fe} \times \dfrac{1 \times 10^3 \text{ g}}{1 \text{ kg}} \times \dfrac{1 \text{ mol Fe}}{55.85 \text{ g}} \times \dfrac{1 \text{ mol FeCO}_3}{1 \text{ mol Fe}} \times \dfrac{115.86 \text{ g}}{1 \text{ mol FeCO}_3} \times \dfrac{1 \text{ kg}}{1 \times 10^3 \text{ g}}$

$= 2.1 \times 10^3 \text{ kg FeCO}_3$

70. a) $1.0 \times 10^3 \text{ kg Pb} \times \dfrac{1 \times 10^3 \text{ g}}{1 \text{ kg}} \times \dfrac{1 \text{ mol Pb}}{207.2 \text{ g}} \times \dfrac{1 \text{ mol PbS}}{1 \text{ mol Pb}} \times \dfrac{239.3 \text{ g}}{1 \text{ mol PbS}} \times \dfrac{1 \text{ kg}}{1 \times 10^3 \text{ g}}$

$= 1.2 \times 10^3 \text{ kg PbS}$

b) $1.0 \times 10^3 \text{ kg Pb} \times \dfrac{1 \times 10^3 \text{ g}}{1 \text{ kg}} \times \dfrac{1 \text{ mol Pb}}{207.2 \text{ g}} \times \dfrac{1 \text{ mol PbCO}_3}{1 \text{ mol Pb}} \times \dfrac{267.2 \text{ g}}{1 \text{ mol PbCO}_3} \times \dfrac{1 \text{ kg}}{1000 \text{ g}}$

$= 1.3 \times 10^3 \text{ kg PbCO}_3$

c) $1.0 \times 10^3 \text{ kg Pb} \times \dfrac{1 \times 10^3 \text{ g}}{1 \text{ kg}} \times \dfrac{1 \text{ mol Pb}}{207.2 \text{ g}} \times \dfrac{1 \text{ mol PbSO}_4}{1 \text{ mol Pb}} \times \dfrac{303.3 \text{ g}}{1 \text{ mol PbSO}_4} \times \dfrac{1 \text{ kg}}{1 \times 10^3 \text{ g}}$

$= 1.5 \times 10^3 \text{ PbSO}_4$

Mass Percent Composition

71. mass % Sr $= \dfrac{2.45 \text{ g Sr}}{2.89 \text{ g SrO}} \times 100\% = 84.8\% \text{ Sr}$

72. mass % Al $= \dfrac{4.78 \text{ g Al}}{6.67 \text{ g Al}_2\text{O}_3} \times 100\% = 71.7\% \text{ Al}$

73. mass % Ca $= \dfrac{0.690 \text{ g Ca}}{1.912 \text{ g CaCl}_2} \times 100\% = 36.1\% \text{ Ca}$

mass % Cl $= \dfrac{1.222 \text{ g Cl}}{1.912 \text{ g CaCl}_2} \times 100\% = 63.91\% \text{ Cl}$

74. mass % C $= \dfrac{0.27 \text{ g C}}{0.45 \text{ Aspirin}} \times 100\% = 6.0 \times 10^1 \% \text{ C}$

mass % H $= \dfrac{0.020 \text{ g H}}{0.45 \text{ Aspirin}} \times 100\% = 4.4\% \text{ H}$

mass % O $= \dfrac{0.16 \text{ g O}}{0.45 \text{ Aspirin}} \times 100\% = 36\% \text{ O}$

75. $\dfrac{\text{mass F}}{28.5 \text{ g CuF}_2} \times 100\% = 37.42\% \text{ F} \Rightarrow \text{mass F} = \dfrac{(37.42\% \text{ F})(28.5 \text{ g CuF}_2)}{100\%} = 10.7 \text{ g}$

76. $\dfrac{4.8 \text{ g Ag}}{\text{mass AgCl}} \times 100\% = 75.27\% \text{ Ag} \Rightarrow \text{mass AgCl} = \dfrac{(4.8 \text{ g Ag})(100\%)}{75.27\% \text{ Ag}} = 6.4 \text{ g}$

77. $\dfrac{3.0 \text{ mg F}}{\text{mass NaF}} \times 100\% = 45.24\% \text{ F} \Rightarrow \text{mass NaF} = \dfrac{(3.0 \text{ mg F})(100\%)}{45.24\% \text{ F}} = 6.6 \text{ mg NaF}$

78. $\dfrac{\overbrace{150 \text{ µg I}}^{\text{assuming 3 sig figs}}}{\text{mass KI}} \times 100\% = 76.45\% \text{ I} \Rightarrow \text{mass KI} = \dfrac{(150 \text{ µg I})(100\%)}{76.45\% \text{ I}} = 196 \text{ µg KI}$

Mass Percent Composition from Chemical Formula

79. Assume 1 mole of each compound and determine the mass % using molar masses.

   a) mass % N = $\dfrac{28.02 \text{ g N}}{44.02 \text{ g N}_2\text{O}} \times 100\% = 63.65\% \text{ N}$

   b) mass % N = $\dfrac{14.01 \text{ g N}}{30.01 \text{ g NO}} \times 100\% = 46.68\% \text{ N}$

   c) mass % N = $\dfrac{14.01 \text{ g N}}{46.01 \text{ g NO}_2} \times 100\% = 30.45\% \text{ N}$

   d) mass % N = $\dfrac{28.02 \text{ g N}}{108.02 \text{ g N}_2\text{O}_5} \times 100\% = 25.94\% \text{ N}$

80. Assume 1 mole of each compound and determine the mass % using molar masses.

   a) mass % C = $\dfrac{24.02 \text{ g C}}{26.04 \text{ g C}_2\text{H}_2} \times 100\% = 92.24\% \text{ C}$

   b) mass % C = $\dfrac{36.03 \text{ g C}}{42.09 \text{ g C}_3\text{H}_6} \times 100\% = 85.60\% \text{ C}$

   c) mass % C = $\dfrac{24.02 \text{ g C}}{30.08 \text{ g C}_2\text{H}_6} \times 100\% = 79.85\% \text{ C}$

   d) mass % C = $\dfrac{24.02 \text{ g C}}{46.08 \text{ g C}_2\text{H}_6\text{O}} \times 100\% = 52.13\% \text{ C}$

81. Assume 1 mole of each compound and determine the mass % using molar masses.

a) mass % C = $\dfrac{24.02 \text{ g C}}{60.06 \text{ g C}_2\text{H}_4\text{O}_2} \times 100\% = 39.99\%$ C

mass % H = $\dfrac{4.04 \text{ g H}}{60.06 \text{ g C}_2\text{H}_4\text{O}_2} \times 100\% = 6.73\%$ H

mass % O = $\dfrac{32.00 \text{ g O}}{60.06 \text{ g C}_2\text{H}_4\text{O}_2} \times 100\% = 53.28\%$ O

b) mass % C = $\dfrac{12.01 \text{ g C}}{46.03 \text{ g CH}_2\text{O}_2} \times 100\% = 26.09\%$ C

mass % H = $\dfrac{2.02 \text{ g H}}{46.03 \text{ g CH}_2\text{O}_2} \times 100\% = 4.39\%$ H

mass % O = $\dfrac{32.00 \text{ g O}}{46.03 \text{ g CH}_2\text{O}_2} \times 100\% = 69.52\%$ O

c) mass % C = $\dfrac{36.03 \text{ g C}}{59.13 \text{ g C}_3\text{H}_9\text{N}} \times 100\% = 60.93\%$ C

mass % H = $\dfrac{9.09 \text{ g H}}{59.13 \text{ g C}_3\text{H}_9\text{N}} \times 100\% = 15.4\%$ H

mass % N = $\dfrac{14.01 \text{ g N}}{59.13 \text{ g C}_3\text{H}_9\text{N}} \times 100\% = 23.69\%$ N

d) mass % C = $\dfrac{48.04 \text{ g C}}{88.18 \text{ g C}_4\text{H}_{12}\text{N}_2} \times 100\% = 54.48\%$ C

mass % H = $\dfrac{12.12 \text{ g H}}{88.18 \text{ g C}_4\text{H}_{12}\text{N}_2} \times 100\% = 13.74\%$ H

mass % N = $\dfrac{28.02 \text{ g N}}{88.18 \text{ g C}_4\text{H}_{12}\text{N}_2} \times 100\% = 31.78\%$ N

82. Assume 1 mole of each compound and determine the mass % using molar masses.

a) mass % Fe = $\dfrac{55.85 \text{ g Fe}}{162.20 \text{ g FeCl}_3} \times 100\% = 34.43\%$ Fe

mass % Cl = $\dfrac{106.35 \text{ g Cl}}{162.20 \text{ g FeCl}_3} \times 100\% = 65.57\%$ Cl

b) mass % Ti = $\dfrac{47.87 \text{ g Ti}}{79.88 \text{ g TiO}_2} \times 100\% = 59.94\%$ Ti

mass % O = $\dfrac{32.00 \text{ g O}}{79.88 \text{ g TiO}_2} \times 100\% = 40.06\%$ O

c) mass % H = $\dfrac{3.03 \text{ g H}}{98.00 \text{ g H}_3\text{PO}_4} \times 100\% = 3.09\%$ H

mass % P = $\dfrac{30.97 \text{ g P}}{98.00 \text{ g H}_3\text{PO}_4} \times 100\% = 31.60\%$ H

mass % O = $\dfrac{64.00 \text{ g O}}{98.00 \text{ g H}_3\text{PO}_4} \times 100\% = 65.31\%$ O

d) mass % H = $\dfrac{1.01 \text{ g H}}{63.02 \text{ g HNO}_3} \times 100\% = 1.60\%$ H

mass % N = $\dfrac{14.01 \text{ g N}}{63.02 \text{ g HNO}_3} \times 100\% = 22.23\%$ N

mass % O = $\dfrac{48.00 \text{ g O}}{63.02 \text{ g HNO}_3} \times 100\% = 76.17\%$ O

83. a) mass % O = $\dfrac{6(16.00 \text{ g O})}{164.09 \text{ g Ca(NO}_3)_2} \times 100\% = 58.50\%$ O

b) mass % O = $\dfrac{4(16.00 \text{ g O})}{151.9 \text{ g FeSO}_4} \times 100\% = 42.13\%$ O

c) mass % O = $\dfrac{2(16.00 \text{ g O})}{44.01 \text{ g CO}_2} \times 100\% = 72.71\%$ O

84. a) mass % Cl = $\dfrac{4(35.45 \text{ g Cl})}{153.8 \text{ g CCl}_4} \times 100\% = 92.19\%$ Cl

b) mass % Cl = $\dfrac{2(35.45 \text{ g Cl})}{143.0 \text{ g Ca(ClO)}_2} \times 100\% = 49.58\%$ Cl

c) mass % Cl = $\dfrac{1(35.45 \text{ g Cl})}{100.5 \text{ g HClO}_4} \times 100\% = 35.27\%$ Cl

85. Hematite: mass % Fe = $\dfrac{111.7 \text{ g Fe}}{159.70 \text{ g Fe}_2\text{O}_3} \times 100\% = 69.94\%$ Fe

   Magnetite: mass % Fe = $\dfrac{167.55 \text{ g Fe}}{231.55 \text{ g Fe}_3\text{O}_4} \times 100\% = 72.36\%$ Fe

   Siderite: mass % Fe = $\dfrac{55.85 \text{ g Fe}}{115.86 \text{ g FeCO}_3} \times 100\% = 48.20\%$ Fe

   The magnetite ore has the highest iron content.

86. NH_3: mass % N = $\dfrac{14.01 \text{ g N}}{17.04 \text{ g NH}_3} \times 100\% = 82.22\%$ N

   $\text{CO(NH}_2)_2$: mass % N = $\dfrac{28.02 \text{ g N}}{60.07 \text{ g CO(NH}_2)_2} \times 100\% = 46.65\%$ N

   NH_4NO_3: mass % N = $\dfrac{28.02 \text{ g N}}{80.06 \text{ g NH}_4\text{NO}_3} \times 100\% = 35.00\%$ N

   $(\text{NH}_4)_2\text{SO}_4$: mass % N = $\dfrac{28.02 \text{ g N}}{132.17 \text{ g (NH}_4)_2\text{SO}_4} \times 100\% = 21.20\%$ N

   The ammonia fertilizer (NH_3) has the highest nitrogen content.

Calculating Empirical Formulas

87. $1.78 \text{ g N} \times \dfrac{1 \text{ mole N}}{14.01 \text{ g}} = 0.127$ moles N

    $4.05 \text{ g O} \times \dfrac{1 \text{ mole O}}{16.00 \text{ g}} = 0.253$ moles O

    $\text{N}_{\frac{0.127}{0.127}}\text{O}_{\frac{0.253}{0.127}} = \text{NO}_2$

88. $2.231 \text{ g Se} \times \dfrac{1 \text{ mole Se}}{78.96 \text{ g}} = 0.02825$ moles Se

    $3.221 \text{ g F} \times \dfrac{1 \text{ mole F}}{19.00 \text{ g}} = 0.1695$ moles F

    $\text{Se}_{\frac{0.02825}{0.02825}}\text{F}_{\frac{0.1695}{0.02825}} = \text{SeF}_6$

89. a) $1.245 \text{ g Ni} \times \dfrac{1 \text{ mole Ni}}{58.69 \text{ g}} = 0.02121$ moles Ni

$5.381 \text{ g I} \times \dfrac{1 \text{ mole I}}{126.90 \text{ g}} = 0.04240$ moles I

$\text{Ni}_{\underset{0.02121}{0.02121}} \text{I}_{\underset{0.02121}{0.04240}} = \text{NiI}_2$

b) $1.443 \text{ g Se} \times \dfrac{1 \text{ mole Se}}{78.96 \text{ g}} = 0.01828$ moles Se

$5.841 \text{ g Br} \times \dfrac{1 \text{ mole Br}}{79.90 \text{ g}} = 0.07310$ moles Br

$\text{Se}_{\underset{0.01828}{0.01828}} \text{Br}_{\underset{0.01828}{0.07310}} = \text{SeBr}_4$

c) $2.128 \text{ g Be} \times \dfrac{1 \text{ mole Be}}{9.01 \text{ g}} = 0.236$ moles Be

$7.557 \text{ g S} \times \dfrac{1 \text{ mole S}}{32.07 \text{ g}} = 0.2356$ moles S

$15.107 \text{ g O} \times \dfrac{1 \text{ mole O}}{16.00 \text{ g}} = 0.9442$ moles O

$\text{Be}_{\underset{0.2356}{0.236}} \text{S}_{\underset{0.2356}{0.2356}} \text{O}_{\underset{0.2356}{0.9442}} = \text{BeSO}_4$

90. a) $2.677 \text{ g Ba} \times \dfrac{1 \text{ mole Ba}}{137.33 \text{ g}} = 0.01949$ moles Ba

$3.115 \text{ g Br} \times \dfrac{1 \text{ mole Br}}{79.90 \text{ g}} = 0.03899$ moles Br

$\text{Ba}_{\underset{0.01949}{0.01949}} \text{Br}_{\underset{0.01949}{0.03899}} = \text{BaBr}_2$

b) $1.651 \text{ g Ag} \times \dfrac{1 \text{ mole Ag}}{107.87 \text{ g}} = 0.01531$ moles Ag

$0.1224 \text{ g O} \times \dfrac{1 \text{ mole O}}{16.00 \text{ g}} = 0.007650$ moles O

$\text{Ag}_{\underset{0.00765}{0.01531}} \text{O}_{\underset{0.00765}{0.00765}} = \text{Ag}_2\text{O}$

c) $0.672 \text{ g Co} \times \dfrac{1 \text{ mole Co}}{58.93 \text{ g}} = 0.01140$ moles Co

$0.569 \text{ g As} \times \dfrac{1 \text{ mole As}}{74.92 \text{ g}} = 0.007595$ moles As

$0.486 \text{ g O} \times \dfrac{1 \text{ mole O}}{16.00 \text{ g}} = 0.0304$ moles O

$\text{Co}_{\underset{0.007594}{0.01140}} \text{As}_{\underset{0.007594}{0.007594}} \text{O}_{\underset{0.007594}{0.03038}} = \text{Co}_{1.50}\text{As}_1\text{O}_4 \Rightarrow 2 \times (\text{Co}_{1.50}\text{As}_1\text{O}_4) = \text{Co}_3\text{As}_2\text{O}_8$

91. Assume a 100 gram sample, percentage composition is then equal to the number of grams of each element.

$$54.50 \text{ g C} \times \frac{1 \text{ mole C}}{12.01 \text{ g}} = 4.538 \text{ moles C}$$

$$13.73 \text{ g H} \times \frac{1 \text{ mole H}}{1.01 \text{ g}} = 13.6 \text{ moles H}$$

$$31.77 \text{ g N} \times \frac{1 \text{ mole N}}{14.01 \text{ g}} = 2.268 \text{ moles N}$$

$$C_{\frac{4.538}{2.268}} H_{\frac{13.6}{2.268}} N_{\frac{2.268}{2.268}} \Rightarrow C_2H_6N$$

92. Assume a 100 gram sample, percentage composition is then equal to the number of grams of each element.

$$37.51 \text{ g C} \times \frac{1 \text{ mole C}}{12.01 \text{ g}} = 3.123 \text{ moles C}$$

$$4.20 \text{ g H} \times \frac{1 \text{ mole g H}}{1.01 \text{ g H}} = 4.16 \text{ moles H}$$

$$58.29 \text{ g O} \times \frac{1 \text{ mole g O}}{16.00 \text{ g O}} = 3.643 \text{ moles O}$$

$$C_{\frac{3.123}{3.123}} H_{\frac{4.16}{3.123}} O_{\frac{3.643}{3.123}} = C_1H_{1.33}O_{1.167}; \text{ Empirical Formula} = 6 \times (C_1H_{1.33}O_{1.167}) = C_6H_8O_7$$

93. Assume a 100 g sample:

a) $62.04 \text{ g C} \times \dfrac{1 \text{ mole C}}{12.01 \text{ g}} = 5.166 \text{ moles C}$

$10.41 \text{ g H} \times \dfrac{1 \text{ mole g H}}{1.01 \text{ g H}} = 10.3 \text{ moles H}$

$27.55 \text{ g O} \times \dfrac{1 \text{ mole g O}}{16.00 \text{ g O}} = 1.722 \text{ moles O}$

$C_{\frac{5.166}{1.722}} H_{\frac{10.3}{1.722}} O_{\frac{1.722}{1.722}} = C_3H_6O$

b) $58.80 \text{ g C} \times \dfrac{1 \text{ mole C}}{12.01 \text{ g}} = 4.896 \text{ moles C}$

$9.87 \text{ g H} \times \dfrac{1 \text{ mole g H}}{1.01 \text{ g H}} = 9.77 \text{ moles H}$

$31.33 \text{ g O} \times \dfrac{1 \text{ mole g O}}{16.00 \text{ g O}} = 1.958 \text{ moles O}$

$C_{\frac{4.896}{1.958}} H_{\frac{9.77}{1.958}} O_{\frac{1.958}{1.958}} = C_{2.5}H_5O_1$; Empirical Formula $= 2 \times (C_{2.5}H_5O_1) = C_5H_{10}O_2$

c) $71.98 \text{ g C} \times \dfrac{1 \text{ mole C}}{12.01 \text{ g}} = 5.993 \text{ moles C}$

$6.71 \text{ g H} \times \dfrac{1 \text{ mole g H}}{1.01 \text{ g H}} = 6.64 \text{ moles H}$

$21.31 \text{ g O} \times \dfrac{1 \text{ mole g O}}{16.00 \text{ g O}} = 1.332 \text{ moles O}$

$C_{\frac{5.993}{1.332}} H_{\frac{6.64}{1.332}} O_{\frac{1.332}{1.332}} = C_{4.50}H_5O_1$; Empirical Formula $= 2 \times (C_{4.50}H_5O_1) = C_9H_{10}O_2$

94. Assume a 100 gram sample, percentage composition is then equal to the number of grams of each element.

a) $63.56 \text{ g C} \times \dfrac{1 \text{ mole C}}{12.01 \text{ g}} = 5.292 \text{ moles C}$

$6.00 \text{ g H} \times \dfrac{1 \text{ mole H}}{1.01 \text{ g H}} = 5.94 \text{ moles H}$

$9.27 \text{ g N} \times \dfrac{1 \text{ mole N}}{14.01 \text{ g N}} = 0.662 \text{ moles N}$

$21.17 \text{ g O} \times \dfrac{1 \text{ mole O}}{16.00 \text{ g O}} = 1.323 \text{ moles O}$

$C_{\frac{5.292}{0.662}} H_{\frac{5.94}{0.662}} N_{\frac{0.662}{0.662}} O_{\frac{1.323}{0.662}} = C_8H_9NO_2$

Empirical Formula = $C_8H_9NO_2$

b) $73.03 \text{ g C} \times \dfrac{1 \text{ mole C}}{12.01 \text{ g}} = 6.081 \text{ moles C}$

$6.13 \text{ g H} \times \dfrac{1 \text{ mole g H}}{1.01 \text{ g H}} = 6.07 \text{ moles H}$

$20.84 \text{ g O} \times \dfrac{1 \text{ mole g O}}{16.00 \text{ g O}} = 1.303 \text{ moles O}$

$C_{\frac{6.081}{1.303}} H_{\frac{6.07}{1.303}} O_{\frac{1.303}{1.303}} = C_{4.667}H_{4.66}O_1$

Empirical Formula = $3 \times (C_{4.667}H_{4.66}O_1) = C_{14}H_{14}O_3$

95. $1.45 \text{ g P} \times \dfrac{1 \text{ mole P}}{30.97 \text{ g P}} = 0.0468 \text{ mole P}$

mass O = $2.57 - 1.45 = 1.12$ g O

$1.12 \text{ g O} \times \dfrac{1 \text{ mole O}}{16.00 \text{ g O}} = 0.0700 \text{ mole O}$

$P_{\frac{0.0468}{0.0468}} O_{\frac{0.0700}{0.0468}} = P_1O_{1.50}$; Empirical Formula = $2 \times (P_1O_{1.50}) = P_2O_3$

96. $2.241 \text{ g Ni} \times \dfrac{1 \text{ mole Ni}}{58.69 \text{ g Ni}} = 0.03818 \text{ mole Ni}$

mass O = $2.852 - 2.241 = 0.611$ g O

$0.611 \text{ g O} \times \dfrac{1 \text{ mole O}}{16.00 \text{ g O}} = 0.0382 \text{ mole O}$

$Ni_{\frac{0.0382}{0.0382}} O_{\frac{0.03819}{0.0382}} = NiO$

97. $0.77 \text{ mg N} \times \dfrac{1 \text{ g}}{1000 \text{ mg}} \times \dfrac{1 \text{ mole N}}{14.01 \text{ g N}} = 5.5 \times 10^{-5} \text{ mole N}$

mass Cl = 6.61 − 0.77 = 5.84 mg Cl

$5.84 \text{ mg Cl} \times \dfrac{1 \text{ g}}{1000 \text{ mg}} \times \dfrac{1 \text{ mole Cl}}{35.45 \text{ g Cl}} = 1.65 \times 10^{-4} \text{ mole Cl}$

$\text{Ni}_{\frac{5.50 \times 10^{-5}}{5.50 \times 10^{-5}}} \text{Cl}_{\frac{1.65 \times 10^{-4}}{5.50 \times 10^{-5}}} = \text{NCl}_3$

98. $45.2 \text{ mg P} \times \dfrac{1 \text{ g}}{1000 \text{ mg}} \times \dfrac{1 \text{ mole P}}{30.97 \text{ g P}} = 1.46 \times 10^{-3} \text{ mole P}$

mass Se = 131.6 − 45.2 = 86.4 mg Se

$86.4 \text{ mg Se} \times \dfrac{1 \text{ g}}{1000 \text{ mg}} \times \dfrac{1 \text{ mole Se}}{78.96 \text{ g}} = 1.09 \times 10^{-3} \text{ mole Se}$

$\text{P}_{\frac{1.46 \times 10^{-3}}{1.09 \times 10^{-3}}} \text{Se}_{\frac{1.09 \times 10^{-3}}{1.09 \times 10^{-3}}} = \text{P}_{1.334} \text{Se}_1$

Empirical Formula = $3 \times (\text{P}_{1.333}\text{Se}_1) = \text{P}_4\text{Se}_3$

Calculating Molecular Formulas

99. $\dfrac{\text{Molar Mass}}{\text{Empirical Mass}} = \text{Multiplier} \Rightarrow \dfrac{56.11}{14.03} = 4 \Rightarrow 4 \times (\text{CH}_2) = \text{C}_4\text{H}_8$

100. $\dfrac{\text{Molar Mass}}{\text{Empirical Mass}} = \text{Multiplier} \Rightarrow \dfrac{219.9}{109.9} = 2 \Rightarrow 2 \times (\text{P}_2\text{O}_3) = \text{P}_4\text{O}_6$

101. $\dfrac{\text{Molar Mass}}{\text{Empirical Mass}} = \text{Multiplier}$

 a) $\dfrac{284.77}{47.46} = 6 \Rightarrow 6 \times (\text{CCl}) = \text{C}_6\text{Cl}_6$

 b) $\dfrac{131.39}{131.38} = 1 \Rightarrow 1 \times (\text{C}_2\text{HCl}_3) = \text{C}_2\text{HCl}_3$

 c) $\dfrac{181.44}{60.48} = 3 \Rightarrow 3 \times (\text{C}_2\text{HCl}) = \text{C}_6\text{H}_3\text{Cl}_3$

102. $\dfrac{\text{Molar Mass}}{\text{Empirical Mass}} = \text{Multiplier}$

 a) $\dfrac{163.26}{163.26} = 1 \Rightarrow 1 \times (\text{C}_{11}\text{H}_{17}\text{N}) = \text{C}_{11}\text{H}_{17}\text{N}$

 b) $\dfrac{186.24}{93.13} = 2 \Rightarrow 2 \times (\text{C}_6\text{H}_7\text{N}) = \text{C}_{12}\text{H}_{14}\text{N}_2$

 c) $\dfrac{312.29}{52.06} = 6 \Rightarrow 6 \times (\text{C}_3\text{H}_2\text{N}) = \text{C}_{18}\text{H}_{12}\text{N}_6$

Cumulative Problems

103. Volume = L³ ⇒ (1.42 cm)³ = 2.86 cm³

mass = density × volume ⇒ (8.96 g Cu/cm³)(2.86 cm³) = 25.6 g Cu

$$25.6 \text{ g Cu} \times \frac{1 \text{ mole Cu}}{63.55 \text{ g}} \times \frac{6.022 \times 10^{23} \text{ Cu atoms}}{1 \text{ mole Cu}} = 2.43 \times 10^{23} \text{ Cu atoms}$$

104. Volume = 4/3 πr³ ⇒ 4/3 π(0.886 cm)³ = 2.91 cm³

mass = density × volume ⇒ (10.5 g Ag/cm³)(2.91 cm³) = 30.6 g Ag

$$30.6 \text{ g Ag} \times \frac{1 \text{ mole Ag}}{107.87 \text{ g}} \times \frac{6.022 \times 10^{23} \text{ Ag atoms}}{1 \text{ mole Ag}} = 1.71 \times 10^{23} \text{ Ag atoms}$$

105. Volume: 1 mL = 1 cm³ ⇒ 0.05 cm³ H₂O

mass = density × volume ⇒ (1.0 g H₂O/cm³)(0.05 cm³) = 0.05 g H₂O

$$0.05 \text{ g H}_2\text{O} \times \frac{1 \text{ mole H}_2\text{O}}{18.02 \text{ g}} \times \frac{6.022 \times 10^{23} \text{ particles}}{1 \text{ mole H}_2\text{O}} = 2 \times 10^{21} \text{ H}_2\text{O molecules}$$

106. mass = density × volume ⇒ (0.788 g C₃H₆O/cm³)(325 cm³) = 256 g C₃H₆O

$$256 \text{ g C}_3\text{H}_6\text{O} \times \frac{1 \text{ mole C}_3\text{H}_6\text{O}}{58.09 \text{ g}} \times \frac{6.022 \times 10^{23} \text{ molecules}}{1 \text{ mole C}_3\text{H}_6\text{O}} =$$

$$2.65 \times 10^{24} \text{ C}_3\text{H}_6\text{O molecules}$$

107.
Substance	Mass	Moles	Number of Particles
Ar	0.018 g	4.5×10^{-4}	2.7×10^{20}
NO₂	8.33×10^{-3} g	1.81×10^{-4}	1.09×10^{20}
K	22.4 mg	5.73×10^{-4}	3.45×10^{20}
C₈H₁₈	3.76 kg	32.9	1.98×10^{25}

108.
Substance	Mass	Moles	Number of Particles
C₆H₁₂O₆	15.8 g	0.0877	5.28×10^{22}
Pb	3.11 g	0.0150	9.04×10^{21}
CF₄	22.5 mg	2.56×10^{-4}	1.54×10^{20}
C	0.466 g	0.0388	2.34×10^{22}

109. a) CuI_2; Formula Mass = 317.35

$$\text{Mass \% Cu} = \frac{63.55}{317.35} \times 100\% = 20.03\% \text{ Cu}$$

$$\text{Mass \% I} = \frac{253.8}{317.35} \times 100\% = 79.97\% \text{ I}$$

b) $NaNO_3$; Formula Mass = 85.00

$$\text{Mass \% Na} = \frac{22.99}{85.00} \times 100\% = 27.05\%$$

$$\text{Mass \% N} = \frac{14.01}{85.00} \times 100\% = 16.48\%$$

$$\text{Mass \% O} = \frac{48.00}{85.00} \times 100\% = 56.47\% \text{ O}$$

c) $PbSO_4$; Formula Mass = 303.3

$$\text{Mass \% Pb} = \frac{207.2}{303.3} \times 100\% = 68.32\% \text{ Pb}$$

$$\text{Mass \% S} = \frac{32.07}{303.3} \times 100\% = 10.57\% \text{ S}$$

$$\text{Mass \% O} = \frac{64.00}{303.3} \times 100\% = 21.10\% \text{ O}$$

d) CaF_2; Formula Mass = 78.08

$$\text{Mass \% Ca} = \frac{40.08}{78.08} \times 100\% = 51.33\% \text{ Ca}$$

$$\text{Mass \% F} = \frac{38.00}{78.08} \times 100\% = 48.67\% \text{ F}$$

110. a) NI_3; Molecular Mass = 394.71

$$\text{Mass \% N} = \frac{14.01}{394.71} \times 100\% = 3.549\% \text{ N}$$

$$\text{Mass \% I} = \frac{380.7}{394.71} \times 100\% = 96.45\% \text{ I}$$

c) PCl_3; Molecular Mass = 137.32

$$\text{Mass \% P} = \frac{30.97}{137.32} \times 100\% = 22.55\% \text{ P}$$

$$\text{Mass \% Cl} = \frac{106.35}{137.32} \times 100\% = 77.45\% \text{ Cl}$$

b) XeF_4; Molecular Mass = 207.29

$$\text{Mass \% Xe} = \frac{131.29}{207.29} \times 100\% = 63.336\% \text{ Xe}$$

$$\text{Mass \% F} = \frac{76.00}{207.29} \times 100\% = 36.66\% \text{ F}$$

d) CO; Molecular Mass = 28.01

$$\text{Mass \% C} = \frac{12.01}{28.01} \times 100\% = 42.88\% \text{ C}$$

$$\text{Mass \% O} = \frac{16.00}{28.01} \times 100\% = 57.12\% \text{ O}$$

111. Step 1: Determine how much Fe_2O_3 would be needed to obtain 1×10^3 kg of iron.

$$1.0\times10^3 \text{ kg Fe} \times \frac{1000 \text{ g}}{1 \text{ kg}} \times \frac{1 \text{ mol Fe}}{55.85 \text{ g}} \times \frac{1 \text{ mol Fe}_2\text{O}_3}{2 \text{ mol Fe}} \times \frac{159.70 \text{ g}}{1 \text{ mol Fe}_2\text{O}_3} \times \frac{1 \text{ kg}}{1000 \text{ g}}$$

$= 1.4\times10^3 \text{ kg Fe}_2\text{O}_3$

Step 2: Based on the ore being 78% Fe_2O_3, determine the amount of rock needed for processing. Recall that 78% Fe_2O_3 can be used as a conversion factor because 78 kg Fe_2O_3 is obtained for every 100 kg of ore mined.

$$1.4\times10^3 \text{ kg Fe}_2\text{O}_3 \times \frac{100 \text{ kg rock}}{78 \text{ kg Fe}_2\text{O}_3} = 1.8\times10^3 \text{ kg rock}$$

112. Step 1: Determine how much pure PbS would be needed to obtain 1×10^3 kg of lead.

$$1.0\times10^3 \text{ kg Pb} \times \frac{1\times10^3 \text{ g}}{1 \text{ kg}} \times \frac{1 \text{ mole Pb}}{207.2 \text{ g}} \times \frac{1 \text{ mole PbS}}{1 \text{ mole Pb}} \times \frac{239.3 \text{ g}}{1 \text{ mol PbS}} \times \frac{1 \text{ kg}}{1\times10^3 \text{ g}}$$

$= 1.2\times10^3 \text{ kg PbS}$

Step 2: Based on the ore being 84% PbS, determine how much rock is needed for processing. Recall that 84% PbS can be used as a conversion factor because 84 kg PbS is obtained for every 100 kg of ore mined.

$$1.2\times10^3 \text{ kg PbS} \times \frac{100 \text{ kg rock}}{84 \text{ kg PbS}} = 1.4\times10^3 \text{ kg rock}$$

113. $\dfrac{12 \text{ kg CHF}_2\text{Cl}}{1 \text{ mo}} \times 12 \text{ mo} \times \dfrac{1000 \text{ g}}{1 \text{ kg}} \times \dfrac{1 \text{ mol CHF}_2\text{Cl}}{86.47 \text{ g}} \times \dfrac{1 \text{ mol Cl}}{1 \text{ mol CHF}_2\text{Cl}} \times \dfrac{1 \text{ kg}}{1000 \text{ g}} = 59 \text{ kg Cl}$

114. $\dfrac{55 \text{ g CF}_2\text{Cl}_2}{1 \text{ mo}} \times 12 \text{ mo} \times \dfrac{1 \text{ mol CF}_2\text{Cl}_2}{120.91 \text{ g}} \times \dfrac{2 \text{ mol Cl}}{1 \text{ mol CF}_2\text{Cl}_2} \times \dfrac{35.45 \text{ g}}{1 \text{ mol Cl}} = 3.9\times10^2 \text{ g Cl}$

115. $1.0 \text{ L H}_2\text{O} \times \dfrac{1000 \text{ cm}^3}{1 \text{ L}} \times \dfrac{1 \text{ g H}_2\text{O}}{1 \text{ cm}^3} \times \dfrac{1 \text{ mol H}_2\text{O}}{18.0 \text{ g}} \times \dfrac{2 \text{ mol H}}{1 \text{ mol H}_2\text{O}} \times \dfrac{1.008 \text{ g}}{1 \text{ mol H}} = 1.1\times10^2 \text{ g H}$

116. $1.0 \text{ kg C}_2\text{H}_6\text{O} \times \dfrac{1000 \text{ g}}{1 \text{ kg}} \times \dfrac{1 \text{ mol C}_2\text{H}_6\text{O}}{46.07 \text{ g}} \times \dfrac{6 \text{ mol H}}{1 \text{ mol C}_2\text{H}_6\text{O}} \times \dfrac{1.01 \text{ g}}{1 \text{ mol H}} = 1.3\times10^2 \text{ g H}$

117.
Formula	Molar Mass	%C (by mass)	%H (by mass)
C_2H_4	28.06	85.60%	14.40%
C_4H_{10}	58.12	82.66%	17.34%
C_4H_8	56.12	85.60%	14.40%
C_3H_8	44.11	81.71%	18.29%

118.
Formula	Name	Molar Mass	%Cr (by mass)	%O (by mass)
Cr_2O_3	Chromium(III) Oxide	152.00	68.42%	31.58%
CrO_2	Chromium(IV) Oxide	84.00	61.90%	38.10%
CrO_3	Chromium(VI) Oxide	100.00	52.00%	48.00%

119. Assume a 100 gram sample, % composition then equals the number of grams of each element.

$$55.80 \text{ g C} \times \frac{1 \text{ mole C}}{12.01 \text{ g}} = 4.646 \text{ moles C}$$

$$7.03 \text{ g H} \times \frac{1 \text{ mole H}}{1.01 \text{ g H}} = 6.96 \text{ moles H}$$

$$37.17 \text{ g O} \times \frac{1 \text{ mole O}}{16.00 \text{ g O}} = 2.323 \text{ moles O}$$

$$C_{\frac{4.646}{2.323}} H_{\frac{6.96}{2.323}} O_{\frac{2.323}{2.323}} = C_2H_3O$$

$$\frac{\text{Molar Mass}}{\text{Empirical Mass}} = \text{Multiplier} \Rightarrow \frac{86.09}{43.04} = 2$$

Molecular Formula = $2 \times (C_2H_3O) = C_4H_6O_2$

120. Assume a 100 gram sample, % composition then equals the number of grams of each element.

$$49.48 \text{ g C} \times \frac{1 \text{ mole C}}{12.01 \text{ g}} = 4.120 \text{ moles C}$$

$$5.19 \text{ g H} \times \frac{1 \text{ mole H}}{1.01 \text{ g H}} = 5.14 \text{ moles H}$$

$$28.85 \text{ g N} \times \frac{1 \text{ mole N}}{14.01 \text{ g N}} = 2.059 \text{ moles N}$$

$$16.48 \text{ g O} \times \frac{1 \text{ mole g O}}{16.00 \text{ g O}} = 1.030 \text{ moles O}$$

$$C_{\frac{4.120}{1.030}} H_{\frac{5.14}{1.030}} N_{\frac{2.059}{1.030}} O_{\frac{1.030}{1.030}} = C_4H_5N_2O$$

$$\frac{\text{Molar Mass}}{\text{Empirical Mass}} = \text{Multiplier} \Rightarrow \frac{194.19}{97.1} = 2$$

Molecular Formula = $2 \times (C_4H_5N_2O) = C_8H_{10}N_4O_2$

121. Assume a 100 gram sample, % composition then equals the number of grams of each element.

$$74.03 \text{ g C} \times \frac{1 \text{ mole C}}{12.01 \text{ g}} = 6.164 \text{ moles C}$$

$$8.70 \text{ g H} \times \frac{1 \text{ mole H}}{1.01 \text{ g H}} = 8.61 \text{ moles H}$$

$$17.27 \text{ g N} \times \frac{1 \text{ mole N}}{14.01 \text{ g N}} = 1.233 \text{ moles N}$$

$$C_{\frac{6.164}{1.233}} H_{\frac{8.630}{1.233}} N_{\frac{1.233}{1.233}} = C_5H_7N$$

$$\frac{\text{Molar Mass}}{\text{Empirical Mass}} = \text{Multiplier} \Rightarrow \frac{162.23}{81.12} = 2$$

Molecular Formula = $2 \times (C_5H_7N) = C_{10}H_{14}N_2$

122. Assume a 100 gram sample, % composition then equals the number of grams of each element.

$$79.37 \text{ g C} \times \frac{1 \text{ mole C}}{12.01 \text{ g}} = 6.609 \text{ moles C}$$

$$8.88 \text{ g H} \times \frac{1 \text{ mole H}}{1.01 \text{ g H}} = 8.79 \text{ moles H}$$

$$11.75 \text{ g O} \times \frac{1 \text{ mole O}}{16.00 \text{ g O}} = 0.7344 \text{ moles O}$$

$$C_{\frac{6.609}{0.7344}} H_{\frac{8.79}{0.7344}} O_{\frac{0.7344}{0.7344}} = C_9H_{12}O$$

$$\frac{\text{Molar Mass}}{\text{Empirical Mass}} = \text{Multiplier} \Rightarrow \frac{272.37}{136.19} = 2$$

Molecular Formula = $2 \times (C_9H_{12}O) = C_{18}H_{24}O_2$

123. The mass of the sample consists of KBr and KI as shown in Equation 1:

Eqn 1: Mass KBr + Mass KI = 5.00 g

The mass of KBr and KI can be calculated by multiplying the moles of each compound by the formula mass of that compound as shown in Equation 2.

Eqn 2: (moles KBr)(FM KBr) + (moles KI)(FM KI) = 5.00 g

The sample contains 1.51 g K, which corresponds to 0.0386193 mol K. (Note: To prevent introducing errors into the calculation, numbers will not be rounded to account for significant figures until the final answer.) Because the sample consists of KBr and KI, which have 1 mole of K per mole of the compound, the following can be written:

Eqn 3: moles KBr + moles KI = 0.0386193

Using Equations 2 and 3, we have the situation of two equations and two unknowns to solve.

$(0.0386193 - \text{mol KI})(119.00) + \text{moles KI}(166.00) = 5.00$

$4.5956967 - 119.00 \text{ mol KI} + 166.00 \text{ mol KI} = 5.00$

$47.00 \text{ mol KI} = 0.4043033$

mol KI = 0.4043033/47.00

mol KI = 0.008602198

moles KBr + 0.008602198 = 0.0386193

mol KBr = 0.03001710

%KI = (0.008602198 mol KI × 166 g/mol KI)/5.00 ×100% = 28.6% KI

%KBr = (0.03001710 mol KBr × 119 g/mol KBr)/5.00 ×100% = 71.4% KBr

124. Given the information in the problem, Equation 1 can be written as:

Eqn. 1: mol CO_2 + mol Ne = 1.75 moles

The total mass of the sample was given as 65.3 g; therefore, Equation 2 can be written as shown:

Eqn. 2: (mole CO_2)(MM CO_2) + (mol Ne) (AM Ne) = 65.3 g

Using Equations 1 and 2, we have the situation of two equations and two unknowns to solve. (Note: To prevent introducing errors into the calculation, numbers will not be rounded to account for significant figures until the final answer.)

mol CO_2 = 1.75 − mol Ne

(mol CO_2)(MM CO_2) + (mol Ne) (AM Ne) = 65.3

(1.75 − mol Ne)(44.01) + (mol Ne) (20.18) = 65.3

77.0175 − 44.01 mol Ne + 20.18 mol Ne = 65.3

−44.01 mol Ne + 20.18 mol Ne = 65.3−77.0175

−23.83 mol Ne = −11.7175

mol Ne = −11.7175/−23.83 = 0.491712

mol CO_2 = 1.75 − 0.491712 = 1.25829

mol % Ne = 0.491712/1.75 × 100% = 28.1%

mol % CO_2 = 1.25829/1.75 × 100% = 71.9%

125. g $C_2H_6S \rightarrow$ mol $C_2H_6S \rightarrow$ mol $SO_2 \rightarrow$ g SO_2

Reaction: $2\ C_2H_6S + 9\ O_2 \rightarrow 4\ CO_2 + 6\ H_2O + 2\ SO_2$

$28.7\text{ g }C_2H_6S \times \dfrac{1\text{ mol }C_2H_6S}{62.15\text{ g }C_2H_6S} \times \dfrac{2\text{ mol }SO_2}{2\text{ mol }C_2H_6S} \times \dfrac{64.07\text{ g }SO_2}{1\text{ mol }SO_2} = 29.6\text{ g }SO_2$

126. g $CH_4S \rightarrow$ mol $CH_4S \rightarrow$ mol $SO_2 \rightarrow$ g SO_2

Reaction: $CH_4S + 3\ O_2 \rightarrow CO_2 + 2\ H_2O + SO_2$

$1.89\text{ g }CH_4S \times \dfrac{1\text{ mol }CH_4S}{48.12\text{ g }CH_4S} \times \dfrac{1\text{ mol }SO_2}{1\text{ mol }CH_4S} \times \dfrac{64.07\text{ g }SO_2}{1\text{ mol }SO_2} = 2.52\text{ g }SO_2$

127. mass ore → mass Fe_2O_3 → mass Fe
10.0 kg Ore × 0.38 = 3.8 kg Fe_2O_3
Fe_2O_3 = 111.7 g Fe/159.7 g Fe_2O_3 = 69.94% Fe
3.8 kg Fe_2O_3 × 0.6994 = 2.7 kg Fe

128. Mass Seawater × 0.035 NaCl = 1.0 g NaCl   Mass Seawater = 28.57 g
Density of Seawater = 1.02 g/ml = 28.57 g/ Volume
Volume Seawater = 28.57/1.02 = 28.0 ml Seawater

Highlight Problems

129. $V = 4/3\pi(7\times 10^8\text{ m})^3 = 1.4\times 10^{27}\text{ m}^3$; convert to $\text{cm}^3 \Rightarrow$

a) $1.4\times 10^{27}\text{ m}^3 \times \dfrac{(100\text{ cm})^3}{(1\text{ m})^3} = 1.4\times 10^{33}\text{ cm}^3$; calculate grams of H $\Rightarrow$

$1.4\times 10^{33}\text{ cm}^3\text{ H} \times \dfrac{1.4\text{ g H}}{1\text{ cm}^3} = 2.0\times 10^{33}\text{ g H}$; convert to moles and atoms $\Rightarrow$

$2.0\times 10^{33}\text{ g H} \times \dfrac{1\text{ mole H}}{1.008\text{ g}} \times \dfrac{6.022\times 10^{23}\text{ atoms}}{1\text{ mole H}} = 1\times 10^{57}\text{ H atoms per star}$

b) $\dfrac{1\times 10^{57}\text{ H atoms}}{1\text{ star}} \times \dfrac{1\times 10^{11}\text{ stars}}{\text{galaxy}} = 1\times 10^{68}\text{ H atoms per galaxy}$

c) $\dfrac{1\times 10^{68}\text{ H atoms}}{1\text{ galaxy}} \times \dfrac{1\times 10^{11}\text{ galaxies}}{\text{universe}} = 1\times 10^{79}\text{ H atoms in the universe}$

130. $100\times 10^6\text{ cars} \times \dfrac{1.1\text{ kg }CF_2Cl_2}{\text{car}} \times 0.25\text{ lost per year} = 2.8\times 10^7\text{ kg }CF_2Cl_2\text{ / year}$

$\dfrac{2.8\times 10^7\text{ kg }CF_2Cl_2}{1\text{ year}} \times \dfrac{1\times 10^3\text{ g}}{1\text{ kg}} \times \dfrac{1\text{ mole }CF_2Cl_2}{120.91\text{ g}} \times \dfrac{2\text{ moles Cl}}{1\text{ mole }CF_2Cl_2} \times \dfrac{35.45\text{ g}}{1\text{ mole Cl}} \times$

$\dfrac{1\text{ kg}}{1\times 10^3\text{ g}} = 1.6\times 10^7\text{ kg Cl per year}$

131. Assume a 100 gram sample, % composition is:

$$95.02 \text{ g C} \times \frac{1 \text{ mole C}}{12.01 \text{ g}} = 7.912 \text{ moles C}$$

$$4.98 \text{ g H} \times \frac{1 \text{ mole H}}{1.01 \text{ g H}} = 4.93 \text{ moles H}$$

$$C_{\frac{7.912}{4.93}} H_{\frac{4.93}{4.93}} = C_{1.6}H_1; \text{ Empirical formula} = 5 \times (C_{1.6}H_1) = C_8H_5$$

$$\frac{\text{Molar Mass}}{\text{Empirical Mass}} = \text{Multiplier} \Rightarrow \frac{202.23}{101.12} = 2;$$

Molecular Formula = $2 \times (C_8H_5) = C_{16}H_{10}$

132. The numerical value of the mole (6.022×10^{23}) is defined as being equal to the number of atoms in exactly 12 g of pure carbon-12.

133. If the mole was defined as the number of carbon atoms in exactly 12 slugs of carbon-12, then the new value of Avogadro's number would be 9.01×10^{27}.

134. $C_{15222}H_{25370}O_{12685}$

$$\frac{162 \text{ g/mol}}{C_6H_{10}O_5 \text{ unit}} \times 2537 \text{ units} = 4.11 \times 10^5 \text{g/mol}$$

empirical formula: $C_6H_{10}O_5$

Data Interpretation and Analysis

135. a) % change = $\frac{0.7 - 0.9}{0.7} \times 100\% = -28.6\% \Rightarrow -3 \times 10^1\%$ (to 1 sig. fig.)

b) % F in NaF = $\frac{19.00}{41.99} \times 100\% = 45.25\%$ F

100.0 lb NaF × 0.4525 = 45.25 lb F

45.25 lb F × $\frac{0.4536 \text{ kg}}{1 \text{ lb}}$ = 20.53 kg F

% F in Na_2SiF_6 = $\frac{6 \times 19.00}{188.1} \times 100\% = 60.60\%$ F

100.0 lb NaF × 0.6060 = 60.60 lb F

60.60 lb F × $\frac{0.4536 \text{ kg}}{1 \text{ lb}}$ = 27.49 kg F

Na_2SiF_6 is the more economical choice.

c) For women:

$\frac{3.1 \text{ mg}}{\text{day}} \times \frac{1 \text{ L}}{0.7 \text{ mg}} = 4$ L/day

For men:

$\frac{3.8 \text{ mg}}{\text{day}} \times \frac{1 \text{ L}}{0.7 \text{ mg}} = 5$ L/day

Chemical Reactions

Questions

1. In a reaction, one or more substances change into different substances. Many examples can be given, such as the neutralization of acid with a base (vinegar + baking soda).

2. If you could see atoms and molecules, you would know a chemical reaction is occurring when:
   1) atoms are combining with other atoms to form a compound.
   2) new molecules are formed.
   3) original compounds decompose.
   4) atoms are exchanging places with atoms in another compound.

3. The following constitute the main evidence that a chemical reaction has occurred:
   1) a color change
   2) formation of a solid
   3) formation of a gas
   4) emission of light
   5) emission of (or absorption of) heat

4. A chemical equation is a way of representing what happens to atoms and molecules during a chemical reaction; that is, what the starting materials (reactants) are and what new materials are formed (products). For example:

$$2\ HCl(aq) + Ba(OH)_2(aq) \rightarrow BaCl_2(aq) + 2H_2O\ (l)$$

The reactants (HCl and $Ba(OH)_2$) are found on the left side of the arrow, and the products ($BaCl_2$ and H_2O) are found on the right side of the arrow. The phase of the reactants and products are given in parentheses. For example, (l) indicates a liquid phase, while (aq) indicates the substances are dissolved in water.

5. a) gas
   b) liquid
   c) solid
   d) aqueous (dissolved in water)

6. To balance a chemical equation, adjust the <u>coefficients</u> as necessary to make the numbers of each type of atom on both sides of the equation equal. Never adjust the <u>subscripts</u> to balance a chemical equation.

7.
	Element	Reactants	Products
a)	Ag	4	4
	O	2	2
	C	1	1

Yes, the equation is balanced.

8. An aqueous solution is a mixture of a substance dissolved in water. Two common examples are saltwater and sugar water.

9. A soluble compound will dissolve in solution in appreciable quantities, while an insoluble compound will hardly dissolve in solution at all.

10. Ionic compounds dissolve in water and dissociate into the ions that make up the compound. For example, NaCl in water will form $Na^+(aq)$ and $Cl^-(aq)$.

11. Polyatomic ions dissolve in water and dissociate into the ions that make up the compound. The polyatomic ion group stays together as one particle. For example, $NaNO_3$ in water will form $Na^+(aq)$ and $NO_3^-(aq)$.

12. A strong electrolyte is a substance that dissociates nearly 100% to form ions when it dissolves in solution, whereas a weak electrolyte will only dissociate to a small extent when dissolved in solution.

13. The solubility rules are as follows:

    Soluble Compounds

    1. Any compound containing Li^+, Na^+, K^+, or NH_4^+.
    2. Any compound containing NO_3^- or $C_2H_3O_2^-$.
    3. Most compounds containing Cl^-, Br^-, or I^-.
       Except with Ag^+, Hg_2^{2+}, or Pb^{2+}, they form insoluble compounds.
    4. Most compounds with SO_4^{2-}.
       Except with Sr^{2+}, Ba^{2+}, Pb^{2+}, or Ca^{2+}, they form insoluble compounds.

    Insoluble Compounds

    1. Most compounds containing OH^- or S^{2-}
       Except with Li^+, Na^+, K^+, or NH_4^+, form soluble compounds.
       Except when S^{2-} is with Sr^{2+}, Ba^{2+}, or Ca^{2+}, form soluble compounds.
       Except when OH^- is with Sr^{2+}, Ba^{2+}, or Ca^{2+}, form slightly soluble compounds.
    2. Most compounds containing CO_3^{2-} or PO_4^{3-}
       Except with Li^+, Na^+, K^+, or NH_4^+, they form soluble compounds.

    These rules are useful because there is no way to predict solubility from looking at the periodic table. These rules allow us to know what compounds will dissolve in water and which ones will not, which is critical when writing chemical equations.

14. A precipitation reaction occurs when two aqueous mixtures containing only soluble compounds are mixed together and an insoluble solid compound forms. An example of a precipitation reaction is:
    $Ba(NO_3)_2(aq) + Na_2SO_4(aq) \rightarrow BaSO_4(s) + 2NaNO_3(aq)$

15. The precipitate in a precipitation reaction will always be the insoluble compound. Otherwise, it would dissolve in water, no solid would form, and there would be no precipitation reaction.

16. A molecular equation shows all the compounds in their complete neutral formula. A complete ionic equation shows all soluble ionic compounds as aqueous ions. The net ionic equation shows only the species that take part in the reaction. The following is an example of each:
    Molecular Equation: $Ba(NO_3)_2(aq) + Na_2SO_4(aq) \rightarrow BaSO_4(s) + 2NaNO_3(aq)$
    Complete Ionic Equation:
    $Ba^{2+}(aq) + 2NO_3^-(aq) + 2Na^+(aq) + SO_4^{2-}(aq) \rightarrow BaSO_4(s) + 2Na^+(aq) + 2NO_3^-(aq)$
    Net Ionic Equation: $Ba^{2+}(aq) + SO_4^{2-}(aq) \rightarrow BaSO_4(s)$

17. An acid–base reaction involves a neutralization of $H^+(aq)$ ions from an acid and $OH^-(aq)$ ions from a base to form water. The counter ions form a salt. Consider the following reaction: $NaOH(aq) + HCl(aq) \rightarrow H_2O(l)$ and $NaCl(aq)$. The base is NaOH, the acid is HCl.

18. The properties of acids include a sour taste, the ability to dissolve some metals, and the tendency to form H^+ in solution. The properties of a base are a bitter taste, a slippery feel, and a tendency to form OH^- in solution.

19. A gas-evolution reaction occurs when one of the products of a chemical reaction is a gas. An example is: $2HCl(aq) + CaS(aq) \rightarrow H_2S(g) + CaCl_2(aq)$.

20. A redox reaction occurs when electrons are exchanged between the reactants. An example is: $2K(s) + Cl_2(g) \rightarrow 2KCl(s)$. The K atom loses electrons while the Cl atom gains electrons.

21. A combustion reaction involves the reaction of a substance with O_2 to form oxygen-containing compounds, often including water.
    An example is: $2C_2H_6(g) + 7O_2(g) \rightarrow 4CO_2(g) + 6H_2O(g)$

22. You can classify chemical reactions in one of two ways: (1) by the type of chemistry occurring during the reaction, such as acid–base chemistry or precipitation chemistry or (2) by study of what happens to atoms during the reaction. We use both methods for classifying reactions, so we can study the similarities and differences between different reactions.

23. In a synthesis reaction, simpler substances combine to form more complex substances, for example: 2Na(s) + Cl$_2$(g) → 2NaCl(s). A decomposition reaction occurs when a complex substance decomposes to form simpler substances; for example: 2H$_2$O(l) → 2H$_2$(g) + O$_2$(g).

24. In a single replacement reaction, one element displaces another in a compound, for example: Zn(s) + CuCl$_2$(aq) → ZnCl$_2$(aq) + Cu(s). In a double displacement reaction, two elements or groups of elements in two different compounds exchange places— for example: AgNO$_3$(aq) + NaCl(aq) → AgCl(s) + NaNO$_3$(aq).

Problems

Evidence of Chemical Reactions

25. a) A chemical reaction because the initial compounds change to form a solid and a color change occurs.
    b) Not a chemical reaction because the initial compound did not change into another substance.
    c) A chemical reaction because the initial compounds change to form a solid.
    d) A chemical reaction because the initial compounds change to form a gas and other new compounds.

26. a) A chemical reaction because new compounds are formed and heat is given off.
    b) Not a chemical reaction because no new compounds were formed.
    c) A chemical reaction because a gas formed when the two reactants were mixed.
    d) Not a chemical reaction because no new compounds were formed.

27. Yes, a chemical reaction has occurred because the bubbles that formed are a new compound formed as a product of the reaction.

28. Yes, a chemical reaction has occurred because the bubbles are a new compound created as the product of the reaction and because the reaction produced heat.

29. Yes, a chemical reaction has occurred because the color change in the hair is due to forming new compounds in the hair itself.

30. No, water boiling is not a chemical reaction because you still have water and there was not a new compound formed by the process.

Writing and Balancing Chemical Equations

31. a)
Pb	1	1
N	2	2
O	6	6
Na	2	2
Cl	2	2

Yes, the equation is balanced.

b)
C	3	3
H	8	8
O	2	10

No, the equation is not balanced.

32. a)
Mg	1	1
S	1	2
Cu	2	2
Cl	4	2

No, the equation is not balanced.

b)
C	12	12
H	28	28
O	38	38

Yes, the equation is balanced.

33. By adding a subscript, the nature of the chemical is changed and the equation no longer accurately represents the chemical reaction it is supposed to describe. The proper method of balancing a reaction is to add coefficients in front of each reactant or product compound. The balanced reaction is: $2H_2O\ (l) \rightarrow 2H_2(g) + O_2(g)$.

34. By changing a subscript, the nature of the chemical is changed and no longer accurately represents the chemical reaction it is supposed to describe. The proper method of balancing a reaction is to add coefficients in front of each reactant or product compounds. The balanced reaction is: $2Al\ (s) + 3Cl_2(g) \rightarrow 2AlCl_3(s)$.

35. a) $PbS(s) + 2HCl(aq) \rightarrow PbCl_2(s) + H_2S(g)$
    b) $CO(g) + 3H_2(g) \rightarrow CH_4(g) + H_2O(l)$
    c) $Fe_2O_3(s) + 3H_2(g) \rightarrow 2Fe(s) + 3H_2O(l)$
    d) $4NH_3(g) + 5O_2(g) \rightarrow 4NO(g) + 6H_2O(g)$

36. a) $2Cu(s) + S(s) \rightarrow Cu_2S(s)$
    b) $2SO_2(g) + O_2(g) \rightarrow 2SO_3(g)$
    c) $4HCl(aq) + MnO_2(s) \rightarrow MnCl_2(aq) + 2H_2O(l) + Cl_2(g)$
    d) $2C_6H_6(l) + 15O_2(g) \rightarrow 12CO_2(g) + 6H_2O(l)$

37. a) $Mg(s) + 2CuNO_3(aq) \rightarrow Mg(NO_3)_2(aq) + 2Cu(s)$
    b) $2N_2O_5(g) \rightarrow 4NO_2(g) + O_2(g)$
    c) $Ca(s) + 2HNO_3(aq) \rightarrow Ca(NO_3)_2(aq) + H_2(g)$
    d) $2CH_3OH(l) + 3O_2(g) \rightarrow 2CO_2(g) + 4H_2O(g)$

38. a) $2C_2H_2(g) + 5O_2(g) \rightarrow 4CO_2(g) + 2H_2O(g)$
    b) $Cl_2(g) + 2KI(aq) \rightarrow I_2(s) + 2KCl(aq)$
    c) $Li_2O(s) + H_2O(l) \rightarrow 2LiOH(aq)$
    d) $2CO(g) + O_2(g) \rightarrow 2CO_2(g)$

39. $2Na(s) + 2H_2O(l) \rightarrow H_2(g) + 2NaOH(aq)$

40. $4Fe(s) + 3O_2(g) \rightarrow 2Fe_2O_3(s)$

41. $2SO_2(g) + O_2(g) + 2H_2O(l) \rightarrow 2H_2SO_4(aq)$

42. $4NO_2(g) + O_2(g) + 2H_2O(l) \rightarrow 4HNO_3(aq)$

43. $V_2O_5(s) + 2H_2(g) \rightarrow V_2O_3(s) + 2H_2O(l)$

44. $2NO_2(g) + 7H_2(g) \rightarrow 2NH_3(g) + 4H_2O(l)$

45. $C_{12}H_{22}O_{11}(aq) + H_2O(l) \rightarrow 4C_2H_5OH(aq) + 4CO_2(g)$

46. $6CO_2(g) + 6H_2O(l) \rightarrow C_6H_{12}O_6(aq) + 6O_2(g)$

47. a) $Na_2S(aq) + Cu(NO_3)_2(aq) \rightarrow 2NaNO_3(aq) + CuS(s)$
    b) $4HCl(aq) + O_2(g) \rightarrow 2H_2O(l) + 2Cl_2(g)$
    c) $2H_2(g) + O_2(g) \rightarrow 2H_2O(l)$
    d) $FeS(s) + 2HCl(aq) \rightarrow FeCl_2(aq) + H_2S(g)$

48. a) $3N_2H_4(l) \rightarrow 4NH_3(g) + N_2(g)$
    b) $3H_2(g) + N_2(g) \rightarrow 2NH_3(g)$
    c) $Cu_2O(s) + C(s) \rightarrow 2Cu(s) + CO(g)$
    d) $H_2(g) + Cl_2(g) \rightarrow 2HCl(g)$

49. a) $BaO_2(s) + H_2SO_4(aq) \rightarrow BaSO_4(s) + H_2O_2(aq)$
    b) $2Co(NO_3)_3(aq) + 3(NH_4)_2S(aq) \rightarrow Co_2S_3(s) + 6NH_4NO_3(aq)$
    c) $Li_2O(s) + H_2O(l) \rightarrow 2LiOH(aq)$
    d) $Hg_2(C_2H_3O_2)_2(aq) + 2KCl(aq) \rightarrow Hg_2Cl_2(s) + 2KC_2H_3O_2(aq)$

50. a) $MnO_2(s) + 4HCl(aq) \rightarrow Cl_2(g) + MnCl_2(aq) + 2H_2O(l)$
    b) $2CO_2(g) + CaSiO_3(s) + H_2O(l) \rightarrow SiO_2(s) + Ca(HCO_3)_2(aq)$
    c) $2Fe(s) + 3S(l) \rightarrow Fe_2S_3(s)$
    d) $3NO_2(g) + H_2O(l) \rightarrow 2HNO_3(aq) + NO(g)$

51. a) $2Rb(s) + 2H_2O(l) \rightarrow 2RbOH(aq) + H_2(g)$
    b) Ok
    c) $2NiS(s) + 3O_2(g) \rightarrow 2NiO(s) + 2SO_2(g)$
    d) $3PbO(s) + 2NH_3(g) \rightarrow 3Pb(s) + N_2(g) + 3H_2O(l)$

52. a) Ok
    b) $4Cr(s) + 3O_2(g) \rightarrow 2Cr_2O_3(s)$
    c) $Al_2S_3(s) + 6H_2O(l) \rightarrow 2Al(OH)_3(s) + 3H_2S(g)$
    d) $Fe_2O_3(s) + 3CO(g) \rightarrow 2Fe(s) + 3CO_2(g)$

53. $C_6H_{12}O_6(aq) + 6O_2(g) \rightarrow 6CO_2(g) + 6H_2O(l)$

54. $C_3H_8(g) + 5O_2(g) \rightarrow 3CO_2(g) + 4H_2O(g)$

55. $2NO(g) + 2CO(g) \rightarrow N_2(g) + 2CO_2(g)$

56. $2NH_3(g) + CO_2(g) \rightarrow CO(NH_2)_2(s) + H_2O(l)$

Solubility

57. a) Soluble: Na^+, $C_2H_3O_2^-$
    b) Soluble: Sn^{2+}, NO_3^-
    c) Insoluble
    d) Soluble: Na^+, PO_4^{3-}

58. a) Soluble: NH_4^+, S^{2-}
    b) Insoluble
    c) Insoluble
    d) Soluble: Pb^{2+}, $C_2H_3O_2^-$

59. Ag^+ with Cl^-, $AgCl$
Ba^{2+} with SO_4^{2-}, $BaSO_4$
Cu^{2+} with CO_3^{2-}, $CuCO_3$
Fe^{3+} with S^{2-}, Fe_2S_3

60. Na^+ with CO_3^{2-}, Na_2CO_3
Sr^{2+} with S^{2-}, SrS
Co^{2+} with SO_4^{2-}, $CoSO_4$
Pb^{2+} with NO_3^-, $Pb(NO_3)_2$

61.
Soluble	Insoluble	Soluble	Insoluble
K_2S	Hg_2I_2	K_2SO_4	$PbSO_4$
BaS	$Cu_3(PO_4)_2$	SrS	$PbCl_2$
NH_4Cl	MgS	Li_2S	Hg_2Cl_2
Na_2CO_3	$CaSO_4$		

62.
Soluble	Insoluble	Soluble	Insoluble
$LiOH$	$CoCO_3$	K_3PO_4	$SrSO_4$
$CaCl_2$	$Cu(OH)_2$	Na_2CO_3	Hg_2Br_2
$Ca(C_2H_3O_2)_2$	$AgCl$	$Pb(NO_3)_2$	$PbBr_2$
		CuI_2	PbI_2

Precipitation Reactions

63. a) no reaction
b) $K_2SO_4(aq) + BaBr_2(aq) \rightarrow BaSO_4(s) + 2\,KBr(aq)$
c) $2NaCl(aq) + Hg_2(C_2H_3O_2)_2(aq) \rightarrow 2NaC_2H_3O_2(aq) + Hg_2Cl_2(s)$
d) no reaction

64. a) $3NaOH(aq) + FeBr_3(aq) \rightarrow 3NaBr(aq) + Fe(OH)_3(s)$
b) $BaCl_2(aq) + 2AgNO_3(aq) \rightarrow 2AgCl(s) + Ba(NO_3)_2(aq)$
c) $Na_2CO_3(aq) + CoCl_2(aq) \rightarrow CoCO_3(s) + 2NaCl(aq)$
d) $K_2S(aq) + BaCl_2(aq) \rightarrow BaS(s) + 2KCl(aq)$

65. a) $Na_2CO_3(aq) + Pb(NO_3)_2(aq) \rightarrow PbCO_3(s) + 2NaNO_3(aq)$
b) $K_2SO_4(aq) + Pb(C_2H_3O_2)_2(aq) \rightarrow PbSO_4(s) + 2KC_2H_3O_2(aq)$
c) $Cu(NO_3)_2(aq) + BaS(aq) \rightarrow CuS(s) + Ba(NO_3)_2(aq)$
d) no reaction

66. a) $2KCl(aq) + Pb(C_2H_3O_2)_2(aq) \rightarrow PbCl_2(s) + 2KC_2H_3O_2(aq)$
b) $Li_2SO_4(aq) + SrCl_2(aq) \rightarrow SrSO_4(s) + 2LiCl(aq)$
c) no reaction
d) $Cr(NO_3)_3(aq) + K_3PO_4(aq) \rightarrow CrPO_4(s) + 3KNO_3(aq)$

67. a) correct
    b) no reaction
    c) correct
    d) $Pb(NO_3)_2(aq) + 2LiCl(aq) \rightarrow 2LiNO_3(aq) + PbCl_2(s)$

68. a) $AgNO_3(aq) + NaCl(aq) \rightarrow NaNO_3(aq) + AgCl(s)$
    b) no reaction
    c) correct
    d) correct

Ionic and Net Ionic Equations

69. Spectator Ions: NO_3^-, K^+

70. Spectator Ions: Na^+, I^-

71. a) Complete Ionic:
     $Ag^+(aq) + NO_3^-(aq) + K^+(aq) + Cl^-(aq) \rightarrow AgCl(s) + K^+(aq) + NO_3^-(aq)$
     Net Ionic: $Ag^+(aq) + Cl^-(aq) \rightarrow AgCl(s)$
    b) Complete Ionic:
     $Ca^{2+}(aq) + S^{2-}(aq) + Cu^{2+}(aq) + 2Cl^-(aq) \rightarrow CuS(s) + Ca^{2+}(aq) + 2Cl^-(aq)$
     Net Ionic: $S^{2-}(aq) + Cu^{2+}(aq) \rightarrow CuS(s)$
    c) Complete Ionic:
     $Na^+(aq) + OH^-(aq) + H^+(aq) + NO_3^-(aq) \rightarrow H_2O(l) + Na^+(aq) + NO_3^-(aq)$
     Net Ionic: $OH^-(aq) + H^+(aq) \rightarrow H_2O(l)$
    d) Complete Ionic:
     $6K^+(aq) + 2PO_4^{3-}(aq) + 3Ni^{2+}(aq) + 6Cl^-(aq) \rightarrow Ni_3(PO_4)_2(s) + 6K^+(aq) + 6Cl^-(aq)$
     Net Ionic: $2PO_4^{3-}(aq) + 3Ni^{2+}(aq) \rightarrow Ni_3(PO_4)_2(s)$

72. a) Complete Ionic:
     $H^+(aq) + I^-(aq) + K^+(aq) + OH^-(aq) \rightarrow H_2O(l) + K^+(aq) + I^-(aq)$
     Net Ionic: $H^+(aq) + OH^-(aq) \rightarrow H_2O(l)$
    b) Complete Ionic:
     $2Na^+(aq) + SO_4^{2-}(aq) + Ca^{2+}(aq) + 2I^-(aq) \rightarrow CaSO_4(s) + 2Na^+(aq) + 2I^-(aq)$
     Net Ionic: $SO_4^{2-}(aq) + Ca^{2+}(aq) \rightarrow CaSO_4(s)$
    c) Complete Ionic:
     $2HC_2H_3O_2(aq) + 2Na^+(aq) + CO_3^{2-}(aq) \rightarrow H_2O(l) + CO_2(g) + 2Na^+(aq) + 2C_2H_3O_2^-(aq)$
     Net Ionic: $2HC_2H_3O_2(aq) + CO_3^{2-}(aq) \rightarrow H_2O(l) + CO_2(g) + 2C_2H_3O_2^-(aq)$
    d) Complete Ionic:
     $NH_4^+(aq) + Cl^-(aq) + Na^+(aq) + OH^-(aq) \rightarrow H_2O(l) + NH_3(g) + Na^+(aq) + Cl^-(aq)$
     Net Ionic: $NH_4^+(aq) + OH^-(aq) \rightarrow H_2O(l) + NH_3(g)$

73. Complete Ionic:
     $Hg_2^{2+}(aq) + 2NO_3^-(aq) + 2Na^+(aq) + 2Cl^-(aq) \rightarrow Hg_2Cl_2(s) + 2Na^+(aq) + NO_3^-(aq)$
     Net Ionic: $Hg_2^{2+}(aq) + 2Cl^-(aq) \rightarrow Hg_2Cl_2(s)$

74. Complete Ionic:
$Pb^{2+}(aq) + 2NO_3^-(aq) + 2K^+(aq) + SO_4^{2-}(aq) \rightarrow PbSO_4(s) + 2K^+(aq) + 2NO_3^-(aq)$
Net Ionic: $Pb^{2+}(aq) + SO_4^{2-}(aq) \rightarrow PbSO_4(s)$

75. a) Complete Ionic: $2Na^+(aq) + CO_3^{2-}(aq) + Pb^{2+}(aq) + 2NO_3^-(aq) \rightarrow$
$PbCO_3(s) + 2Na^+(aq) + 2NO_3^-(aq)$
Net Ionic: $CO_3^{2-}(aq) + Pb^{2+}(aq) \rightarrow PbCO_3(s)$
b) Complete Ionic: $2K^+(aq) + SO_4^{2-}(aq) + Pb^{2+}(aq) + 2C_2H_3O_2^-(aq) \rightarrow$
$PbSO_4(s) + 2K^+(aq) + 2C_2H_3O_2^-(aq)$
Net Ionic: $SO_4^{2-}(aq) + Pb^{2+}(aq) \rightarrow PbSO_4(s)$
c) Complete Ionic: $Cu^{2+}(aq) + 2NO_3^-(aq) + Ba^{2+}(aq) + S^{2-}(aq) \rightarrow$
$CuS(s) + Ba^{2+}(aq) + 2NO_3^-(aq)$
Net Ionic: $Cu^{2+}(aq) + S^{2-}(aq) \rightarrow CuS(s)$
d) No Reaction

76. a) Complete Ionic: $2K^+(aq) + 2Cl^-(aq) + Pb^{2+}(aq) + 2C_2H_3O_2^-(aq) \rightarrow$
$PbCl_2(s) + 2K^+(aq) + 2C_2H_3O_2^-(aq)$
Net Ionic: $2Cl^-(aq) + Pb^{2+}(aq) \rightarrow PbCl_2(s)$
b) Complete Ionic: $2Li^+(aq) + SO_4^{2-}(aq) + Sr^{2+}(aq) + 2Cl^-(aq) \rightarrow$
$SrSO_4(s) + 2Li^+(aq) + 2Cl^-(aq)$
Net Ionic: $SO_4^{2-}(aq) + Sr^{2+}(aq) \rightarrow SrSO_4(s)$
c) No Reaction
d) Complete Ionic: $Cr^{3+}(aq) + 3NO_3^-(aq) + 3K^+(aq) + PO_4^{3-}(aq) \rightarrow$
$CrPO_4(s) + 3K^+(aq) + 3NO_3^-(aq)$
Net Ionic: $Cr^{3+}(aq) + PO_4^{3-}(aq) \rightarrow CrPO_4(s)$

Acid–Base and Gas-Evolution Reactions

77. Molecular: $HCl(aq) + KOH(aq) \rightarrow KCl(aq) + H_2O(l)$
Net Ionic: $H^+(aq) + OH^-(aq) \rightarrow H_2O(l)$

78. Molecular: $2HNO_3(aq) + Ca(OH)_2(aq) \rightarrow Ca(NO_3)_2(aq) + 2H_2O(l)$
Net Ionic: $H^+(aq) + OH^-(aq) \rightarrow H_2O(l)$

79. a) $2HCl(aq) + Ba(OH)_2(aq) \rightarrow 2H_2O(l) + BaCl_2(aq)$
b) $H_2SO_4(aq) + 2KOH(aq) \rightarrow 2H_2O(l) + K_2SO_4(aq)$
c) $HClO_4(aq) + NaOH(aq) \rightarrow H_2O(l) + NaClO_4(aq)$

80. a) $2HC_2H_3O_2(aq) + Ca(OH)_2(aq) \rightarrow 2H_2O(l) + Ca(C_2H_3O_2)_2(aq)$
b) $HBr(aq) + LiOH(aq) \rightarrow LiBr(aq) + H_2O(l)$
c) $H_2SO_4(aq) + Ba(OH)_2(aq) \rightarrow 2H_2O(l) + BaSO_4(s)$

81. a) $HBr(aq) + NaHCO_3(aq) \rightarrow CO_2(g) + H_2O(l) + NaBr(aq)$
b) $NH_4I(aq) + KOH(aq) \rightarrow H_2O(l) + NH_3(g) + KI(aq)$
c) $2HNO_3(aq) + K_2SO_3(aq) \rightarrow SO_2(g) + H_2O(l) + 2KNO_3(aq)$
d) $2HI(aq) + Li_2S(aq) \rightarrow H_2S(g) + 2LiI(aq)$

82. a) $2HClO_4(aq) + K_2CO_3(aq) \rightarrow CO_2(g) + H_2O(l) + 2KClO_4(aq)$
    b) $HC_2H_3O_2(aq) + LiHSO_3 \rightarrow SO_2(g) + H_2O(l) + LiC_2H_3O_2(aq)$
    c) $(NH_4)_2SO_4(aq) + Ca(OH)_2(aq) \rightarrow CaSO_4(s) + 2NH_3(g) + 2H_2O(l)$
    d) $2HCl(aq) + ZnS(s) \rightarrow H_2S(g) + ZnCl_2(aq)$

Oxidation–Reduction and Combustion

83. b) metal reacting with a nonmetal
    d) transfer of electrons between reactants

84. a) transfer of electrons between reactants
    b) substance reacting with elemental oxygen
    d) metal reacting with a nonmetal

85. a) $2C_2H_6(g) + 7O_2(g) \rightarrow 4CO_2(g) + 6H_2O(g)$
    b) $2Ca(s) + O_2(g) \rightarrow 2CaO(s)$
    c) $2C_3H_8O(l) + 9O_2(g) \rightarrow 6CO_2(g) + 8H_2O(g)$
    d) $2C_4H_{10}S(l) + 15O_2(g) \rightarrow 8CO_2(g) + 10H_2O(g) + 2SO_2(g)$

86. a) $S(s) + O_2(g) \rightarrow SO_2(g)$
    b) $C_7H_{16}(l) + 11O_2(g) \rightarrow 7CO_2(g) + 8H_2O(g)$
    c) $C_4H_{10}O(l) + 6O_2(g) \rightarrow 4CO_2(g) + 5H_2O(g)$
    d) $CS_2(l) + 3O_2(g) \rightarrow CO_2(g) + 2SO_2(g)$

87. a) $2Ag(s) + Br_2(g) \rightarrow 2AgBr(s)$
    b) $2K(s) + Br_2(g) \rightarrow 2KBr(s)$
    c) $2Al(s) + 3Br_2(g) \rightarrow 2AlBr_3(s)$
    d) $Ca(s) + Br_2(g) \rightarrow CaBr_2(s)$

88. a) $Zn(s) + Cl_2(g) \rightarrow ZnCl_2(s)$
    b) $2Ga(s) + 3Cl_2(g) \rightarrow 2GaCl_3(s)$
    c) $2Rb(s) + Cl_2(g) \rightarrow 2RbCl(s)$
    d) $Mg(s) + Cl_2(g) \rightarrow MgCl_2(s)$

Classifying Chemical Reactions by What Atoms Do

89. a) double displacement
    b) synthesis
    c) single displacement
    d) decomposition

90. a) decomposition
    b) synthesis
    c) single displacement
    d) double displacement

91. a) synthesis
    b) decomposition
    c) synthesis

92. a) double displacement
    b) synthesis
    c) synthesis

Cumulative Problems

93. a) Complete Ionic: $2Na^+(aq) + 2I^-(aq) + Hg_2^{2+}(aq) + 2NO_3^-(aq) \rightarrow$
    $Hg_2I_2(s) + 2Na^+(aq) + 2NO_3^-(aq)$
    Net Ionic: $2I^-(aq) + Hg_2^{2+}(aq) \rightarrow Hg_2I_2(s)$
    b) Complete Ionic: $2H^+(aq) + 2ClO_4^-(aq) + Ba^{2+}(aq) + 2OH^-(aq) \rightarrow$
    $2H_2O(l) + 2ClO_4^-(aq) + Ba^{2+}(aq)$
    Net Ionic: $H^+(aq) + OH^-(aq) \rightarrow H_2O(l)$
    c) No Reaction
    d) Complete Ionic: $2H^+(aq) + 2Cl^-(aq) + 2Li^+(aq) + CO_3^{2-}(aq) \rightarrow$
    $CO_2(g) + H_2O(l) + 2Li^+(aq) + 2Cl^-(aq)$
    Net Ionic: $2H^+(aq) + CO_3^{2-}(aq) \rightarrow CO_2(g) + H_2O(l)$

94. a) Complete Ionic: $Li^+(aq) + Cl^-(aq) + Ag^+(aq) + NO_3^-(aq) \rightarrow$
    $AgCl(s) + NO_3^-(aq) + Li^+(aq)$
    Net Ionic: $Cl^-(aq) + Ag^+(aq) \rightarrow AgCl(s)$
    b) Complete Ionic: $2H^+(aq) + SO_4^{2-}(aq) + 2Li^+(aq) + SO_3^{2-}(aq) \rightarrow$
    $SO_2(g) + H_2O(l) + 2Li^+(aq) + SO_4^{2-}(aq)$
    Net Ionic: $2H^+(aq) + SO_3^{2-}(aq) \rightarrow SO_2(g) + H_2O(l)$
    c) Complete Ionic: $2HC_2H_3O_2(aq) + Ca^{2+}(aq) + 2OH^-(aq) \rightarrow$
    $2H_2O(l) + 2C_2H_3O_2^-(aq) + Ca^{2+}(aq)$
    Net Ionic: $2HC_2H_3O_2(aq) + 2OH^-(aq) \rightarrow 2H_2O(l) + 2C_2H_3O_2^-(aq)$
    d) No Reaction

95. a) No Reaction
    b) No Reaction
    c) Complete Ionic: $H^+(aq) + NO_3^-(aq) + K^+(aq) + HSO_3^-(aq) \rightarrow$
    $SO_2(g) + H_2O(l) + NO_3^-(aq) + K^+(aq)$
    Net Ionic: $H^+(aq) + HSO_3^-(aq) \rightarrow SO_2(g) + H_2O(l)$
    d) Complete Ionic: $Mn^{3+}(aq) + 3Cl^-(aq) + 3K^+(aq) + PO_4^{3-}(aq) \rightarrow$
    $MnPO_4(s) + 3Cl^-(aq) + 3K^+(aq)$
    Net Ionic: $Mn^{3+}(aq) + PO_4^{3-}(aq) \rightarrow MnPO_4(s)$

96. a) No Reaction
    b) No Reaction
    c) Complete Ionic: $Cr^{3+}(aq) + 3NO_3^-(aq) + 3Li^+(aq) + 3OH^-(aq) \rightarrow$
    $\qquad Cr(OH)_3(s) + 3NO_3^-(aq) + Li^+(aq)$
    Net Ionic: $Cr^{3+}(aq) + 3OH^-(aq) \rightarrow Cr(OH)_3(s)$
    d) Complete Ionic: $2H^+(aq) + 2Cl^-(aq) + Hg_2^{2+}(aq) + 2NO_3^-(aq) \rightarrow$
    $\qquad Hg_2Cl_2(s) + 2NO_3^-(aq) + 2H^+(aq)$
    Net Ionic: $2Cl^-(aq) + Hg_2^{2+}(aq) \rightarrow Hg_2Cl_2(s)$

97. a) acid–base; $KOH(aq) + HC_2H_3O_2(aq) \rightarrow H_2O(l) + KC_2H_3O_2(aq)$
    b) gas–evolution/acid–base: $2HBr(aq) + K_2CO_3(aq) \rightarrow H_2O(l) + CO_2(g) + 2KBr(aq)$
    c) synthesis: $2H_2(g) + O_2(g) \rightarrow 2H_2O(g)$
    d) precipitation: $2NH_4Cl(aq) + Pb(NO_3)_2(aq) \rightarrow PbCl_2(s) + 2NH_4NO_3(aq)$

98. a) No Reaction
    b) combustion: $2C_5H_{12}O(l) + 15O_2(g) \rightarrow 12H_2O(g) + 10CO_2(g)$
    c) gas evolution: $2NH_4Cl(aq) + Ca(OH)_2(aq) \rightarrow 2NH_3(g) + 2H_2O(l) + CaCl_2(aq)$
    d) precipitation: $SrS(aq) + CuSO_4(aq) \rightarrow SrSO_4(s) + CuS(s)$

99. a) oxidation–reduction, single–displacement
    b) acid–base, gas evolution
    c) gas evolution, double–displacement
    d) precipitation, double–displacement

100. a) precipitation, double–displacement
     b) synthesis, oxidation–reduction
     c) oxidation–reduction, single–displacement
     d) gas evolution, oxidation–reduction, single–displacement

101. For calcium chloride:
     Molecular: $3CaCl_2(aq) + 2Na_3PO_4(aq) \rightarrow Ca_3(PO_4)_2(s) + 6NaCl(aq)$
     Complete: $3Ca^{2+}(aq) + 6Cl^-(aq) + 6Na^+(aq) + 2PO_4^{3-}(aq) \rightarrow$
     $\qquad Ca_3(PO_4)_2(s) + 6Na^+(aq) + 6Cl^-(aq)$
     Net Ionic: $3Ca^{2+}(aq) + 2PO_4^{3-}(aq) \rightarrow Ca_3(PO_4)_2(s)$

     For magnesium nitrate:
     Molecular: $3Mg(NO_3)_2(aq) + 2Na_3PO_4(aq) \rightarrow Mg_3(PO_4)_2(s) + 6NaNO_3(aq)$
     Complete: $3Mg^{2+}(aq) + 6NO_3^-(aq) + 6Na^+(aq) + 2PO_4^{3-}(aq) \rightarrow$
     $\qquad Mg_3(PO_4)_2(s) + 6Na^+(aq) + 6NO_3^-(aq)$
     Net Ionic: $3Mg^{2+}(aq) + 2PO_4^{3-}(aq) \rightarrow Mg_3(PO_4)_2(s)$

102. For nitric acid:
   Complete: $2H^+(aq) + 2NO_3^-(aq) + Ca^{2+}(aq) + CO_3^{2-}(aq) \rightarrow$
   $CO_2(g) + H_2O(l) + Ca^{2+}(aq) + 2NO_3^-(aq)$
   Net Ionic: $2H^+(aq) + CO_3^{2-}(aq) \rightarrow CO_2(g) + H_2O(l)$

   For sulfuric acid:
   Complete: $2H^+(aq) + SO_4^{2-}(aq) + Ca^{2+}(aq) + CO_3^{2-}(aq) \rightarrow$
   $CO_2(g) + H_2O(l) + CaSO_4(s)$
   Net Ionic: $2H^+(aq) + SO_4^{2-}(aq) + Ca^{2+}(aq) + CO_3^{2-}(aq) \rightarrow$
   $CO_2(g) + H_2O(l) + CaSO_4(s)$

   You would know the neutralization process was working by the formation of carbon dioxide gas bubbles in solution.

103. a) $Pb^{2+}(aq) + 2Cl^-(aq) \rightarrow PbCl_2(s)$
    b) $Ca^{2+}(aq) + SO_4^{2-}(aq) \rightarrow CaSO_4(s)$
    c) $Ag^+(aq) + Cl^-(aq) \rightarrow AgCl(s)$
    d) $Hg_2^{2+}(aq) + 2Cl^-(aq) \rightarrow Hg_2Cl_2(s)$

104. a) $Hg_2^{2+}(aq) + 2Cl^-(aq) \rightarrow Hg_2Cl_2(s)$
    b) $Ca^{2+}(aq) + SO_4^{2-}(aq) \rightarrow CaSO_4(s)$
    c) $Ba^{2+}(aq) + SO_4^{2-}(aq) \rightarrow BaSO_4(s)$
    d) $Ag^+(aq) + Cl^-(aq) \rightarrow AgCl(s)$

105. $2K_3PO_4(aq) + 3Ca^{2+}(aq) \rightarrow Ca_3(PO_4)_2(s) + 6K^+(aq)$

   $0.112 \text{ mol } K_3PO_4 \times \dfrac{3 \text{ mol } Ca^{2+}}{2 \text{ mol } K_3PO_4} = 0.168 \text{ mol } Ca^{2+}$

   $0.168 \text{ mol } Ca^{2+} \times \dfrac{40.08 \text{ g}}{1 \text{ mol } Ca^{2+}} = 6.73 \text{ g } Ca^{2+}$

106. $Na_2CO_3(aq) + Mg^{2+}(aq) \rightarrow MgCO_3(s) + 2Na^+(aq)$

   $0.0877 \text{ mol } Na_2CO_3 \times \dfrac{1 \text{ mol } Mg^{2+}}{1 \text{ mol } Na_2CO_3} \times \dfrac{24.31 \text{ g}}{1 \text{ mol } Mg^{2+}} = 2.13 \text{ g } Mg^{2+}$

107. Assuming lead(II) ions in solution: $Pb^{2+}(aq) + 2NaCl(aq) \rightarrow PbCl_2(s) + 2Na^+(aq)$

   $0.133 \text{ g } Pb^{2+} \times \dfrac{1 \text{ mol } Pb^{2+}}{207.2 \text{ g}} \times \dfrac{2 \text{ mol } NaCl}{1 \text{ mol } Pb^{2+}} = 0.00128 \text{ mol } NaCl$

   $0.133 \text{ g } Pb^{2+} \times \dfrac{1 \text{ mol } Pb^{2+}}{207.2 \text{ g}} \times \dfrac{2 \text{ mol } NaCl}{1 \text{ mol } Pb^{2+}} \times \dfrac{58.44 \text{ g}}{1 \text{ mol } NaCl} = 0.0750 \text{ g } NaCl$

108. $Ag^+(aq) + KCl(aq) \rightarrow AgCl(s) + K^+(aq)$

$$1.77 \text{ g } Ag^+ \times \frac{1 \text{ mol } Ag^+}{107.87 \text{ g}} \times \frac{1 \text{ mol } KCl}{1 \text{ mol } Ag^+} = 0.0164 \text{ mol } KCl$$

$$1.77 \text{ g } Ag^+ \times \frac{1 \text{ mol } Ag^+}{107.87 \text{ g}} \times \frac{1 \text{ mol } KCl}{1 \text{ mol } Ag^+} \times \frac{74.55 \text{ g}}{1 \text{ mol } KCl} = 1.22 \text{ g } KCl$$

Highlight Problem

109. The first figure is based on a chemical reaction. You can see the changes in the arrangement of atoms within the compounds before and after detonation.

110. a) Molecular: $Fe(NO_3)_3(aq) + 3NaOH(aq) \rightarrow Fe(OH)_3(s) + 3NaNO_3(aq)$
   Complete: $Fe^{3+}(aq) + 3NO_3^-(aq) + 3Na^+(aq) + 3OH^-(aq) \rightarrow$
   $Fe(OH)_3(s) + 3Na^+(aq) + 3NO_3^-(aq)$
   Net Ionic: $Fe^{3+}(aq) + 3OH^-(aq) \rightarrow Fe(OH)_3(s)$
   b) Molecular: $CoCl_2(aq) + 2KOH(aq) \rightarrow Co(OH)_2(s) + 2KCl(aq)$
   Complete: $Co^{2+}(aq) + 2Cl^-(aq) + 2K^+(aq) + 2OH^-(aq) \rightarrow$
   $Co(OH)_2(s) + 2K^+(aq) + 2OH^-(aq)$
   Net Ionic: $Co^{2+}(aq) + 2OH^-(aq) \rightarrow Co(OH)_2(s)$
   c) Molecular: $AgNO_3(aq) + NaI(aq) \rightarrow AgI(s) + NaNO_3(aq)$
   Complete: $Ag^+(aq) + NO_3^-(aq) + Na^+(aq) + I^-(aq) \rightarrow$
   $AgI(s) + Na^+(aq) + NO_3^-(aq)$
   Net Ionic: $Ag^+(aq) + I^-(aq) \rightarrow AgI(s)$

111. Individual group answers will differ.

112. Individual group answers will differ; see the solubility rules to verify answers in Figure 7.7.

113. Specific examples will vary; representative reactions are shown here:
   Precipitation: Reactions that form solid substances in water.
   $AgNO_3(aq) + NaCl(aq) \rightarrow AgCl(s) + NaNO_3(aq)$
   Acid–base: Reactions that form water upon mixing of an acid and a base.
   $HCl(aq) + NaOH(aq) \rightarrow H_2O(l) + NaCl(aq)$
   Gas evolution: Reactions that evolve a gas.
   $NH_4Cl(aq) + NaOH(aq) \rightarrow H_2O(l) + NH_3(g) + NaCl(aq)$
   Redox (noncombustion): Reactions that involve the transfer of electrons
   $Zn(s) + CuCl_2(aq) \rightarrow Cu(s) + ZnCl_2(aq)$
   Combustion: Type of redox reaction that provides energy
   $CH_4(g) + O_2(g) \rightarrow CO_2(g) + 2H_2O(g)$

Data Interpretation and Analysis

114. a) Ag^+, Ca^{2+}, Cu^{2+}
    NaCl added: No precipitate, so Ag^+ is not present.
    Na_2SO_4 added: Precipitate formed, so Ca^{2+} is present.
    Na_2CO_3 added: Precipitate formed, so Cu^{2+} is present.
   b) Hg_2^{2+}, Ba^{2+}, Fe^{2+}
    KCl added: Precipitate formed, so Hg_2^{2+} is present.
    K_2SO_4 added: No precipitate formed, so Ba^{2+} is not present.
    K_2CO_3 added: Precipitate formed, so Fe^{2+} is present.

Quantities in Chemical Reactions 8

Questions

1. Reaction stoichiometry is important because it allows us to calculate how much reactant should be used to form a desired amount of a product. It can be thought of as a recipe for a desired compound. Just think: if we had no way to determine the amount of raw ingredients needed to make a drug in the pharmaceutical industry, there would be no way to produce sufficient amounts at a reasonable cost.

2. a) 1 mol N_2 : 3 mol H_2 : 2 mol NH_3

   b) 2 molecules $N_2 \times \dfrac{3 \text{ molecules } H_2}{1 \text{ molecule } N_2} = 6$ molecules H_2

   c) 2 mol $N_2 \times \dfrac{3 \text{ mol } H_2}{1 \text{ mol } N_2} = 6$ mol H_2

3. $\dfrac{2 \text{ moles NaCl}}{1 \text{ mole } Cl_2}$

4. Coefficients in a balanced chemical reaction represent molecules or moles, NOT grams. "2 moles of Na react with 1 mole of Cl_2 to form 2 moles of NaCl."

5. mass reactant → moles reactant → moles product → mass product

6. 7 cups noodles $\times \dfrac{4 \text{ servings pasta}}{2 \text{ cups noodles}} = 14$ servings pasta

   27 tomatoes $\times \dfrac{4 \text{ servings pasta}}{12 \text{ tomatoes}} = 9$ servings pasta

   9 garlic cloves $\times \dfrac{4 \text{ servings pasta}}{3 \text{ garlic cloves}} = 12$ servings pasta

   The tomatoes are the limiting reactant, 9 servings of pasta can be made.

7. The limiting reactant is the reactant that runs out first and, therefore, determines the maximum amount of product that can be formed in a reaction.

8. The theoretical yield is the amount of product that can be made based on the limiting reactant.

9. The amount of product that is actually produced by a reaction is often less than the theoretical maximum amount. The percent yield of a reaction is the actual yield/theoretical yield × 100%.

10. You start by determining which reactant was the limiting reactant. To do this, you convert each reactant into moles of the product. The reactant that produces the smallest amount of moles of product is the limiting reactant. Finally, you would determine the grams of the product produced by converting the moles of product, from the limiting reactant, to grams.

11. d) A and B are present in a mass ratio of 1:2. Because the reaction requires 2 moles of B for every mole of A, then A will be the limiting reagent only if the molar mass of A > molar mass of B. This will result in a molar ratio that is less than 1:2, even though the reactants are present in a 1:2 mass ratio.

12. The limiting reactant was A because it was completely used up. B was present in excess; some remained when the reaction was complete.

13. The enthalpy of reaction (ΔH_{rxn}) is the amount of thermal energy (or heat) that flows when a reaction occurs at constant pressure. This quantity is important as it allows one to calculate the amount of thermal energy produced or consumed by a chemical reaction given a set of specific conditions.

14. If thermal energy flows out of the reaction and into the surroundings (as in an exothermic reaction), then ΔH_{rxn} is negative. If, by contrast, thermal energy flows into the reaction and out of the surroundings (as in an endothermic reaction), then ΔH_{rxn} is positive.

Problems

Mole-to-Mole Conversions

15. a) $2 \text{ mol A} \times \dfrac{1 \text{ mol C}}{1 \text{ mol A}} = 2 \text{ mol C}$ b) $2 \text{ mol B} \times \dfrac{1 \text{ mol C}}{2 \text{ mol B}} = 1 \text{ mol C}$

    c) $3 \text{ mol A} \times \dfrac{1 \text{ mol C}}{1 \text{ mol A}} = 3 \text{ mol C}$ d) $3 \text{ mol B} \times \dfrac{1 \text{ mol C}}{2 \text{ mol B}} = 1.5 \text{ mol C}$

16. a) $6 \text{ mol A} \times \dfrac{3 \text{ mol B}}{2 \text{ mol A}} = 9 \text{ mol B}$ b) $2 \text{ mol A} \times \dfrac{3 \text{ mol B}}{2 \text{ mol A}} = 3 \text{ mol B}$

    c) $7 \text{ mol A} \times \dfrac{3 \text{ mol B}}{2 \text{ mol A}} = 10.5 \text{ mol B}$ d) $11 \text{ mol A} \times \dfrac{3 \text{ mol B}}{2 \text{ mol A}} = 16.5 \text{ mol B}$

17. a) $1.3 \text{ mol N}_2\text{O}_5 \times \dfrac{4 \text{ mol NO}_2}{2 \text{ mol N}_2\text{O}_5} = 2.6 \text{ mol NO}_2$

    b) $5.8 \text{ mol N}_2\text{O}_5 \times \dfrac{4 \text{ mol NO}_2}{2 \text{ mol N}_2\text{O}_5} = 11.6 \text{ mol NO}_2$

    c) $4.45 \times 10^3 \text{ mol N}_2\text{O}_5 \times \dfrac{4 \text{ mol NO}_2}{2 \text{ mol N}_2\text{O}_5} = 8.90 \times 10^3 \text{ mol NO}_2$

    d) $1.006 \times 10^{-3} \text{ mol N}_2\text{O}_5 \times \dfrac{4 \text{ mol NO}_2}{2 \text{ mol N}_2\text{O}_5} = 2.012 \times 10^{-3} \text{ mol NO}_2$

18. a) $5.3 \text{ mol } N_2H_4 \times \dfrac{4 \text{ mol } NH_3}{3 \text{ mol } N_2H_4} = 7.1 \text{ mol } NH_3$

b) $2.28 \text{ mol } N_2H_4 \times \dfrac{4 \text{ mol } NH_3}{3 \text{ mol } N_2H_4} = 3.04 \text{ mol } NH_3$

c) $5.8 \times 10^{-2} \text{ mol } N_2H_4 \times \dfrac{4 \text{ mol } NH_3}{3 \text{ mol } N_2H_4} = 7.7 \times 10^{-2} \text{ mol } NH_3$

d) $9.76 \times 10^7 \text{ mol } N_2H_4 \times \dfrac{4 \text{ mol } NH_3}{3 \text{ mol } N_2H_4} = 1.30 \times 10^8 \text{ mol } NH_3$

19. $2 \text{ molecules } SO_2 \times \dfrac{2 \text{ molecules } H_2S}{1 \text{ molecule } SO_2} = 4 \text{ molecules } H_2S$ (choice c)

20. $6 \text{ molecules } F_2 \times \dfrac{2 \text{ molecules } ClF_3}{3 \text{ molecules } F_2} = 4 \text{ molecules } ClF_3$ (choice b)

21. a) $1.75 \text{ mol } H_2 \times \dfrac{2 \text{ mol } HCl}{1 \text{ mol } H_2} = 3.50 \text{ mol } HCl$

b) $1.75 \text{ mol } O_2 \times \dfrac{2 \text{ mol } H_2O}{1 \text{ mol } O_2} = 3.50 \text{ mol } H_2O$

c) $1.75 \text{ mol } Na \times \dfrac{1 \text{ mol } Na_2O_2}{2 \text{ mol } Na} = 0.875 \text{ mol } Na_2O_2$

d) $1.75 \text{ mol } O_2 \times \dfrac{2 \text{ mol } SO_3}{3 \text{ mol } O_2} = 1.17 \text{ mol } SO_3$

22. a) $0.112 \text{ mol } O_2 \times \dfrac{2 \text{ mol } CaO}{1 \text{ mol } O_2} = 0.224 \text{ mol } CaO$

b) $0.112 \text{ mol } O_2 \times \dfrac{2 \text{ mol } Fe_2O_3}{3 \text{ mol } O_2} = 0.0747 \text{ mol } Fe_2O_3$

c) $0.112 \text{ mol } K \times \dfrac{2 \text{ mol } K_2O}{4 \text{ mol } K} = 0.0560 \text{ mol } K_2O$

d) $0.112 \text{ mol } O_2 \times \dfrac{2 \text{ mol } Al_2O_3}{3 \text{ mol } O_2} = 0.0747 \text{ mol } Al_2O_3$

23. a) $2.4 \text{ mol PbS} \times \dfrac{2 \text{ mol PbO}}{2 \text{ mol PbS}} = 2.4 \text{ mol PbO}$

$2.4 \text{ mol PbS} \times \dfrac{2 \text{ mol SO}_2}{2 \text{ mol PbS}} = 2.4 \text{ mol SO}_2$

b) $2.4 \text{ mol O}_2 \times \dfrac{2 \text{ mol PbO}}{3 \text{ mol O}_2} = 1.6 \text{ mol PbO}$

$2.4 \text{ mol O}_2 \times \dfrac{2 \text{ mol SO}_2}{3 \text{ mol O}_2} = 1.6 \text{ mol SO}_2$

c) $5.3 \text{ mol PbS} \times \dfrac{2 \text{ mol PbO}}{2 \text{ mol PbS}} = 5.3 \text{ mol PbO}$

$5.3 \text{ mol PbS} \times \dfrac{2 \text{ mol SO}_2}{2 \text{ mol PbS}} = 5.3 \text{ mol SO}_2$

d) $5.3 \text{ mol O}_2 \times \dfrac{2 \text{ mol PbO}}{3 \text{ mol O}_2} = 3.5 \text{ mol PbO}$

$5.3 \text{ mol O}_2 \times \dfrac{2 \text{ mol SO}_2}{3 \text{ mol O}_2} = 3.5 \text{ mol SO}_2$

24. a) $4.6 \text{ mol C}_3\text{H}_8 \times \dfrac{3 \text{ mol CO}_2}{1 \text{ mol C}_3\text{H}_8} = 13.8 \text{ mol CO}_2 \Rightarrow 14 \text{ mol CO}_2$

$4.6 \text{ mol C}_3\text{H}_8 \times \dfrac{4 \text{ mol H}_2\text{O}}{1 \text{ mol C}_3\text{H}_8} = 18.4 \text{ mol H}_2\text{O} \Rightarrow 18 \text{ mol H}_2\text{O}$

b) $4.6 \text{ mol O}_2 \times \dfrac{3 \text{ mol CO}_2}{5 \text{ mol O}_2} = 2.76 \text{ mol CO}_2 \Rightarrow 2.8 \text{ mol CO}_2$

$4.6 \text{ mol O}_2 \times \dfrac{4 \text{ mol H}_2\text{O}}{5 \text{ mol O}_2} = 3.68 \text{ mol H}_2\text{O} \Rightarrow 3.7 \text{ mol H}_2\text{O}$

c) $0.0558 \text{ mol C}_3\text{H}_8 \times \dfrac{3 \text{ mol CO}_2}{1 \text{ mol C}_3\text{H}_8} = 0.167 \text{ mol CO}_2$

$0.0558 \text{ mol C}_3\text{H}_8 \times \dfrac{4 \text{ mol H}_2\text{O}}{1 \text{ mol C}_3\text{H}_8} = 0.223 \text{ mol H}_2\text{O}$

d) $0.0558 \text{ mol O}_2 \times \dfrac{3 \text{ mol CO}_2}{5 \text{ mol O}_2} = 0.0335 \text{ mol CO}_2$

$0.0558 \text{ mol O}_2 \times \dfrac{4 \text{ mol H}_2\text{O}}{5 \text{ mol O}_2} = 0.0446 \text{ mol H}_2\text{O}$

25.

mol N_2H_4	mol N_2O_4	mol N_2	mol H_2O
4	2	6	8
6	3	9	12
4	2	6	8
11	5.5	16.5	22
3	1.5	4.5	6
8.26	4.13	12.4	16.5

26.

mol SiO$_2$	mol C	mol SiC	mol CO
2	6	2	4
3	9	3	6
5	15	5	10
3.2	9.5	3.2	6.4
3.2	9.6	3.2	6.4

27. $2\,C_4H_{10}(g) + 13\,O_2(g) \rightarrow 8\,CO_2(g) + 10\,H_2O(g)$

$$4.9 \text{ mol } C_4H_{10} \times \frac{13 \text{ mol } O_2}{2 \text{ mol } C_4H_{10}} = 32 \text{ mol } O_2$$

28. $2\,HC_2H_3O_2(aq) + Ca(OH)_2(aq) \rightarrow 2\,H_2O(l) + Ca(C_2H_3O_2)_2(aq)$

$$1.07 \text{ mol } HC_2H_3O_2 \times \frac{1 \text{ mol } Ca(OH)_2}{2 \text{ mol } HC_2H_3O_2} = 0.535 \text{ mol } Ca(OH)_2$$

29. a) $Pb(s) + 2\,AgNO_3(aq) \rightarrow Pb(NO_3)_2(aq) + 2\,Ag(s)$

b) $9.3 \text{ mol Pb} \times \dfrac{2 \text{ mol AgNO}_3}{1 \text{ mol Pb}} = 19 \text{ mol AgNO}_3$

c) $28.4 \text{ mol Pb} \times \dfrac{2 \text{ mol Ag}}{1 \text{ mol Pb}} = 56.8 \text{ mol Ag}$

30. a) $2\,Al(s) + 3\,H_2SO_4(aq) \rightarrow Al(SO_4)_3(aq) + 3\,H_2(g)$

b) $8.3 \text{ mol Al} \times \dfrac{3 \text{ mol } H_2SO_4}{2 \text{ mol Al}} = 12 \text{ mol } H_2SO_4$

c) $0.341 \text{ mol Al} \times \dfrac{3 \text{ mol } H_2}{2 \text{ mol Al}} = 0.512 \text{ mol } H_2$

Mass-to-Mass Conversions

31. a) $2.13 \text{ g HgO} \times \dfrac{1 \text{ mol HgO}}{216.59 \text{ g}} \times \dfrac{1 \text{ mol } O_2}{2 \text{ mol HgO}} \times \dfrac{32.00 \text{ g}}{1 \text{ mol } O_2} = 0.157 \text{ g } O_2$

b) $6.77 \text{ g HgO} \times \dfrac{1 \text{ mol HgO}}{216.59 \text{ g}} \times \dfrac{1 \text{ mol } O_2}{2 \text{ mol HgO}} \times \dfrac{32.00 \text{ g}}{1 \text{ mol } O_2} = 0.500 \text{ g } O_2$

c) $1.55 \times 10^3 \text{ g HgO} \times \dfrac{1 \text{ mol HgO}}{216.59 \text{ g}} \times \dfrac{1 \text{ mol } O_2}{2 \text{ mol HgO}} \times \dfrac{32.00 \text{ g}}{1 \text{ mol } O_2} = 115 \text{ g } O_2$

d) $3.87 \times 10^{-3} \text{ g HgO} \times \dfrac{1 \text{ mol HgO}}{216.59 \text{ g}} \times \dfrac{1 \text{ mol } O_2}{2 \text{ mol HgO}} \times \dfrac{32.00 \text{ g}}{1 \text{ mol } O_2} = 2.86 \times 10^{-4} \text{ g } O_2$

32. a) $2.72 \text{ g KClO}_3 \times \dfrac{1 \text{ mol KClO}_3}{122.55 \text{ g}} \times \dfrac{3 \text{ mol O}_2}{2 \text{ mol KClO}_3} \times \dfrac{32.00 \text{ g}}{1 \text{ mol O}_2} = 1.07 \text{ g O}_2$

b) $0.361 \text{ g KClO}_3 \times \dfrac{1 \text{ mol KClO}_3}{122.55 \text{ g}} \times \dfrac{3 \text{ mol O}_2}{2 \text{ mol KClO}_3} \times \dfrac{32.00 \text{ g}}{1 \text{ mol O}_2} = 0.141 \text{ g O}_2$

c) $8.36 \times 10^4 \text{ g KClO}_3 \times \dfrac{1 \text{ mol KClO}_3}{122.55 \text{ g}} \times \dfrac{3 \text{ mol O}_2}{2 \text{ mol KClO}_3} \times \dfrac{32.00 \text{ g}}{1 \text{ mol O}_2} = 3.27 \times 10^4 \text{ g O}_2$

d) $0.0224 \text{ g KClO}_3 \times \dfrac{1 \text{ mol KClO}_3}{122.55 \text{ g}} \times \dfrac{3 \text{ mol O}_2}{2 \text{ mol KClO}_3} \times \dfrac{32.00 \text{ g}}{1 \text{ mol O}_2} = 8.77 \times 10^{-3} \text{ g O}_2$

33. a) $2.4 \text{ g Cl}_2 \times \dfrac{1 \text{ mol Cl}_2}{70.90 \text{ g}} \times \dfrac{2 \text{ mol NaCl}}{1 \text{ mol Cl}_2} \times \dfrac{58.44 \text{ g}}{1 \text{ mol NaCl}} = 4.0 \text{ g NaCl}$

b) $2.4 \text{ g CaO} \times \dfrac{1 \text{ mol CaO}}{56.08 \text{ g}} \times \dfrac{1 \text{ mol CaCO}_3}{1 \text{ mol CaO}} \times \dfrac{100.09 \text{ g}}{1 \text{ mol CaCO}_3} = 4.3 \text{ g CaCO}_3$

c) $2.4 \text{ g Mg} \times \dfrac{1 \text{ mol Mg}}{24.31 \text{ g}} \times \dfrac{2 \text{ mol MgO}}{2 \text{ mol Mg}} \times \dfrac{40.31 \text{ g}}{1 \text{ mol MgO}} = 4.0 \text{ g MgO}$

d) $2.4 \text{ g Na}_2\text{O} \times \dfrac{1 \text{ mol Na}_2\text{O}}{61.98 \text{ g}} \times \dfrac{2 \text{ mol NaOH}}{1 \text{ mol Na}_2\text{O}} \times \dfrac{40.00 \text{ g}}{1 \text{ mol NaOH}} = 3.1 \text{ g NaOH}$

34. a) $17.8 \text{ g Ca} \times \dfrac{1 \text{ mol Ca}}{40.08 \text{ g}} \times \dfrac{1 \text{ mol CaCl}_2}{1 \text{ mol Ca}} \times \dfrac{110.98 \text{ g}}{1 \text{ mol CaCl}_2} = 49.3 \text{ g CaCl}_2$

b) $17.8 \text{ g Br}_2 \times \dfrac{1 \text{ mol Br}_2}{159.80 \text{ g}} \times \dfrac{2 \text{ mol KBr}}{1 \text{ mol Br}_2} \times \dfrac{119.00 \text{ g}}{1 \text{ mol KBr}} = 26.5 \text{ g KBr}$

c) $17.8 \text{ g O}_2 \times \dfrac{1 \text{ mol O}_2}{32.00 \text{ g}} \times \dfrac{2 \text{ mol Cr}_2\text{O}_3}{3 \text{ mol O}_2} \times \dfrac{152.00 \text{ g}}{1 \text{ mol Cr}_2\text{O}_3} = 56.4 \text{ g Cr}_2\text{O}_3$

d) $17.8 \text{ g Sr} \times \dfrac{1 \text{ mol Sr}}{87.62 \text{ g}} \times \dfrac{2 \text{ mol SrO}}{2 \text{ mol Sr}} \times \dfrac{103.62 \text{ g}}{1 \text{ mol SrO}} = 21.1 \text{ g SrO}$

35. a) $4.7 \text{ g Al} \times \dfrac{1 \text{ mol Al}}{26.98 \text{ g}} \times \dfrac{1 \text{ mol Al}_2\text{O}_3}{2 \text{ mol Al}} \times \dfrac{101.96 \text{ g}}{1 \text{ mol Al}_2\text{O}_3} = 8.9 \text{ g Al}_2\text{O}_3$

$4.7 \text{ g Al} \times \dfrac{1 \text{ mol Al}}{26.98 \text{ g}} \times \dfrac{2 \text{ mol Fe}}{2 \text{ mol Al}} \times \dfrac{55.85 \text{ g}}{1 \text{ mol Fe}} = 9.7 \text{ g Fe}$

b) $4.7 \text{ g Fe}_2\text{O}_3 \times \dfrac{1 \text{ mol Fe}_2\text{O}_3}{159.70 \text{ g}} \times \dfrac{1 \text{ mol Al}_2\text{O}_3}{1 \text{ mol Fe}_2\text{O}_3} \times \dfrac{101.96 \text{ g}}{1 \text{ mol Al}_2\text{O}_3} = 3.0 \text{ g Al}_2\text{O}_3$

$4.7 \text{ g Fe}_2\text{O}_3 \times \dfrac{1 \text{ mol Fe}_2\text{O}_3}{159.70 \text{ g}} \times \dfrac{2 \text{ mol Fe}}{1 \text{ mol Fe}_2\text{O}_3} \times \dfrac{55.85 \text{ g}}{1 \text{ mol Fe}} = 3.3 \text{ g Fe}$

36. a) $10.8 \text{ g HCl} \times \dfrac{1 \text{ mol HCl}}{36.46 \text{ g}} \times \dfrac{2 \text{ mol NaCl}}{2 \text{ mol HCl}} \times \dfrac{58.44 \text{ g}}{1 \text{ mol NaCl}} = 17.3 \text{ g NaCl}$

$10.8 \text{ g HCl} \times \dfrac{1 \text{ mol HCl}}{36.46 \text{ g}} \times \dfrac{1 \text{ mol H}_2\text{O}}{2 \text{ mol HCl}} \times \dfrac{18.02 \text{ g}}{1 \text{ mol H}_2\text{O}} = 2.67 \text{ g H}_2\text{O}$

$10.8 \text{ g HCl} \times \dfrac{1 \text{ mol HCl}}{36.46 \text{ g}} \times \dfrac{1 \text{ mol CO}_2}{2 \text{ mol HCl}} \times \dfrac{44.01 \text{ g}}{1 \text{ mol CO}_2} = 6.52 \text{ g CO}_2$

b) $10.8 \text{ g Na}_2\text{CO}_3 \times \dfrac{1 \text{ mol Na}_2\text{CO}_3}{105.99 \text{ g}} \times \dfrac{2 \text{ mol NaCl}}{1 \text{ mol Na}_2\text{CO}_3} \times \dfrac{58.44 \text{ g}}{1 \text{ mol NaCl}} = 11.9 \text{ g NaCl}$

$10.8 \text{ g Na}_2\text{CO}_3 \times \dfrac{1 \text{ mol Na}_2\text{CO}_3}{105.99 \text{ g}} \times \dfrac{1 \text{ mol H}_2\text{O}}{1 \text{ mol Na}_2\text{CO}_3} \times \dfrac{18.02 \text{ g}}{1 \text{ mol H}_2\text{O}} = 1.84 \text{ g H}_2\text{O}$

$10.8 \text{ g Na}_2\text{CO}_3 \times \dfrac{1 \text{ mol Na}_2\text{CO}_3}{105.99 \text{ g}} \times \dfrac{1 \text{ mol CO}_2}{1 \text{ mol Na}_2\text{CO}_3} \times \dfrac{44.01 \text{ g}}{1 \text{ mol CO}_2} = 4.48 \text{ g CO}_2$

37.

Mass CH_4	Mass O_2	Mass CO_2	Mass H_2O
0.645 g	2.57 g	1.77 g	1.45 g
22.32 g	89.00 g	61.20 g	50.12 g
5.041 g	20.10 g	13.82 g	11.32 g
1.07 g	4.28 g	2.94 g	2.41 g
3.18 kg	12.7 kg	8.72 kg	7.14 kg
8.57×10^2 kg	3.42×10^3 kg	2.35×10^3 kg	1.92×10^3 kg

38.

Mass C_4H_{10}	Mass O_2	Mass CO_2	Mass H_2O
0.310 g	1.11 g	0.939 g	0.481 g
5.22 g	18.7 g	15.8 g	8.09 g
3.342 g	11.96 g	10.12 g	5.180 g
5.83 g	20.9 g	17.7 g	9.04 g
232 mg	0.830 g	702 mg	0.360 g
39.0 mg	139 mg	118 mg	60.4 mg

39. a) $2.5 \text{ g NaOH} \times \dfrac{1 \text{ mol NaOH}}{40.00 \text{ g}} \times \dfrac{1 \text{ mol HCl}}{1 \text{ mol NaOH}} \times \dfrac{36.46 \text{ g}}{1 \text{ mol HCl}} = 2.3 \text{ g HCl}$

b) $2.5 \text{ g Ca(OH)}_2 \times \dfrac{1 \text{ mol Ca(OH)}_2}{74.10 \text{ g}} \times \dfrac{2 \text{ mol HNO}_3}{1 \text{ mol Ca(OH)}_2} \times \dfrac{63.02 \text{ g}}{1 \text{ mol HNO}_3}$

$= 4.3 \text{ g HNO}_3$

c) $2.5 \text{ g KOH} \times \dfrac{1 \text{ mol KOH}}{56.11 \text{ g}} \times \dfrac{1 \text{ mol H}_2\text{SO}_4}{2 \text{ mol KOH}} \times \dfrac{98.09 \text{ g}}{1 \text{ mol H}_2\text{SO}_4} = 2.2 \text{ g H}_2\text{SO}_4$

40. a) $17.3 \text{ g Pb(NO}_3)_2 \times \dfrac{1 \text{ mol Pb(NO}_3)_2}{331.2 \text{ g}} \times \dfrac{2 \text{ mol KI}}{1 \text{ mol Pb(NO}_3)_2} \times \dfrac{166.00 \text{ g}}{1 \text{ mol KI}} = 17.3 \text{ g KI}$

b) $17.3 \text{ g CuCl}_2 \times \dfrac{1 \text{ mol CuCl}_2}{134.45 \text{ g}} \times \dfrac{1 \text{ mol Na}_2\text{CO}_3}{1 \text{ mol CuCl}_2} \times \dfrac{105.99 \text{ g}}{1 \text{ mol Na}_2\text{CO}_3} = 13.6 \text{ g Na}_2\text{CO}_3$

c) $17.3 \text{ g Sr(NO}_3)_2 \times \dfrac{1 \text{ mol Sr(NO}_3)_2}{211.64 \text{ g}} \times \dfrac{1 \text{ mol K}_2\text{SO}_4}{1 \text{ mol Sr(NO}_3)_2} \times \dfrac{174.27 \text{ g}}{1 \text{ mol K}_2\text{SO}_4}$
$= 14.2 \text{ g K}_2\text{SO}_4$

41. $22.5 \text{ g Al} \times \dfrac{1 \text{ mol Al}}{26.98 \text{ g}} \times \dfrac{3 \text{ mol H}_2\text{SO}_4}{2 \text{ mol Al}} \times \dfrac{98.09 \text{ g}}{1 \text{ mol H}_2\text{SO}_4} = 123 \text{ g H}_2\text{SO}_4$

$22.5 \text{ g Al} \times \dfrac{1 \text{ mol Al}}{26.98 \text{ g}} \times \dfrac{3 \text{ mol H}_2}{2 \text{ mol Al}} \times \dfrac{2.02 \text{ g}}{1 \text{ mol H}_2} = 2.53 \text{ g H}_2$

42. $2.8 \text{ g Fe} \times \dfrac{1 \text{ mol Fe}}{55.85 \text{ g}} \times \dfrac{2 \text{ mol HCl}}{1 \text{ mol Fe}} \times \dfrac{36.46 \text{ g}}{1 \text{ mol HCl}} = 3.7 \text{ g HCl}$

$2.8 \text{ g Fe} \times \dfrac{1 \text{ mol Fe}}{55.85 \text{ g}} \times \dfrac{1 \text{ mol H}_2}{1 \text{ mol Fe}} \times \dfrac{2.02 \text{ g}}{1 \text{ mol H}_2} = 0.10 \text{ g H}_2$

Limiting Reactant, Theoretical Yield, and Percent Yield

43. a) $2 \text{ mol A} \times \dfrac{3 \text{ mol C}}{2 \text{ mol A}} = 3 \text{ mol C}$

$5 \text{ mol B} \times \dfrac{3 \text{ mol C}}{4 \text{ mol B}} = 3.75 \text{ mol C}$ ∴ The limiting reactant is A.

b) $1.8 \text{ mol A} \times \dfrac{3 \text{ mol C}}{2 \text{ mol A}} = 2.7 \text{ mol C}$

$4 \text{ mol B} \times \dfrac{3 \text{ mol C}}{4 \text{ mol B}} = 3 \text{ mol C}$ ∴ The limiting reactant is A.

c) $3 \text{ mol A} \times \dfrac{3 \text{ mol C}}{2 \text{ mol A}} = 4.5 \text{ mol C}$

$4 \text{ mol B} \times \dfrac{3 \text{ mol C}}{4 \text{ mol B}} = 3 \text{ mol C}$ ∴ The limiting reactant is B.

d) $22 \text{ mol A} \times \dfrac{3 \text{ mol C}}{2 \text{ mol A}} = 33 \text{ mol C}$

$40 \text{ mol B} \times \dfrac{3 \text{ mol C}}{4 \text{ mol B}} = 30 \text{ mol C}$ ∴ The limiting reactant is B.

44. a) $1 \text{ mol A} \times \dfrac{1 \text{ mol C}}{1 \text{ mol A}} = 1 \text{ mol C}$

$4 \text{ mol B} \times \dfrac{1 \text{ mol C}}{3 \text{ mol B}} = 1.3 \text{ mol C}$ ∴ The limiting reactant is A.

b) $2 \text{ mol A} \times \dfrac{1 \text{ mol C}}{1 \text{ mol A}} = 2 \text{ mol C}$

$3 \text{ mol B} \times \dfrac{1 \text{ mol C}}{3 \text{ mol B}} = 1 \text{ mol C}$ ∴ The limiting reactant is B.

c) $0.5 \text{ mol A} \times \dfrac{1 \text{ mol C}}{1 \text{ mol A}} = 0.5 \text{ mol C}$

$1.6 \text{ mol B} \times \dfrac{1 \text{ mol C}}{3 \text{ mol B}} = 0.53 \text{ mol C}$ ∴ The limiting reactant is A.

d) $24 \text{ mol A} \times \dfrac{1 \text{ mol C}}{1 \text{ mol A}} = 24 \text{ mol C}$

$75 \text{ mol B} \times \dfrac{1 \text{ mol C}}{3 \text{ mol B}} = 25 \text{ mol C}$ ∴ The limiting reactant is A.

45. a) $1 \text{ mol A} \times \dfrac{3 \text{ mol C}}{1 \text{ mol A}} = 3 \text{ mol C}$

$1 \text{ mol B} \times \dfrac{3 \text{ mol C}}{2 \text{ mol B}} = 1.5 \text{ mol C}$

The limiting reactant is B. Therefore, the theoretical yield of C is 1.5 mol.

b) $2 \text{ mol A} \times \dfrac{3 \text{ mol C}}{1 \text{ mol A}} = 6 \text{ mol C}$

$2 \text{ mol B} \times \dfrac{3 \text{ mol C}}{2 \text{ mol B}} = 3 \text{ mol C}$

The limiting reactant is B. Therefore, the theoretical yield of C is 3 mol.

c) $1 \text{ mol A} \times \dfrac{3 \text{ mol C}}{1 \text{ mol A}} = 3 \text{ mol C}$

$3 \text{ mol B} \times \dfrac{3 \text{ mol C}}{2 \text{ mol B}} = 4.5 \text{ mol C}$

The limiting reactant is A. Therefore, the theoretical yield of C is 3 mol.

d) $32 \text{ mol A} \times \dfrac{3 \text{ mol C}}{1 \text{ mol A}} = 96 \text{ mol C}$

$68 \text{ mol B} \times \dfrac{3 \text{ mol C}}{2 \text{ mol B}} = 102 \text{ mol C}$

The limiting reactant is A. Therefore, the theoretical yield of C is 96 mol.

46. a) $2 \text{ mol A} \times \dfrac{2 \text{ mol C}}{2 \text{ mol A}} = 2 \text{ mol C}$

$4 \text{ mol B} \times \dfrac{2 \text{ mol C}}{3 \text{ mol B}} = 2.7 \text{ mol C}$

The limiting reactant is A. Therefore, the theoretical yield of C is 2 mol.

b) $3 \text{ mol A} \times \dfrac{2 \text{ mol C}}{2 \text{ mol A}} = 3 \text{ mol C}$

$3 \text{ mol B} \times \dfrac{2 \text{ mol C}}{3 \text{ mol B}} = 2 \text{ mol C}$

The limiting reactant is B. Therefore, the theoretical yield of C is 2 mol.

c) $5 \text{ mol A} \times \dfrac{2 \text{ mol C}}{2 \text{ mol A}} = 5 \text{ mol C}$

$6 \text{ mol B} \times \dfrac{2 \text{ mol C}}{3 \text{ mol B}} = 4 \text{ mol C}$

The limiting reactant is B. Therefore, the theoretical yield of C is 4 mol.

d) $4 \text{ mol A} \times \dfrac{2 \text{ mol C}}{2 \text{ mol A}} = 4 \text{ mol C}$

$5 \text{ mol B} \times \dfrac{2 \text{ mol C}}{3 \text{ mol B}} = 3.3 \text{ mol C}$

The limiting reactant is B. Therefore, the theoretical yield of C is 3.3 mol.

47. a) $1 \text{ mol K} \times \dfrac{2 \text{ mol KCl}}{2 \text{ mol K}} = 1 \text{ mol KCl}$

$1 \text{ mol Cl}_2 \times \dfrac{2 \text{ mol KCl}}{1 \text{ mol Cl}_2} = 2 \text{ mol KCl}$ ∴ The limiting reactant is K.

b) $1.8 \text{ mol K} \times \dfrac{2 \text{ mol KCl}}{2 \text{ mol K}} = 1.8 \text{ mol KCl}$

$1 \text{ mol Cl}_2 \times \dfrac{2 \text{ mol KCl}}{1 \text{ mol Cl}_2} = 2 \text{ mol KCl}$ ∴ The limiting reactant is K.

c) $2.2 \text{ mol K} \times \dfrac{2 \text{ mol KCl}}{2 \text{ mol K}} = 2.2 \text{ mol KCl}$

$1 \text{ mol Cl}_2 \times \dfrac{2 \text{ mol KCl}}{1 \text{ mol Cl}_2} = 2 \text{ mol KCl}$ ∴ The limiting reactant is Cl_2.

d) $14.6 \text{ mol K} \times \dfrac{2 \text{ mol KCl}}{2 \text{ mol K}} = 14.6 \text{ mol KCl}$

$7.8 \text{ mol Cl}_2 \times \dfrac{2 \text{ mol KCl}}{1 \text{ mol Cl}_2} = 15.6 \text{ mol KCl}$ ∴ The limiting reactant is K.

48. a) $1 \text{ mol Cr} \times \dfrac{2 \text{ mol Cr}_2\text{O}_3}{4 \text{ mol Cr}} = 0.5 \text{ mol Cr}_2\text{O}_3$

$1 \text{ mol O}_2 \times \dfrac{2 \text{ mol Cr}_2\text{O}_3}{3 \text{ mol O}_2} = 0.66 \text{ mol Cr}_2\text{O}_3 \quad \therefore$ The limiting reactant is Cr.

b) $4 \text{ mol Cr} \times \dfrac{2 \text{ mol Cr}_2\text{O}_3}{4 \text{ mol Cr}} = 2 \text{ mol Cr}_2\text{O}_3$

$2.5 \text{ mol O}_2 \times \dfrac{2 \text{ mol Cr}_2\text{O}_3}{3 \text{ mol O}_2} = 1.7 \text{ mol Cr}_2\text{O}_3 \quad \therefore$ Limiting reactant is O_2.

c) $12 \text{ mol Cr} \times \dfrac{2 \text{ mol Cr}_2\text{O}_3}{4 \text{ mol Cr}} = 6 \text{ mol Cr}_2\text{O}_3$

$10 \text{ mol O}_2 \times \dfrac{2 \text{ mol Cr}_2\text{O}_3}{3 \text{ mol O}_2} = 6.7 \text{ mol Cr}_2\text{O}_3 \quad \therefore$ Limiting reactant is Cr.

d) $14.8 \text{ mol Cr} \times \dfrac{2 \text{ mol Cr}_2\text{O}_3}{4 \text{ mol Cr}} = 7.4 \text{ mol Cr}_2\text{O}_3$

$10.3 \text{ mol O}_2 \times \dfrac{2 \text{ mol Cr}_2\text{O}_3}{3 \text{ mol O}_2} = 6.9 \text{ mol Cr}_2\text{O}_3 \quad \therefore$ Limiting reactant is O_2.

49. a) $2 \text{ mol Mn} \times \dfrac{2 \text{ mol MnO}_3}{2 \text{ mol Mn}} = 2 \text{ mol MnO}_3$

$2 \text{ mol O}_2 \times \dfrac{2 \text{ mol MnO}_3}{3 \text{ mol O}_2} = 1.3 \text{ mol MnO}_3$

The limiting reactant is O_2, $\therefore$ the theoretical yield of MnO_3 is 1.3 mol.

b) $4.8 \text{ mol Mn} \times \dfrac{2 \text{ mol MnO}_3}{2 \text{ mol Mn}} = 4.8 \text{ mol MnO}_3$

$8.5 \text{ mol O}_2 \times \dfrac{2 \text{ mol MnO}_3}{3 \text{ mol O}_2} = 5.7 \text{ mol MnO}_3$

The limiting reactant is Mn, $\therefore$ the theoretical yield of MnO_3 is 4.8 mol.

c) $0.114 \text{ mol Mn} \times \dfrac{2 \text{ mol MnO}_3}{2 \text{ mol Mn}} = 0.114 \text{ mol MnO}_3$

$0.161 \text{ mol O}_2 \times \dfrac{2 \text{ mol MnO}_3}{3 \text{ mol O}_2} = 0.107 \text{ mol MnO}_3$

The limiting reactant is O_2, $\therefore$ the theoretical yield of MnO_3 is 0.107 mol.

d) $27.5 \text{ mol Mn} \times \dfrac{2 \text{ mol MnO}_3}{2 \text{ mol Mn}} = 27.5 \text{ mol MnO}_3$

$43.8 \text{ mol O}_2 \times \dfrac{2 \text{ mol MnO}_3}{3 \text{ mol O}_2} = 29.2 \text{ mol MnO}_3$

The limiting reactant is Mn, $\therefore$ the theoretical yield of MnO_3 is 27.5 mol.

50. a) $2 \text{ mol Ti} \times \dfrac{1 \text{ mol TiCl}_4}{1 \text{ mol Ti}} = 2 \text{ mol TiCl}_4$

$2 \text{ mol Cl}_2 \times \dfrac{1 \text{ mol TiCl}_4}{2 \text{ mol Cl}_2} = 1 \text{ mol TiCl}_4$

The limiting reactant is Cl_2, $\therefore$ the theoretical yield of TiCl_4 is 1 mol.

b) $5 \text{ mol Ti} \times \dfrac{1 \text{ mol TiCl}_4}{1 \text{ mol Ti}} = 5 \text{ mol TiCl}_4$

$9 \text{ mol Cl}_2 \times \dfrac{1 \text{ mol TiCl}_4}{2 \text{ mol Cl}_2} = 4.5 \text{ mol TiCl}_4$

The limiting reactant is Cl_2, $\therefore$ the theoretical yield of TiCl_4 is 4.5 mol.

c) $0.483 \text{ mol Ti} \times \dfrac{1 \text{ mol TiCl}_4}{1 \text{ mol Ti}} = 0.483 \text{ mol TiCl}_4$

$0.911 \text{ mol Cl}_2 \times \dfrac{1 \text{ mol TiCl}_4}{2 \text{ mol Cl}_2} = 0.456 \text{ mol TiCl}_4$

The limiting reactant is Cl_2, $\therefore$ the theoretical yield of TiCl_4 is 0.456 mol.

d) $12.4 \text{ mol Ti} \times \dfrac{1 \text{ mol TiCl}_4}{1 \text{ mol Ti}} = 12.4 \text{ mol TiCl}_4$

$15.8 \text{ mol Cl}_2 \times \dfrac{1 \text{ mol TiCl}_4}{2 \text{ mol Cl}_2} = 7.90 \text{ mol TiCl}_4$

The limiting reactant is Cl_2, $\therefore$ the theoretical yield of TiCl_4 is 7.90 mol.

51. $9 \text{ mol A} \times \dfrac{2 \text{ mol C}}{3 \text{ mol A}} = 6 \text{ mol C}$

$8 \text{ mol B} \times \dfrac{2 \text{ mol C}}{4 \text{ mol B}} = 4 \text{ mol C}$ $\therefore$ B is the limiting reagent.

Final reaction vessel would contain:

    4 mol C (theoretical yield)

    0 mol B (the limiting reagent)

    Excess mol A = Starting moles − Consumed moles

        Starting moles = 9 moles A

        Consumed moles = $8 \text{ mol B} \times \dfrac{3 \text{ mol A}}{4 \text{ mol B}} = 6 \text{ mol A}$

    Excess mol A = 9 mol A − 6 mol A = 3 mol A

52. $5 \text{ mol S} \times \dfrac{2 \text{ mol SO}_3}{2 \text{ mol S}} = 5 \text{ mol SO}_3$

$9 \text{ mol O}_2 \times \dfrac{2 \text{ mol SO}_3}{3 \text{ mol O}_2} = 6 \text{ mol SO}_3 \quad \therefore \text{ S is the limiting reagent.}$

Final reaction vessel would contain:

$\quad$ 5 mol SO_3 (theoretical yield)

$\quad$ 0 mol S (the limiting reagent)

Excess mol O_2 = Starting moles − Consumed moles

$\quad$ Starting moles = 9 moles O_2

$\quad$ Consumed moles = $5 \text{ mol S} \times \dfrac{3 \text{ mol O}_2}{2 \text{ mol S}} = 7.5 \text{ mol O}_2$

$\quad$ Excess mol A = $9 \text{ mol O}_2 - 7.5 \text{ mol O}_2 = 1.5 \text{ mol O}_2$

53. a) Because there is only 1 O_2 molecule and it only requires 4 of the 7 available HCl molecules, 2 Cl_2 molecules are formed.
    b) There are only 6 available HCl molecules; however, 12 molecules are needed to fully react with the 3 available O_2 molecules. Therefore, only 3 Cl_2 molecules are formed.
    c) There are only 4 HCl molecules available and excess O_2 molecules; therefore, 2 Cl_2 molecules are formed.

54. a) There are 3 O_2 molecules that only require 2 of the 3 CH_3OH molecules present to form 2 CO_2 molecules.
    b) There is only 1 CH_3OH molecule present and excess amounts of O_2 molecules; therefore, 1 molecule of CO_2 could be formed.
    c) There are 3 O_2 molecules present, which only require 2 of the 4 CH_3OH molecules present to form 2 molecules of CO_2.

55. a) $1.0 \text{ g Li} \times \dfrac{1 \text{ mol Li}}{6.94 \text{ g}} \times \dfrac{2 \text{ mol LiF}}{2 \text{ mol Li}} = 0.14 \text{ mol LiF}$

$\quad 1.0 \text{ g F}_2 \times \dfrac{1 \text{ mol F}_2}{38.00 \text{ g}} \times \dfrac{2 \text{ mol LiF}}{1 \text{ mol F}_2} = 0.053 \text{ mol LiF} \quad \therefore \text{ Limiting reactant is F}_2.$

b) $10.5 \text{ g Li} \times \dfrac{1 \text{ mol Li}}{6.94 \text{ g}} \times \dfrac{2 \text{ mol LiF}}{2 \text{ mol Li}} = 1.51 \text{ mol LiF}$

$\quad 37.2 \text{ g F}_2 \times \dfrac{1 \text{ mol F}_2}{38.00 \text{ g}} \times \dfrac{2 \text{ mol LiF}}{1 \text{ mol F}_2} = 1.96 \text{ mol LiF} \quad \therefore \text{ Limiting reactant is Li.}$

c) $2.85 \times 10^3 \text{ g Li} \times \dfrac{1 \text{ mol Li}}{6.94 \text{ g}} \times \dfrac{2 \text{ mol LiF}}{2 \text{ mol Li}} = 411 \text{ mol LiF}$

$\quad 6.79 \times 10^3 \text{ g F}_2 \times \dfrac{1 \text{ mol F}_2}{38.00 \text{ g}} \times \dfrac{2 \text{ mol LiF}}{1 \text{ mol F}_2} = 357 \text{ mol LiF} \quad \therefore \text{ Limiting reactant is F}_2.$

56. a) $1.0 \text{ g Al} \times \dfrac{1 \text{ mol Al}}{26.98 \text{ g}} \times \dfrac{2 \text{ mol Al}_2\text{O}_3}{4 \text{ mol Al}} = 0.019 \text{ mol Al}_2\text{O}_3$

$1.0 \text{ g O}_2 \times \dfrac{1 \text{ mol O}_2}{32.00 \text{ g}} \times \dfrac{2 \text{ mol Al}_2\text{O}_3}{3 \text{ mol O}_2} = 0.021 \text{ mol Al}_2\text{O}_3$

The limiting reactant is Al.

b) $2.2 \text{ g Al} \times \dfrac{1 \text{ mol Al}}{26.98 \text{ g}} \times \dfrac{2 \text{ mol Al}_2\text{O}_3}{4 \text{ mol Al}} = 0.041 \text{ mol Al}_2\text{O}_3$

$1.8 \text{ g O}_2 \times \dfrac{1 \text{ mol O}_2}{32.00 \text{ g}} \times \dfrac{2 \text{ mol Al}_2\text{O}_3}{3 \text{ mol O}_2} = 0.038 \text{ mol Al}_2\text{O}_3$

The limiting reactant is O_2.

c) $0.353 \text{ g Al} \times \dfrac{1 \text{ mol Al}}{26.98 \text{ g}} \times \dfrac{2 \text{ mol Al}_2\text{O}_3}{4 \text{ mol Al}} = 6.54 \times 10^{-3} \text{ mol Al}_2\text{O}_3$

$0.482 \text{ g O}_2 \times \dfrac{1 \text{ mol O}_2}{32.00 \text{ g}} \times \dfrac{2 \text{ mol Al}_2\text{O}_3}{3 \text{ mol O}_2} = 1.00 \times 10^{-2} \text{ mol Al}_2\text{O}_3$

The limiting reactant is Al.

57. a) $1.0 \text{ g Al} \times \dfrac{1 \text{ mol Al}}{26.98 \text{ g}} \times \dfrac{2 \text{ mol AlCl}_3}{2 \text{ mol Al}} \times \dfrac{133.33 \text{ g}}{1 \text{ mol AlCl}_3} = 4.9 \text{ g AlCl}_3$

$1.0 \text{ g Cl}_2 \times \dfrac{1 \text{ mol Cl}_2}{70.90 \text{ g}} \times \dfrac{2 \text{ mol AlCl}_3}{3 \text{ mol Cl}_2} \times \dfrac{133.33 \text{ g}}{1 \text{ mol AlCl}_3} = 1.3 \text{ g AlCl}_3$

The limiting reactant is Cl_2, the theoretical yield is 1.3 g of $AlCl_3$.

b) $5.5 \text{ g Al} \times \dfrac{1 \text{ mol Al}}{26.98 \text{ g}} \times \dfrac{2 \text{ mol AlCl}_3}{2 \text{ mol Al}} \times \dfrac{133.33 \text{ g}}{1 \text{ mol AlCl}_3} = 27 \text{ g AlCl}_3$

$19.8 \text{ g Cl}_2 \times \dfrac{1 \text{ mol Cl}_2}{70.90 \text{ g}} \times \dfrac{2 \text{ mol AlCl}_3}{3 \text{ mol Cl}_2} \times \dfrac{133.33 \text{ g}}{1 \text{ mol AlCl}_3} = 24.8 \text{ g AlCl}_3$

The limiting reactant is Cl_2, the theoretical yield is 24.8 g of $AlCl_3$.

c) $0.439 \text{ g Al} \times \dfrac{1 \text{ mol Al}}{26.98 \text{ g}} \times \dfrac{2 \text{ mol AlCl}_3}{2 \text{ mol Al}} \times \dfrac{133.33 \text{ g}}{1 \text{ mol AlCl}_3} = 2.17 \text{ g AlCl}_3$

$2.29 \text{ g Cl}_2 \times \dfrac{1 \text{ mol Cl}_2}{70.90 \text{ g}} \times \dfrac{2 \text{ mol AlCl}_3}{3 \text{ mol Cl}_2} \times \dfrac{133.33 \text{ g}}{1 \text{ mol AlCl}_3} = 2.87 \text{ g AlCl}_3$

The limiting reactant is Al, the theoretical yield is 2.17 g $AlCl_3$.

58. a) $1.0 \text{ g Ti} \times \dfrac{1 \text{ mol Ti}}{47.88 \text{ g}} \times \dfrac{1 \text{ mol TiF}_4}{1 \text{ mol Ti}} \times \dfrac{123.88 \text{ g}}{1 \text{ mol TiF}_4} = 2.6 \text{ g TiF}_4$

$1.0 \text{ g F}_2 \times \dfrac{1 \text{ mol F}_2}{38.00 \text{ g}} \times \dfrac{1 \text{ mol TiF}_4}{2 \text{ mol F}_2} \times \dfrac{123.88 \text{ g}}{1 \text{ mol TiF}_4} = 1.6 \text{ g TiF}_4$

The limiting reactant is F_2, ∴ the theoretical yield of TiF_4 is 1.6 g.

b) $4.8 \text{ g Ti} \times \dfrac{1 \text{ mol Ti}}{47.88 \text{ g}} \times \dfrac{1 \text{ mol TiF}_4}{1 \text{ mol Ti}} \times \dfrac{123.88 \text{ g}}{1 \text{ mol TiF}_4} = 12 \text{ g TiF}_4$

$3.2 \text{ g F}_2 \times \dfrac{1 \text{ mol F}_2}{38.00 \text{ g}} \times \dfrac{1 \text{ mol TiF}_4}{2 \text{ mol F}_2} \times \dfrac{123.88 \text{ g}}{1 \text{ mol TiF}_4} = 5.2 \text{ g TiF}_4$

The limiting reactant is F_2, ∴ the theoretical yield of TiF_4 is 5.2 g.

c) $0.388 \text{ g Ti} \times \dfrac{1 \text{ mol Ti}}{47.88 \text{ g}} \times \dfrac{1 \text{ mol TiF}_4}{1 \text{ mol Ti}} \times \dfrac{123.88 \text{ g}}{1 \text{ mol TiF}_4} = 1.00 \text{ g TiF}_4$

$0.341 \text{ g F}_2 \times \dfrac{1 \text{ mol F}_2}{38.00 \text{ g}} \times \dfrac{1 \text{ mol TiF}_4}{2 \text{ mol F}_2} \times \dfrac{123.88 \text{ g}}{1 \text{ mol TiF}_4} = 0.556 \text{ g TiF}_4$

The limiting reactant is F_2, ∴ the theoretical yield of TiF_4 is 0.556 g.

59. Percent Yield $= \dfrac{18.5}{24.8} \times 100\% = 74.6\%$

60. Percent Yield $= \dfrac{0.104}{0.118} \times 100\% = 88.1\%$

61. $14.4 \text{ g CaO} \times \dfrac{1 \text{ mol CaO}}{56.08 \text{ g}} \times \dfrac{1 \text{ mol CaCO}_3}{1 \text{ mol CaO}} \times \dfrac{100.09 \text{ g}}{1 \text{ mol CaCO}_3} = 25.7 \text{ g CaCO}_3$

$13.8 \text{ g CO}_2 \times \dfrac{1 \text{ mol CO}_2}{44.01 \text{ g}} \times \dfrac{1 \text{ mol CaCO}_3}{1 \text{ mol CO}_2} \times \dfrac{100.09 \text{ g}}{1 \text{ mol CaCO}_3} = 31.4 \text{ g CaCO}_3$

The limiting reactant is CaO.

The theoretical yield is 25.7 g $CaCO_3$.

The percent yield $= \dfrac{19.4}{25.7} \times 100\% = 75.5\%$.

62. $61.5 \text{ g SO}_3 \times \dfrac{1 \text{ mol SO}_3}{80.07 \text{ g}} \times \dfrac{1 \text{ mol H}_2\text{SO}_4}{1 \text{ mol SO}_3} \times \dfrac{98.09 \text{ g}}{1 \text{ mol H}_2\text{SO}_4} = 75.3 \text{ g H}_2\text{SO}_4$

$11.2 \text{ g H}_2\text{O} \times \dfrac{1 \text{ mol H}_2\text{O}}{18.02 \text{ g}} \times \dfrac{1 \text{ mol H}_2\text{SO}_4}{1 \text{ mol H}_2\text{O}} \times \dfrac{98.09 \text{ g}}{1 \text{ mol H}_2\text{SO}_4} = 61.0 \text{ g H}_2\text{SO}_4$

The limiting reactant is H_2O.

The theoretical yield is 61.0 g H_2SO_4.

The percent yield $= \dfrac{54.9}{61.0} \times 100\% = 90.0\%$.

63. $11.2 \text{ g NiS}_2 \times \dfrac{1 \text{ mol NiS}_2}{122.83 \text{ g}} \times \dfrac{2 \text{ mol NiO}}{2 \text{ mol NiS}_2} \times \dfrac{74.69 \text{ g}}{1 \text{ mol NiO}} = 6.81 \text{ g NiO}$

$5.43 \text{ g O}_2 \times \dfrac{1 \text{ mol O}_2}{32.00 \text{ g}} \times \dfrac{2 \text{ mol NiO}}{5 \text{ mol O}_2} \times \dfrac{74.69 \text{ g}}{1 \text{ mol NiO}} = 5.07 \text{ g NiO}$

The limiting reactant is O_2.

The theoretical yield is 5.07 g NiO.

The percent yield $= \dfrac{4.86}{5.07} \times 100\% = 95.9\%$.

64. $63.1 \text{ g HCl} \times \dfrac{1 \text{ mol HCl}}{36.46 \text{ g}} \times \dfrac{2 \text{ mol Cl}_2}{4 \text{ mol HCl}} \times \dfrac{70.90 \text{ g}}{1 \text{ mol Cl}_2} = 61.4 \text{ g Cl}_2$

$17.2 \text{ g O}_2 \times \dfrac{1 \text{ mol O}_2}{32.00 \text{ g}} \times \dfrac{2 \text{ mol Cl}_2}{1 \text{ mol O}_2} \times \dfrac{70.90 \text{ g}}{1 \text{ mol Cl}_2} = 76.2 \text{ g Cl}_2$

The limiting reactant is HCl.

The theoretical yield is 61.4 g Cl_2.

The percent yield $= \dfrac{49.3}{61.4} \times 100\% = 80.3\%$.

65. $135.8 \text{ g NaCl} \times \dfrac{1 \text{ mol NaCl}}{58.44 \text{ g}} \times \dfrac{1 \text{ mol PbCl}_2}{2 \text{ mol NaCl}} \times \dfrac{278.1 \text{ g}}{1 \text{ mol PbCl}_2} = 323.1 \text{ g PbCl}_2$

$195.7 \text{ g Pb}^{2+} \times \dfrac{1 \text{ mol Pb}^{2+}}{207.2 \text{ g}} \times \dfrac{1 \text{ mol PbCl}_2}{1 \text{ mol Pb}^{2+}} \times \dfrac{278.1 \text{ g}}{1 \text{ mol PbCl}_2} = 262.7 \text{ g PbCl}_2$

The limiting reactant is Pb^{2+}.

The theoretical yield is 262.7 g $PbCl_2$.

The percent yield $= \dfrac{252.4}{262.7} \times 100\% = 96.1\%$.

66. $10.1 \text{ g Mg} \times \dfrac{1 \text{ mol Mg}}{24.31 \text{ g}} \times \dfrac{2 \text{ mol MgO}}{2 \text{ mol Mg}} \times \dfrac{40.31 \text{ g}}{1 \text{ mol MgO}} = 16.7 \text{ g MgO}$

$10.5 \text{ g O}_2 \times \dfrac{1 \text{ mol O}_2}{32.00 \text{ g}} \times \dfrac{2 \text{ mol MgO}}{1 \text{ mol O}_2} \times \dfrac{40.31 \text{ g}}{1 \text{ mol MgO}} = 26.5 \text{ g MgO}$

The limiting reactant is Mg.

The theoretical yield is 16.7 g MgO.

The percent yield is $\dfrac{11.9 \text{ g MgO}}{16.7 \text{ g MgO}} \times 100 = 71.3\%$.

67. $10.0 \text{ g TiO}_2 \times \dfrac{1 \text{ mol TiO}_2}{79.87 \text{ g}} \times \dfrac{1 \text{ mol Ti}}{1 \text{ mole TiO}_2} \times \dfrac{47.87 \text{ g}}{1 \text{ mol Ti}} = 5.99 \text{ g Ti}$

$10.0 \text{ g C} \times \dfrac{1 \text{ mol C}}{12.01 \text{ g}} \times \dfrac{1 \text{ mol Ti}}{2 \text{ mole C}} \times \dfrac{47.87 \text{ g}}{1 \text{ mol Ti}} = 19.9 \text{ g Ti}$

∴ C is excess reagent, TiO_2 is the limiting reagent.

Upon completion the reaction vessel would contain

    0 g TiO_2 (the limiting reagent)

    5.99 g Ti

$10.0 \text{ g TiO}_2 \times \dfrac{1 \text{ mol TiO}_2}{79.866 \text{ g}} \times \dfrac{2 \text{ mol CO}}{1 \text{ mole TiO}_2} \times \dfrac{28.01 \text{ g}}{1 \text{ mol CO}} = 7.01 \text{ g CO}$

Mass of unused carbon can be calculated using the law of conservation of mass.
20.0 g of reactants, 13.00 g products (5.99 g Ti + 7.01 g CO); therefore,
7.0 g carbon left.

68. $27.5 \text{ g N}_2\text{H}_4 \times \dfrac{1 \text{ mol N}_2\text{H}_4}{32.06 \text{ g}} \times \dfrac{3 \text{ mol N}_2}{2 \text{ mole N}_2\text{H}_4} \times \dfrac{28.02 \text{ g}}{1 \text{ mol N}_2} = 36.1 \text{ g N}_2$

$74.9 \text{ g N}_2\text{O}_4 \times \dfrac{1 \text{ mol N}_2\text{O}_4}{92.02 \text{ g}} \times \dfrac{3 \text{ mol N}_2}{1 \text{ mole N}_2\text{O}_4} \times \dfrac{28.02 \text{ g}}{1 \text{ mol N}_2} = 68.4 \text{ g N}_2$

∴ N_2H_4 is limiting reagent, N_2O_4 is the excess reagent.

Upon completion, the reaction vessel would contain

    0 g N_2H_4 (the limiting reagent)

    36.1 g N_2

$27.5 \text{ g N}_2\text{H}_4 \times \dfrac{1 \text{ mol N}_2\text{H}_4}{32.06 \text{ g}} \times \dfrac{4 \text{ mol H}_2\text{O}}{2 \text{ mole N}_2\text{H}_4} \times \dfrac{18.02 \text{ g}}{1 \text{ mol H}_2\text{O}} = 30.9 \text{ g H}_2\text{O}$

Mass of unused N_2O_4 can be calculated using the law of conservation of mass
102.4 g of reactants, 67.0 g products; therefore, 35.4 g of unreacted N_2O_4.

Enthalpy and Stoichiometry ΔH_{rxn}

69. a) exothermic, $-\Delta H$
    b) endothermic, $+\Delta H$
    c) exothermic, $-\Delta H$

70. a) endothermic, $+\Delta H$
    b) exothermic, $-\Delta H$
    c) endothermic, $+\Delta H$

71. a) $1 \text{ mol A} \times \dfrac{-55 \text{ kJ}}{1 \text{ mol A}} = -55 \text{ kJ}$

b) $2 \text{ mol A} \times \dfrac{-55 \text{ kJ}}{1 \text{ mol A}} = -110 \text{ kJ} \Rightarrow -1.1 \times 10^2 \text{ kJ}$

c) $1 \text{ mol B} \times \dfrac{-55 \text{ kJ}}{2 \text{ mol B}} = -27.5 \text{ kJ} \Rightarrow -28 \text{ kJ}$

d) $2 \text{ mol B} \times \dfrac{-55 \text{ kJ}}{2 \text{ mol B}} = -55 \text{ kJ}$

72. a) $2 \text{ mol A} \times \dfrac{-125 \text{ kJ}}{2 \text{ mol A}} = -125 \text{ kJ}$

b) $3 \text{ mol A} \times \dfrac{-125 \text{ kJ}}{2 \text{ mol A}} = -187.5 \text{ kJ} \Rightarrow -188 \text{ kJ}$

c) $3 \text{ mol B} \times \dfrac{-125 \text{ kJ}}{3 \text{ mol B}} = -125 \text{ kJ}$

d) $5 \text{ mol B} \times \dfrac{-125 \text{ kJ}}{3 \text{ mol B}} = -208 \text{ kJ}$

73. $155 \text{ g C}_3\text{H}_6\text{O} \times \dfrac{1 \text{ mol C}_3\text{H}_6\text{O}}{58.09 \text{ g C}_3\text{H}_6\text{O}} \times \dfrac{-1790 \text{ kJ}}{1 \text{ mol C}_3\text{H}_6\text{O}} = -4.78 \times 10^3 \text{ kJ}$

74. $237 \text{ g CH}_4 \times \dfrac{1 \text{ mol CH}_4}{16.05 \text{ g CH}_4} \times \dfrac{-802.3 \text{ kJ}}{1 \text{ mol CH}_4} = -1.18 \times 10^4 \text{ kJ}$

75. $-1.55 \times 10^3 \text{ kJ} \times \dfrac{1 \text{ mol C}_8\text{H}_{18}}{-5074.1 \text{ kJ}} \times \dfrac{114.26 \text{ g}}{1 \text{ mol C}_8\text{H}_{18}} = 34.9 \text{ g C}_8\text{H}_{18}$

76. $175 \text{ kJ} \times \dfrac{1 \text{ mol H}_2\text{O}}{44.01 \text{ kJ}} \times \dfrac{18.02 \text{ g}}{1 \text{ mol H}_2\text{O}} = 71.7 \text{ g H}_2\text{O}$

Cumulative Problems

77. You can estimate about how much of each reactant is present: approximately 1 mole N_2, 4–5 moles O_2, and 2 moles of H_2O. Looking at the reaction, you need 2 moles of N_2 to 5 moles of O_2 to 2 moles of H_2O; therefore, N_2 is the limiting reagent.

78. You can estimate about how much of each reactant is present: approximately 1 mole CO and 1 mole O_2. Looking at the balanced reaction you need 2 moles of CO for every 1 mole of O_2; therefore, CO is the limiting reagent.

79. $Ba^{2+}(aq) + Na_2SO_4(aq) \rightarrow BaSO_4(s) + 2\ Na^+(aq)$

$258\ mg\ BaSO_4 \times \dfrac{1\ g}{1000\ mg} \times \dfrac{1\ mol\ BaSO_4}{233.40\ g} \times \dfrac{1\ mol\ Ba^{2+}}{1\ mol\ BaSO_4} \times \dfrac{137.33\ g}{1\ mol\ Ba^{2+}}$

$= 0.152\ g\ Ba^{2+}$

80. $Ag^+(aq) + KCl(aq) \rightarrow AgCl(s) + K^+(aq)$

$0.212\ mg\ AgCl \times \dfrac{1\ g}{1000\ mg} \times \dfrac{1\ mol\ AgCl}{143.32\ g} \times \dfrac{1\ mol\ Ag^+}{1\ mol\ AgCl} \times \dfrac{107.87\ g}{1\ mol\ Ag^+} \times \dfrac{1000\ mg}{1\ g}$

$= 0.160\ mg\ Ag^+$

81. $NaHCO_3(aq) + HCl(aq) \rightarrow H_2O(l) + CO_2(g) + NaCl(aq)$

$3.5\ g\ NaHCO_3 \times \dfrac{1\ mol\ NaHCO_3}{84.01\ g} \times \dfrac{1\ mol\ HCl}{1\ mol\ NaHCO_3} \times \dfrac{36.46\ g}{1\ mol\ HCl} = 1.5\ g\ HCl$

82. $CaCO_3(aq) + 2\ HCl(aq) \rightarrow H_2O(l) + CO_2(g) + CaCl_2(aq)$

$5.8\ g\ HCl \times \dfrac{1\ mol\ HCl}{36.46\ g} \times \dfrac{1\ mol\ CaCO_3}{2\ mol\ HCl} \times \dfrac{100.09\ g}{1\ mol\ CaCO_3} = 8.0\ g\ CaCO_3$

83. $2\ C_8H_{18}(l) + 25\ O_2(g) \rightarrow 18\ H_2O(l) + 16\ CO_2(g)$

$1.0\ kg\ C_8H_{18} \times \dfrac{1 \times 10^3\ g}{1\ kg} \times \dfrac{1\ mol\ C_8H_{18}}{114.26\ g} \times \dfrac{16\ mol\ CO_2}{2\ mol\ C_8H_{18}} \times \dfrac{44.01\ g}{1\ mol\ CO_2} \times \dfrac{1\ kg}{1 \times 10^3\ g}$

$= 3.1\ kg\ CO_2$

84. $C_3H_8(l) + 5\ O_2(g) \rightarrow 4\ H_2O(l) + 3\ CO_2(g)$

$18.9\ L\ C_3H_8 \times \dfrac{1 \times 10^3\ mL}{1\ L} \times \dfrac{0.621\ g}{1\ mL} \times \dfrac{1\ mol\ C_3H_8}{44.11\ g} \times \dfrac{3\ mol\ CO_2}{1\ mol\ C_3H_8} \times \dfrac{44.01\ g}{1\ mol\ CO_2} \times$

$\dfrac{1\ kg}{1 \times 10^3\ g} = 35.1\ kg\ CO_2$

85. $3\ CaCl_2(aq) + 2\ Na_3PO_4(aq) \rightarrow Ca_3(PO_4)_2(s) + 6\ NaCl(aq)$

$4.8\ g\ CaCl_2 \times \dfrac{1\ mol\ CaCl_2}{110.98\ g} \times \dfrac{2\ mol\ Na_3PO_4}{3\ mol\ CaCl_2} \times \dfrac{163.94\ g}{1\ mol\ Na_3PO_4} = 4.7\ g\ Na_3PO_4$

86. $Mg^{2+}(aq) + 2\ NaOH(aq) \rightarrow Mg(OH)_2(s) + 2\ Na^+(aq)$

$88.4\ mg\ Mg^{+2} \times \dfrac{1\ g}{1 \times 10^3\ mg} \times \dfrac{1\ mol\ Mg^{2+}}{24.31\ g} \times \dfrac{2\ mol\ NaOH}{1\ mol\ Mg^{2+}} \times \dfrac{40.00\ g}{1\ mol\ NaOH}$

$= 0.291\ g\ NaOH$

87. $Zn(s) + 2 HCl(aq) \rightarrow H_2(g) + ZnCl_2(aq)$

$$14.5 \text{ g } H_2 \times \frac{1 \text{ mol } H_2}{2.02 \text{ g}} \times \frac{1 \text{ mol } Zn}{1 \text{ mol } H_2} \times \frac{65.39 \text{ g}}{1 \text{ mol } Zn} = 4.69 \times 10^2 \text{ g } Zn$$

88. $2 Na_2O_2(s) + 2 H_2O(l) \rightarrow 4 NaOH(aq) + O_2(g)$

$$35.23 \text{ g } Na_2O_2 \times \frac{1 \text{ mol } Na_2O_2}{77.98 \text{ g}} \times \frac{1 \text{ mol } O_2}{2 \text{ mol } Na_2O_2} \times \frac{32.00 \text{ g}}{1 \text{ mol } O_2} = 7.229 \text{ g } O_2$$

89. $2 NH_4NO_3(s) \rightarrow 2 N_2(g) + O_2(g) + 4 H_2O(g)$

$$1.00 \text{ kg } NH_4NO_3 \times \frac{1000 \text{ g}}{1 \text{ kg}} \times \frac{1 \text{ mol } NH_4NO_3}{80.06 \text{ g}} \times \frac{1 \text{ mol } O_2}{2 \text{ mol } NH_4NO_3} \times \frac{32.00 \text{ g}}{1 \text{ mol } O_2}$$
$$= 2.00 \times 10^2 \text{ g } O_2$$

90. $2 KClO_3(s) \rightarrow 3 O_2(g) + 2 KCl(s)$

$$45.8 \text{ g } KClO_3 \times \frac{1 \text{ mol } KClO_3}{122.55 \text{ g}} \times \frac{3 \text{ mol } O_2}{2 \text{ mol } KClO_3} \times \frac{32.00 \text{ g}}{1 \text{ mol } O_2} = 17.9 \text{ g } O_2$$

91. $5.00 \text{ mL } C_4H_6O_3 \times \frac{1.08 \text{ g}}{1 \text{ mL}} \times \frac{1 \text{ mol } C_4H_6O_3}{102.10 \text{ g}} \times \frac{1 \text{ mol } C_9H_8O_4}{1 \text{ mol } C_4H_6O_3} \times \frac{180.17 \text{ g}}{1 \text{ mol } C_9H_8O_4}$
$= 9.53 \text{ g } C_9H_8O_4$

$2.08 \text{ g } C_7H_6O_3 \times \frac{1 \text{ mol } C_7H_6O_3}{138.13 \text{ g}} \times \frac{1 \text{ mol } C_9H_8O_4}{1 \text{ mol } C_7H_6O_3} \times \frac{180.17 \text{ g}}{1 \text{ mol } C_9H_8O_4} = 2.71 \text{ g } C_9H_8O_4$

The limiting reagent is salicylic acid, $C_7H_6O_3$.

The theoretical yield of aspirin, $C_9H_8O_4$, is then 2.71 g.

The percent yield $= \frac{2.01}{2.71} \times 100\% = 74.2\%$.

92. $C_2H_5OH(l) + 3 O_2(g) \rightarrow 3 H_2O(g) + 2 CO_2(g)$

$3.8 \text{ mL } C_2H_5OH \times \frac{0.789 \text{ g}}{1 \text{ mL}} \times \frac{1 \text{ mol } C_2H_5OH}{46.08 \text{ g}} \times \frac{3 \text{ mol } H_2O}{1 \text{ mol } C_2H_5OH} \times \frac{18.02 \text{ g}}{1 \text{ mol } H_2O}$
$= 3.52 \text{ g } H_2O$

$12.5 \text{ g } O_2 \times \frac{1 \text{ mol } O_2}{32.00 \text{ g}} \times \frac{3 \text{ mol } H_2O}{3 \text{ mol } O_2} \times \frac{18.02 \text{ g}}{1 \text{ mol } H_2O} = 7.04 \text{ g } H_2O$

The limiting reagent is ethanol, C_2H_5OH.

The theoretical yield of water is then 3.52 g.

The percent yield $= \frac{3.10}{3.52} \times 100\% = 88.1\%$.

93. $68.2 \text{ kg NH}_3 \times \dfrac{1\times 10^3 \text{g}}{1 \text{ kg}} \times \dfrac{1 \text{ mol NH}_3}{17.04 \text{ g}} \times \dfrac{1 \text{ mol CH}_4\text{N}_2\text{O}}{2 \text{ mol NH}_3} \times \dfrac{60.07 \text{ g}}{1 \text{ mol CH}_4\text{N}_2\text{O}} \times \dfrac{1 \text{ kg}}{1000 \text{ g}}$

$= 1.20 \times 10^2 \text{ kg CH}_4\text{N}_2\text{O}$

$105 \text{ kg CO}_2 \times \dfrac{1\times 10^3 \text{g}}{1 \text{ kg}} \times \dfrac{1 \text{ mol CO}_2}{44.01 \text{ g}} \times \dfrac{1 \text{ mol CH}_4\text{N}_2\text{O}}{1 \text{ mol CO}_2} \times \dfrac{60.07 \text{ g}}{1 \text{ mol CH}_4\text{N}_2\text{O}} \times \dfrac{1 \text{ kg}}{1000 \text{ g}}$

$= 143 \text{ kg CH}_4\text{N}_2\text{O}$

The limiting reagent is ammonia, NH_3.

The theoretical yield of urea is then 1.20×10^2 kg.

The percent yield $= \dfrac{87.5}{1.20 \times 10^2} \times 100\% = 72.9\%$.

94. $SiO_2(l) + 2\ C(s) \rightarrow Si(l) + 2\ CO(g)$

$52.8 \text{ kg SiO}_2 \times \dfrac{1000 \text{ g}}{1 \text{ kg}} \times \dfrac{1 \text{ mol SiO}_2}{60.09 \text{ g}} \times \dfrac{1 \text{ mol Si}}{1 \text{ mol SiO}_2} \times \dfrac{28.09 \text{ g}}{1 \text{ mol Si}} \times \dfrac{1 \text{ kg}}{1000 \text{ g}} = 24.7 \text{ kg Si}$

$25.8 \text{ kg C} \times \dfrac{1\times 10^3 \text{g}}{1 \text{ kg}} \times \dfrac{1 \text{ mol C}}{12.01 \text{ g}} \times \dfrac{1 \text{ mol Si}}{2 \text{ mol C}} \times \dfrac{28.09 \text{ g}}{1 \text{ mol Si}} \times \dfrac{1 \text{ kg}}{1\times 10^3 \text{g}} = 30.2 \text{ kg Si}$

The limiting reactant is SiO_2, the theoretical yield of silicon is 24.7 kg.

The percent yield $= \dfrac{22.4}{24.7} \times 100\% = 90.7\%$.

95. $\dfrac{0.550 \text{ mg Pb}}{\text{L}} \times \dfrac{1 \text{ g}}{1000 \text{ mg}} \times 5.0 \text{ L} \times \dfrac{1 \text{ mol Pb}}{207.2 \text{ g}} \times \dfrac{1 \text{ mol C}_4\text{H}_6\text{O}_4\text{S}_2}{1 \text{ mol Pb}} \times \dfrac{182.2 \text{ g}}{1 \text{ mol C}_4\text{H}_6\text{O}_4\text{S}_2}$

$= 0.00242$ g, or 2.42 mg, of Succimer

96. $4.4 \text{ g O}_2 \times \dfrac{1 \text{ mol O}_2}{32.00 \text{ g}} \times \dfrac{4 \text{ mol KO}_2}{3 \text{ mol O}_2} \times \dfrac{71.10 \text{ g}}{1 \text{ mol KO}_2} = 13 \text{ g KO}_2$

97. If cooking is 10% efficient, then 1.6×10^3 kJ × 10 = 1.6×10^4 kJ of heat must be produced by the barbeque.

$-1.6 \times 10^4 \text{ kJ} \times \dfrac{3 \text{ mol CO}_2}{-2044 \text{ kJ}} \times \dfrac{44.01 \text{ g}}{1 \text{ mol CO}_2} = 1.0 \times 10^3 \text{ g CO}_2$

98. $-5.00 \times 10^2 \text{ kJ} \times \dfrac{1 \text{ mol CO}_2}{-393.5 \text{ kJ}} \times \dfrac{44.01 \text{ g}}{1 \text{ mol CO}_2} = 55.9 \text{ g CO}_2$

Highlight Problems

99. The balloon in (b) has a 2:1 ratio of reactants that matches the reaction stoichiometry.

100. $HCl(aq) + NaOH(aq) \rightarrow H_2O(l) + NaCl(aq)$
Beaker (b) will neutralize the HCl.

101. Solution is based on the assumption that all numbers have 3 significant digits (this was not stated in the text of the problem).
$2\ C_8H_{18}(l) + 25\ O_2(g) \rightarrow 18\ H_2O(l) + 16\ CO_2(g)$

$9.0 \times 10^{12}\ kg\ C_8H_{18} \times \dfrac{1 \times 10^3\ g}{1\ kg} \times \dfrac{1\ mol\ C_8H_{18}}{114.26\ g} \times \dfrac{16\ mol\ CO_2}{2\ mol\ C_8H_{18}} \times \dfrac{44.01\ g}{1\ mol\ CO_2} \times \dfrac{1\ kg}{1000\ g}$

$= 2.77 \times 10^{13}\ kg\ CO_2\ /\ year$

$(2.77 \times 10^{13}\ kg\ CO_2\ /\ year)(X\ years) = 3.00 \times 10^{15}\ kg\ CO_2 \Rightarrow$
$X = 108\ years$

102. $CaCO_3(s) + H_2SO_4(aq) \rightarrow H_2O(l) + CO_2(g) + CaSO_4(aq)$

$\dfrac{5.0 \times 10^{-3}\ g\ H_2SO_4}{1\ L} \times 5.2 \times 10^9\ L \times \dfrac{1\ mol\ H_2SO_4}{98.09\ g} \times \dfrac{1\ mol\ CaCO_3}{1\ mol\ H_2SO_4} \times \dfrac{100.09\ g}{1\ mol\ CaCO_3}$

$\times \dfrac{1\ kg}{1000\ g} = 2.7 \times 10^4\ kg\ CaCO_3$

103. $2\ C_8H_{18}(l) + 25\ O_2(g) \rightarrow 16\ CO_2(g) + 18\ H_2O(g)$

$55\ L\ C_8H_{18} \times \dfrac{1000\ mL}{1\ L} \times \dfrac{0.70\ g}{1\ mL} \times \dfrac{1\ mol\ C_8H_{18}}{114.2\ g} \times \dfrac{25\ mol\ O_2}{2\ mol\ C_8H_{18}} \times \dfrac{25\ L}{1\ mol\ O_2} =$

$1.05 \times 10^5\ L\ O_2$

$\dfrac{Volume\ of\ O_2}{Volume\ of\ air} = 0.20 \Rightarrow Volume\ of\ air = \dfrac{1.05 \times 10^5}{0.20} = 5.3 \times 10^5\ L\ of\ air$

104. Answers will vary for each group.

105. a) $C_3H_8(g) + 5O_2(g) \rightarrow 3CO_2(g) + 4H_2O(g)$
b) $C_3H_8(g) + 5O_2(g) \rightarrow 3CO_2(g) + 4H_2O(g)$
c) $-5.0\ MJ \times \dfrac{1 \times 10^6\ J}{1\ MJ} \times \dfrac{1\ kJ}{1000\ J} \times \dfrac{1\ mol\ C_3H_8}{-2219\ kJ} \times \dfrac{44.1\ g}{1\ mol\ C_3H_8} = 99\ g\ C_3H_8$
d) $99\ g\ C_3H_8 \times \dfrac{1\ cm^3}{2.01\ g} \times \dfrac{1\ L}{1000\ cm^3} \times \dfrac{\$0.67}{1\ L} = \$0.033$

Data Interpretation and Analysis

106. a) A is the limiting reagent for experiments 1–3 because the mass of C increases proportionally each time the amount of A is increased.
　b) B is the limiting reagent for experiments 4–7 because each time A increases, it does not have an effect on the mass of C produced. B is the limiting reagent in experiment 8 because the amount of C produced increased proportionately with the increase in B.
　c) 2A + B
　d) 2C
　e) 84.4% average percent yield

Electrons in Atoms and the Periodic Table

9

Questions

1. Both the Bohr model and the quantum-mechanical model were developed in the early 1900s. These models explain how electrons are arranged within the atomic structure and how the electrons affect the chemical and physical properties of each element.

2. Light is a form of electromagnetic radiation, a type of energy that travels through space at a constant speed of 3.0×10^8 m/s.

3. White light actually contains the wavelengths of all visible colors (red, orange, yellow, green, blue, indigo, and violet). We see a certain color of light only when the specific wavelength that corresponds to that color is present.

4. When we look at a blue object, it appears blue because the object reflects the wavelength corresponding to blue and absorbs all of the other wavelengths from the white light that illuminated the object.

5. The wavelengths of light and energy are inversely related—the shorter the wavelength, the greater the energy. Wavelength and frequency are inversely related—the shorter the wavelength, the higher the frequency.

6. The sun, stars, and some unstable atomic nuclei have the ability to generate gamma rays.

7. X-rays are used to image internal bones and organs.

8. It is best to avoid excess exposure to gamma rays and X-rays because these photons have enough energy to destroy biological molecules in your body, increasing your risk for cancer. Occasional medical X-rays do not pose a threat; however, continual exposure to them does.

9. The ultraviolet rays, which cause sunburn and suntan, also have the ability to damage biological molecules. UV rays are linked to skin cancer, cataracts, and premature skin wrinkling.

10. All warm bodies emit infrared radiation. This fact is the basis for night vision technology.

11. Only certain molecules, such as water, can absorb microwaves. This explains why your food, which contains water, is heated while your plate, which doesn't contain water, does not heat.

12. Radio waves are the type of electromagnetic radiation used in communication devices, including cell phones.

13. The Bohr model for hydrogen places the single electron in a circular orbit around the nucleus. The orbit of the electron is quantized. That is, it has a fixed energy at a specific fixed distance from the nucleus.

14. An emission spectrum is the pattern of specific wavelengths of light that are emitted by specific elements when they are in an excited state. The Bohr model explains emission spectra at specific wavelengths as the result of an electron in a high-energy orbital relaxing down to a lower-energy state, with the excess energy given off as a photon of light with a quantized amount of energy.

15. The Bohr model orbit is a circular orbit that maps the exact path an electron would make around a nucleus. The quantum-mechanical model, however, has an orbital that is best described as a region that has the greatest probability for the location of an electron.

16. The ground state is the lowest energy state of an atom in which all electrons are in the lowest possible energy orbital. The excited state of an atom is when an electron occupies an unstable higher energy orbital.

17. The e^- has wave-particle duality, which means the path of an electron is not predictable. The motion of a baseball is predictable. A probability map shows a statistical, reproducible pattern of where the electron is located.

18. The quantum-mechanical orbitals have "fuzzy" boundaries because they represent regions in space in which the electron is statistically most likely to be located. The farther from the nucleus, the lower the probability of finding the electron, which produces the "fuzzy" effect.

19. The four subshells are *s*, *p*, *d*, and *f*. The *s* subshell contains 1 orbital, the *p* subshell 3 orbitals, the *d* subshell 5 orbitals, and the *f* subshell 7 orbitals. The maximum numbers of electrons are 2, 6, 10, and 14, respectively.

20. 1*s*, 2*s*, 2*p*, 3*s*, 3*p*, 4*s*, 3*d*, 4*p*, 5*s*, 4*d*

21. The Pauli exclusion principle states that the maximum number of electrons in an orbital is 2. The 2 electrons must have opposite spin direction to occupy the same orbital. This is important in assigning electron configurations, as it allows you to determine how and where the electrons should be assigned.

22. Hund's rule states that, when filling orbitals of equal energy levels, you need to place a single electron in each orbital with the same spin direction before you add a second electron of opposite spin to any orbitals.

23. [Ne] represents $1s^2 2s^2 2p^6$
    [Kr] represents $1s^2 2s^2 2p^6 3s^2 3p^6 4s^2 3d^{10} 4p^6$

24. Valence electrons are the *s* and *p* electrons in the outermost shell (highest principal quantum number *n*). The core electrons are the inner electrons.

25.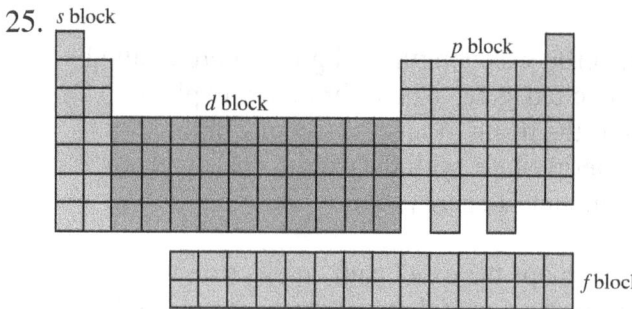

26. The quantum-mechanical model helps explain the chemical properties of the atoms, that is, their tendency to gain or lose electrons and how many electrons they gain or lose. It also explains periodic trends such as ionization energies and atomic radius.

27. Group 1 elements form +1 ions because the electron configuration of the ion will match that of a noble gas. The group 7 elements form −1 ions for the same reason.

28. a) In general, ionization energies increase going across a row in the periodic table from left to right because it is easier to remove electrons from metals (on the left) to obtain a closed-shell electronic configuration. Nonmetals (on the right) prefer to gain electrons to obtain the closed-shell configuration, so removing an electron would be expected to involve greater energy. Ionization energy decreases going down a column (group or family) because the valence electrons are in higher-energy shells farther away from the nucleus and are easier to remove to form ions.
   b) Atomic size increases down a group because additional shells or layers of orbitals are added. Atomic size decreases across a period from left to right as the increase in positive charge on the nucleus exerts a stronger attraction on the electronic orbitals, pulling them closer to the nucleus and, therefore, shrinking the size of the atom.
   c) Metallic character decreases as you move across a period from left to right and increases as you move down a group. This is consistent with the trends for ionization energy, since metals tend to lose electrons to form cations. The lower the ionization energy (i.e., the easier it is to form cations), the more metallic the element.

Problems

Wavelength, Energy, and Frequency of Electromagnetic Radiation

29. a) $1.0 \text{ ft} \times \dfrac{12 \text{ in}}{1 \text{ ft}} \times \dfrac{2.54 \text{ cm}}{1 \text{ in}} \times \dfrac{1 \text{ m}}{100 \text{ cm}} \times \dfrac{1 \text{ s}}{3.00 \times 10^8 \text{ m}} \times \dfrac{1 \text{ ns}}{1 \times 10^{-9} \text{s}} = 1.0 \text{ ns}$

b) $2462 \text{ mi} \times \dfrac{5280 \text{ ft}}{1 \text{ mi}} \times \dfrac{12 \text{ in}}{1 \text{ ft}} \times \dfrac{2.54 \text{ cm}}{1 \text{ in}} \times \dfrac{1 \text{ m}}{100 \text{ cm}} \times \dfrac{1 \text{ s}}{3.00 \times 10^8 \text{ m}} \times \dfrac{1 \text{ ms}}{1 \times 10^{-3} \text{s}} = 13.21 \text{ ms}$

c) $4.5 \times 10^9 \text{ km} \times \dfrac{1000 \text{ m}}{1 \text{ km}} \times \dfrac{1 \text{ s}}{3.00 \times 10^8 \text{ m}} \times \dfrac{1 \text{ min}}{60 \text{ s}} \times \dfrac{1 \text{ hr}}{60 \text{ min}} = 4.2 \text{ hr (4 hr 12 min)}$

30. a) $1.0 \text{ s} \times \dfrac{3.00 \times 10^8 \text{ m}}{1 \text{ s}} = 3.0 \times 10^8 \text{ m}$

b) $1.0 \text{ day} \times \dfrac{24 \text{ hr}}{1 \text{ day}} \times \dfrac{60 \text{ min}}{1 \text{ hr}} \times \dfrac{60 \text{ s}}{1 \text{ min}} \times \dfrac{3.00 \times 10^8 \text{ m}}{1 \text{ s}} = 2.6 \times 10^{13} \text{ m}$

c) $1.0 \text{ yr} \times \dfrac{365 \text{ day}}{1 \text{ yr}} \times \dfrac{24 \text{ hr}}{1 \text{ day}} \times \dfrac{60 \text{ min}}{1 \text{ hr}} \times \dfrac{60 \text{ s}}{1 \text{ min}} \times \dfrac{3.00 \times 10^8 \text{ m}}{1 \text{ s}} = 9.5 \times 10^{15} \text{ m}$

31. c) infrared

32. d) ultraviolet

33. radio waves < microwaves < infrared < ultraviolet

34. gamma rays > visible light > microwaves > radio waves

35. Two of the following three: gamma rays, X-rays, or ultraviolet rays.

36. microwaves and radio waves

37. a) radio waves < infrared < X-rays
    b) radio waves < infrared < X-rays
    c) X-rays < infrared < radio waves

38. a) gamma rays > visible > microwaves
    b) gamma rays > visible > microwaves
    c) microwaves > visible > gamma rays

The Bohr Model

39. Bohr orbits have fixed <u>energies</u> and fixed <u>distances</u>.

40. When an electron moves from a lower-energy orbit to a higher-energy orbit, energy is absorbed. When an electron relaxes from a higher orbit to a lower orbit, a photon of light is emitted that corresponds to the difference in orbit energy levels.

41. The smaller the wavelength, the larger the energy of the photon. Therefore, the 410 nm wavelength corresponds to the $n = 6$ to $n = 2$ transition and the 434 nm wavelength corresponds to the $n = 5$ to $n = 2$ transition.

42. The smaller the wavelength, the larger the energy of the photon. Therefore, the 486 nm wavelength corresponds to the $n = 4$ to $n = 2$ transition, and the 656 nm wavelength corresponds to the $n = 3$ to $n = 2$ transition.

The Quantum-Mechanical Model

43. The 2s and 3p would have the same shape as the 1s and 2p. The only difference is that they would be larger in size.

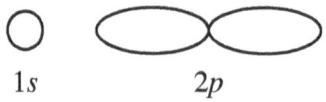

1s   2p

44. The 4d orbital will have the same shape as the 3d orbital. The only difference is that it would be larger in size.

3d$_{yz}$, 3d$_{xy}$, 3d$_{xz}$, 3d$_{x^2-y^2}$ all share the clover leaf shape   3d$_{z^2}$ Shape

45. On average, the 2s e⁻ is closer to the nucleus because it is in a smaller orbital.

46. On average, the 4p e⁻ is farther from the nucleus because it is in a larger orbital.

47. The transition with the smallest energy difference will produce the longer wavelength. This would correspond to the 2p to 1s transition.

48. The transition with the smallest energy difference will produce the longer wavelength. This would correspond to the 3p to 2s transition.

Electron Configurations

49. a) Sr: $1s^2 2s^2 2p^6 3s^2 3p^6 4s^2 3d^{10} 4p^6 5s^2$
    b) Ge: $1s^2 2s^2 2p^6 3s^2 3p^6 4s^2 3d^{10} 4p^2$
    c) Li: $1s^2 2s^1$
    d) Kr: $1s^2 2s^2 2p^6 3s^2 3p^6 4s^2 3d^{10} 4p^6$

50. a) N: $1s^2 2s^2 2p^3$
    b) Mg: $1s^2 2s^2 2p^6 3s^2$
    c) Ar: $1s^2 2s^2 2p^6 3s^2 3p^6$
    d) Se: $1s^2 2s^2 2p^6 3s^2 3p^6 4s^2 3d^{10} 4p^4$

51. a) He: [↑↓] 0 unpaired
        $1s$
    b) B: [↑↓] [↑↓] [↑][ ][ ] 1 unpaired
        $1s$ $2s$ $2p$
    c) Li: [↑↓] [↑] 1 unpaired
        $1s$ $2s$
    d) N: [↑↓] [↑↓] [↑][↑][↑] 3 unpaired
        $1s$ $2s$ $2p$

52. a) F: [↑↓] [↑↓] [↑↓][↑↓][↑] 1 unpaired
        $1s$ $2s$ $2p$
    b) C: [↑↓] [↑↓] [↑][↑][ ] 2 unpaired
        $1s$ $2s$ $2p$
    c) Ne: [↑↓] [↑↓] [↑↓][↑↓][↑↓] 0 unpaired
        $1s$ $2s$ $2p$
    d) Be: [↑↓] [↑↓] 0 unpaired
        $1s$ $2s$

53. a) [Ar]$4s^2 3d^{10} 4p^1$
    b) [Ar]$4s^2 3d^{10} 4p^3$
    c) [Kr]$5s^1$
    d) [Kr]$5s^2 4d^{10} 5p^2$

54. a) [Kr]$5s^2 4d^{10} 5p^4$
    b) [Ar]$4s^2 3d^{10} 4p^5$
    c) [Kr]$5s^2 4d^{10} 5p^5$
    d) [Xe]$6s^1$

55. a) Zn: [Ar]$4s^2 3d^{10}$
    b) Cu: [Ar]$4s^1 3d^{10}$
    c) Zr: [Kr]$5s^2 4d^2$
    d) Fe: [Ar]$4s^2 3d^6$

56. a) Mn: [Ar]$4s^2 3d^5$
    b) Ti: [Ar]$4s^2 3d^2$
    c) Cd: [Kr]$4s^2 4d^{10}$
    d) V: [Ar]$4s^2 3d^3$

Valence Electrons and Core Electrons

57. Valence electrons are underlined.
    a) Kr: $1s^2 2s^2 2p^6 3s^2 3p^6 \underline{4s^2} 3d^{10} \underline{4p^6}$
    b) Ge: $1s^2 2s^2 2p^6 3s^2 3p^6 \underline{4s^2} 3d^{10} \underline{4p^2}$
    c) Cl: $1s^2 2s^2 2p^6 \underline{3s^2} \underline{3p^5}$
    d) Sr: $1s^2 2s^2 2p^6 3s^2 3p^6 4s^2 3d^{10} 4p^6 \underline{5s^2}$

58. Valence electrons are underlined.
    a) Sb: $1s^2 2s^2 2p^6 3s^2 3p^6 4s^2 3d^{10} 4p^6 \underline{5s^2} 4d^{10} \underline{5p^3}$
    b) N: $1s^2 \underline{2s^2 2p^3}$
    c) B: $1s^2 \underline{2s^2 2p^1}$
    d) K: $1s^2 2s^2 2p^6 3s^2 3p^6 \underline{4s^1}$

59. a) Br: [↑↓] [↑↓][↑↓][↑] 1 unpaired electron
          4s   4p

    b) Kr: [↑↓] [↑↓][↑↓][↑↓] 0 unpaired electron
          4s   4p

    c) Na: [↑] [ ][ ][ ] 1 unpaired electron
          3s   3p

    d) In: [↑↓] [↑][ ][ ] 1 unpaired electron
          5s   5p

60. a) Ne: [↑↓] [↑↓][↑↓][↑↓] 0 unpaired electron
          2s   2p

    b) I: [↑↓] [↑↓][↑↓][↑] 1 unpaired electron
         5s   5p

    c) Sr: [↑↓] [ ][ ][ ] 0 unpaired electron
          5s   5p

    d) Ge: [↑↓] [↑][↑][ ] 2 unpaired electrons
          4s   4p

61. a) 6
    b) 6
    c) 7
    d) 1

62. a) 2
    b) 3
    c) 2
    d) 6

Electron Configurations and the Periodic Table

63. a) ns^1
    b) ns^2
    c) ns^2np^3
    d) ns^2np^5

64. a) ns^2np^1
    b) ns^2np^2
    c) ns^2np^4
    d) ns^2np^6

65. a) [Ne]$3s^23p^1$
    b) [He]$2s^2$
    c) [Kr]$5s^24d^{10}5p^1$
    d) [Kr]$5s^24d^2$

66. a) Tl: [Xe]$6s^24f^{14}5d^{10}6p^1$
    b) Co: [Ar]$4s^23d^7$
    c) Ba: [Xe]$6s^2$
    d) Sb: [Kr]$5s^24d^{10}5p^3$

67. a) Sr: [Kr]$5s^2$
    b) Y: [Kr]$5s^24d^1$
    c) Ti: [Ar]$4s^23d^2$
    d) Te: [Kr]$5s^24d^{10}5p^4$

68. a) [Ar]$4s^23d^{10}4p^4$
    b) [Kr]$5s^24d^{10}5p^2$
    c) [Xe]$6s^24f^{14}5d^{10}6p^2$
    d) [Kr]$5s^24d^{10}$

69. a) 2
    b) 3
    c) 5
    d) 6

70. a) 6
    b) 10
    c) 0
    d) 10

71. Period 1 consists of two elements within the 1s subshell. The s subshell has a maximum of two elements. Period 2 consists of eight elements within the 2s and 2p subshells. The s subshell has a maximum of two elements, and the p subshell has a maximum of six elements for a combined total of eight elements.

72. Period 3 consists of eight elements within the 3s and 3p subshells. The s subshell has a maximum of two elements, and the p subshell has a maximum of six elements for a combined total of eight elements. Period 4 consists of 18 elements within the 4s, 3d, and 4p subshells. The s subshell has a maximum of two elements, the d subshell has a maximum of ten elements, and the p subshell has a maximum of six elements for a combined total of 18 elements.

73. a) aluminum
    b) sulfur
    c) argon
    d) magnesium

74. a) arsenic or vanadium
    b) selenium
    c) vanadium
    d) krypton

75. a) chlorine
    b) gallium
    c) iron
    d) rubidium

76. a) sodium
    b) cadmium
    c) barium
    d) antimony

Periodic Trends

77. a) As
    b) Br
    c) cannot determine based on periodic properties alone
    d) S

78. a) Al
    b) Cl
    c) Ge
    d) S

79. Pb < Sn < Te < S < Cl

80. In < Ga < Si < N < F

81. a) In
    b) Si
    c) Pb
    d) C

82. a) Sn
    b) Ga
    c) cannot determine based on periodic properties alone
    d) Sn

83. F < S < Si < Ge < Ca < Rb

84. S < Se < Sb < Pb < Cs

85. a) Sr
    b) Bi
    c) cannot determine based on periodic properties alone
    d) As

86. a) Pb
    b) K
    c) cannot determine based on periodic properties alone
    d) Sn

87. S < Se < Sb < In < Ba < Fr

88. N < P < Si < Al < Ga < Sr

Cumulative Problems

89. When $n = 3$, there can be $3s$, $3p$, and $3d$ subshells. The s has 2 e⁻, p has 6 e⁻, and the d has 10 e⁻ for a total of 18 electrons.

90. When $n = 4$, there can be $4s$, $4p$, $4d$, and $4f$ subshells. The s has 2 e⁻, p has 6 e⁻, the d has 10 e⁻, and the f has 14 e⁻ for a total of 32 electrons.

91. The alkali metals all share the [ ]ns^2 electron configuration. By losing 2 electrons to form the 2+ ion, the electron configuration of the ion becomes the same as a noble gas.

92. The group 16 elements starting with oxygen all share the [ ]ns^2np^4 electron configuration. By gaining 2 electrons to form the 2⁻ ion, the electron configuration of the ion becomes the same as a noble gas.

93. a) $1s^2 2s^2 2p^6 3s^2 3p^6$
    b) $1s^2 2s^2 2p^6 3s^2 3p^6$
    c) $1s^2 2s^2 2p^6 3s^2 3p^6$
    d) $1s^2 2s^2 2p^6 3s^2 3p^6 4s^2 3d^{10} 4p^6$
    All electron configurations are isoelectronic with noble gases.

94. a) $1s^22s^22p^6$
    b) $1s^22s^22p^63s^23p^6$
    c) $1s^2$
    d) $1s^22s^22p^6$
    All electron configurations are isoelectronic with noble gases.

95. Metals are on the left side of the periodic table because they tend to give up electrons to form positive ions, which assume an electron configuration identical to that of a noble gas. Nonmetals are on the right side of the periodic table because they tend to gain electrons to form negative ions, which assume an electron configuration identical to that of a noble gas. The metalloids are the boundary region between the metals and nonmetals.

96. The trends shown in Figure 4.14 reflect the tendency of elements to form ions that will have an identical electron configuration to that of a noble gas.

97. a) There is a maximum of 6p and 2s electrons for any principal quantum number.
    $1s^22s^22p^63s^23p^3$
    b) There is not a 2d subshell.
    $1s^22s^22p^63s^23p^2$
    c) There is not a 1p subshell.
    $1s^22s^22p^3$
    d) There is a maximum of 6p electrons for any principal quantum number.
    $1s^22s^22p^63s^23p^3$

98. a) There is a maximum of 6p and 2s electrons for any principal quantum number.
    $1s^22s^22p^63s^23p^64s^2$
    b) The 4s orbital follows the 3p orbital.
    $1s^22s^22p^63s^23p^64s^23d^8$
    c) The 2s orbital follows the 1s orbital.
    $1s^22s^22p^6$
    d) The 3d orbital follows the 4s orbital.
    $1s^22s^22p^63s^23p^64s^23d^{10}4p^3$

99. The electron configuration of bromine shows that it is one electron short of having the same electron configuration of argon, a noble gas. Therefore, bromine is very reactive because it wants to obtain a full outer shell of electrons as the noble gases already have. Krypton is a noble gas that already has a full outer shell; therefore there is no advantage to undergoing any reactions.

100. The electron configuration of potassium shows that it has 1 electron more than the electron configuration of argon, a noble gas. Therefore, potassium is very reactive because it wants to lose 1 electron in order to obtain a full outer shell of electrons as the noble gases already have. Argon is a noble gas that already has a full outer shell; therefore, there is no advantage to undergoing any reactions.

101. Oxidation is the loss of an electron and is related to ionization energy. The trend for ionization energy is that it decreases as you move downward and toward the left side of the periodic table. Therefore, K is the most easily oxidized.

102. Reduction is the gain of an electron and is related to electron affinity. The trend for electron affinity is that it increases as you move up and to the right side of the periodic table. Therefore, Cl is the most easily reduced.

103. $E = \dfrac{hc}{\lambda} \Rightarrow \lambda = \dfrac{hc}{E} = \dfrac{(6.626 \times 10^{-34} \text{ J} \cdot \text{s})(3.00 \times 10^8 \text{ m/s})}{3.0 \times 10^{-19} \text{ J}} = 6.6 \times 10^{-7} \text{ m}$

104. $E = \dfrac{hc}{\lambda} \Rightarrow \lambda = \dfrac{hc}{E} = \dfrac{(6.626 \times 10^{-34} \text{ J} \cdot \text{s})(3.00 \times 10^8 \text{ m/s})}{4.1 \times 10^{-19} \text{ J}} = 4.8 \times 10^{-7} \text{ m}$

105. $1.496 \times 10^8 \text{ km} \times \dfrac{1 \times 10^3 \text{ m}}{1 \text{ km}} \times \dfrac{1 \text{ s}}{3.00 \times 10^8 \text{ m}} = 498.7 \text{ s} = 8.31 \text{ min} = 8 \text{ min } 19 \text{ sec}$

106. $1 \text{ light yr} = \dfrac{3.00 \times 10^8 \text{ m}}{\text{s}} \times \dfrac{1 \text{ km}}{1 \times 10^3 \text{ m}} \times \dfrac{60 \text{ s}}{1 \text{ min}} \times \dfrac{60 \text{ min}}{1 \text{ hr}} \times \dfrac{24 \text{ hrs}}{1 \text{ day}} \times \dfrac{365 \text{ days}}{1 \text{ yr}}$

$= 9.46 \times 10^{12}$ km per year

$4.3 \text{ light yr} \times \dfrac{9.46 \times 10^{12} \text{ km}}{1 \text{ light year}} = 4.1 \times 10^{13} \text{ km}$

107. a) $\lambda = \dfrac{h}{mv} = \dfrac{(6.626 \times 10^{-34} \text{ J} \cdot \text{s})}{4.59 \times 10^{-5} \text{ kg} \times 95 \, m/s} = 1.5 \times 10^{-31} \text{ m}$

b) $\lambda = \dfrac{h}{mv} = \dfrac{(6.626 \times 10^{-34} \text{ J} \cdot \text{s})}{9.109 \times 10^{-31} \text{ kg} \times 3.88 \times 10^6 \text{ m/s}} = 1.87 \times 10^{-10} \text{ m}$

The wavelength of an e⁻ is a measurable value. The wavelength of a golf-ball is so small that it is not a practical value to even attempt to measure.

108. $E = \dfrac{hc}{\lambda} = \dfrac{(6.626 \times 10^{-34} \text{ J} \cdot \text{s})(3.00 \times 10^8 \text{ m/s})}{632 \text{ nm} \times \dfrac{1 \text{ m}}{1 \times 10^9 \text{ nm}}} \times \dfrac{6.022 \times 10^{23} \text{ photons}}{1 \text{ mol photons}} = 1.89 \times 10^5 \text{ J}$

109. The dips in ionization energy occur when the electron being removed occupies a more stable location than the following electron of the following element. This occurs for the full *s* orbital (2A) and when the *p* orbital is half full (5A), illustrating the stability associated with full and half-full orbitals.

110. The ionization energy steadily increases as consecutive electrons are removed as it becomes more difficult to remove electrons from positively charged cations. An unusually large jump in ionization energy occurs when the ion is isoelectronic with a noble gas, as this is a particularly stable electron configuration. For magnesium, after the second ionization step it is isoelectronic with neon; therefore, the third ionization energy would be very large. For sodium, the second ionization energy would be very large as it becomes isoelectronic with neon after the first ionization energy.

Highlight Problems

111. a) $E = \dfrac{hc}{\lambda} = \dfrac{(6.626 \times 10^{-34}\,J \cdot s)(3.00 \times 10^{8}\,m/s)}{1500\,nm \times \dfrac{1\,m}{1 \times 10^{9}\,nm}} = 1.3 \times 10^{-19}$ J per photon

$\dfrac{1.3 \times 10^{-19}\,J}{photon} \times \dfrac{1\,kJ}{1 \times 10^{3}\,J} \times \dfrac{6.022 \times 10^{23}\,photons}{1\,mole} = 8.0 \times 10^{1}$ kJ/mol, No.

b) $E = \dfrac{hc}{\lambda} = \dfrac{(6.626 \times 10^{-34}\,J \cdot s)(3.00 \times 10^{8}\,m/s)}{500\,nm \times \dfrac{1\,m}{1 \times 10^{9}\,nm}} = 3.98 \times 10^{-19}$ J per photon

$\dfrac{3.98 \times 10^{-19}\,J}{photon} \times \dfrac{1\,kJ}{1 \times 10^{3}\,J} \times \dfrac{6.022 \times 10^{23}\,photons}{1\,mole} = 239$ kJ/mol, No.

c) $E = \dfrac{hc}{\lambda} = \dfrac{(6.626 \times 10^{-34}\,J \cdot s)(3.00 \times 10^{8}\,m/s)}{150\,nm \times \dfrac{1\,m}{1 \times 10^{9}\,nm}} = 1.3 \times 10^{-18}$ J per photon

$\dfrac{1.3 \times 10^{-18}\,J}{photon} \times \dfrac{1\,kJ}{1 \times 10^{3}\,J} \times \dfrac{6.022 \times 10^{23}\,photons}{1\,mole} = 800$ kJ/mol, Yes.

112. When Einstein said, "God does not play dice with the universe," he was referring to the idea that there must be some order and predictability to the universe. The quantum-mechanical model, however, introduces the idea that there is an unpredictable nature to the fundamental particles that make up the entire universe.

113.

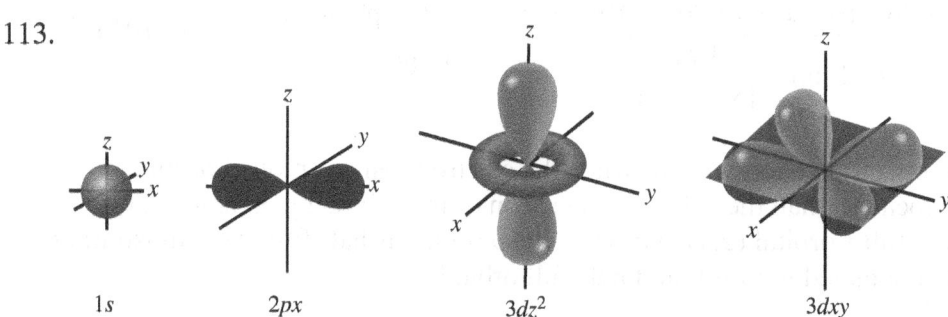

1s   2px   3dz²   3dxy

114. Answers will vary.

115. Answers will vary.

116. Atomic size decreases as you move to the right across a period and increases as you move down a column in the periodic table.

    Ionization energy increases as you move to the right across a period and decreases as you move down a column in the periodic table.

    Metallic character decreases as you move to the right across a period and increases as you move down a column in the periodic table.

Data Interpretation and Analysis

117. a) Ionization energy is positive as removing an electron from an atom is an endothermic process. Electron affinities are negative as they represent an exothermic release of energy as there is a release of energy.
    b) The ionization energy increases as you go from left to right across the periodic table. This trend is because the size of the atoms is getting smaller and the removal of an electron from a smaller atom requires more energy because there is a greater effective nuclear charge to overcome in removing the electron.
    c) Mg: [Ne]$3s^2$ Al: [Ne]$3s^23p^1$ P: [Ne]$3s^23p^3$ S: [Ne]$3s^23p^4$
    The removal of the electron from aluminum results in an ion that has a filled $3s$ orbital. Filled orbital subshells are more stable than other configurations. Sulfur is also an exception as the removal of the electron creates a half-filled $3p$ subshell. Half-filled subshells are also more stable than other configurations.
    d) The electron affinity becomes more negative (releases more energy) as you move from left to right across the periodic table. There is greater release of energy as the added electron is closer to the nucleus of smaller elements, and there is a greater effective nuclear charge.
    e) Si: [Ne]$3s^23p^2$ P: [Ne]$3s^23p^3$
    The electron affinity for Si is more exothermic than P because the addition of an electron for silicon will create a half-filled $3p$ subshell, whereas adding another electron to phosphorus will result in having to pair two electrons together in one of the p orbitals. The creation of half-filled subshells is more stable; therefore, there is a greater release of energy.
    f) Removing one electron from Na: +500 kJ/mol
       Adding one electron to Cl: −350 kJ/mol
       Net energy change   +150 kJ/mol, endothermic

Chemical Bonding 10

Questions

1. Bonding theories are important because they predict how atoms bond together to form compounds. They predict what combinations of atoms form compounds and what combinations do not. Bonding theories can explain the shapes of molecules, which in turn determine many of their physical and chemical properties.

2. Ne = $1s^2 2s^2 2p^6$, 8 valence electrons
   Ar = $1s^2 2s^2 2p^6 3s^2 3p^6$, 8 valence electrons

3. An octet is an atom that contains 8 valance electrons. A duet is a pair of valence electrons. A chemical bond is the sharing or transfer of electrons to form a stable electron configuration.

4. An ionic bond is formed as a result of the transfer of electrons from one atom (or group of atoms) to another. The oppositely charged particles that are formed are then attracted to each other. A covalent bond exists when the electrons are shared between two atoms.

5. The Lewis structure for potassium has 1 valence electron, while the Lewis structure for monoatomic chlorine has 7 valence electrons. From these structures we can determine that if potassium gives up the 1 valence electron to chlorine, K^+ and Cl^- are formed. Therefore, the formula must be KCl.

6. A pair of electrons shared between two atoms is called a bonding pair. A pair of electrons located only on one atom is called a lone pair.

7. A double bond is shorter and stronger than a single bond. A triple bond is shorter and even stronger than a double bond or a single bond.

8. a) Write the correct skeletal structure for the molecule.
   b) Calculate the total number of electrons for the Lewis structure by summing the valence electrons of each atom in the molecule.
   c) Distribute the electrons among the atoms, giving octets (or duets for hydrogen) to as many atoms as possible.
   d) If an atom lacks an octet, form double or triple bonds, as needed.

9. Add up the valence electrons from each atom that is forming the molecule.

10. Sum the valence electrons as you did for molecules, but add 1 electron for every −1 charge and subtract 1 electron for every +1 charge.

11. The octet rule has exceptions because the theory is not 100% accurate; however, it does work well in a majority of cases. Some exceptions to the rules are compounds that have odd numbers of valence electrons, boron compounds that tend to form with only 6 valence electrons, and some compounds that have more than 8 valence electrons.

12. Resonance structures of molecules are the average of one or more equivalent Lewis structures and represent the true structure of the molecule. Resonance structures are needed because the equivalent Lewis structures are not an accurate representation of the molecule.

13. The valence shell electron pair repulsion (VSEPR) theory states that molecular shape is dictated by the fact that electron pairs, whether lone pairs or bonding pairs, repel each other and try to assume an optimum maximum distance from each other.

14. a) linear
    b) trigonal planar
    c) tetrahedral

15. a) 180°
    b) 120°
    c) 109.5°

16. In the VSEPR theory, the electron geometry refers to the shape of the electron pairs, both lone and bonding pairs. The molecular geometry is concerned only with the shape of the molecule itself, which means that only the placement of atoms as a result of the electron geometry is taken into account.

17. Electronegativity is the ability of an atom to attract electrons toward itself in a covalent bond.

18. The periodic trend in electronegativity is that it increases as you move up and to the right in the periodic table (excluding the noble gases). Therefore, fluorine is the most electronegative element with an electronegativity value of 4.0.

19. A polar covalent bond is a covalent bond in which electrons are not equally shared between the two atoms.

20. A dipole moment is the development of a slightly positive end and a slightly negative end within a bond due to the uneven sharing of electrons.

21. When you try to mix a polar liquid with a nonpolar liquid, they will separate to form two distinct regions.

22. The two requirements of a polar molecule are:
    1. They must have polar bonds.
    2. The polar bonds must be distributed asymmetrically; otherwise they cancel out.

Problems

Writing Lewis Structures for Elements

23. a) $1s^2 \underline{2s^2 2p^3}$, ·N̈:

    b) $1s^2 \underline{2s^2 2p^2}$, ·C̈·

    c) $1s^2 2s^2 2p^6 \underline{3s^2 3p^5}$, :C̈l·

    d) $1s^2 2s^2 2p^6 \underline{3s^2 3p^6}$, :Är:

24. a) $1s^2 \underline{2s^1}$, ·Li

    b) $1s^2 2s^2 2p^6 \underline{3s^2 3p^3}$, :Ṗ·

    c) $1s^2 \underline{2s^2 2p^5}$, :F̈·

    d) $1s^2 \underline{2s^2 2p^6}$, :N̈e:

25. a) :Ï·

    b) :S̈·

    c) ·Ġe·

    d) ·Ċa

26. a) :K̈r:

    b) :Ṗ·

    c) ·Ḃ·

    d) ·Na

27. :Ẍ: Halogens tend to gain 1 electron in reactions.

28. M· Alkali metals tend to lose 1 electron in reactions.

29. ·M· Alkaline earth metals tend to lose 2 electrons in reactions.

30. :Ẍ: Group 6 elements tend to gain 2 electrons in reactions.

31. a) Al^{3+}
    b) Mg^{2+}
    c) $\left[:\ddot{Se}: \right]^{2-}$
    d) $\left[:\ddot{N}: \right]^{3-}$

32. a) Sr^{2+}
    b) $\left[:\ddot{S}: \right]^{2-}$
    c) Li^+
    d) $\left[:\ddot{Cl}: \right]^{3-}$

33. a) Kr
    b) Ne
    c) Kr
    d) Xe

34. a) Kr
    b) Xe
    c) Kr
    d) Ne

Lewis Structures for Ionic Compounds

35. a) covalent
    b) ionic
    c) covalent
    d) ionic

36. a) covalent
    b) covalent
    c) ionic
    d) ionic

37. a) Na$^+$ [:F̈:]$^-$
    b) Ca^{2+} [:Ö:]$^{2-}$
    c) [:B̈r:]$^-$ Sr^{2+} [:B̈r:]$^-$
    d) K$^+$ [:Ö:]$^{2-}$ K$^+$

38. a) Sr^{2+} [:Ö:]$^{2-}$
    b) Li$^+$ [:S̈:]$^{2-}$ Li$^+$
    c) [:Ï:]$^-$ Ca^{2+} [:Ï:]$^-$
    d) Rb$^+$ [:F̈:]$^-$

39. a) CaS
    b) MgBr$_2$
    c) CsI
    d) Ca$_3$N$_2$

40. a) Al$_2$S$_3$
    b) Na$_2$S
    c) SrSe
    d) BaF$_2$

41. a) [:F̈:]$^-$Mg^{2+}[:F̈:]$^-$
    b) Mg^{2+} [:Ö:]$^{2-}$
    c) Mg^{2+}[:N̈:]$^{3-}$Mg^{2+}[:N̈:]$^{3-}$Mg^{2+}

42. a) [:F̈:]$^-$Al^{3+}[:F̈:]$^-$
       [:F̈:]$^-$
    b) [:Ö:]$^{2-}$Al^{3+}[:Ö:]$^{2-}$Al^{3+}[:Ö:]$^{2-}$
    c) Al^{3+}[:N̈:]$^{3-}$

43. a) Cs$^+$ [:C̈l:]$^-$
    b) Ba^{2+} [:Ö:]$^{2-}$
    c) [:Ï:]$^-$ Ca^{2+} [:Ï:]$^-$

44. a) Na$^+$ [:Ö:]$^{2-}$ Na$^+$
    b) Mg^{2+}[:Ö:]$^{2-}$
    c) Li$^+$ [:S̈:]$^{2-}$ Li$^+$

Lewis Structures for Covalent Compounds

45. a) A single hydrogen atom has 1 valence electron. When 2 hydrogen atoms share a single valence electron with the other, they each get a duet, a stable configuration for hydrogen.
   b) A single iodine atom has 7 valence electrons. When 2 iodine atoms share a single valence electron with the other, they each get an octet.
   c) A single nitrogen atom has 5 valence electrons. When 2 nitrogen atoms share 3 valence electrons with the other, they each get an octet.
   d) A single oxygen atom has 6 valence electrons. When 2 oxygen atoms share a pair of electrons with the other, they each get an octet.

46. Each H atom in a compound can contribute 1 valence e⁻, a S atom contributes 6 valence e⁻. When H_2S forms, each H atom shares its one e⁻ with S and in doing so has achieved a stable duet. The S atom started with 6 valence electrons, and upon forming bonds with H atoms, it achieves a stable octet. HS and H_3S would not be stable as S would not have an octet in either formula.

47. a) H—P̈—H
       |
       H
   b) :C̈l—S̈—C̈l:
   c) :F̈—F̈:
   d) H—Ï:

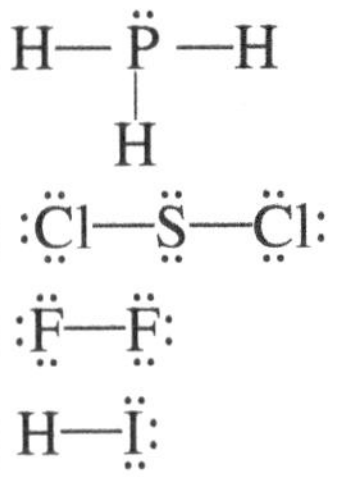

48. a) H—C—H
       | |
       H   H
   b) :F̈—N̈—F̈:
          |
         :F̈:
   c) :F̈—Ö—F̈:
   d) H—Ö—H

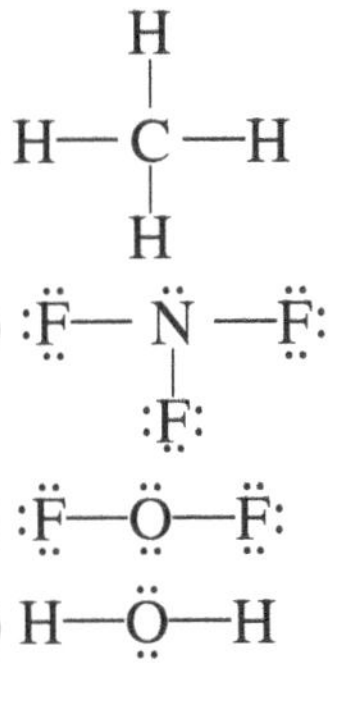

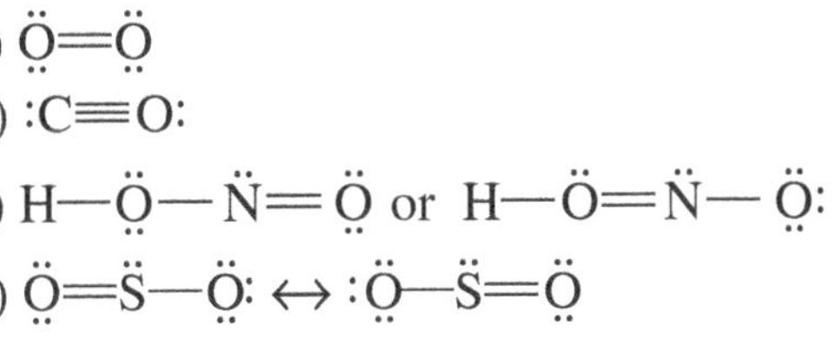

49. a) Ö=Ö
   b) :C≡O:
   c) H—Ö—N̈=Ö   or   H—Ö=N̈—Ö:
   d) Ö=S̈—Ö: ↔ :Ö—S̈=Ö

50. a) 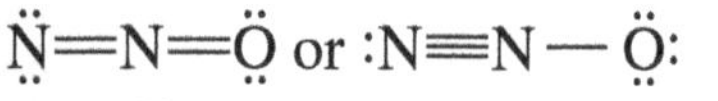

b) 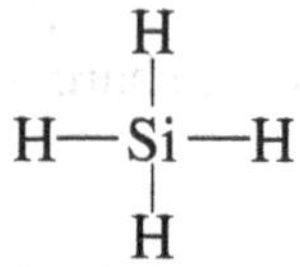

c) 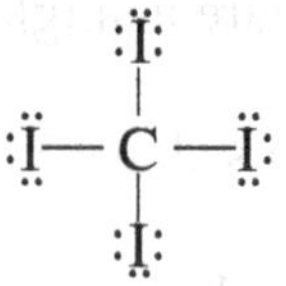

d) 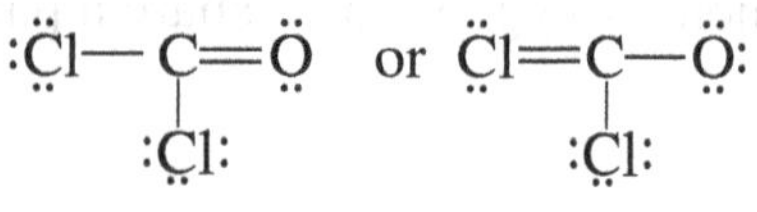

51. a) 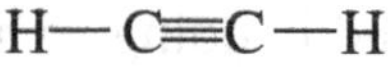

b) 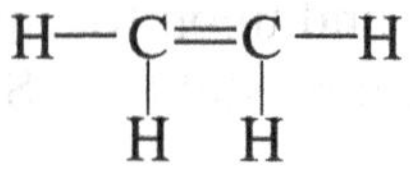

c) 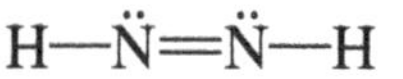

d) 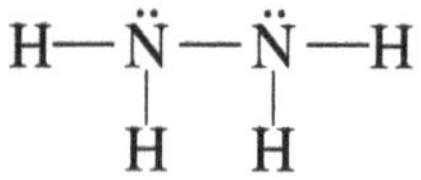

52. a) 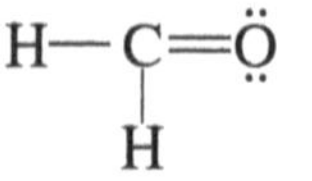

b) 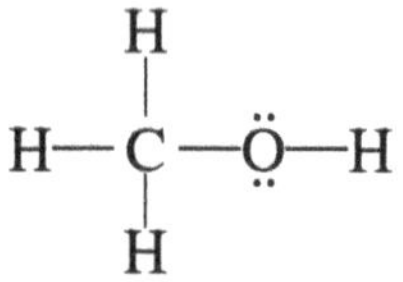

c) 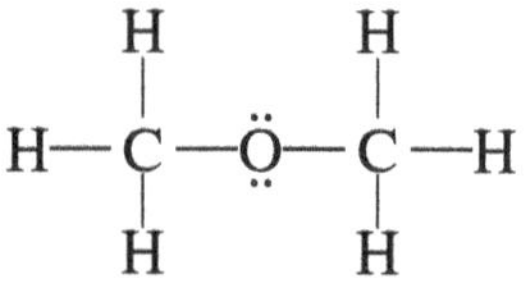

d) 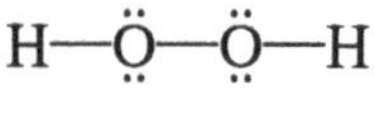

53. a) 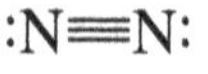

b) 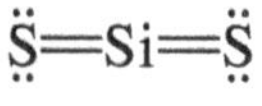

c) 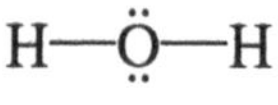

d) 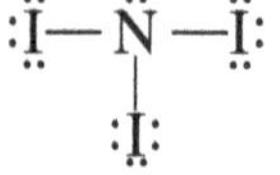

54. a) H—N̈—H
          |
          H

b) :C̈l—Ö—C̈l:

c)
```
      :O:
      ‖
H—C—Ö—H
```

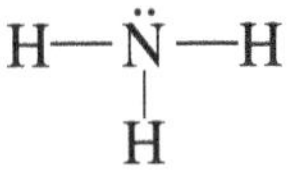

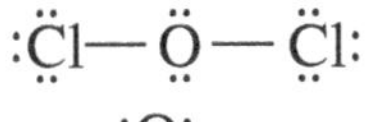

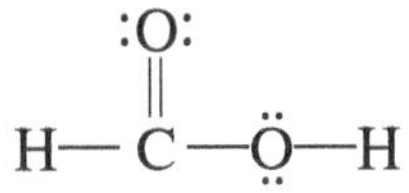

d) H—B̈r:

55. a) Ö=S̈e—Ö: ↔ :Ö—S̈e=Ö

b) $\left[\begin{array}{c} Ö=C—Ö: \\ | \\ :Ö: \end{array} \right]^{2-}$ ↔ $\left[\begin{array}{c} :Ö—C=Ö \\ | \\ :Ö: \end{array} \right]^{2-}$ ↔ $\left[\begin{array}{c} :Ö—C—Ö: \\ ‖ \\ :O: \end{array} \right]^{2-}$

c) $\left[:C̈l—Ö: \right]^{-}$

d) $\left[:Ö—C̈l—Ö: \right]^{-}$

56. a) $\left[\begin{array}{c} :Ö—C̈l—Ö: \\ | \\ :Ö: \end{array} \right]^{-}$

b) $\left[\begin{array}{c} :Ö: \\ | \\ :Ö—Cl—Ö: \\ | \\ :O: \end{array} \right]^{-}$

c) $\left[\begin{array}{c} Ö=N—Ö: \\ | \\ :O: \end{array} \right]^{-}$ ↔ $\left[\begin{array}{c} :Ö—N=Ö \\ | \\ :O: \end{array} \right]^{-}$ ↔ $\left[\begin{array}{c} :Ö—N—Ö: \\ ‖ \\ :O: \end{array} \right]^{-}$

d)
```
Ö=S—Ö:        :Ö—S=Ö        :Ö—S—Ö:
   |             |              ‖
  :O:           :O:            :O:
```
↔ ↔

57. a) $\left[\begin{array}{c} :Ö: \\ | \\ :Ö—P—Ö: \\ | \\ :O: \end{array} \right]^{3-}$

b) $\left[:C≡N: \right]^{-}$

c) $\left[Ö=N̈—Ö: \right]^{-}$ ↔ $\left[:Ö—N̈=Ö \right]^{-}$

d) $\left[\begin{array}{c} :Ö—S̈—Ö: \\ | \\ :O: \end{array} \right]^{2-}$

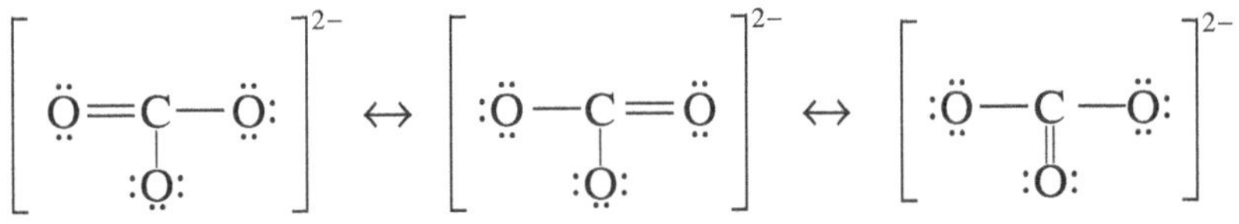

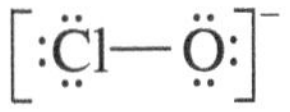

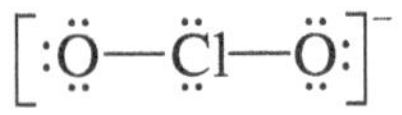

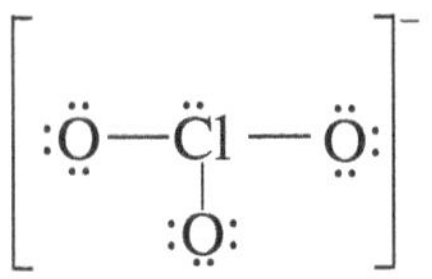

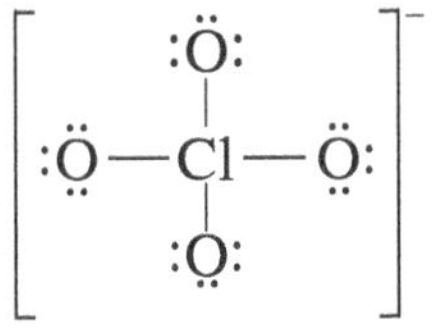

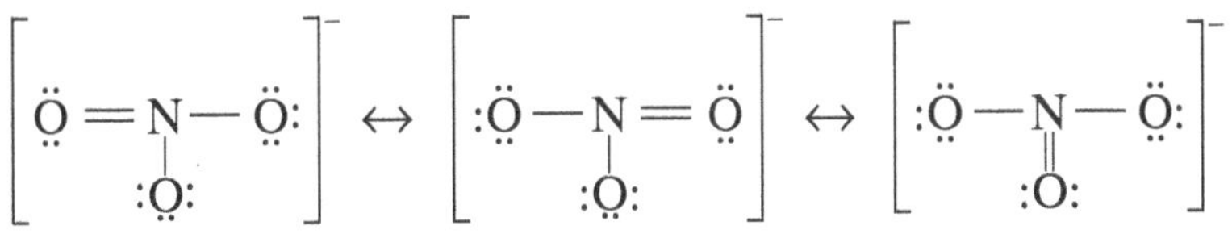

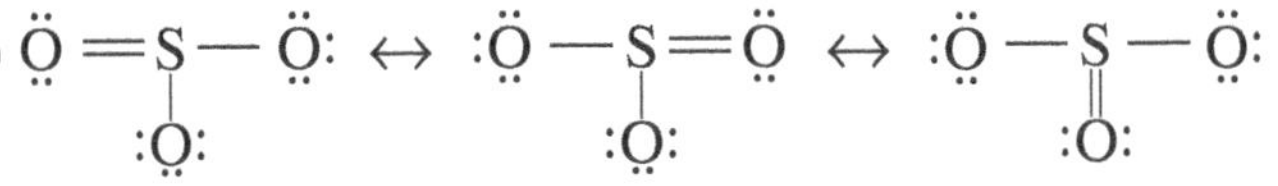

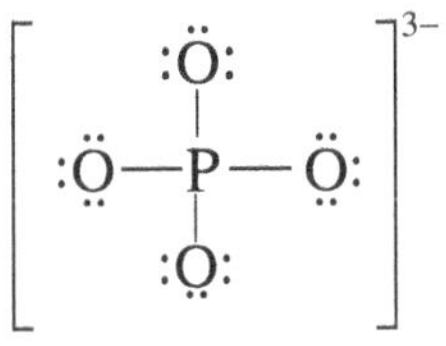

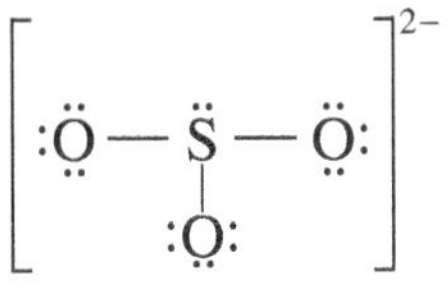

58. a) 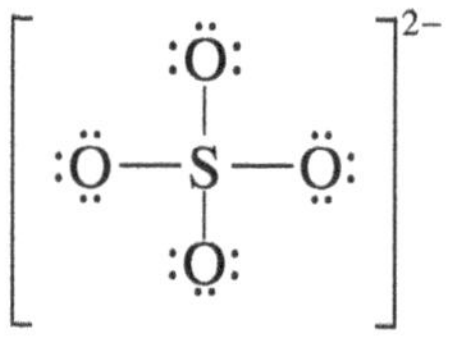 $\left[\begin{array}{c} \ddot{\text{O}}\text{:} \\ | \\ \text{:}\ddot{\text{O}}-\text{S}-\ddot{\text{O}}\text{:} \\ | \\ \text{:}\ddot{\text{O}}\text{:} \end{array} \right]^{2-}$

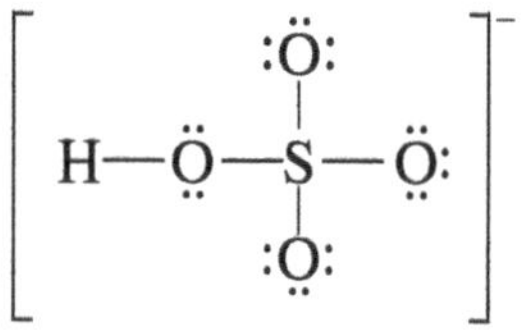

b) $\left[\begin{array}{c} \ddot{\text{O}}\text{:} \\ | \\ \text{H}-\ddot{\text{O}}-\text{S}-\ddot{\text{O}}\text{:} \\ | \\ \text{:}\ddot{\text{O}}\text{:} \end{array} \right]^{-}$

c) $\left[\begin{array}{c} \text{H} \\ | \\ \text{H}-\text{N}-\text{H} \\ | \\ \text{H} \end{array} \right]^{+}$

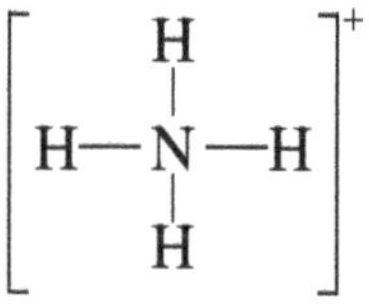

d) 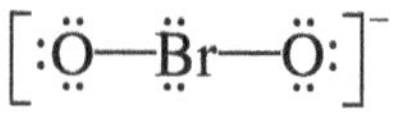 $\left[\text{:}\ddot{\text{O}}-\ddot{\text{Br}}-\ddot{\text{O}}\text{:} \right]^{-}$

59. a) 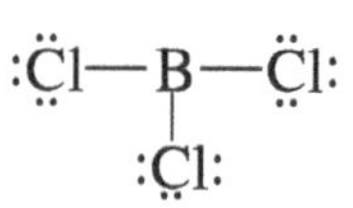 $\text{:}\ddot{\text{Cl}}-\text{B}-\ddot{\text{Cl}}\text{:}$
$\phantom{\text{a) :Cl-B}}|$
$\phantom{\text{a) :Cl-B}}\text{:}\ddot{\text{Cl}}\text{:}$

b) $\ddot{\text{O}}=\dot{\text{N}}-\ddot{\text{O}}\text{:} \leftrightarrow \text{:}\ddot{\text{O}}-\dot{\text{N}}=\ddot{\text{O}}$

c) $\text{H}-\text{B}-\text{H}$
$\phantom{\text{c) H-B}}|$
$\phantom{\text{c) H-B}}\text{H}$

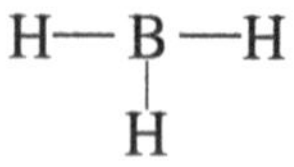

60. a) 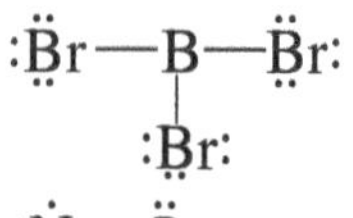 $\text{:}\ddot{\text{Br}}-\text{B}-\ddot{\text{Br}}\text{:}$
$\phantom{\text{a) :Br-B}}|$
$\phantom{\text{a) :Br-B}}\text{:}\ddot{\text{Br}}\text{:}$

b) $\text{:}\dot{\text{N}}=\ddot{\text{O}}$

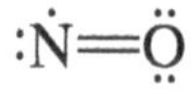

Predicting the Shapes of Molecules

61. a) 4
    b) 4
    c) 2
    d) 4

62. a) 4
    b) 4
    c) 4
    d) 4

63. a) 2 bonding groups, 2 lone pairs
    b) 3 bonding groups, 1 lone pair
    c) 2 bonding groups, 0 lone pair
    d) 4 bonding groups, 0 lone pair

64. a) 4 bonding groups, 0 lone pairs
    b) 2 bonding groups, 2 lone pairs
    c) 2 bonding groups, 2 lone pairs
    d) 3 bonding groups, 1 lone pair

65. a) tetrahedral
    b) trigonal planar
    c) linear
    d) trigonal planar

66. a) linear
    b) trigonal planar
    c) tetrahedral
    d) trigonal planar

67. a) 109.5°
    b) 120°
    c) 180°
    d) 120°

68. a) 180°
    b) 120°
    c) 109.5°
    d) 120°

69. a) electron geometry = linear
       molecular geometry = linear
    b) electron geometry = trigonal planar
       molecular geometry = bent
    c) electron geometry = tetrahedral
       molecular geometry = bent
    d) electron geometry = tetrahedral
       molecular geometry = trigonal pyramidal

70. a) electron geometry = linear
       molecular geometry = linear
    b) electron geometry = trigonal planar
       molecular geometry = trigonal planar
    c) electron geometry = tetrahedral
       molecular geometry = tetrahedral

71. a) 180°
    b) 120°
    c) 109.5°
    d) 109.5°

72. a) 180°
    b) 120°
    c) 109.5°

73. a) electron geometry = linear
       molecular geometry = linear
    b) Both nitrogen atoms have identical electron and molecular geometry.
       electron geometry = trigonal planar
       molecular geometry = bent
    c) Both nitrogen atoms have identical electron and molecular geometry.
       electron geometry = tetrahedral
       molecular geometry = trigonal pyramidal

74. a) The carbon atom:
       electron geometry = tetrahedral
       molecular geometry = tetrahedral
       The oxygen atom:
       electron geometry = tetrahedral
       molecular geometry = bent
    b) Both carbon atoms:
       electron geometry = tetrahedral
       molecular geometry = tetrahedral
       The oxygen atom:
       electron geometry = tetrahedral
       molecular geometry = bent
    c) Both oxygen atoms:
       electron geometry = tetrahedral
       molecular geometry = bent

75. a) trigonal planar
    b) bent
    c) trigonal planar
    d) tetrahedral

76. a) tetrahedral
    b) bent
    c) bent
    d) tetrahedral

Electronegativity and Polarity

77. a) 1.2
    b) 1.8
    c) 2.8

78. a) 4.0
    b) 2.5
    c) 2.5

79. Cl > Si > Ga > Ca > Rb

80. Cs < Ba < Si < N < F

81. a) 2.8 − 1.2 = 1.6, polar covalent
    b) 4.0 − 1.6 = 2.4, ionic
    c) 2.8 − 2.8 = 0,   pure covalent
    d) 3.5 − 1.8 = 1.7, polar covalent

82. a) 3.0 − 0.8 = 2.2, ionic
    b) 3.0 − 3.0 = 0,   pure covalent
    c) 2.5 − 2.5 = 0,   pure covalent
    d) 3.0 − 2.5 = 0.5, polar covalent

83. | Atom 1 (E.N.) | Atom 2 (E.N.) | E.N. Difference |
    |---|---|---|
    | H (2.1) | H (2.1) | 0.0 |
    | I (2.5) | Cl (3.0) | 0.5 |
    | H (2.1) | Br (2.8) | 0.7 |
    | C (2.5) | O (3.5) | 1.0 |

    Trend: H_2 < ICl < HBr < CO

84. | Atom 1 (E.N.) | Atom 2 (E.N.) | E.N. Difference |
    |---|---|---|
    | H (2.1) | Cl (3.0) | 0.9 |
    | N (3.0) | O (3.5) | 0.5 |
    | H (2.1) | I (2.5) | 0.4 |
    | F (4.0) | F (4.0) | 0.0 |

    Trend: HCl > NO > HI > F_2

85. a) polar, different electronegativities
    b) nonpolar, same electronegativities
    c) nonpolar, same electronegativities
    d) polar, different electronegativities

86. a) nonpolar, same electronegativities
    b) polar, different electronegativities
    c) polar, different electronegativities
    d) nonpolar, same electronegativities

87. a) 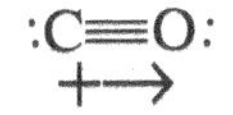 :C≡O:
       $\overset{+\longrightarrow}{}$

    b) nonpolar

    c) nonpolar

    d) 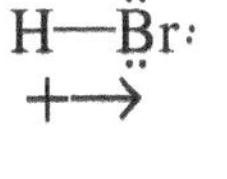 H—B̈r:
       $\overset{+\longrightarrow}{}$

88. a) nonpolar

   b) :N═Ö·
      +⟶

   c) H─C̈l:
      +⟶

   d) nonpolar

89. a) nonpolar: The C and S atoms have identical electronegativities.
   b) polar
   c) nonpolar: The C and H atoms have different electronegativities, but symmetry cancels the effect.
   d) polar

90. a) polar
   b) polar
   c) polar
   d) nonpolar: The C and O atoms have different electronegativities, but symmetry cancels the effect.

91. a) nonpolar: The B and H atoms have nearly identical electronegativities, so the bonds are not sufficiently polar, and the symmetry of the molecule would cancel out any polarity effects.
   b) polar
   c) nonpolar: The C and H atoms have a small difference in electronegativities (0.4), and symmetry cancels out any polarity effects.
   d) polar

92. a) polar
   b) polar
   c) nonpolar: The C and F atoms have different electronegativities, but symmetry cancels the effect.
   d) polar

Cumulative Problems

93. a) Ca: $1s^2 2s^2 2p^6 3s^2 3p^6 \underline{4s^2}$, ·Ca·

   b) Ga: $1s^2 2s^2 2p^6 3s^2 3p^6 \underline{4s^2} 3d^{10} \underline{4p^1}$, ·Ġa·

   c) As: $1s^2 2s^2 2p^6 3s^2 3p^6 \underline{4s^2} 3d^{10} \underline{4p^3}$, ·Ȧs:

   d) I: $1s^2 2s^2 2p^6 3s^2 3p^6 4s^2 3d^{10} 4p^6 \underline{5s^2} 4d^{10} \underline{5p^5}$, :Ï:

94. a) Rb: $1s^2 2s^2 2p^6 3s^2 3p^6 4s^2 3d^{10} 4p^6 \underline{5s^1}$, Rb·

   b) Ge: $1s^2 2s^2 2p^6 3s^2 3p^6 \underline{4s^2} 3d^{10} \underline{4p^2}$, ·Ġe·

   c) Kr: $1s^2 2s^2 2p^6 3s^2 3p^6 \underline{4s^2} 3d^{10} \underline{4p^6}$, :K̈r:

   d) Se: $1s^2 2s^2 2p^6 3s^2 3p^6 \underline{4s^2} 3d^{10} \underline{4p^4}$, ·S̈e:

95. a) ionic, $K^+ \; [:\!\ddot{\underset{..}{S}}\!:]^{2-} \; K^+$

   b) covalent,
   $$H-\underset{\underset{:\ddot{F}:}{|}}{C}=\ddot{O}$$

   c) ionic, $Mg^{2+} \; [:\!\ddot{\underset{..}{Se}}\!:]^{2-}$

   d) covalent,
   $$:\!\ddot{B}r-\underset{\underset{:\ddot{B}r:}{|}}{P}-\ddot{B}r$$

96. a) covalent, $H-C\equiv N:$

   b) covalent, $:\!\ddot{\underset{..}{Cl}}\!-\!\ddot{\underset{..}{F}}\!:$

   c) ionic, $[:\!\ddot{\underset{..}{I}}\!:]^- \; Mg^{2+} \; [:\!\ddot{\underset{..}{I}}\!:]^-$

   d) ionic, $Ca^{2+} \; [:\!\ddot{\underset{..}{S}}\!:]^{2-}$

97.
$$:\!\ddot{\underset{..}{Cl}}\!-\!\underset{\underset{:\ddot{\underset{..}{Cl}}:}{||}}{\overset{:\ddot{O}:}{C}}\!-\!\ddot{\underset{..}{Cl}}\!: \qquad :\!\ddot{\underset{..}{Cl}}\!\underset{C}{\diagup}\!\overset{:\!\!:\!\overset{||}{O}\!:}{}\!\overset{\diagdown}{}\!\ddot{\underset{..}{Cl}}\!:$$

Polar   Trigonal Planar

98.
$$H-\underset{\underset{H}{|}}{\overset{\overset{H}{|}}{C}}-\overset{:\ddot{O}:}{\overset{||}{C}}-H \qquad H\text{···}\overset{\overset{H}{|}}{\underset{\underset{H}{\blacktriangle}}{C}}-C\underset{H}{\overset{O}{\diagup \diagdown}}$$

Polar

99.
$$H-\underset{\underset{H}{|}}{\overset{\overset{H}{|}}{C}}-\overset{:\ddot{O}:}{\overset{||}{C}}-\ddot{\underset{..}{O}}H \qquad H\text{···}\overset{\overset{H}{|}}{\underset{\underset{H}{\blacktriangle}}{C}}-C\underset{OH}{\overset{O}{\diagup \diagdown}}$$

Polar

100. The shape of each carbon atom is trigonal planar.

$$\underset{HC\underset{\underset{C}{||}}{\diagdown}}{HC\overset{\overset{H}{C}}{\diagup}}\underset{\underset{H}{\diagup}}{\overset{CH}{\diagdown}}CH \quad \longleftrightarrow \quad \underset{HC\underset{\underset{C}{||}}{\diagdown}}{HC\overset{\overset{H}{C}}{\diagup}}\underset{\underset{H}{\diagup}}{\overset{CH}{\diagdown}}CH$$

101. $H:\!\ddot{\underset{..}{Cl}}\!: + Na^+[:\!\ddot{\underset{..}{O}}\!-H]^- \rightarrow H-\ddot{\underset{..}{O}}-H + Na^+[:\!\ddot{\underset{..}{Cl}}\!:]^-$

102. $\left[:\ddot{O}-N=\ddot{O}:\atop{|\atop:\ddot{O}:}\right]^{-}$ Pb^{2+} $\left[:\ddot{O}-N=\ddot{O}:\atop{|\atop:\ddot{O}:}\right]^{-}$ $+ 2Li^{+}\left[:\ddot{\underset{..}{C}}l:\right]^{-} \rightarrow$

$\left[:\ddot{\underset{..}{C}}l:\right]^{-}Pb^{2+}\left[:\ddot{\underset{..}{C}}l:\right]^{-} + 2Li^{+}\left[:\ddot{O}-N=\ddot{O}:\atop{|\atop:\ddot{O}:}\right]^{-}$

103. $K\cdot$, $:\ddot{\underset{..}{C}}l-\ddot{\underset{..}{C}}l: \Rightarrow [K]^{+}\left[:\ddot{\underset{..}{C}}l:\right]^{-}$, K was oxidized and Cl_2 was reduced

104. $Ca:$, $:\ddot{\underset{..}{B}}r-\ddot{\underset{..}{B}}r: \Rightarrow [:\ddot{\underset{..}{B}}r:]^{-}[Ca]^{2+}[:\ddot{\underset{..}{B}}r:]^{-}$ Ca was oxidized and Br_2 was reduced

105. a) $K^{+}\left[:\ddot{O}-H\right]^{-}$

b) $K^{+}\left[:\ddot{O}-N=\ddot{O}\atop{|\atop:\ddot{O}:}\right]^{-} \leftrightarrow K^{+}\left[\ddot{O}=N-\ddot{O}:\atop{|\atop:\ddot{O}:}\right]^{-} \leftrightarrow K^{+}\left[:\ddot{O}-N-\ddot{O}:\atop{|\atop:\ddot{O}:}\right]^{-}$

c) $Li^{+}\left[:\ddot{I}-\ddot{O}:\right]^{-}$

d) $Ba^{2+}\left[:\ddot{O}-C=\ddot{O}\atop{|\atop:\ddot{O}:}\right]^{2-} \leftrightarrow Ba^{2+}\left[\ddot{O}=C-\ddot{O}:\atop{|\atop:\ddot{O}:}\right]^{2-} \leftrightarrow Ba^{2+}\left[:\ddot{O}-C-\ddot{O}:\atop{|\atop:\ddot{O}:}\right]^{2-}$

106. a) $Rb^{+}\left[:\ddot{O}-\ddot{I}-\ddot{O}:\right]^{-}$

b) $\left[:\ddot{O}-H\right]^{-}Ca^{2+}\left[:\ddot{O}-H\right]^{-}$

c) $\left[H-\underset{\underset{H}{|}}{\overset{\overset{H}{|}}{N}}-H\right]^{+}\left[:\ddot{\underset{..}{C}}l:\right]^{-}$

d) $[:C\equiv N:]^{-}Sr^{2+}[:C\equiv N:]^{-}$

107. a)
```
      :F:
   :F: |
    \P—F:
   :F:/ |
       :F:
```

b)
```
      :F:
      |
:F—S—F:
      |
      :F:
```

c)
```
      :F:
      |
:F—Se—F:
      |
      :F:
```

108. a)
```
      :F:
   :F. |
    \Cl—F:
   :F:/ |
       :F:
```

b)
```
       :F:
   :F. | .F:
    \ S /
   :F:/ | \F:
       :F:
```

c)
```
       :F:
   :F. |
    \I—F:
   :F:/ |
       :F:
```

109. $\dfrac{46.02 \text{ g}}{1 \text{ mol}} \times \dfrac{26.10\% \text{ C}}{100} = 12.01 \text{ g} \Rightarrow 1 \text{ mol C}$ Formula = CH_2O_2

$\dfrac{46.02 \text{ g}}{1 \text{ mol}} \times \dfrac{4.38\% \text{ H}}{100} = 2.02 \text{ g} \Rightarrow 2 \text{ mol H}$

$\dfrac{46.02 \text{ g}}{1 \text{ mol}} \times \dfrac{69.52\% \text{ O}}{100} = 32.00 \text{ g} \Rightarrow 2 \text{ mol O}$

```
      :O:
      ||
  H—C—Ö—H
```

110. $\dfrac{42.04 \text{ g}}{1 \text{ mol}} \times \dfrac{28.57\% \text{ C}}{100} = 12.01 \text{ g C} \Rightarrow 1 \text{ mol C}$

$\dfrac{42.04 \text{ g}}{1 \text{ mol}} \times \dfrac{4.80\% \text{ H}}{100} = 2.02 \text{ g H} \Rightarrow 2 \text{ mol H}$

$\dfrac{42.04 \text{ g}}{1 \text{ mol}} \times \dfrac{66.64\% \text{ N}}{100} = 28.02 \text{ g N} \Rightarrow 2 \text{ mol N}$; s ∴ Formula = CH_2N_2

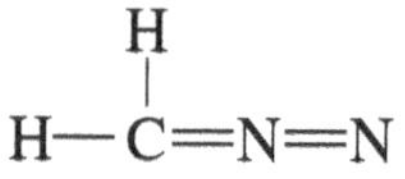

```
       H
       |
   H—C=N=N
```

111. The best possible structure for HOO is H—Ö—Ö·; this is not a stable molecule as the second oxygen atom does not have an octet. The electron geometry is tetrahedral and a bent molecular geometry.

112. The best possible structure for CH_3 is
```
       H
       |
   H—C—H
       ·
```
; this is not a stable molecule as the carbon atom does not have a complete octet. The electron geometry is tetrahedral and a trigonal pyramidal molecular geometry.

Highlight Problems

113. 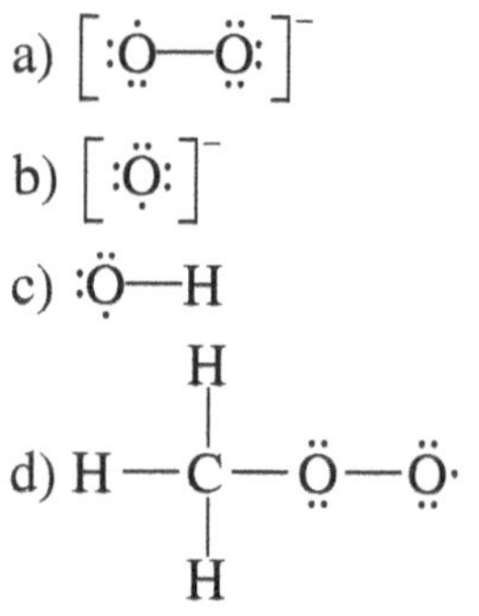 a) $\left[:\ddot{\text{O}}\text{—}\ddot{\text{O}}: \right]^-$

   b) $\left[:\ddot{\text{O}}: \right]^-$

   c) :Ö—H

   d)
```
       H
       |
   H—C—Ö—Ö·
       |
       H
```

114. 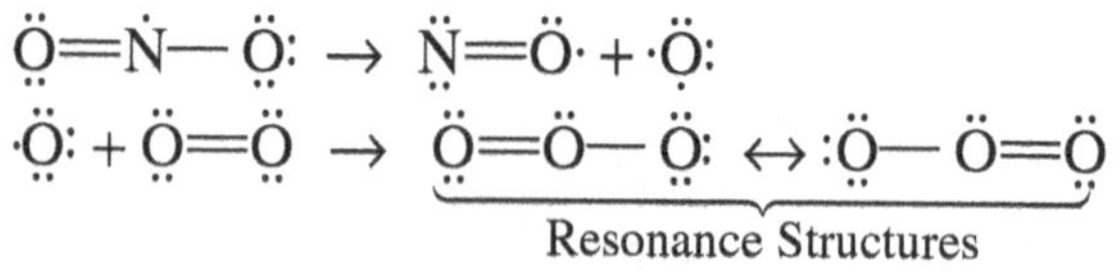 Ö=Ṅ—Ö: → Ṅ=Ö· + ·Ö:

   ·Ö: + Ö=Ö → $\underbrace{\text{Ö=Ö—Ö:} \leftrightarrow :\text{Ö—Ö=Ö}}_{\text{Resonance Structures}}$

115. a) incorrect; The structure should have tetrahedral electron geometry and a bent molecular geometry.

   H—S̈e—H

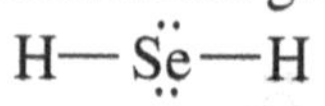

   b) correct

   c) incorrect; The structure should have tetrahedral electron geometry and a trigonal pyramidal geometry.
```
   :C̈l—P̈—C̈l:
       |
       :C̈l:
```

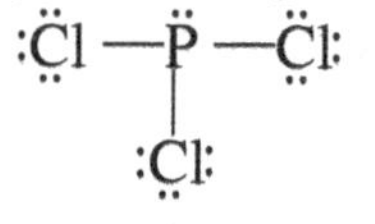

   d) correct

116. 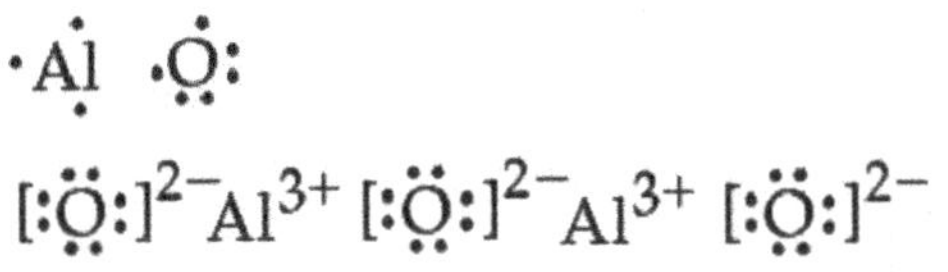 ·Ȧl· ·Ö:

[:Ö:]²⁻ Al³⁺ [:Ö:]²⁻ Al³⁺ [:Ö:]²⁻

117. Lewis Structure rules:
 1. Write the correct skeletal structure for the molecule.
 2. Calculate the total number of electrons for the Lewis structure by summing the valence electrons of each atom in the molecule. If you are writing a Lewis structure for a polyatomic ion, you must consider the charge of the ion when calculating the total number of electrons.
 3. Distribute the electrons among the atoms, giving octets (duets for hydrogen) to as many atoms as possible. Terminal atoms get electrons first, then central atoms.
 4. If any atoms lack an octet, form double or triple bonds as necessary to give them octets.

118. (a) 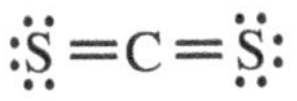 :S̈=C=S̈:

(b) Linear :S̈=C=S̈:

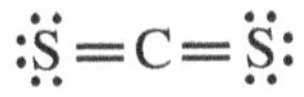

(c) Nonpolar bonds

(d) Nonpolar molecule

(a) 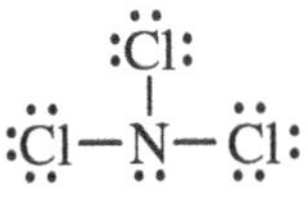 :C̈l—N̈—C̈l: with :C̈l: above N

(b) Trigonal pyramidal

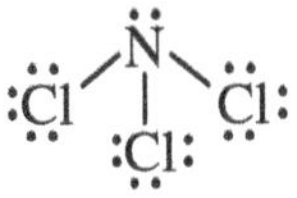

(c) Nonpolar bonds

(d) Polar molecule

(a) 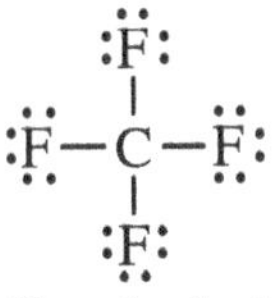 :F̈—C—F̈: with :F̈: above and :F̈: below

(b) Tetrahedral

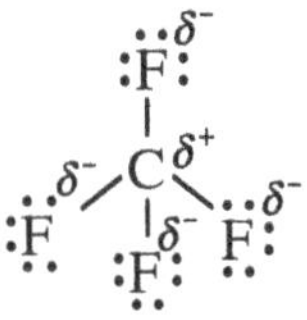

(c) Polar bonds

(d) Nonpolar molecule

(a) 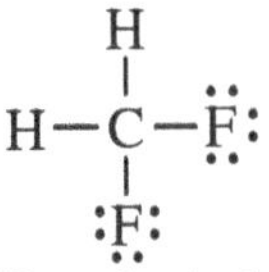 H—C—F̈: with H above and :F̈: below

(b) Tetrahedral

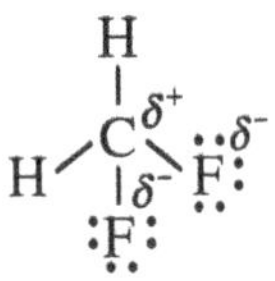

(c) Polar and nonpolar bonds

(c) Polar molecule

Copyright © 2018 Pearson Education, Inc.

Data Interpretation and Analysis

119.

NO$_2$	:Ö=Ṅ−Ö:	134
NO$_2^+$	[:Ö=N−Ö:]$^+$	180
NO$_2^-$	[:Ö=N̈−Ö:]$^-$	115

The bond angle for NO$_2^+$ corresponds to 180 degrees, as there are two electron regions around the central nitrogen atom. The NO$_2$ molecule has an additional electron present on the nitrogen atom, which repels the other two bonding regions causing the oxygen atoms to move closer together—from 180 degrees to 134 degrees. The nitrite ion, NO$_2^-$, has a lone pair of electrons, and the addition of the extra electron causes more repulsion between the regions, causing the oxygen atoms to move closer together from 134 degrees to 115 degrees.

Gases   11

Questions

1. Pressure is the force per unit area that is caused by gaseous molecules as they collide with a surface.

2. A straw works by creating a pressure difference between the inside and outside of the straw. This is done by creating a low pressure (sucking) on the inside of the straw. The pressure on the outside of the straw remains the same but is now higher than the inside. This forces the liquid up the inside of the straw and into your mouth. The upper limit of a straw is 10.3 m, or 34 ft.

3. The kinetic molecular theory assumptions:
   - Gas particles are in constant motion and move in a straight line.
   - Gas particles do not interact with each other (no attraction or repulsion). When particles collide, they bounce back like perfect billiard balls.
   - There is a large distance between gas particles compared to the size of each particle.
   - The average kinetic energy (energy of motion) is proportional to the temperature of the gas.

4.
   - Gases are compressible. This is because of the large distance between gas particles.
   - Gases assume the shape and volume of their container. This is because gas particles are in constant, straight-line motion.
   - Gases have low densities as compared with solids and liquids. This is because there is a large distance between particles that is filled with only empty space.

5. Ear pain experienced during changes in altitude is caused by air cavities within our ear. When the external pressure lowers, the air within the cavity is at a higher pressure. This causes the eardrum to bulge outward, which is the source of pain.

6. The main units to measure pressure are the atmosphere (atm), the pascal (Pa), the millimeters of mercury (mm Hg), and the torr.

7. Boyle's law describes the relationship between the pressure and volume of a gas. It states that pressure and volume are inversely related, or $V \propto 1/P$. As pressure increases, volume decreases (n and T being constant). When the volume of a gas sample is decreased, the same number of gas particles is crowded into a smaller volume, causing more collisions with the walls of the container and therefore increasing the pressure.

Copyright © 2018 Pearson Education, Inc.

8. When a diver is more than a few meters beneath the surface, the air pressure in the lungs is greater than the air pressure at the water's surface. As the diver ascends to the surface, the volume of air in the lungs will expand as the external pressure decreases.

9. Extra long snorkels would not work because the pressure of the air on the surface is at about 1 atm of pressure while the air in a diver's lung is at high pressure (10 m depth = 2 atm pressure). If you connected the diver with a snorkel at the surface, the high pressure in the lungs would force all of the air out to the surface, the opposite of what you want!

10. Charles's law describes the relationship between the temperature and volume of a gas. The volume of a gas is directly proportional to the absolute temperature (in Kelvin), or $V \propto T$. The volume of a given amount of gas increases as temperature increases (assuming constant P). If the temperature of a gas sample is increased, the gas particles move faster, and if the pressure is to remain constant, the volume must increase.

11. The gas molecules in a hot-air balloon are moving faster and, according to Charles's law, will occupy a larger volume than the same amount of gas at ambient temperature. As the volume occupied by a given amount of gas increases, the density of the gas decreases and the balloon rises due to the buoyancy effect. The air inside the balloon is less dense than the air outside the balloon.

12. The combined gas law, $\dfrac{P_1 V_1}{T_1} = \dfrac{P_2 V_2}{T_2}$, is used when three variables are changed and n is constant.

13. Avogadro's law states that the volume of a gas and the amount of the gas are directly proportional. If the number of gas particles increases at constant pressure and temperature, the particles must occupy more volume.

14. The ideal gas law is $PV = nRT$. The ideal gas law is useful because you can calculate any of the variables given the other three (R is a constant). The ideal gas law can also be used to calculate the molar mass of a gas.

15. The ideal gas law is most accurate under the conditions of low pressure and high temperature. It breaks down at high pressures and low temperatures. This breakdown occurs because the gases are no longer acting according to the kinetic molecular theory. The gas particles start interacting (attraction and repulsion), and the distance between gas particles is no longer large.

16. Partial pressure is the pressure of a single gas in a mixture of two or more gases.

17. Dalton's law of partial pressure states that the total pressure exerted by a gas is equal to the sum of the partial pressures of the various component gases in the mixture: $P_{total} = P_1 + P_2 + P_3 + \ldots$.

18. Hypoxia is a medical condition caused by low levels of O_2. Mild hypoxia leads to dizziness, headache, and shortness of breath. Severe hypoxia can lead to unconsciousness and even death. Oxygen toxicity is a condition caused by increased oxygen levels in body tissues, which can cause muscle twitching, tunnel vision, and convulsions.

19. Deep-sea divers use a helium-oxygen mixture to prevent oxygen toxicity and nitrogen narcosis.

20. When a gas is collected over water, it is not pure because some of the water molecules will have evaporated and your sample will contain water vapor.

21. The vapor pressure of a liquid is equal to the equilibrium partial pressure of the gas in contact with the liquid. As temperature increases, the vapor pressure of a liquid increases.

22. Standard temperature and pressure (STP) are 273 K and 1 atm. The molar volume of a gas: 1 mole of any gas has a volume of 22.4 L at conditions of STP.

Problems

Converting Between Pressure Units

23. a) $1277 \text{ mm Hg} \times \dfrac{1 \text{ atm}}{760 \text{ mm Hg}} = 1.680 \text{ atm}$

    b) $2.38 \times 10^5 \text{ Pa} \times \dfrac{1 \text{ atm}}{101,325 \text{ Pa}} = 2.35 \text{ atm}$

    c) $127 \text{ psi} \times \dfrac{1 \text{ atm}}{14.7 \text{ psi}} = 8.64 \text{ atm}$

    d) $455 \text{ torr} \times \dfrac{1 \text{ atm}}{760 \text{ torr}} = 0.599 \text{ atm}$

24. a) $921 \text{ torr} \times \dfrac{1 \text{ atm}}{760 \text{ torr}} = 1.21 \text{ atm}$

    b) $4.8 \times 10^4 \text{ Pa} \times \dfrac{1 \text{ atm}}{101,325 \text{ Pa}} = 0.47 \text{ atm}$

    c) $87.5 \text{ psi} \times \dfrac{1 \text{ atm}}{14.7 \text{ psi}} = 5.95 \text{ atm}$

    d) $34.22 \text{ in Hg} \times \dfrac{1 \text{ atm}}{29.92 \text{ in Hg}} = 1.144 \text{ atm}$

25. a) $2.3 \text{ atm} \times \dfrac{760 \text{ torr}}{1 \text{ atm}} = 1.7 \times 10^3 \text{ torr}$

b) $4.7 \times 10^{-2} \text{ atm} \times \dfrac{760 \text{ mm Hg}}{1 \text{ atm}} = 36 \text{ mm Hg}$

c) $24.8 \text{ psi} \times \dfrac{760 \text{ mm Hg}}{14.7 \text{ psi}} = 1.28 \times 10^3 \text{ mm Hg}$

d) $32.84 \text{ in Hg} \times \dfrac{760 \text{ torr}}{29.92 \text{ in Hg}} = 834.2 \text{ torr}$

26. a) $1.06 \text{ atm} \times \dfrac{760 \text{ mm Hg}}{1 \text{ atm}} = 806 \text{ mm Hg}$

b) $95{,}422 \text{ Pa} \times \dfrac{760 \text{ mm Hg}}{101{,}325 \text{ Pa}} = 715.72 \text{ mm Hg}$

c) $22.3 \text{ psi} \times \dfrac{760 \text{ torr}}{14.7 \text{ psi}} = 1.15 \times 10^3 \text{ torr}$

d) $35.78 \text{ in Hg} \times \dfrac{760 \text{ mm Hg}}{29.92 \text{ in Hg}} = 908.9 \text{ mm Hg}$

27.

Pascals	Atmospheres	mm Hg	Torr	psi
882	0.00870	6.62	6.62	0.128
5.65×10^4	0.558	424	424	8.20
1.71×10^5	1.69	1.28×10^3	1.28×10^3	24.8
1.02×10^5	1.01	764	764	14.8
3.32×10^4	0.328	249	249	4.82

28.

Pascals	Atmospheres	mm Hg	Torr	psi
1.94×10^5	1.91	1.45×10^3	1.45×10^3	28.1
1.15×10^4	0.113	86.3	86.3	1.67
9.62×10^4	0.949	721	721	14.0
1.45×10^4	0.143	109	109	2.11
2.69×10^5	2.65	2.01×10^3	2.01×10^3	38.9

29. a) $24.9 \text{ in Hg} \times \dfrac{1 \text{ atm}}{29.92 \text{ in Hg}} = 0.832 \text{ atm}$

b) $24.9 \text{ in Hg} \times \dfrac{760 \text{ mm Hg}}{29.92 \text{ in Hg}} = 632 \text{ mm Hg}$

c) $24.9 \text{ in Hg} \times \dfrac{14.7 \text{ psi}}{29.92 \text{ in Hg}} = 12.2 \text{ psi}$

d) $24.9 \text{ in Hg} \times \dfrac{101{,}325 \text{ Pa}}{29.92 \text{ in Hg}} = 8.43 \times 10^4 \text{ Pa}$

30. a) $235 \text{ mm Hg} \times \dfrac{760 \text{ torr}}{760 \text{ mm Hg}} = 235 \text{ torr}$

b) $235 \text{ mm Hg} \times \dfrac{14.7 \text{ psi}}{760 \text{ mm Hg}} = 4.55 \text{ psi}$

c) $235 \text{ mm Hg} \times \dfrac{29.92 \text{ in Hg}}{760 \text{ mm Hg}} = 9.25 \text{ in Hg}$

d) $235 \text{ mm Hg} \times \dfrac{1 \text{ atm}}{760 \text{ mm Hg}} = 0.309 \text{ atm}$

31. a) $31.85 \text{ in Hg} \times \dfrac{760 \text{ mm Hg}}{29.92 \text{ in Hg}} = 809.0 \text{ mm Hg}$

b) $31.85 \text{ in Hg} \times \dfrac{1 \text{ atm}}{29.92 \text{ in Hg}} = 1.065 \text{ atm}$

c) $31.85 \text{ in Hg} \times \dfrac{760 \text{ torr}}{29.92 \text{ in Hg}} = 809.0 \text{ torr}$

d) $31.85 \text{ in Hg} \times \dfrac{101{,}325 \text{ Pa}}{29.92 \text{ in Hg}} \times \dfrac{1 \text{ kPa}}{1000 \text{ Pa}} = 107.9 \text{ kPa}$

32. a) $652 \text{ mm Hg} \times \dfrac{760 \text{ torr}}{760 \text{ mm Hg}} = 652 \text{ torr}$

b) $652 \text{ mm Hg} \times \dfrac{1 \text{ atm}}{760 \text{ mm Hg}} = 0.858 \text{ atm}$

c) $652 \text{ mm Hg} \times \dfrac{29.92 \text{ in Hg}}{760 \text{ mm Hg}} = 25.7 \text{ in Hg}$

d) $652 \text{ mm Hg} \times \dfrac{14.7 \text{ psi}}{760 \text{ mm Hg}} = 12.6 \text{ psi}$

Simple Gas Laws

33. $P_1 = 705 \text{ mm Hg}$, $V_1 = 3.95 \text{ L}$, $V_2 = 5.38 \text{ L}$, $P_2 = ?$; $P_1V_1 = P_2V_2$
(705 mm Hg)(3.95 L) = (P_2)(5.38 L)
$P_2 = \dfrac{(705 \text{ mm Hg})(3.95 \text{ L})}{(5.38 \text{ L})} = 518 \text{ mm Hg}$

34. $P_1 = 1.65 \text{ atm}$, $V_1 = 22.8 \text{ L}$, $V_2 = 10.7 \text{ L}$, $P_2 = ?$; $P_1V_1 = P_2V_2$
(1.65 atm)(22.8 L) = (P_2)(10.7 L)
$P_2 = \dfrac{(1.65 \text{ atm})(22.8 \text{ L})}{(10.7 \text{ L})} = 3.52 \text{ atm}$

35. $P_1 = 1.0$ atm, $V_1 = 6.3$ L, $V_2 = ?$ L, $P_2 = 3.5$ atm; $P_1V_1 = P_2V_2$

$$(1.0 \text{ atm})(6.3 \text{ L}) = (V_2)(3.5 \text{ atm}) \Rightarrow V_2 = \frac{(1.0 \text{ atm})(6.3 \text{ L})}{(3.5 \text{ atm})} = 1.8 \text{ L}$$

36. $P_1 = 5.5$ atm, $V_1 = 5.2$ L, $V_2 = ?$ L, $P_2 = 1.0$ atm; $P_1V_1 = P_2V_2$

$$(5.5 \text{ atm})(5.2 \text{ L}) = (V_2)(1.0 \text{ atm}) \Rightarrow V_2 = \frac{(5.5 \text{ atm})(5.2 \text{ L})}{(1.0 \text{ atm})} = 29 \text{ L}$$

37.
P1	V1	P2	V2
755 mm Hg	2.85 L	885 mm Hg	<u>2.43 L</u>
<u>9.35 atm</u>	1.33 L	4.32 atm	2.88 L
192 mm Hg	382 mL	<u>152 mm Hg</u>	482 mL
2.11 atm	<u>226 mL</u>	3.82 atm	125 mL

38.
P1	V1	P2	V2
<u>2.40 atm</u>	1.90 L	4.19 atm	1.09 L
755 mm Hg	118 mL	709 mm Hg	<u>126 mL</u>
2.75 atm	6.75 mL	<u>0.373 atm</u>	49.8 mL
343 torr	<u>17.5 L</u>	683 torr	8.79 L

39. $T_1 = 299$ K, $V_1 = 3.2$ L, $T_2 = 376$ K, $V_2 = ?$; $\frac{V_1}{T_1} = \frac{V_2}{T_2}$

$$\frac{3.2 \text{ L}}{299 \text{ K}} = \frac{V_2}{376 \text{ K}} \Rightarrow V_2 = \frac{(3.2 \text{ L})(376 \text{ K})}{(299 \text{ K})} = 4.0 \text{ L}$$

40. $T_1 = 298$ K, $V_1 = 2.7$ L, $T_2 = 77$ K, $V_2 = ?$; $\frac{V_1}{T_1} = \frac{V_2}{T_2}$

$$\frac{2.7 \text{ L}}{298 \text{ K}} = \frac{V_2}{77 \text{ K}} \Rightarrow T_2 = \frac{(2.7 \text{ L})(77 \text{ K})}{(298 \text{ K})} = 0.70 \text{ L}$$

41. $T_1 = 22 + 273 = 295$ K, $V_1 = 48.3$ mL, $T_2 = 87 + 273 = 3.60 \times 10^2$ K, $V_2 = ?$; $\frac{V_1}{T_1} = \frac{V_2}{T_2}$

$$\frac{48.3 \text{ mL}}{295 \text{ K}} = \frac{V_2}{3.60 \times 10^2 \text{ K}} \Rightarrow V_2 = \frac{(48.3 \text{ mL})(3.60 \times 10^2 \text{ K})}{(295 \text{ K})} = 58.9 \text{ mL}$$

42. $T_1 = 95.3 + 273 = 368$ K, $V_1 = 1.55$ mL, $T_2 = 0.0 + 273 = 273$ K, $V_2 = ?$; $\frac{V_1}{T_1} = \frac{V_2}{T_2}$

$$\frac{1.55 \text{ mL}}{368 \text{ K}} = \frac{V_2}{273 \text{ K}} \Rightarrow V_2 = \frac{(1.55 \text{ mL})(273 \text{ K})}{(368 \text{ K})} = 1.15 \text{ mL}$$

43.
V1	T1	V2	T2
1.08 L	25.4 °C	1.33 L	__94.5 °C__
__58.9 mL__	77 K	228 mL	298 K
115 cm³	__12.5 °C__	119 cm³	22.4 °C
232 L	18.5 °C	__294 L__	96.2 °C

44.
V1	T1	V2	T2
119 L	10.5 °C	__162 L__	112.3 °C
__75.4 mL__	135 K	176 mL	315 K
2.11 L	15.4 °C	2.33 L	__44.9 °C__
15.4 cm³	__−45.7 °C__	19.2 cm³	10.4 °C

45. $n_1 = 0.12$ mol, $V_1 = 2.55$ L, $n_2 = 0.32$ mol, $V_2 = ?$; $\dfrac{V_1}{n_1} = \dfrac{V_2}{n_2}$

$\dfrac{2.55 \text{ L}}{0.12 \text{ mol}} = \dfrac{V_2}{0.32 \text{ mol}} \Rightarrow V_2 = \dfrac{(2.55 \text{ L})(0.32 \text{ mol})}{0.12 \text{ mol}} = 6.8 \text{ L}$

46. $n_1 = 0.48$ mol, $V_1 = 11.7$ L, $n_2 = 0.72$ mol, $V_2 = ?$; $\dfrac{V_1}{n_1} = \dfrac{V_2}{n_2}$

$\dfrac{11.7 \text{ L}}{0.48 \text{ mol}} = \dfrac{V_2}{0.72 \text{ mol}} \Rightarrow V_2 = \dfrac{(11.7 \text{ L})(0.72 \text{ mol})}{0.48 \text{ mol}} = 18 \text{ L}$

47. $n_1 = 0.128$ mol, $V_1 = 2.76$ L, $n_2 = 0.128 + 0.073 = 0.201$ mol, $V_2 = ?$; $\dfrac{V_1}{n_1} = \dfrac{V_2}{n_2}$

$\dfrac{2.76 \text{ L}}{0.128 \text{ mol}} = \dfrac{V_2}{0.201 \text{ mol}} \Rightarrow V_2 = \dfrac{(2.76 \text{ L})(0.201 \text{ mol})}{(0.128 \text{ mol})} = 4.33 \text{ L}$

48. $n_1 = 0.87$ mol, $V_1 = 334$ mL, $n_2 = 0.87 + 0.22 = 1.09$ mol, $V_2 = ?$; $\dfrac{V_1}{n_1} = \dfrac{V_2}{n_2}$

$\dfrac{334 \text{ mL}}{0.87 \text{ mol}} = \dfrac{V_2}{1.09 \text{ mol}} \Rightarrow V_2 = \dfrac{(334 \text{ mL})(1.09 \text{ mol})}{(0.87 \text{ mol})} = 4.2 \times 10^2 \text{ mL}$

49.
V1	n1	V2	n2
38.5 mL	1.55×10^{-3} mol	49.4 mL	__1.99×10^{-3} mol__
__8.03 L__	1.37 mol	26.8 L	4.57 mol
11.2 L	0.628 mol	__15.7 L__	0.881 mol
422 mL	__0.0109 mol__	671 mL	0.0174 mol

50.
V1	n1	V2	n2
25.2 L	5.05 mol	__15.1 L__	3.03 mol
__499 mL__	1.10 mol	414 mL	0.913 mol
8.63 L	0.0018 mol	10.9 L	__0.0023 mol__
53 mL	__1.1×10^{-3} mol__	13 mL	2.61×10^{-4} mol

Combined Gas Law

51. $P_1 = 725$ mm Hg, $V_1 = 28.4$ L, $T_1 = 305$ K, $V_2 = 14.8$ L, $T_2 = 375$ K, $P_2 = ?$

$$\frac{P_1 V_1}{T_1} = \frac{P_2 V_2}{T_2} \Rightarrow \frac{(725 \text{ mm Hg})(28.4 \text{ L})}{(305 \text{ K})} = \frac{P_2(14.8 \text{ L})}{(375 \text{ K})}$$

$$P_2 = \frac{(725 \text{ mm Hg})(28.4 \text{ L})(375 \text{ K})}{(305 \text{ K})(14.8 \text{ L})} = 1.71 \times 10^3 \text{ mm Hg}$$

52. $P_1 = 1.32$ atm, $V_1 = 218$ mL, $T_1 = 298$ K, $P_2 = 1.55$ atm, $V_2 = ?$, $T_2 = 335$ K,

$$\frac{P_1 V_1}{T_1} = \frac{P_2 V_2}{T_2} \Rightarrow \frac{(1.32 \text{ atm})(218 \text{ mL})}{(298 \text{ K})} = \frac{(1.55 \text{ atm})V_2}{(335 \text{ K})}$$

$$V_2 = \frac{(1.32 \text{ atm})(218 \text{ mL})(335 \text{ K})}{(298 \text{ K})(1.55 \text{ atm})} = 209 \text{ mL}$$

53. $P_1 = 1.0$ atm, $V_1 = 2.8$ L, $T_1 = 34 + 273 = 307$ K,
$P_2 = 3.5$ atm, $V_2 = ?$, $T_2 = 18 + 273 = 291$ K

$$\frac{P_1 V_1}{T_1} = \frac{P_2 V_2}{T_2} \Rightarrow \frac{(1.0 \text{ atm})(2.8 \text{ L})}{(307 \text{ K})} = \frac{(3.5 \text{ atm})(V_2)}{(291 \text{ K})} \Rightarrow$$

$$V_2 = \frac{(1.0 \text{ atm})(2.8 \text{ L})(291 \text{ K})}{(307 \text{ K})(3.5 \text{ atm})} = 0.76 \text{ L}$$

54. $P_1 = 765$ mm Hg, $V_1 = 585$ mL, $T_1 = 25 + 273 = 298$ K,
$P_2 = 442$ mm Hg, $V_2 = ?$, $T_2 = 5 + 273 = 278$ K

$$\frac{P_1 V_1}{T_1} = \frac{P_2 V_2}{T_2} \Rightarrow \frac{(765 \text{ mm Hg})(585 \text{ mL})}{(298 \text{ K})} = \frac{(442 \text{ mm Hg})(V_2)}{(278 \text{ K})} \Rightarrow$$

$$V_2 = \frac{(765 \text{ mm Hg})(585 \text{ mL})(278 \text{ K})}{(298 \text{ K})(442 \text{ mm Hg})} = 945 \text{ mL}$$

55. $P_1 = 735$ mm Hg, $T_1 = 28 + 273 = 301$ K, $P_2 = ?$, $T_2 = 86 + 273 = 359$ K

$$\frac{P_1}{T_1} = \frac{P_2}{T_2} \Rightarrow \frac{(735 \text{ mm Hg})}{(301 \text{ K})} = \frac{(P_2)}{(359 \text{ K})} \Rightarrow P_2 = \frac{(735 \text{ mm Hg})(359 \text{ K})}{(301 \text{ K})} = 877 \text{ mm Hg}$$

56. $P_1 = 44$ psi, $T_1 = 11 + 273 = 284$ K, $P_2 = ?$, $T_2 = 37 + 273 = 3.10 \times 10^2$ K

$$\frac{P_1}{T_1} = \frac{P_2}{T_2} \Rightarrow \frac{(44 \text{ psi})}{(284 \text{ K})} = \frac{(P_2)}{(310 \text{ K})} \Rightarrow P_2 = \frac{(44 \text{ psi})(3.10 \times 10^2 \text{ K})}{(284 \text{ K})} = 48 \text{ psi}$$

Rise in Pressure $= 48 - 44 = 4$ psi

57.
P1	V1	T1	P2	V2	T2
121 atm	1.58 L	12.2 °C	1.54 atm	133 L	32.3 °C
721 torr	141 mL	135 K	801 torr	152 mL	<u>162 K</u>
5.51 atm	0.879 L	22.1 °C	<u>4.86 atm</u>	1.05 L	38.3 °C

58.
P1	V1	T1	P2	V2	T2
1.01 atm	<u>0.27 L</u>	2.7 °C	0.54 atm	0.58 L	42.3 °C
123 torr	41.5 mL	<u>45.8 K</u>	626 torr	36.5 mL	205 K
<u>0.412 atm</u>	1.879 L	20.8 °C	0.412 atm	2.05 L	48.1 °C

The Ideal Gas Laws

59. P = 1.25 atm, V = ?, n = 0.255 mol, T = 305 K, PV = nRT

$$(1.25 \text{ atm}) V = (0.255 \text{ mol})(0.0821 \frac{L \cdot atm}{mol \cdot K})(305 \text{ K})$$

$$V = \frac{(0.255 \text{ mol})(0.0821 \frac{L \cdot atm}{mol \cdot K})(305 \text{ K})}{(1.25 \text{ atm})} = 5.11 \text{ L}$$

60. P = ?, V = 20.0 L, n = 0.683 mol, T = 325 K, PV = nRT

$$P(20.0 \text{ L}) = (0.683 \text{ mol})(0.0821 \frac{L \cdot atm}{mol \cdot K})(325 \text{ K})$$

$$P = \frac{(0.683 \text{ mol})(0.0821 \frac{L \cdot atm}{mol \cdot K})(325 \text{ K})}{(20.0 \text{ L})} = 0.911 \text{ atm}$$

61. P = 1.8 atm, V = 28.5 L, n = ?, T = 298 K, PV = nRT

$$(1.8 \text{ atm})(28.5 \text{ L}) = (n)(0.0821 \frac{L \cdot atm}{mol \cdot K})(298 \text{ K}) \Rightarrow$$

$$n = \frac{(1.8 \text{ atm})(28.5 \text{ L})}{(0.0821 \frac{L \cdot atm}{mol \cdot K})(298 \text{ K})} = 2.1 \text{ mol}$$

62. P = 1.3 atm, V = 11.8 L, n = 0.52 mol, T = ?, PV = nRT

$$(1.3 \text{ atm})(11.8 \text{ L}) = (0.52 \text{ mol})(0.0821 \frac{L \cdot atm}{mol \cdot K})(T) \Rightarrow$$

$$T = \frac{(1.3 \text{ atm})(11.8 \text{ L})}{(0.52 \text{ mol})(0.0821 \frac{L \cdot atm}{mol \cdot K})} = 3.6 \times 10^2 \text{ K}$$

63. $P = 43.2 \text{ psi} \times \dfrac{1 \text{ atm}}{14.7 \text{ psi}} = 2.94 \text{ atm}$, $V = 11.8$ L, $n = ?$, $T = 25 + 273 = 298$ K

$PV = nRT \Rightarrow (2.94 \text{ atm})(11.8 \text{ L}) = (n)(0.0821 \dfrac{\text{L} \cdot \text{atm}}{\text{mol} \cdot \text{K}})(298 \text{ K}) \Rightarrow$

$n = \dfrac{(2.94 \text{ atm})(11.8 \text{ L})}{(0.0821 \dfrac{\text{L} \cdot \text{atm}}{\text{mol} \cdot \text{K}})(298 \text{ K})} = 1.42 \text{ mol}$

64. $P = ?$ mm Hg, $V = 0.214$ L, $n = 0.0115$ mol, $T = 45 + 273 = 318$ K

$PV = nRT \Rightarrow (P)(0.215 \text{ L}) = (0.0115 \text{ mol})(0.0821 \dfrac{\text{L} \cdot \text{atm}}{\text{mol} \cdot \text{K}})(318 \text{ K}) \Rightarrow$

$P = \dfrac{(0.0115 \text{ mol})(0.0821 \dfrac{\text{L} \cdot \text{atm}}{\text{mol} \cdot \text{K}})(318 \text{ K})}{(0.215 \text{ L})} = 1.40 \text{ atm}$

$1.40 \text{ atm} \times \dfrac{760 \text{ mm Hg}}{1 \text{ atm}} = 1.06 \times 10^3 \text{ mm Hg}$

65.

P	V	n	T
1.05 atm	1.19 L	0.112 mol	<u>136 K</u>
112 torr	<u>40.9 L</u>	0.241 mol	304 K
<u>1.49 atm</u>	28.5 mL	1.74×10^{-3} mol	25.4 °C
0.559 atm	0.439 L	<u>0.0117 mol</u>	255 K

66.

P	V	n	T
2.39 atm	1.21 L	<u>0.172 mol</u>	205 K
512 torr	<u>26.9 L</u>	0.741 mol	298 K
0.433 atm	0.192 L	0.0131 mol	<u>77.3 K</u>
<u>6.80 atm</u>	20.2 mL	5.71×10^{-3} mol	20.4 °C

67. $P = 24.1 \text{ psi} \times \dfrac{1 \text{ atm}}{14.7 \text{ psi}} = 1.64 \text{ atm}$, $V = 3.5$ L, $n = ?$, $T = 25 + 273 = 298$ K

$PV = nRT \Rightarrow (1.64 \text{ atm})(3.5 \text{ L}) = (n)(0.0821 \dfrac{\text{L} \cdot \text{atm}}{\text{mol} \cdot \text{K}})(298 \text{ K}) \Rightarrow$

$n = \dfrac{(1.64 \text{ atm})(3.5 \text{ L})}{(0.0821 \dfrac{\text{L} \cdot \text{atm}}{\text{mol} \cdot \text{K}})(298 \text{ K})} = 0.23 \text{ mol}$

68. $P = 47.1 \text{ psi} \times \dfrac{1 \text{ atm}}{14.7 \text{ psi}} = 3.20 \text{ atm}$, $V = 4.8 \text{ L}$, $n = ?$, $T = 25 + 273 = 298 \text{ K}$

$PV = nRT \Rightarrow (3.20 \text{ atm})(4.8 \text{ L}) = (n)(0.0821 \dfrac{\text{L} \cdot \text{atm}}{\text{mol} \cdot \text{K}})(298 \text{ K}) \Rightarrow$

$n = \dfrac{(3.20 \text{ atm})(4.8 \text{ L})}{(0.0821 \dfrac{\text{L} \cdot \text{atm}}{\text{mol} \cdot \text{K}})(298 \text{ K})} = 0.63 \text{ mol}$

69. $P = 745 \text{ mm Hg} \times \dfrac{1 \text{ atm}}{760 \text{ mm Hg}} = 0.980 \text{ atm}$, $V = 248 \text{ mL} \times \dfrac{1 \text{ L}}{1000 \text{ mL}} = 0.248 \text{ L}$,

$n = ?$, $T = 28 + 273 = 301 \text{ K}$

$PV = nRT \Rightarrow (0.980 \text{ atm})(0.248 \text{ L}) = (n)(0.0821 \dfrac{\text{L} \cdot \text{atm}}{\text{mol} \cdot \text{K}})(301 \text{ K}) \Rightarrow$

$n = \dfrac{(0.980 \text{ atm})(0.248 \text{ L})}{(0.0821 \dfrac{\text{L} \cdot \text{atm}}{\text{mol} \cdot \text{K}})(301 \text{ K})} = 0.00983 \text{ mol}$

molar mass $= \dfrac{0.433 \text{ g}}{0.00983 \text{ mol}} = 44.0 \text{ g/mol}$

70. $P = 721 \text{ mm Hg} \times \dfrac{1 \text{ atm}}{760 \text{ mm Hg}} = 0.949 \text{ atm}$, $V = 113 \text{ mL} \times \dfrac{1 \text{ L}}{1000 \text{ mL}} = 0.113 \text{ L}$,

$n = ?$, $T = 32 + 273 = 305 \text{ K}$

$PV = nRT \Rightarrow (0.949 \text{ atm})(0.113 \text{ L}) = (n)(0.0821 \dfrac{\text{L} \cdot \text{atm}}{\text{mol} \cdot \text{K}})(305 \text{ K}) \Rightarrow$

$n = \dfrac{(0.949 \text{ atm})(0.113 \text{ L})}{(0.0821 \dfrac{\text{L} \cdot \text{atm}}{\text{mol} \cdot \text{K}})(305 \text{ K})} = 0.00428 \text{ mol}$

molar mass $= \dfrac{0.171 \text{ g}}{0.00428 \text{ mol}} = 40.0 \text{ g/mol}$

71. $P = 886 \text{ torr} \times \dfrac{1 \text{ atm}}{760 \text{ torr}} = 1.17 \text{ atm}$, $V = 224 \text{ mL} \times \dfrac{1 \text{ L}}{1000 \text{ mL}} = 0.224 \text{ L}$,

$n = ?$, $T = 55 + 273 = 328 \text{ K}$

$PV = nRT \Rightarrow (1.17 \text{ atm})(0.224 \text{ L}) = (n)(0.0821 \dfrac{\text{L} \cdot \text{atm}}{\text{mol} \cdot \text{K}})(328 \text{ K}) \Rightarrow$

$n = \dfrac{(1.17 \text{ atm})(0.224 \text{ L})}{(0.0821 \dfrac{\text{L} \cdot \text{atm}}{\text{mol} \cdot \text{K}})(328 \text{ K})} = 0.00973 \text{ mol}$

molar mass $= \dfrac{38.8 \text{ mg} \times \dfrac{1 \text{ g}}{1000 \text{ mg}}}{0.00973 \text{ mol}} = 3.99 \text{ g/mol}$

72. $P = 753 \text{ mm Hg} \times \dfrac{1 \text{ atm}}{760 \text{ torr}} = 0.991 \text{ atm}$, $V = 117 \text{ mL} \times \dfrac{1 \text{ L}}{1000 \text{ mL}} = 0.117 \text{ L}$,
    $n = ?$, $T = 85 + 273 = 358 \text{ K}$

    $PV = nRT \Rightarrow (0.991 \text{ atm})(0.117 \text{ L}) = (n)(0.0821 \dfrac{\text{L} \cdot \text{atm}}{\text{mol} \cdot \text{K}})(358 \text{ K}) \Rightarrow$

    $n = \dfrac{(0.991 \text{ atm})(0.117 \text{ L})}{(0.0821 \dfrac{\text{L} \cdot \text{atm}}{\text{mol} \cdot \text{K}})(358 \text{ K})} = 0.00394 \text{ mol}$

    molar mass $= \dfrac{0.555 \text{ g}}{0.00394 \text{ mol}} = 141 \text{ g/mol}$

Partial Pressure

73. $P_{tot} = P_1 + P_2 + P_3 + ... \Rightarrow P_{tot} = 217 \text{ torr} + 106 \text{ torr} + 248 \text{ torr} = 571 \text{ torr}$

74. $P_{tot} = P_1 + P_2 + P_3 + ... \Rightarrow$
    $P_{tot} = 422 \text{ mm Hg} + 102 \text{ mm Hg} + 165 \text{ mm Hg} + 52 \text{ mm Hg} = 741 \text{ mm Hg}$

75. $P_{tot} = P_{He} + P_{O_2} \Rightarrow 11.0 \text{ atm} = P_{He} + 0.30 \text{ atm} \Rightarrow P_{He} = 11.0 - 0.30 = 10.7 \text{ atm}$

76. $P_{tot} = P_{He} + P_{N_2} + P_{O_2} \Rightarrow 752 \text{ mm Hg} = 234 + 197 + P_{O_2} \Rightarrow$
    $P_{O_2} = 752 - 234 - 197 = 321 \text{ mm Hg}$

77. $P_{tot} = P_{H_2} + P_{H_2O} \Rightarrow 732 \text{ mm Hg} = P_{H_2} + 31.8$
    $P_{H_2} = 732 - 31.8 = 7.00 \times 10^2 \text{ mm Hg}$

78. $P_{tot} = P_{O_2} + P_{H_2O} \Rightarrow 753 \text{ mm Hg} = P_{O_2} + 23.8$
    $P_{O_2} = 753 - 23.8 = 729 \text{ mm Hg}$

79. $P_{N_2} = 1.12 \text{ atm} \times 0.78 = 0.87 \text{ atm}$ $P_{O_2} = 1.12 \text{ atm} \times 0.22 = 0.25 \text{ atm}$

80. $P_{CO_2} = 758 \text{ mm Hg} \times 0.00038 = 0.29 \text{ mm Hg}$

81. $P_{O_2} = 8.5 \text{ atm} \times 0.040 = 0.34 \text{ atm}$

82. If the diver is using compressed air, the partial pressure is:
    $P_{O_2} = 11 \text{ atm} \times 0.22 = 2.4 \text{ atm}$.

    Thus, the diver would be in danger of oxygen toxicity.

Molar Volume

83. a) $22.5 \text{ mol Cl}_2 \times \dfrac{22.4 \text{ L}}{1 \text{ mol Cl}_2} = 504 \text{ L}$

    b) $3.6 \text{ mol N}_2 \times \dfrac{22.4 \text{ L}}{1 \text{ mol N}_2} = 81 \text{ L}$

    c) $2.2 \text{ mol He} \times \dfrac{22.4 \text{ L}}{1 \text{ mol He}} = 49 \text{ L}$

    d) $27 \text{ mol CH}_4 \times \dfrac{22.4 \text{ L}}{1 \text{ mol CH}_4} = 6.0 \times 10^2 \text{ L}$

84. a) $21.2 \text{ mol N}_2\text{O} \times \dfrac{22.4 \text{ L}}{1 \text{ mol N}_2\text{O}} = 475 \text{ L}$

    b) $0.215 \text{ mol CO} \times \dfrac{22.4 \text{ L}}{1 \text{ mol CO}} = 4.82 \text{ L}$

    c) $0.364 \text{ mol CO}_2 \times \dfrac{22.4 \text{ L}}{1 \text{ mol CO}_2} = 8.15 \text{ L}$

    d) $8.6 \text{ mol C}_2\text{H}_6 \times \dfrac{22.4 \text{ L}}{1 \text{ mol C}_2\text{H}_6} = 1.9 \times 10^2 \text{ L}$

85. a) $73.9 \text{ g N}_2 \times \dfrac{1 \text{ mol N}_2}{28.02 \text{ g}} \times \dfrac{22.4 \text{ L}}{1 \text{ mol N}_2} = 59.1 \text{ L}$

    b) $42.9 \text{ g O}_2 \times \dfrac{1 \text{ mol O}_2}{32.00 \text{ g}} \times \dfrac{22.4 \text{ L}}{1 \text{ mol O}_2} = 30.0 \text{ L}$

    c) $148 \text{ g NO}_2 \times \dfrac{1 \text{ mol NO}_2}{46.01 \text{ g}} \times \dfrac{22.4 \text{ L}}{1 \text{ mol NO}_2} = 72.1 \text{ L}$

    d) $245 \text{ mg CO}_2 \times \dfrac{1 \text{ g}}{1000 \text{ mg}} \times \dfrac{1 \text{ mol CO}_2}{44.01 \text{ g}} \times \dfrac{22.4 \text{ L}}{1 \text{ mol CO}_2} = 0.125 \text{ L}$

86. a) $48.9 \text{ g He} \times \dfrac{1 \text{ mol He}}{4.003 \text{ g}} \times \dfrac{22.4 \text{ L}}{1 \text{ mol He}} = 274 \text{ L}$

    b) $45.2 \text{ g Xe} \times \dfrac{1 \text{ mol Xe}}{131.29 \text{ g}} \times \dfrac{22.4 \text{ L}}{1 \text{ mol Xe}} = 7.71 \text{ L}$

    c) $48.2 \text{ mg Cl}_2 \times \dfrac{1 \text{ g}}{1000 \text{ mg}} \times \dfrac{1 \text{ mol Cl}_2}{70.90 \text{ g}} \times \dfrac{22.4 \text{ L}}{1 \text{ mol Cl}_2} = 0.0152 \text{ L}$

    d) $3.83 \text{ kg SO}_2 \times \dfrac{1000 \text{ g}}{1 \text{ kg}} \times \dfrac{1 \text{ mol SO}_2}{64.07 \text{ g}} \times \dfrac{22.4 \text{ L}}{1 \text{ mol SO}_2} = 1.34 \times 10^3 \text{ L}$

87. a) $178 \text{ mL CO}_2 \times \dfrac{1 \text{ L}}{1000 \text{ mL}} \times \dfrac{1 \text{ mol CO}_2}{22.4 \text{ L}} \times \dfrac{44.01 \text{ g}}{1 \text{ mol CO}_2} = 0.350 \text{ g}$

b) $155 \text{ mL O}_2 \times \dfrac{1 \text{ L}}{1000 \text{ mL}} \times \dfrac{1 \text{ mol O}_2}{22.4 \text{ L}} \times \dfrac{32.00 \text{ g}}{1 \text{ mol O}_2} = 0.221 \text{ g}$

c) $1.25 \text{ L SF}_6 \times \dfrac{1 \text{ mol SF}_6}{22.4 \text{ L}} \times \dfrac{146.07 \text{ g}}{1 \text{ mol SF}_6} = 8.15 \text{ g}$

88. a) $5.82 \text{ L NO} \times \dfrac{1 \text{ mol NO}}{22.4 \text{ L}} \times \dfrac{30.01 \text{ g}}{1 \text{ mol NO}} = 7.80 \text{ g}$

b) $0.324 \text{ L N}_2 \times \dfrac{1 \text{ mol N}_2}{22.4 \text{ L}} \times \dfrac{28.02 \text{ g}}{1 \text{ mol N}_2} = 0.405 \text{ g}$

c) $139 \text{ cm}^3 \text{ Ar} \times \dfrac{1 \text{ L}}{1000 \text{ cm}^3} \times \dfrac{1 \text{ mol Ar}}{22.4 \text{ L}} \times \dfrac{39.95 \text{ g}}{1 \text{ mol Ar}} = 0.248 \text{ g}$

Gases in Chemical Reactions

89. $n = 1.07 \text{ mol C} \times \dfrac{1 \text{ mol H}_2}{1 \text{ mol C}} = 1.07 \text{ mol H}_2$, $P = 1.0 \text{ atm}$, $T = 315 \text{ K}$, $V = ?$

$PV = nRT \Rightarrow (1.0 \text{ atm})(V) = (1.07 \text{ mol})(0.0821 \dfrac{\text{L} \cdot \text{atm}}{\text{mol} \cdot \text{K}})(315 \text{ K}) \Rightarrow$

$V = \dfrac{(1.07 \text{ mol})(0.0821 \dfrac{\text{L} \cdot \text{atm}}{\text{mol} \cdot \text{K}})(315 \text{ K})}{(1.0 \text{ atm})} = 28 \text{ L}$

90. $P = 0.988 \text{ atm}$, $V = 1.3 \text{ L}$, $n = ?$, $T = 325 \text{ K}$

$PV = nRT \Rightarrow (0.988 \text{ atm})(1.3 \text{ L}) = (n)(0.0821 \dfrac{\text{L} \cdot \text{atm}}{\text{mol} \cdot \text{K}})(325 \text{ K}) \Rightarrow$

$n = \dfrac{(0.988 \text{ atm})(1.3 \text{ L})}{(0.0821 \dfrac{\text{L} \cdot \text{atm}}{\text{mol} \cdot \text{K}})(325 \text{ K})} = 0.048 \text{ mol O}_2$

$0.048 \text{ moles O}_2 \times \dfrac{2 \text{ mol H}_2\text{O}}{1 \text{ mol O}_2} = 0.096 \text{ mol H}_2\text{O}$

91. $P = 748 \text{ mm Hg} \times \dfrac{1 \text{ atm}}{760 \text{ mm Hg}} = 0.984 \text{ atm}$, $V = ?$, $T = 86 + 273 = 359 \text{ K}$

$n = 0.55 \text{ mol CH}_3\text{OH} \times \dfrac{2 \text{ mol H}_2}{1 \text{ mol CH}_3\text{OH}} = 1.1 \text{ mol H}_2$

$PV = nRT \Rightarrow (0.984 \text{ atm})(V) = (1.1 \text{ mol})(0.0821 \dfrac{\text{L} \cdot \text{atm}}{\text{mol} \cdot \text{K}})(359 \text{ K}) \Rightarrow$

$V = \dfrac{(1.1 \text{ mol})(0.0821 \dfrac{\text{L} \cdot \text{atm}}{\text{mol} \cdot \text{K}})(359 \text{ K})}{(0.984 \text{ atm})} = 33 \text{ L of H}_2$

The molar ratio of CO to H_2 is 1:2, we can therefore divide the volume of H_2 in half to determine CO. Volume of CO = 33/2 = 17 L

Warning: *This only works for gases! It does not apply to liquids or solids.*

92. $P = 782 \text{ mm Hg} \times \dfrac{1 \text{ atm}}{760 \text{ mm Hg}} = 1.03 \text{ atm}$, $V = ?$, $T = 25 + 273 = 298 \text{ K}$

$n = 2.4 \text{ mol Al} \times \dfrac{3 \text{ mol O}_2}{4 \text{ mol Al}} = 1.8 \text{ mol O}_2$

$PV = nRT \Rightarrow (1.03 \text{ atm})(V) = (1.8 \text{ mol})(0.0821 \dfrac{\text{L} \cdot \text{atm}}{\text{mol} \cdot \text{K}})(298 \text{ K}) \Rightarrow$

$V = \dfrac{(1.8 \text{ mol})(0.0821 \dfrac{\text{L} \cdot \text{atm}}{\text{mol} \cdot \text{K}})(298 \text{ K})}{(1.03 \text{ atm})} = 43 \text{ L}$

93. $P = 892 \text{ torr} \times \dfrac{1 \text{ atm}}{760 \text{ torr}} = 1.17 \text{ atm}$, $V = ?$, $T = 95 + 273 = 368 \text{ K}$

$n = 18.5 \text{ g Al} \times \dfrac{1 \text{ mol Al}}{26.98 \text{ g}} \times \dfrac{1 \text{ mol N}_2}{2 \text{ mol Al}} = 0.343 \text{ mol N}_2$

$PV = nRT \Rightarrow (1.17 \text{ atm})(V) = (0.343 \text{ mol})(0.0821 \dfrac{\text{L} \cdot \text{atm}}{\text{mol} \cdot \text{K}})(368 \text{ K}) \Rightarrow$

$V = \dfrac{(0.343 \text{ mol})(0.0821 \dfrac{\text{L} \cdot \text{atm}}{\text{mol} \cdot \text{K}})(368 \text{ K})}{(1.17 \text{ atm})} = 8.86 \text{ L}$

94. $P = 687 \text{ torr} \times \dfrac{1 \text{ atm}}{760 \text{ torr}} = 0.904 \text{ atm}, V = ?, T = 35 + 273 = 308 \text{ K}$

$n = 28 \text{ g NaCl} \times \dfrac{1 \text{ mol NaCl}}{58.44 \text{ g}} \times \dfrac{1 \text{ mol Cl}_2}{2 \text{ mol NaCl}} = 0.24 \text{ mol Cl}_2$

$PV = nRT \Rightarrow (0.904 \text{ atm})(V) = (0.24 \text{ mol})(0.0821 \dfrac{\text{L} \cdot \text{atm}}{\text{mol} \cdot \text{K}})(308 \text{ K}) \Rightarrow$

$V = \dfrac{(0.24 \text{ mol})(0.0821 \dfrac{\text{L} \cdot \text{atm}}{\text{mol} \cdot \text{K}})(308 \text{ K})}{(0.904 \text{ atm})} = 6.7 \text{ L}$

95. $24.8 \text{ L H}_2 \times \dfrac{1 \text{ mol H}_2}{22.4 \text{ L}} \times \dfrac{2 \text{ mol NH}_3}{3 \text{ mol H}_2} \times \dfrac{17.04 \text{ g}}{1 \text{ mol NH}_3} = 12.6 \text{ g NH}_3$

96. $58.5 \text{ mL N}_2 \times \dfrac{1 \text{ L}}{1000 \text{ mL}} \times \dfrac{1 \text{ mol N}_2}{22.4 \text{ L}} \times \dfrac{6 \text{ mol Li}}{1 \text{ mol N}_2} \times \dfrac{6.94 \text{ g}}{1 \text{ mol Li}} = 0.109 \text{ g Li}$

97. $156.8 \text{ mL O}_2 \times \dfrac{1 \text{ L}}{1000 \text{ mL}} \times \dfrac{1 \text{ mol O}_2}{22.4 \text{ L}} \times \dfrac{2 \text{ mol Ca}}{1 \text{ mol O}_2} \times \dfrac{40.08 \text{ g}}{1 \text{ mol Ca}} = 0.561 \text{ g Ca}$

98. $14.8 \text{ L O}_2 \times \dfrac{1 \text{ mol O}_2}{22.4 \text{ L}} \times \dfrac{2 \text{ mol MgO}}{1 \text{ mol O}_2} \times \dfrac{40.31 \text{ g}}{1 \text{ mol MgO}} = 53.3 \text{ g MgO}$

Cumulative Problems

99. $P = 1.00 \text{ atm}, V = ?, n = 1.00 \text{ mol}, T = 273 \text{ K}$

$V = \dfrac{(1.00 \text{ mol})(0.0821 \dfrac{\text{L} \cdot \text{atm}}{\text{mol} \cdot \text{K}})(273 \text{ K})}{(1.00 \text{ atm})} = 22.4 \text{ L}$

100. $28.0 \text{ g N}_2 \times \dfrac{1 \text{ mol N}_2}{28.02 \text{ g}} = 1.00 \text{ mol N}_2 \quad 4.00 \text{ g He} \times \dfrac{1 \text{ mol He}}{4.00 \text{ g}} = 1.00 \text{ mol He}$

$V_{\text{either gas}} = (1.00 \text{ mol})(0.0821 \dfrac{\text{L} \cdot \text{atm}}{\text{mol} \cdot \text{K}}) \dfrac{(T)}{(P)}$

As long as both gases are at the same T and P, the volumes will be identical.

101. $P = 267 \text{ torr} \times \dfrac{1 \text{ atm}}{760 \text{ torr}} = 0.351 \text{ atm}$, $V = 255 \text{ mL} \times \dfrac{1 \text{ L}}{1000 \text{ mL}} = 0.255 \text{ L}$,

n = ?, T = 25 + 273 = 298 K, mass = 143.289 − 143.187 = 0.102 g

$PV = nRT \Rightarrow (0.351 \text{ atm})(0.255 \text{ L}) = (n)(0.0821 \dfrac{\text{L} \cdot \text{atm}}{\text{mol} \cdot \text{K}})(298 \text{ K})$

$n = \dfrac{(0.351 \text{ atm})(0.255 \text{ L})}{(0.0821 \dfrac{\text{L} \cdot \text{atm}}{\text{mol} \cdot \text{K}})(298 \text{ K})} = 0.00366 \text{ moles}$

molar mass = $\dfrac{0.102 \text{ g}}{0.00366 \text{ moles}} = 27.9 \text{ g/mol}$

102. $P = 768 \text{ torr} \times \dfrac{1 \text{ atm}}{760 \text{ torr}} = 1.01 \text{ atm}$, $V = 118 \text{ mL} \times \dfrac{1 \text{ L}}{1000 \text{ mL}} = 0.118 \text{ L}$,

n = ?, T = 35 + 273 = 308 K, mass = 97.171 − 97.129 = 0.042 g

$PV = nRT \Rightarrow (0.101 \text{ atm})(0.118 \text{ L}) = (n)(0.0821 \dfrac{\text{L} \cdot \text{atm}}{\text{mol} \cdot \text{K}})(308 \text{ K})$

$n = \dfrac{(1.01 \text{ atm})(0.118 \text{ L})}{(0.0821 \dfrac{\text{L} \cdot \text{atm}}{\text{mol} \cdot \text{K}})(308 \text{ K})} = 4.71 \times 10^{-4} \text{ moles}$

molar mass (g/mol) = $\dfrac{\text{g of gas}}{\text{moles of gas}} = \dfrac{0.042 \text{ g}}{4.71 \times 10^{-4} \text{ mol}} = 89 \text{ g/mol}$

The molar mass of He is 4.00 g/mol ∴ the same is not pure He.

103. $P = 556 \text{ mm Hg} \times \dfrac{1 \text{ atm}}{760 \text{ mm Hg}} = 0.732 \text{ atm}$, $V = 158 \text{ mL} \times \dfrac{1 \text{ L}}{1000 \text{ mL}} = 0.158 \text{ L}$,

n = ?, T = 25 + 273 = 298 K, mass = 0.275 g

$PV = nRT \Rightarrow (0.732 \text{ atm})(0.158 \text{ L}) = (n)(0.0821 \dfrac{\text{L} \cdot \text{atm}}{\text{mol} \cdot \text{K}})(298 \text{ K})$

$n = \dfrac{(0.732 \text{ atm})(0.158 \text{ L})}{(0.0821 \dfrac{\text{L} \cdot \text{atm}}{\text{mol} \cdot \text{K}})(298 \text{ K})} = 0.00473 \text{ moles}$

molar mass = $\dfrac{0.275 \text{ g}}{0.00473 \text{ moles}} = 58.1 \text{ g/mol}$

In 1 mole of compound,

mass C = $58.1 \times 0.8266 = 48.0 \text{ g} \Rightarrow 48.0 \text{ g} \times \dfrac{1 \text{ mole C}}{12.01 \text{ g}} = 4 \text{ mol C}$

mass H = $58.1 \times 0.1734 = 10.1 \text{ g} \Rightarrow 10.1 \text{ g} \times \dfrac{1 \text{ mol H}}{1.01 \text{ g}} = 10 \text{ mol H}$

Molecular formula: C_4H_{10}

104. $n = 258 \text{ mL} \times \dfrac{1 \text{ L}}{1000 \text{ mL}} \times \dfrac{1 \text{ mol}}{22.4 \text{ L}} = 0.0115 \text{ mol}$; molar mass $= \dfrac{0.646 \text{ g}}{0.0115 \text{ mol}} = 56.2 \dfrac{\text{g}}{\text{mol}}$

In 1 mole of compound:

$\text{mass C} = 56.2 \times 0.8563 = 48.1 \text{ g} \Rightarrow 48.1 \text{ g} \times \dfrac{1 \text{ mol C}}{12.01 \text{ g}} = 4.00 \text{ mol C}$

$\text{mass H} = 56.2 \times 0.1437 = 8.08 \text{ g} \Rightarrow 8.08 \text{ g} \times \dfrac{1 \text{ mol H}}{1.01 \text{ g}} = 8.00 \text{ mol H}$

Molecular formula: C_4H_8

105. $P_{tot} = P_{H_2O} + P_{H_2} \Rightarrow 748 \text{ mm Hg} = 23.8 + P_{H_2} \Rightarrow P_{H_2} = 748 - 23.8 = 724 \text{ mm Hg}$

$P = 724 \text{ mm Hg} \times \dfrac{1 \text{ atm}}{760 \text{ mm Hg}} = 0.953 \text{ atm}, T = 25 + 273 = 298 \text{ K}, n = ?$

$V = 325 \text{ mL} \times \dfrac{1 \text{ L}}{1000 \text{ mL}} = 0.325 \text{ L}$

$PV = nRT \Rightarrow (0.953 \text{ atm})(0.325 \text{ L}) = (n)(0.0821 \text{ L} \cdot \text{atm/mol} \cdot \text{K})(298 \text{ K})$

$n = \dfrac{(0.953 \text{ atm})(0.325 \text{ L})}{(0.0821 \text{ L} \cdot \text{atm/mol} \cdot \text{K})(298 \text{ K})} = 0.0127 \text{ mol H}_2$

$0.0127 \text{ mol H}_2 \times \dfrac{1 \text{ mol Zn}}{1 \text{ mol H}_2} \times \dfrac{65.39 \text{ g}}{1 \text{ mol Zn}} = 0.830 \text{ g Zn}$

106. $P_{tot} = P_{H_2O} + P_{O_2} \Rightarrow 745 \text{ mm Hg} = 55.3 + P_{O_2} \Rightarrow P_{O_2} = 745 - 55.3 = 6.90 \times 10^2 \text{ mm Hg}$

$P = 6.90 \times 10^2 \text{ mm Hg} \times \dfrac{1 \text{ atm}}{760 \text{ mm Hg}} = 0.908 \text{ atm}, T = 40 + 273 = 313 \text{ K}, V = ?$

$n = 24.78 \text{ g NiO} \times \dfrac{1 \text{ mol NiO}}{74.69 \text{ g}} \times \dfrac{1 \text{ mol O}_2}{2 \text{ mol NiO}} = 0.166 \text{ mol O}_2$

$PV = nRT \Rightarrow (0.908 \text{ atm})(V) = (0.166 \text{ mol})(0.0821 \text{ L} \cdot \text{atm/mol} \cdot \text{K})(313 \text{ K})$

$V = \dfrac{(0.166 \text{ mol})(0.0821 \text{ L} \cdot \text{atm/mol} \cdot \text{K})(313 \text{ K})}{(0.908 \text{ atm})} = 4.70 \text{ L}$

107. $P_{tot} = P_{H_2O} + P_{H_2} \Rightarrow$ 748 torr = 55.3 + $P_{H_2} \Rightarrow P_{H_2}$ = 748 − 55.3 = 693 torr

$P = 693 \text{ mm Hg} \times \dfrac{1 \text{ atm}}{760 \text{ mm Hg}} = 0.912$ atm

T = 40 + 273 = 313 K, V = 1.78 L, n = ?

$PV = nRT \Rightarrow (0.912 \text{ atm})(1.78 \text{ L}) = (n)(0.0821 \text{ L} \cdot \text{atm/mol} \cdot \text{K})(313 \text{ K})$

$n = \dfrac{(0.912 \text{ atm})(1.78 \text{ L})}{(0.0821 \text{ L} \cdot \text{atm/mol} \cdot \text{K})(313 \text{ K})} = 0.0632 \text{ mol } H_2$

$0.0632 \text{ mol } H_2 \times \dfrac{2.02 \text{ g}}{1 \text{ mol } H_2} = 0.128 \text{ g } H_2$

108. $P_{tot} = P_{H_2O} + P_{O_2} \Rightarrow$ 697 torr = 23.8 + $P_{O_2} \Rightarrow P_{O_2}$ = 697 − 23.8 = 673 torr

$P = 673 \text{ torr} \times \dfrac{1 \text{ atm}}{760 \text{ mm Hg}} = 0.886$ atm,

T = 25 + 273 = 298 K, n = ?

$V = 235 \text{ mL} \times \dfrac{1 \text{ L}}{1000 \text{ mL}} = 0.235$ L

$PV = nRT \Rightarrow (0.886 \text{ atm})(0.235 \text{ L}) = (n)(0.0821 \text{ L} \cdot \text{atm/mol} \cdot \text{K})(298 \text{ K})$

$n = \dfrac{(0.886 \text{ atm})(0.235 \text{ L})}{(0.0821 \text{ L} \cdot \text{atm/mol} \cdot \text{K})(298 \text{ K})} = 0.00851 \text{ mol } O_2$

$0.00851 \text{ mol } O_2 \times \dfrac{32.00 \text{ g}}{1 \text{ mol } O_2} = 0.272 \text{ g } O_2$

109. $P_{tot} = P_{H_2O} + P_{O_2} \Rightarrow$ 752 mm Hg = 23.8 + $P_{O_2} \Rightarrow P_{O_2}$ = 752 − 23.8 = 728 mm Hg

$P = 728 \text{ torr} \times \dfrac{1 \text{ atm}}{760 \text{ mm Hg}} = 0.958$ atm,

T = 25 + 273 = 298 K, V = ?

$n = 15.8 \text{ g Ag} \times \dfrac{1 \text{ mol Ag}}{107.9 \text{ g}} \times \dfrac{1 \text{ mol } O_2}{4 \text{ mol Ag}} = 0.0366$ mol

$PV = nRT \Rightarrow (0.958 \text{ atm})(V) = (0.0366)(0.0821 \text{ L} \cdot \text{atm/mol} \cdot \text{K})(298 \text{ K})$

$V = \dfrac{(0.0366)(0.0821 \text{ L} \cdot \text{atm/mol} \cdot \text{K})(298 \text{ K})}{(0.958 \text{ atm})} = 0.935$ L

110. $2.45 \text{ kg CO} \times \dfrac{1000 \text{ g}}{1 \text{ kg}} \times \dfrac{1 \text{ mol CO}}{28.01 \text{ g}} = 87.5 \text{ mol CO}$

$87.5 \text{ mol CO} \times \dfrac{1 \text{ mol CO}_2}{1 \text{ mol CO}} \times \dfrac{22.4 \text{ L}}{1 \text{ mol CO}_2} = 1.96 \times 10^3 \text{ L of CO}_2$

$87.5 \text{ mol CO} \times \dfrac{1 \text{ mol H}_2}{1 \text{ mol CO}} \times \dfrac{22.4 \text{ L}}{1 \text{ mol H}_2} = 1.96 \times 10^3 \text{ L of H}_2$

Total volume (L) of products:

$1.96 \times 10^3 \text{ L} + 1.96 \times 10^3 \text{ L} = 3.92 \times 10^3 \text{ L}$

111. $HCl(aq) + NaHCO_3(s) \rightarrow CO_2(g) + NaCl(aq) + H_2O(l)$

$T = 22.7 + 273 = 296 \text{ K}, \quad V = 28.2 \text{ mL} \times \dfrac{1 \text{ L}}{1000 \text{ mL}} = 0.0282 \text{ L}$

$PV = nRT \Rightarrow (0.954 \text{ atm})(0.0282 \text{ L}) = (n)(0.0821 \dfrac{\text{L} \cdot \text{atm}}{\text{mol} \cdot \text{K}})(295.9 \text{ K}) \Rightarrow$

$n = \dfrac{(0.954 \text{ atm})(0.0282 \text{ L})}{(0.0821 \dfrac{\text{L} \cdot \text{atm}}{\text{mol} \cdot \text{K}})(295.9 \text{ K})} = 1.11 \times 10^{-3} \text{ mol CO}_2$

$1.11 \times 10^{-3} \text{ mol CO}_2 \times \dfrac{1 \text{ mol NaHCO}_3}{1 \text{ mol CO}_2} \times \dfrac{84.01 \text{ g}}{1 \text{ mol NaHCO}_3}$

$= 0.0933 \text{ g NaHCO}_3$

112. $2 \text{ HCl}(aq) + K_2S(s) \rightarrow H_2S(g) + 2 \text{ KCl}(aq)$

$T = 25.8 + 273 = 299 \text{ K}, \quad V = 42.9 \text{ mL} \times \dfrac{1 \text{ L}}{1000 \text{ mL}} = 0.0429 \text{ L}$

$PV = nRT \Rightarrow (752 \text{ torr} \times \dfrac{1 \text{ atm}}{760 \text{ torr}})(0.0429 \text{ L}) = (n)(0.0821 \dfrac{\text{L} \cdot \text{atm}}{\text{mol} \cdot \text{K}})(299 \text{ K}) \Rightarrow$

$n = \dfrac{(0.989 \text{ atm})(0.0429 \text{ L})}{(0.0821 \dfrac{\text{L} \cdot \text{atm}}{\text{mol} \cdot \text{K}})(299 \text{ K})} = 1.73 \times 10^{-3} \text{ mol H}_2S$

$1.73 \times 10^{-3} \text{ mol H}_2S \times \dfrac{1 \text{ mol K}_2S}{1 \text{ mol H}_2S} \times \dfrac{110.27 \text{ g}}{1 \text{ mol K}_2S} = 0.191 \text{ g K}_2S$

113. a) $285.5 \text{ mL SO}_2 \times \dfrac{1 \text{ L}}{1000 \text{ mL}} \times \dfrac{1 \text{ mol SO}_2}{22.4 \text{ L}} \times \dfrac{2 \text{ mol SO}_3}{2 \text{ mol SO}_2} = 0.0127 \text{ mol SO}_3$

$158.9 \text{ mL O}_2 \times \dfrac{1 \text{ L}}{1000 \text{ mL}} \times \dfrac{1 \text{ mol O}_2}{22.4 \text{ L}} \times \dfrac{2 \text{ mol SO}_3}{1 \text{ mol O}_2} = 0.0142 \text{ mol SO}_3$

The limiting reactant is SO_2.

The theoretical yield is 0.0127 mol SO_3.

b) $187.2 \text{ mL SO}_3 \times \dfrac{1 \text{ L}}{1000 \text{ mL}} \times \dfrac{1 \text{ mol SO}_3}{22.4 \text{ L}} = 0.00836 \text{ mol SO}_3$

Percent Yield $= \dfrac{\text{Actual}}{\text{Theoretical}} \times 100\% \Rightarrow \dfrac{0.00836}{0.0127} \times 100\% = 65.8\%$

114. a) $88.6 \text{ L H}_2 \times \dfrac{1 \text{ mol H}_2}{22.4 \text{ L}} \times \dfrac{4 \text{ mol PH}_3}{6 \text{ mol H}_2} = 2.64 \text{ mol PH}_3$

$158.3 \text{ g P}_4 \times \dfrac{1 \text{ mol P}_4}{123.9 \text{ g}} \times \dfrac{4 \text{ mol PH}_3}{1 \text{ mol P}_4} = 5.11 \text{ mol PH}_3$

The limiting reactant is H_2.

The theoretical yield is 2.64 mol PH_3.

b) $48.3 \text{ L PH}_3 \times \dfrac{1 \text{ mol PH}_3}{22.4 \text{ L}} = 2.16 \text{ mol PH}_3$

Percent Yield $= \dfrac{\text{Actual}}{\text{Theoretical}} \times 100\% \Rightarrow \dfrac{2.16}{2.64} \times 100\% = 81.8\%$

115. a) $12.8 \text{ L NO}_2 \times \dfrac{1 \text{ mol NO}_2}{22.4 \text{ L}} \times \dfrac{2 \text{ mol HNO}_3}{3 \text{ mol NO}_2} = 0.381 \text{ mol HNO}_3$

$14.9 \text{ g H}_2\text{O} \times \dfrac{1 \text{ mol H}_2\text{O}}{18.02 \text{ g}} \times \dfrac{2 \text{ mol HNO}_3}{1 \text{ mol H}_2\text{O}} = 1.65 \text{ mol HNO}_3$

Limiting reactant: NO_2

Mass $= 0.381 \text{ mol HNO}_3 \times \dfrac{63.02 \text{ g}}{1 \text{ mol HNO}_3} = 24.0 \text{ g HNO}_3$

b) Percent Yield $= \dfrac{\text{Actual}}{\text{Theoretical}} \times 100\% \Rightarrow \dfrac{14.8}{24.0} \times 100\% = 61.7\%$

116. a) $P = 632 \text{ mm Hg} \times \dfrac{1 \text{ atm}}{760 \text{ mm Hg}} = 0.832 \text{ atm}$, $V = 84.8 \text{ L}$, $n = ?$,

   $T = 35 + 273 = 308 \text{ K}$

   $PV = nRT \Rightarrow (0.832 \text{ atm})(84.8 \text{ L}) = (n)(0.0821 \dfrac{\text{L} \cdot \text{atm}}{\text{mol} \cdot \text{K}})(308 \text{ K})$

   $n = \dfrac{(0.832 \text{ atm})(84.8 \text{ L})}{(0.0821 \dfrac{\text{L} \cdot \text{atm}}{\text{mol} \cdot \text{K}})(308 \text{ K})} = 2.79 \text{ mol O}_2$

   $2.79 \text{ mol O}_2 \times \dfrac{2 \text{ mol NO}_2}{1 \text{ mol O}_2} = 5.58 \text{ mol NO}_2$

   $158.2 \text{ g NO} \times \dfrac{1 \text{ mol NO}}{30.01 \text{ g}} \times \dfrac{2 \text{ mol NO}_2}{2 \text{ mol NO}} = 5.27 \text{ mol NO}_2$

   The limiting reagent is NO.

   b) $P = 0.832 \text{ atm}$, $V = 97.3 \text{ L}$, $n = ?$, $T = 308 \text{ K}$

   $PV = nRT \Rightarrow (0.832 \text{ atm})(97.3 \text{ L}) = (n)(0.0821 \dfrac{\text{L} \cdot \text{atm}}{\text{mol} \cdot \text{K}})(308 \text{ K})$

   $n = \dfrac{(0.832 \text{ atm})(97.3 \text{ L})}{(0.0821 \dfrac{\text{L} \cdot \text{atm}}{\text{mol} \cdot \text{K}})(308 \text{ K})} = 3.20 \text{ mol NO}_2$

   $\text{Percent Yield} = \dfrac{\text{Actual}}{\text{Theoretical}} \times 100\% \Rightarrow \dfrac{3.20}{5.27} \times 100\% = 60.7\%$

117. $11.83 \text{ g (NH}_4)_2\text{CO}_3 \times \dfrac{1 \text{ mol (NH}_4)_2\text{CO}_3}{96.11 \text{ g}} = 0.123 \text{ mol (NH}_4)_2\text{CO}_3$

   $0.123 \text{ mol (NH}_4)_2\text{CO}_3 \times \dfrac{4 \text{ mol gas}}{1 \text{ mol (NH}_4)_2\text{CO}_3} = 0.492 \text{ mol gas}$

   $T = 22 + 273 = 295 \text{ K}$

   $PV = nRT \Rightarrow (1.02 \text{ atm})(V) = (0.492 \text{ mol})(0.0821 \dfrac{\text{L} \cdot \text{atm}}{\text{mol} \cdot \text{K}})(295 \text{ K}) \Rightarrow$

   $V = \dfrac{(0.492 \text{ mol})(0.0821 \dfrac{\text{L} \cdot \text{atm}}{\text{mol} \cdot \text{K}})(295 \text{ K})}{(1.02 \text{ atm})} = 11.7 \text{ L}$

118. $1.55 \text{ kg NH}_4\text{NO}_3 \times \dfrac{1000 \text{ g}}{1 \text{ kg}} \times \dfrac{1 \text{ mol NH}_4\text{NO}_3}{80.06 \text{ g}} = 19.4 \text{ mol NH}_4\text{NO}_3$

$19.4 \text{ mol NH}_4\text{NO}_3 \times \dfrac{7 \text{ mol gas}}{2 \text{ mol NH}_4\text{NO}_3} = 67.9 \text{ mol gas}$

$PV = nRT \Rightarrow (\dfrac{748 \text{ mm Hg}}{760 \text{ mm Hg/atm}})(V) = (67.9 \text{ mol})(0.0821 \dfrac{\text{L} \cdot \text{atm}}{\text{mol} \cdot \text{K}})(298 \text{ K}) \Rightarrow$

$V = \dfrac{(67.8 \text{ mol})(0.0821 \dfrac{\text{L} \cdot \text{atm}}{\text{mol} \cdot \text{K}})(298 \text{ K})}{(0.984 \text{ atm})} = 1.69 \times 10^3 \text{ L}$

119. $235 \text{ mg He} \times \dfrac{1 \text{ g}}{1000 \text{ mg}} \times \dfrac{1 \text{ mol He}}{4.00 \text{ g}} = 0.0588 \text{ mol He}$

$325 \text{ mg Ne} \times \dfrac{1 \text{ g}}{1000 \text{ mg}} \times \dfrac{1 \text{ mol Ne}}{20.18 \text{ g}} = 0.0161 \text{ mol Ne}$

Total Moles = 0.0588 + 0.0161 = 0.0749 mol

$P_{He} = \dfrac{0.0588 \text{ mol He}}{0.0749 \text{ mol Total}} \times 453 \text{ torr} = 356 \text{ torr}$

120. $4.33 \text{ g CO}_2 \times \dfrac{1 \text{ mol CO}_2}{44.01 \text{ g}} = 0.0984 \text{ mol CO}_2$

$3.11 \text{ g CH}_4 \times \dfrac{1 \text{ mol CH}_4}{16.05 \text{ g}} = 0.194 \text{ mol CH}_4$

Total Moles = 0.0984 + 0.198 = 0.292 mol

$P_{CO_2} = \dfrac{0.0984 \text{ mol CO}_2}{0.292 \text{ mol Total}} \times 1.09 \text{ atm} = 0.367 \text{ atm}$

121. Initially, the flask is filled with equal pressures of SO_2 and O_2, for a total pressure of 0.20 atm. According to the reaction, 2 moles of SO_2 react with a single mole of O_2; therefore, the SO_2 will react completely, leaving one-half of the O_2 unreacted (0.05 atm). The 0.10 atm of SO_2 that reacted forms an equal amount of SO_3 gas; therefore, the final pressure of the vessel would be the unreacted O_2 and the newly formed SO_3, which is equal to 0.15 atm.

122. From the balanced reaction: CO reacts with 2× amount of H_2. The 112 torr of CO would react with 224 torr of the H_2 to produce 112 torr of the CH_3OH. As there is an excess of H_2 (282 torr), there would be 58 torr of H_2 remaining after the reaction. The reaction did not go to completion; therefore, the amount of CH_3OH actually produced is set equal to X, the amount of leftover CO = 112 − X, the amount of leftover H_2 = 58 + (224 − 2X). The final pressure is then equal to the sum of all gases: 196 torr = X + (112 − X) + (58 + (224 − 2X)). Solving X = 99. The % yield of the reaction is therefore 99/112 × 100% = 88.4%.

Highlight Problems

123. Choice c will have the greatest pressure because it has the highest number of gas particles that can collide with the container walls to create pressure.

124. The sketch should show the same number of gas particles in half the original volume. The particles are undergoing a greater number of collisions with the walls as the pressure has increased to 4 atm.

$$\frac{P_1V_1}{T_1} = \frac{P_2V_2}{T_2} \Rightarrow \frac{(1 \text{ atm})(1 \text{ L})}{(298 \text{ K})} = \frac{(P_2)(0.5 \text{ L})}{(523 \text{ K})} \Rightarrow$$

$$P_2 = \frac{(1 \text{ atm})(1 \text{ L})(523 \text{ K})}{(298 \text{ K})(0.5 \text{ L})} = 4 \text{ atm}$$

125. $11.8 \text{ L} \times \frac{1 \text{ mol N}_2}{22.4 \text{ L}} \times \frac{2 \text{ mol NaN}_3}{3 \text{ mol N}_2} \times \frac{65.02 \text{ g}}{1 \text{ mol NaN}_3} = 22.8 \text{ g NaN}_3$

126. $P = 125 \text{ psi} \times \frac{1 \text{ atm}}{14.7 \text{ psi}} = 8.50 \text{ atm}$, $V = 855 \text{ mL} \times \frac{1 \text{ L}}{1000 \text{ mL}} = 0.855 \text{ L}$,

    $n = ?$, $T = 25 + 273 = 298 \text{ K}$

    $PV = nRT \Rightarrow (8.50 \text{ atm})(0.855 \text{ L}) = (n)(0.0821 \frac{\text{L} \cdot \text{atm}}{\text{mol} \cdot \text{K}})(298 \text{ K})$

    $n = \frac{(8.50 \text{ atm})(0.855 \text{ L})}{(0.0821 \frac{\text{L} \cdot \text{atm}}{\text{mol} \cdot \text{K}})(298 \text{ K})} = 0.297 \text{ mol gas in tires}$

    mass of air: $0.297 \text{ mol} \times \frac{28.8 \text{ g Air}}{1 \text{ mol Air}} = 8.55 \text{ g air}$

    mass of He: $0.297 \text{ mol} \times \frac{4.00 \text{ g He}}{1 \text{ mol He}} = 1.19 \text{ g He}$

127. $\frac{V_1}{T_1} = \frac{V_2}{T_2} \Rightarrow \frac{(2.95 \text{ L})}{(298 \text{ K})} = \frac{(V_2)}{(77 \text{ K})} \Rightarrow V_2 = \frac{(77 \text{ K})(2.95 \text{ L})}{(298 \text{ K})} = 0.76 \text{ L}$

    The difference in the volumes (0.15 L) is due to the fact that gas behavior is no longer ideal at extremely low temperatures.

128. $T = 1155 + 273 = 1428 \text{ K}$

    $\frac{P_1}{T_1} = \frac{P_2}{T_2} \Rightarrow \frac{(755 \text{ mm Hg})}{(298 \text{ K})} = \frac{(P_2)}{(1428 \text{ K})} \Rightarrow$

    $P_2 = \frac{(755 \text{ mm Hg})(1428 \text{ K})}{(298 \text{ K})} = 3.62 \times 10^3 \text{ mm Hg}$

129.

Variables Related	Name of Law	Proportionality Expression	Equality Expression	Held Constant
V, P	Boyle's law	$V \propto 1/P$	$P_1V_1 = P_2V_2$	n, T
V, T	Charles's law	$V \propto T$	$V_1/T_1 = V_2/T_2$	n, P
V, n	Avogadro's law	$V \propto n$	$V_1/n_1 = V_2/n_2$	P, T
P, T	Gay-Lussac's law	$P \propto T$	$P_1/T_1 = P_2/T_2$	V, n
P, V, T	combined gas law	$V \propto T/P$	$P_1V_1/T_1 = P_2V_2/T_2$	n

130. $PV = nRT$

$P = 1.2$ atm

$V = ?$

$n = 10.4 \text{ g CO}_2 \times \dfrac{1 \text{ mol CO}_2}{44.01 \text{ g}} = 0.236 \text{ mol CO}_2$

$R = 0.0821$ L·atm/mol·K

$T = 29 \,°C + 273 = 302$ K

$V = \dfrac{nRT}{P} = \dfrac{(0.236 \text{ mol})(0.0821 \text{ L·atm/mol·K})(302 \text{ K})}{1.2 \text{ atm}} = 4.9$ L

131. Calculate moles of hydrogen gas produced from 14.22 g Al:

$14.22 \text{ g Al} \times \dfrac{1 \text{ mol Al}}{26.98 \text{ g}} \times \dfrac{3 \text{ mol H}_2}{2 \text{ mol Al}} = 0.7906 \text{ mol H}_2$

Calculate volume of hydrogen gas using the ideal gas law:

$PV = nRT$

$P = 749 \text{ mm Hg} \times \dfrac{1 \text{ atm}}{760 \text{ mm Hg}} = 0.986$ atm

$V = ?$

$n = 0.7906$ mol H$_2$

$R = 0.0821$ L·atm/mol·K

$T = 31 \,°C + 273 = 304$ K

$V = \dfrac{nRT}{P} = \dfrac{(0.7906 \text{ mol})(0.0821 \text{ L·atm/mol·K})(304 \text{ K})}{0.986 \text{ atm}} = 20.0$ L

The vapor pressure of water at 30 °C is 31.8 mm Hg (Table 11.4). The volume of hydrogen gas is 20.0 L. This problem assumed that there is no water vapor present; however, if water vapor was present, the volume would be identical as gases mix perfectly.

Data Interpretation and Analysis

132. (a) No
    (b) Inverse relationship (when one was high, the other was low)
    (c) 1:1
    (d) During the day, ozone is being produced by traffic as well as NO (not shown on the graph), which will begin to react together as the day proceeds. Toward the end of the day, traffic comes to a minimum, and as such the production of ozone and nitrogen monoxide trails off. However, the reaction between the two gases starts to produce nitrogen dioxide, which continues into the evening. Thus, the ozone levels decrease as the nitrogen dioxide increases.

Liquids, Solids, and Intermolecular Forces

12

Questions

1. Intermolecular forces are attractive forces that occur between individual molecules. Living organisms depend on intermolecular forces not only for taste but also for many other physiological processes. For example, intermolecular forces help determine the shapes of protein molecules and are central to DNA, the inheritable molecules that serve as blueprints for life.

2. The surface tension of water pulls the water into a spherical shape because it minimizes the surface area to volume ratio.

3. The relative magnitude of the intermolecular forces to thermal energy determines whether a substance is a solid, liquid, or gas.

4. Properties of Liquids
   - High densities in comparison to gases.
     - Liquids have high densities in comparison to gases because the atoms or molecules that compose liquids are much closer together.
   - Indefinite shape; they assume the shape of their container.
     - Liquids assume the shape of their containers because the atoms or molecules that compose them are free to flow.
   - Definite volume; they are not easily compressed.
     - Liquids are not easily compressed because the molecules or atoms that compose them are in close contact—they cannot be pushed closer together.

5. Properties of Solids
   - High densities in comparison to gases.
     - Solids have high densities in comparison to gases because the atoms or molecules that compose solids are also close together.
   - Definite shape; they do not assume the shape of their container.
     - The molecules or atoms that compose solids are fixed in place.
   - Definite volume; they are not easily compressed.
     - Solids have a definite volume and cannot be compressed because the molecules or atoms composing them are in close contact.
   - May be crystalline (ordered) or amorphous (disordered).
     - Solids may be crystalline, in which case the atoms or molecules that compose them arrange themselves in a well-ordered, three-dimensional array, or they may be amorphous, in which case the atoms or molecules that compose them have no long-range order.

6. In an amorphous solid, the solid particles have no long-range order. In a crystalline solid, the solid particles arrange themselves in a well-ordered, three-dimensional array.

7. Surface tension is the tendency of liquids to minimize their surface area due to the interaction of molecules between each other. The surface tension increases when intermolecular forces increase.

8. Viscosity is a liquid's resistance to flow and depends on intermolecular forces and molecular shape. The viscosity of a liquid increases with increasing intermolecular forces.

9. Evaporation occurs when a liquid undergoes a physical change to a gas. Condensation is the physical change that takes place when a gaseous substance changes to a liquid form.

10. A glass of water evaporates more slowly than water spilled on the table because the spilled water has a larger surface area for evaporation.

11. The process of evaporation below the boiling point only occurs at the surface of the liquid. At the boiling point, there is sufficient thermal energy that molecules within the interior of the liquid can break free and enter into the gas phase.

12. The boiling point is the temperature at which the vapor pressure of a liquid is equal to the pressure above the solution. The normal boiling point is the boiling point of a liquid when the external pressure is 1 atmosphere.

13. The intermolecular forces in acetone are weaker; therefore, it evaporates faster and is the more volatile compound.

14. Condensation, the opposite process of evaporation, corresponds to the physical process of a gaseous compound being converted into the liquid state. Dynamic equilibrium is when the condensation process is proceeding at exactly the same rate as the evaporation process in a system.

15. Vapor pressure is the partial pressure of a gas in dynamic equilibrium with its liquid. The vapor pressure of a compound increases with increasing temperature and decreases with increasing strength of the intermolecular forces.

16. You feel cooler when you sweat because the process of evaporation is endothermic. That means that heat from your body is absorbed by the sweat as it evaporates, thereby cooling you.

17. The process of condensation is the opposite of evaporation in that it is exothermic. This means that energy is released from the gas as it forms a liquid. Steam at 100 °C releases more heat into your hand due to condensation than water at the same temperature.

18. When a liquid boils, there is sufficient thermal energy in a system that the liquid molecules in the interior of the solution can go into the gas phase. These gaseous molecules form bubbles in solution that float to the surface.

19. The process of freezing is exothermic, which releases heat from the water into the freezer. If the freezer cannot remove the excess heat, the freezer will start to warm up and the water will not freeze.

20. The ice absorbs heat from the liquid in order to melt, which lowers the temperature of the liquid.

21. Melting ice is endothermic as the ice must absorb energy from the surroundings. The sign of ΔH is positive. Freezing water is an exothermic process as energy is released from the water into the surroundings and the sign of ΔH is negative.

22. The boiling of water is endothermic as the water must absorb energy from the surroundings. The sign of ΔH is positive. Condensation of steam is an exothermic process as energy is released from the water into the surroundings and the sign of ΔH is negative.

23. Dispersion force (aka London force) is a type of intermolecular force present between all molecules and atoms. This type of intermolecular force arises from fluctuations in the spatial distribution of electrons in the atom or molecule that causes a temporary positive and negative charge within a particle. The dispersion force increases with increasing molar mass.

24. Dipole–dipole force exists in all polar molecules. The polar molecules have a permanent dipole (a positive and negative end) that interacts with neighboring dipoles. Polar compounds will have higher melting and boiling points than nonpolar compounds of similar molar mass.

25. The hydrogen bond occurs when hydrogen bonds to F, O, or N (very electronegative), which in turn pulls the shared electrons in the bond away from hydrogen. Hydrogen becomes very positive and forms a hydrogen bond to the lone pair electrons on F, O, or N of another molecule. Compounds that can form hydrogen bonds will have higher melting and boiling points than other compounds that do not form hydrogen bonds.

26. The ion–dipole force occurs in mixtures of ionic compounds and polar compounds. An important group of mixtures would be aqueous solutions of ionic compounds.

27. Dispersion forces < dipole–dipole < hydrogen bond < ion–dipole

28. Molecular solids are those solids that are composed of molecules, such as ice. This type of solid is held together by intermolecular forces (dispersion, dipole–dipole, hydrogen bonding).

29. Molecular solids as a whole tend to have low to moderately low melting points relative to other types of solids; however, strong molecular forces can increase their melting points relative to each other.

30. Ionic solids are solids composed of the smallest electrically neutral collection of cations and anions that compose a compound (i.e., the formula unit). Ionic solids are held together by electrostatic attraction.

31. Ionic solids tend to have much higher melting points relative to the melting points of other types of solids.

32. Atomic solids are solids whose composite units are atoms of a single element. Atomic solids can be broken down into three categories: covalent, nonbonding, and metallic. Covalent atomic solids have high melting points. Nonbonding atomic solids have very low melting points. Metallic atomic solids have a range of low and high melting points.

33. Water is unique because it has a low molar mass and is a liquid at room temperature, which is not the case with other low molar mass compounds. Additionally, while most compounds contract during the freezing process, water expands. This is the reason ice floats on water.

34. If ice was denser than water, it would sink to the bottom of liquid water. This would allow lakes to freeze solid, destroying all aquatic life.

Problems

Evaporation, Condensation, Melting, and Freezing

35. The second beaker of 55 mL in a 12-cm diameter dish will evaporate more quickly because evaporation occurs at the surface and the 12-cm dish has a larger surface area.

36. The first dish has a larger diameter, resulting in a surface area approximately three times larger than the second dish.

37. Acetone will feel cooler because it has weaker intermolecular forces, which make it more volatile. It will evaporate faster and, therefore, will remove more heat from your hand.

38. Water is volatile and will evaporate from your skin, providing a cooling effect. The oil is essentially nonvolatile and will not evaporate.

39. The temperature will increase from −5 °C to the melting point at 0 °C. The temperature will not increase until all of the ice has melted. After the ice has completely melted, the temperature will increase until it reaches room temperature (25 °C).

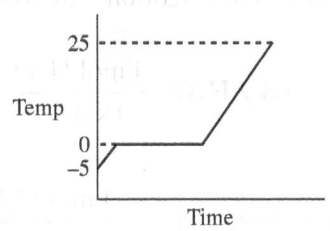

40. The temperature will increase from 25 °C until it reaches the boiling point at 100 °C. The temperature of the water will not increase from the boiling point until all of the water has been converted to steam. The temperature of the steam could then continue to rise if it did not escape from the system.

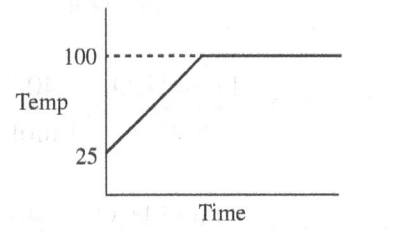

41. The steam would cause a more severe burn because steam condensation is an exothermic process and this excess heat would increase the severity of the burn.

42. Towns near the sea have milder nighttime temperatures because water has a relatively high heat capacity (about four times higher) compared to the land. At night when the temperature of the air decreases, heat that is stored within the water is transferred into the colder air, resulting in warmer air. Towns that are inland lack a source of heat at night and therefore have colder temperatures.

43. The water in the bag is undergoing the freezing process, which releases heat (exothermic) into the rest of the cooler.

44. The water is undergoing the freezing process, which releases heat (exothermic) into the freezer. The freezer must run extensively to remove this excess heat.

45. The ice chest full of ice at 0 °C would be colder because in order to melt the ice, heat from the cooler would be absorbed (endothermic). The ice chest full of water at 0 °C would be warmer because the freezing process releases heat (exothermic) and would warm the cooler.

46. The 50 g of water will warm up faster than the 50 g ice/water mixture because the heat absorbed in the ice/water mixture goes to melting the ice before it would go to increasing the water temperature.

47. The boiling point occurs when the vapor pressure of the liquid is equal to the external pressure. Because of the altitude of Denver (1 mile above sea level), the atmospheric pressure is less than 1 atm. Therefore, the vapor pressure will match the atmospheric pressure at a lower temperature.

48. The boiling point occurs when the vapor pressure of the liquid is equal to the external pressure. Because pressure decreases with increased altitude, the atmospheric pressure on Mount Everest is significantly less than 1 atm.

Heat of Vaporization and Heat of Fusion

49. $33.8 \text{ g H}_2\text{O} \times \dfrac{1 \text{ mol H}_2\text{O}}{18.02 \text{ g}} \times \dfrac{40.7 \text{ kJ}}{1 \text{ mol H}_2\text{O}} = 76.3 \text{ kJ}$

50. $43.9 \text{ g C}_3\text{H}_6\text{O} \times \dfrac{1 \text{ mol C}_3\text{H}_6\text{O}}{58.09 \text{ g}} \times \dfrac{29.1 \text{ kJ}}{1 \text{ mol C}_3\text{H}_6\text{O}} = 22.0 \text{ kJ}$

51. $2.8 \text{ g H}_2\text{O} \times \dfrac{1 \text{ mol H}_2\text{O}}{18.02 \text{ g}} \times \dfrac{40.7 \text{ kJ}}{1 \text{ mol H}_2\text{O}} = 6.3 \text{ kJ}$

52. $4.86 \text{ g H}_2\text{O} \times \dfrac{1 \text{ mol H}_2\text{O}}{18.02 \text{ g}} \times \dfrac{40.7 \text{ kJ}}{1 \text{ mol H}_2\text{O}} = 11.0 \text{ kJ}$

53. $4.25 \text{ g H}_2\text{O} \times \dfrac{1 \text{ mol H}_2\text{O}}{18.02 \text{ g}} \times \dfrac{44.0 \text{ kJ}}{1 \text{ mol H}_2\text{O}} = 10.4 \text{ kJ}$

54. $65.6 \text{ g C}_3\text{H}_8\text{O} \times \dfrac{1 \text{ mol C}_3\text{H}_8\text{O}}{60.11 \text{ g}} \times \dfrac{45.4 \text{ kJ}}{1 \text{ mol C}_3\text{H}_6\text{O}} = 49.5 \text{ kJ}$

55. $835 \text{ kJ} \times \dfrac{1 \text{ mol H}_2\text{O}}{40.7 \text{ kJ}} \times \dfrac{18.02 \text{ g}}{1 \text{ mol H}_2\text{O}} = 3.70 \times 10^2 \text{ g H}_2\text{O}$

56. $1078 \text{ kJ} \times \dfrac{1 \text{ mol H}_2\text{O}}{40.7 \text{ kJ}} \times \dfrac{18.02 \text{ g}}{1 \text{ mol H}_2\text{O}} \times \dfrac{1 \text{ mL}}{1.0 \text{ g}} \times \dfrac{1 \text{ L}}{1000 \text{ ml}} = 0.48 \text{ L H}_2\text{O}$

57. $37.4 \text{ g H}_2\text{O} \times \dfrac{1 \text{ mol H}_2\text{O}}{18.02 \text{ g}} \times \dfrac{6.02 \text{ kJ}}{1 \text{ mol H}_2\text{O}} = 12.5 \text{ kJ}$

58. $23.9 \text{ g C}_4\text{H}_{10}\text{O} \times \dfrac{1 \text{ mol C}_4\text{H}_{10}\text{O}}{74.14 \text{ g}} \times \dfrac{7.27 \text{ kJ}}{1 \text{ mol C}_4\text{H}_{10}\text{O}} = 2.34 \text{ kJ}$

59. $34.2 \text{ g H}_2\text{O} \times \dfrac{1 \text{ mol H}_2\text{O}}{18.02 \text{ g}} \times \dfrac{6.02 \text{ kJ}}{1 \text{ mol H}_2\text{O}} = 11.4 \text{ kJ}$

60. $2.55 \text{ kg C}_4\text{H}_{10}\text{O} \times \dfrac{1000 \text{ g}}{1 \text{ kg}} \times \dfrac{1 \text{ mol C}_4\text{H}_{10}\text{O}}{74.14 \text{ g}} \times \dfrac{7.27 \text{ kJ}}{1 \text{ mol C}_4\text{H}_{10}\text{O}} = 2.50 \times 10^2 \text{ kJ}$

61. Heat water from 28.0 °C to 100.0 °C:

$$2.55 \text{ g H}_2\text{O} \times 4.18 \text{ J/g} \cdot °\text{C} \times (100 °\text{C} - 28 °\text{C}) \times \frac{1 \text{ kJ}}{1000 \text{ J}} = 0.77 \text{ kJ}$$

Vaporize water at 100 °C:

$$2.55 \text{ g H}_2\text{O} \times \frac{1 \text{ mol H}_2\text{O}}{18.02 \text{ g}} \times \frac{40.7 \text{ kJ}}{1 \text{ mol H}_2\text{O}} = 5.76 \text{ kJ}$$

Total: 0.77 kJ + 5.76 kJ = 6.53 kJ

62. Warm ice from −12 °C to 0 °C:

$$5.88 \text{ g H}_2\text{O} \times 2.09 \text{ J/g} \cdot °\text{C} \times (0 °\text{C} - 12.0 °\text{C}) \times \frac{1 \text{ kJ}}{1000 \text{ J}} = 0.147 \text{ kJ}$$

Melt the ice at 0 °C:

$$5.88 \text{ g H}_2\text{O} \times \frac{1 \text{ mol H}_2\text{O}}{18.02 \text{ g}} \times \frac{6.02 \text{ kJ}}{1 \text{ mol H}_2\text{O}} = 1.96 \text{ kJ}$$

Warm ice from 0 °C to 25 °C:

$$5.88 \text{ g H}_2\text{O} \times 4.18 \text{ J/g} \cdot °\text{C} \times (25.0 °\text{C} - 0 °\text{C}) \times \frac{1 \text{ kJ}}{1000 \text{ J}} = 0.614 \text{ kJ}$$

Total: 0.147 kJ + 1.96 kJ + 0.614 kJ = 2.73 kJ

Intermolecular Forces

63. a) dispersion
    b) dispersion
    c) dispersion, dipole–dipole
    d) dispersion, hydrogen bond, dipole–dipole

64. a) dispersion, dipole–dipole
    b) dispersion, dipole–dipole, hydrogen bond
    c) dispersion
    d) dispersion

65. a) dispersion, dipole–dipole
    b) dispersion, dipole–dipole, hydrogen bond
    c) dispersion
    d) dispersion

66. a) dispersion
    b) dispersion, dipole–dipole
    c) dispersion, dipole–dipole, hydrogen bond
    d) dispersion

67. Ion–dipole forces form between the ions and the water molecules. Hydrogen bonding would also be present between the water molecules.

68. Ion–dipole forces form between the ions and the water molecules. Hydrogen bonding would also be present between the water molecules.

69. Choice d would have the highest boiling point because it has a larger molar mass, which indicates stronger dispersion forces.

70. The only intermolecular force present in noble gases is dispersion force, which increases with increasing molecular weight. Hence, radon will have the highest boiling point.

71. The CH_3OH compound will be a liquid at room temperature because it will have hydrogen bonding, dipole–dipole, and dispersion intermolecular forces, while CH_3SH would contain only dipole–dipole and dispersion intermolecular forces.

72. The CH_3CH_2OH compound will be a liquid at room temperature because it will have hydrogen bonding, dipole–dipole, and dispersion intermolecular forces, while CH_3OCH_3 would contain only dipole–dipole and dispersion intermolecular forces.

73. The first liquid that would start to form is NH_3 because of its ability to form hydrogen bonds. CH_4 cannot form hydrogen bonds, so condensation would not start until a much lower temperature was reached.

74. CS_2 is a liquid at room temperature, and CO_2 remains in the gas phase because of the greater dispersion force present in CS_2 as compared to CO_2.

75. No, $CH_3CH_2CH_2CH_2CH_3$ is nonpolar where H_2O is polar and they are not miscible.

76. Yes, CH_3OH and H_2O are both polar compounds; therefore, they are miscible.

77. a) No, CCl_4 is nonpolar and H_2O is polar.
b) Yes, both compounds are nonpolar.
c) Yes, both compounds are polar.

78. a) Yes, both compounds are nonpolar.
b) No, CBr_4 is nonpolar and H_2O is polar.
c) No, Cl_2 is nonpolar and H_2O is polar.

Types of Solids

79. a) atomic
b) molecular
c) ionic
d) atomic

80. a) ionic
b) molecular
c) atomic
d) molecular

81. a) molecular
    b) ionic
    c) molecular
    d) molecular

82. a) molecular
    b) covalent atomic
    c) ionic
    d) metallic atomic

83. Choice c because ionic compounds tend to have high melting points due to the strong electrostatic attraction present in the solid.

84. Choice a because covalent atomic solids tend to have the highest melting point due to the strong covalent bond network that goes through the entire solid.

85. a) Ti(s) has a higher melting point because metallic atomic solids have stronger metallic bonds compared to the weak dispersion force that holds together nonbonding atomic solids like Ne(s).
    b) H_2O(s) has a higher melting point because of hydrogen bonds, which are not found in H_2S.
    c) Xe(s) has a higher melting point because of stronger dispersion forces found in larger atoms.
    d) NaCl(s) has a higher melting point because electrostatic attraction in ionic solids is stronger than intermolecular forces in covalent solids.

86. a) Fe(s) has a higher melting point because metallic bonds are stronger than the intermolecular forces in molecular solids.
    b) KCl(s) has a higher melting point because the electrostatic attraction in ionic solids is stronger than the intermolecular forces in molecular solids.
    c) TiO_2(s) has a higher melting point because the electrostatic attraction in ionic solids is stronger than the intermolecular forces in molecular solids.

87. Ne < SO_2 < NH_3 < H_2O < NaF

88. KF > CH_3OH > CO_2 > Ne

Cumulative Problems

89. a) $78 \text{ g H}_2\text{O} \times \dfrac{1 \text{ mol H}_2\text{O}}{18.02 \text{ g}} \times \dfrac{6.02 \text{ kJ}}{1 \text{ mol H}_2\text{O}} \times \dfrac{1000 \text{ J}}{1 \text{ kJ}} = 2.6 \times 10^4 \text{ J}$

b) $78 \text{ g H}_2\text{O} \times \dfrac{1 \text{ mol H}_2\text{O}}{18.02 \text{ g}} \times \dfrac{6.02 \text{ kJ}}{1 \text{ mol H}_2\text{O}} = 26 \text{ kJ}$

c) $78 \text{ g H}_2\text{O} \times \dfrac{1 \text{ mol H}_2\text{O}}{18.02 \text{ g}} \times \dfrac{6.02 \text{ kJ}}{1 \text{ mol H}_2\text{O}} \times \dfrac{1000 \text{ J}}{1 \text{ kJ}} \times \dfrac{1 \text{ cal}}{4.18 \text{ J}} = 6.2 \times 10^3 \text{ cal}$

d) $78 \text{ g H}_2\text{O} \times \dfrac{1 \text{ mol H}_2\text{O}}{18.02 \text{ g}} \times \dfrac{6.02 \text{ kJ}}{1 \text{ mol H}_2\text{O}} \times \dfrac{1000 \text{ J}}{1 \text{ kJ}} \times \dfrac{1 \text{ cal}}{4.18 \text{ J}} \times \dfrac{1 \text{ Cal}}{1000 \text{ cal}} = 6.2 \text{ Cal}$

90. a) $145 \text{ g H}_2\text{O} \times \dfrac{1 \text{ mol H}_2\text{O}}{18.02 \text{ g}} \times \dfrac{6.02 \text{ kJ}}{1 \text{ mol H}_2\text{O}} \times \dfrac{1000 \text{ J}}{1 \text{ kJ}} = 4.84 \times 10^4 \text{ J}$

b) $145 \text{ g H}_2\text{O} \times \dfrac{1 \text{ mol H}_2\text{O}}{18.02 \text{ g}} \times \dfrac{6.02 \text{ kJ}}{1 \text{ mol H}_2\text{O}} = 48.4 \text{ kJ}$

c) $145 \text{ g H}_2\text{O} \times \dfrac{1 \text{ mol H}_2\text{O}}{18.02 \text{ g}} \times \dfrac{6.02 \text{ kJ}}{1 \text{ mol H}_2\text{O}} \times \dfrac{1000 \text{ J}}{1 \text{ kJ}} \times \dfrac{1 \text{ cal}}{4.18 \text{ J}} = 1.16 \times 10^4 \text{ cal}$

d) $145 \text{ g H}_2\text{O} \times \dfrac{1 \text{ mol H}_2\text{O}}{18.02 \text{ g}} \times \dfrac{6.02 \text{ kJ}}{1 \text{ mol H}_2\text{O}} \times \dfrac{1000 \text{ J}}{1 \text{ kJ}} \times \dfrac{1 \text{ cal}}{4.18 \text{ J}} \times \dfrac{1 \text{ Cal}}{1000 \text{ cal}} = 11.6 \text{ Cal}$

91. $8.5 \text{ g H}_2\text{O} \times \dfrac{1 \text{ mol H}_2\text{O}}{18.02 \text{ g}} \times \dfrac{6.02 \text{ kJ}}{1 \text{ mol H}_2\text{O}} \times \dfrac{1000 \text{ J}}{1 \text{ kJ}} = 2.8 \times 10^3 \text{ J}$

$2.8 \times 10^3 \text{ J} = 255 \text{ g} \times 4.18 \text{ J/g} \cdot °\text{C} \times \Delta \text{T}$

$\Delta \text{T} = \dfrac{2.8 \times 10^3 \text{ J}}{255 \text{ g} \times 4.18 \text{ J/g} \cdot °\text{C}} = 2.7 \, °\text{C}$

92. $14.7 \text{ g H}_2\text{O} \times \dfrac{1 \text{ mol H}_2\text{O}}{18.02 \text{ g}} \times \dfrac{6.02 \text{ kJ}}{1 \text{ mol H}_2\text{O}} \times \dfrac{1000 \text{ J}}{1 \text{ kJ}} = 4.91 \times 10^3 \text{ J}$

$4.91 \times 10^3 \text{ J} = 324 \text{ g} \times 4.18 \text{ J/g} \cdot °\text{C} \times \Delta \text{T}$

$\Delta \text{T} = \dfrac{4.91 \times 10^3 \text{ J}}{324 \text{ g} \times 4.18 \text{ J/g} \cdot °\text{C}} = 3.63 \, °\text{C}$

93. $q = 352 \text{ g} \times 4.18 \text{ J/g} \cdot °\text{C} \times 25 \, °\text{C} = 3.68 \times 10^4 \text{ J}$

$3.68 \times 10^4 \text{ J} \times \dfrac{1 \text{ kJ}}{1000 \text{ J}} \times \dfrac{1 \text{ mol H}_2\text{O}}{6.02 \text{ kJ}} \times \dfrac{18.02 \text{ g}}{1 \text{ mol H}_2\text{O}} = 1.1 \times 10^2 \text{ g H}_2\text{O}$

94. $q = 55.8 \text{ g} \times 4.18 \text{ J/g} \cdot {}°\text{C} \times 55.0 \, {}°\text{C} = 1.28 \times 10^4 \text{ J}$

$1.28 \times 10^4 \text{ J} \times \dfrac{1 \text{ kJ}}{1000 \text{ J}} \times \dfrac{1 \text{ mol H}_2\text{O}}{6.02 \text{ kJ}} \times \dfrac{18.02 \text{ g}}{1 \text{ mol H}_2\text{O}} = 38.4 \text{ g H}_2\text{O}$

95. Cooling Steam: $18.02 \text{ g H}_2\text{O} \times 1.84 \text{ J/g} \cdot {}°\text{C} \times 45 \, {}°\text{C} \times \dfrac{1 \text{ kJ}}{1000 \text{ J}} = 1.5 \text{ kJ}$

Condensing Steam: $1 \text{ mol H}_2\text{O} \times 40.7 \text{ kJ/mol} = 40.7 \text{ kJ}$

Cooling Water: $18.02 \text{ g H}_2\text{O} \times 4.18 \text{ J/g} \cdot {}°\text{C} \times 1.00 \times 10^2 \, {}°\text{C} \times \dfrac{1 \text{ kJ}}{1000 \text{ J}} = 7.53 \text{ kJ}$

Freezing Water: $1 \text{ mol H}_2\text{O} \times \dfrac{6.02 \text{ kJ}}{1 \text{ mol H}_2\text{O}} = 6.02 \text{ kJ}$

Cooling Ice: $18.02 \text{ g H}_2\text{O} \times 2.09 \text{ J/g} \cdot {}°\text{C} \times 50.0 \, {}°\text{C} \times \dfrac{1 \text{ kJ}}{1000 \text{ J}} = 1.88 \text{ kJ}$

Total = 1.5 + 40.7 + 7.53 + 6.02 + 1.88 = 57.6 kJ

96. $10.0 \text{ g H}_2\text{O} \times \dfrac{1 \text{ mol}}{18.02 \text{ g H}_2\text{O}} = 0.555 \text{ mol H}_2\text{O}$

Warming Ice: $10.0 \text{ g H}_2\text{O} \times 2.09 \text{ J/g} \cdot {}°\text{C} \times 10.0 \, {}°\text{C} \times \dfrac{1 \text{ kJ}}{1000 \text{ J}} = 0.209 \text{ kJ}$

Melting Ice: $0.555 \text{ mol H}_2\text{O} \times \dfrac{6.02 \text{ kJ}}{1 \text{ mol H}_2\text{O}} = 3.34 \text{ kJ}$

Warming Water: $10.0 \text{ g H}_2\text{O} \times 4.18 \text{ J/g} \cdot {}°\text{C} \times 1.00 \times 10^2 \, {}°\text{C} \times \dfrac{1 \text{ kJ}}{1000 \text{ J}} = 4.18 \text{ kJ}$

Vaporizing Water: $0.555 \text{ mol H}_2\text{O} \times 40.7 \text{ kJ/mol} = 22.6 \text{ kJ}$

Warming Steam: $10.0 \text{ g H}_2\text{O} \times 1.84 \text{ J/g} \cdot {}°\text{C} \times 10.0 \, {}°\text{C} \times \dfrac{1 \text{ kJ}}{1000 \text{ J}} = 0.184 \text{ kJ}$

Total = 0.209 + 3.34 + 4.18 + 22.6 + 0.184 = 30.5 kJ

97. a) H—S̈e—H Bent Geometry, Dispersion & Dipole–Dipole Forces

b) Ö=S̈—Ö: Bent Geometry, Dispersion & Dipole–Dipole Forces

c) H—C—C̈l: Tetrahedral Geometry, Dispersion & Dipole–Dipole Forces
(with :C̈l: above and :C̈l: below the central C, and H on the left)

d) Ö=C=Ö Linear Geometry, Dispersion Forces

98. a) :Cl—B—Cl: Trigonal Planar Geometry, Dispersion Forces
       |
       :Cl:

   b) H—C=Ö Trigonal Planar Geometry, Dispersion & Dipole–Dipole Forces
        |
        H

   c) :S̈=C=S̈: Linear Geometry, Dispersion Forces

   d) :Cl—N̈—Cl: Trigonal Pyramidal Geometry, Dispersion &
        | Dipole–Dipole Forces
        :Cl:

99. $Na^+[:\ddot{F}:]^-$ $Mg^{2+}[:\ddot{O}:]^{2-}$

    Because the strength of a +2 to −2 attraction is stronger than a +1 to −1, MgO has the higher melting point.

100. $K^+[:\ddot{F}:]^-$ $Ca^{2+}[:\ddot{O}:]^{2-}$ Because the strength of a +2 to −2 attraction is stronger than a +1 to −1, CaO will have the higher melting point.

101. As the molecular weight increases from Cl to I, the greater the London dispersion forces present, which will increase the boiling point as observed. However, HF is the only compound listed that has the ability to form hydrogen bonds, which explains the anomaly in the trend.

102. The larger the compound, the stronger the London dispersion forces present between molecules. The trends would show that the larger the molecule, the higher the boiling point of the compound. Water is an exception because it has the ability to form hydrogen bonds, whereas the others cannot.

103. Heat lost (Bulk water) + Heat gained (Melting ice, Warming water from ice) = 0

$$4.18 \frac{J}{g \cdot °C} \times 550.0 \text{ g } H_2O \times (T_f - 28.0\,°C) + 23.5 \text{ g } H_2O \times \frac{1 \text{ mol } H_2O}{18.02 \text{ g}} \times \frac{6.02 \times 10^3 \text{ J}}{1 \text{ mol } H_2O}$$

Heat lost by water | Heat of fusion of ice (melting the ice)

$$+ 4.18 \frac{J}{g \cdot °C} \times 23.5 \text{ g } H_2O \times (T_f - 0\,°C) = 0$$

The water from the newly melted ice cube warms up

$(2.30 \times 10^3\, T_f - 6.44 \times 10^4) + (7.85 \times 10^3) + (98.2\, T_f - 0) = 0$

$2.40 \times 10^3\, T_f - 5.65 \times 10^4 = 0 \Rightarrow T_f = \frac{5.65 \times 10^4}{2.40 \times 10^3} = 23.5\,°C$

104. $\underbrace{\text{Heat lost}}_{\substack{\text{Heat of vaporization} \\ \text{Water formed} \\ \text{by steam condensing}}} + \underbrace{\text{Heat gained}}_{\text{Water}} = 0$

$\underbrace{1.10 \text{ g H}_2\text{O} \times \dfrac{1 \text{ mol H}_2\text{O}}{18.02 \text{ g}} \times \dfrac{-4.07 \times 10^4 \text{ J}}{1 \text{ mol H}_2\text{O}}}_{\text{Heat of vaporization (condensing the steam)}} +$

$\underbrace{4.18 \dfrac{\text{J}}{\text{g} \cdot {}^\circ\text{C}} \times 1.10 \text{ g H}_2\text{O} \times (T_f - 100.0 \, {}^\circ\text{C})}_{\text{Heat lost from condensation water cooling}} +$

$\underbrace{4.18 \dfrac{\text{J}}{\text{g} \cdot {}^\circ\text{C}} \times 38.5 \text{ g H}_2\text{O} \times (T_f - 27 \, {}^\circ\text{C}) = 0}_{\text{Heat gained by the water}}$

$(-2.48 \times 10^3) + (4.60 \, T_f - 4.60 \times 10^2) + (161 \, T_f - 4.35 \times 10^3) = 0$

$166 \, T_f - 7.29 \times 10^3 = 0 \Rightarrow T_f = \dfrac{7.29 \times 10^3}{166} = 43.9 \, {}^\circ\text{C}$

Highlight Problems

105. Interior molecules have the most neighbors. The surface molecule is more likely to evaporate. The number of neighbors would change; however, surface molecules will always have less than interior molecules and will always be more likely to evaporate.

106. Soap works because one end of the soap molecule is polar, which likes to be in water, and the other end is nonpolar, which can dissolve the grease.

107. a) This is a valid criticism because ice displaces more volume than the liquid water that makes it up. The melting of ice in a cup would actually result in a small decrease in volume.
    b) The melting of ice from the continent would increase ocean levels because this water is not currently in the ocean itself.

108. Rubbing alcohol is a volatile compound due to weak intermolecular forces. Because it is a volatile compound, evaporation can occur rapidly and evaporation is an endothermic process (absorbs heat from the surroundings). The heat that is absorbed comes from your skin, which you sense as the cold feeling.

109. Acetone
   (a) 58.08 g/mol
   (b)

   $$\underset{H_3C}{}\overset{O\,\delta-}{\underset{}{\overset{\|}{C}}}\,\delta+\underset{CH_3}{}$$

   (c) Polar, nonsymmetric distribution of polar bonds
   (d) No hydrogen bonding, lacks H bonded to O, N, or F
   (e) Dipole–dipole
   (f) Intermediate boiling point
   (g) The strength of London dispersion forces increases with increasing molar mass.

   Butane
   (a) 58.12 g/mol
   (b)

   $$H_3C\diagdown\underset{}{CH_2}\diagdown\underset{}{CH_2}\diagdown CH_3$$

   (c) Nonpolar, no polar bonds present
   (d) No hydrogen bonding, lacks H bonded to O, N, or F
   (e) London dispersion forces
   (f) Lowest boiling point
   (g) The strength of London dispersion forces increases with increasing molar mass.

   1-propanol
   (a) 60.09 g/mol
   (b)

   $$H_3C\diagdown CH_2\diagdown\underset{\delta+}{CH_2}\diagdown\overset{\delta-}{OH}$$

   (c) Polar, nonsymmetric distribution of polar bonds
   (d) Hydrogen bonding can occur as O is bonded directly to H
   (e) Hydrogen bonding
   (f) Highest boiling point
   (g) The strength of London dispersion forces increases with increasing molar mass.

110. Carbon monoxide has a boiling point of –191.5 °C, and carbon dioxide has a boiling point of –57 °C. Carbon monoxide is a polar compound indicating dipole–dipole intermolecular forces, and carbon dioxide is a nonpolar molecule indicating London dispersion forces. Typically, dipole–dipole forces are stronger than London dispersion forces; however, that is when the molecules in question have a similar molar mass. The larger molecular mass of carbon dioxide provides greater strength to the dispersion forces present and therefore a higher boiling point.

111. Triad Table of Atomic Masses

Element 1 (amu)	Element 2 (amu)	Element 3 (amu)	Average Mass of 1 and 3
Oxygen 16.00	Sulfur 32.07	Selenium 78.97	47.49
Chlorine 35.45	Bromine 79.90	Iodine 126.9	81.18
Lithium 6.94	Sodium 22.99	Potassium 39.10	23.02

The melting point of each compound is determined by the strength of the dominant intermolecular force present, which would increase as a function of the molar mass of the compound. Therefore, the trend should hold true.

Triad Table of Melting Points

Compound 1	Compound 2	Compound 3	Average mp of 1 and 3
HCl –114.2 °C	HBr –87 °C	HI –50.8 °C	–82.5 °C

112. Assuming the heat of vaporization of sweat is the same as water at 25 °C and that the density of sweat is 1 g/mL:

$$100\,\frac{J}{s} \times \frac{60\,s}{1\,min} \times \frac{60\,min}{1\,hr} \times \frac{1\,kJ}{1000\,J} \times \frac{1\,mol\,H_2O}{44.0\,kJ} \times \frac{18.02\,g}{1\,mol\,H_2O} \times \frac{1\,mL\,H_2O}{1\,g} = 147\,\frac{mL\,H_2O}{hr}$$

Data Interpretation and Analysis

113. (a) Pentane and hexane have boiling points greater than 25 °C.
   (b) As the hydrocarbons get larger, so do the boiling and melting point values. The boiling point trend is more regular than the melting point trend.
   (c) Dispersion intermolecular forces exist in the hydrocarbons. The larger the molecule, the stronger the dispersion force; therefore, more energy is required to overcome the forces during boiling.
   (d) See graph.
   (e) approx. 95 °C

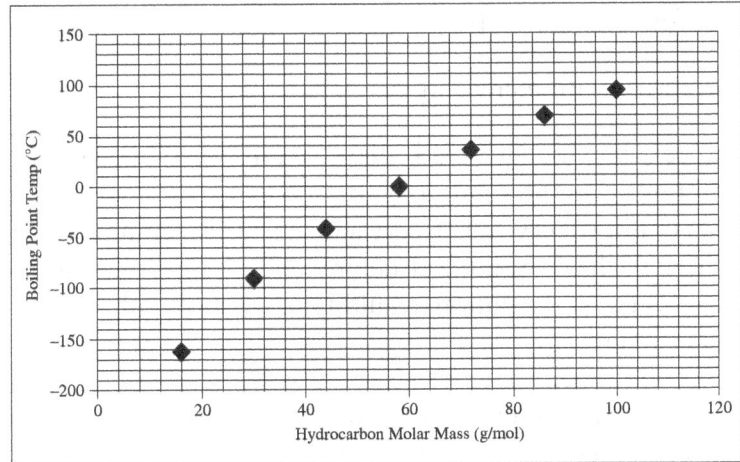

Solutions 13

Questions

1. A solution is a homogeneous mixture of two or more substances. Some examples include air, seawater, soda water, and brass.

2. An aqueous solution exists when a solid, liquid, or a gas is dissolved in water.

3. The solvent is the major component of the solution. The solute is the minor component in the solution. An example is soda pop where carbon dioxide and sugar are solutes and water is the solvent. Another example is saltwater, in which salt is the solute and water is the solvent.

4. "Like dissolves like" is a general rule for predicting solubility. If the solute and solvent display similar intermolecular forces (i.e., dispersion, dipole–dipole, H-bonding), then they will be able to interact and dissolve in each other. In other words, polar solutes will dissolve in polar solvents, and nonpolar solutes will dissolve in nonpolar solvents.

5. Solubility is the amount of a compound (grams) that will dissolve in a specified amount of a solvent.

6. a) The additional solute does not dissolve.
   b) The additional solute will dissolve.
   c) The additional solute will cause the excess solute to come out of solution.

7. A strong electrolyte solution is one that will conduct electricity due to the dissociation of ionic species in solution. A nonelectrolyte solution will not conduct electricity due to the absence of ions in solution. Ionic compounds tend to form strong electrolytes, and molecular compounds form nonelectrolyte solutions.

8. The solubility of gases decreases with increasing temperature.

9. Recrystallization is the process in which a solid is dissolved in a suitable solvent, usually at elevated temperatures. As the solution cools, it becomes saturated, and the excess solid begins to reform crystals. Recrystallization is a common way to purify a solid. The formation of the crystalline structure tends to reject impurities when regrown slowly from a saturated solution, resulting in crystals with fewer impurities than the original crystals.

10. Rock candy is made from a saturated solution of sugar and a string placed into the solution. As the solution cools, it becomes saturated and the sugar crystals come out of solution and crystalize on the string.

11. The bubbles that form in water when it is heated to a temperature lower than the boiling point are dissolved gases that are coming out of solution. The solubility of gases decreases as a function of increased temperatures.

12. The term "flat" soda pop refers to the absence of carbonation (the carbon dioxide has been removed). This occurs faster with warm soda pop because the solubility of gases decreases with increasing temperature.

13. The solubility of a gas increases with increasing pressure and decreases with decreasing pressure. In soda pop, the pressure is provided by a large amount of carbon dioxide gas that is pumped into the can before sealing it. When the can is opened, the pressure is released and the solubility of carbon dioxide decreases, resulting in bubbling.

14. A dilute solution is one that contains a small amount of solute relative to the solvent. A concentrated solution is one that contains a large amount of solute relative to the solvent.

15. Mass percent is the number of grams of solute per 100 g of solution. Molarity (M) is defined as the number of moles of solute per liter of solution.

16. A stock solution is a concentrated solution from which you prepare a more diluted solution.

17. The addition of a nonvolatile solute will lower the freezing point and raise the boiling point of a solution relative to that of the pure solvent.

18. Colligative properties are those properties that depend only on how much solute is added, not the type of solute. Colligative properties include boiling point elevation, freezing point depression, and osmotic pressure.

19. Molality is a common unit of concentration defined as the moles of solute dissolved per kilogram of solvent.

20. Osmosis is the flow of solvent through a semipermeable membrane from a region of lower solute concentration to a region of higher solute concentration.

21. By drinking seawater, the one survivor created a region of high salt concentration on the outside of the cells in his body. This caused osmosis of water out of the cells in the body into the saltwater he had consumed, causing more severe dehydration.

22. Intravenous fluids are always saline solutions that match the osmotic pressure of bodily fluids so that they do not cause water to move either into or out of the cells. If pure water were administered intravenously, cells would swell and possibly burst open as water entered them.

Problems

Solutions

23. a) not a solution
    b) not a solution
    c) solution
    d) solution (Sterling silver is a solid solution of 925 parts Ag and 75 parts Cu.)

24. a) solution
    b) solution
    c) not a solution
    d) solution

25. a) solvent: water, solute: salt
    b) solvent: water, solute: sugar
    c) solvent: water, solute: carbon dioxide

26. a) solvent: water, solute: ethyl alcohol
    b) solvent: water, solute: oxygen
    c) solvent: water, solute: ethylene glycol

27. a) hexane, ethyl ether, or toluene
    b) water, acetone, or methyl alcohol
    c) hexane, ethyl ether, or toluene
    d) water, acetone, or methyl alcohol

28. a) water, acetone, or methyl alcohol
    b) water, acetone, or methyl alcohol
    c) hexane, ethyl ether, or toluene
    d) water, acetone, or methyl alcohol

Solid Dissolved in Water

29. The dissolved particles are the cations and anions that make up the ionic solute. This solution is referred to as a strong electrolyte.

30. The dissolved particles are the intact molecules. This solution is referred to as a nonelectrolyte.

31. From the graph, the solubility of NaCl at 25 °C is ~35 g NaCl/100 g H_2O. A concentration of 35 g NaCl/100 g H_2O is saturated.

32. From the graph, the solubility of KNO_3 at 25 °C is ~35 g KNO_3/100 g H_2O. A concentration of 28 g KNO_3/100 g H_2O is unsaturated.

33. From the graph, the solubility of KNO₃ at 40 °C is ~62 g/100 g H₂O. A concentration of 42 g/100 g H₂O is below the solubility limit, so the solution is initially unsaturated. As the temperature cools from 40 to 0 °C, the solubility limit decreases. At a temperature of ~28 °C, the solubility of KNO₃ drops below the value of 42 g/100 g H₂O. Below this temperature, the excess KNO₃ will recrystallize as the solution cools.

34. When the solution cools to ~ 50 °C, the solution will be saturated. Further cooling will result in the recrystallization of KCl as the excess solute comes out of solution.

35. a) $\dfrac{30.0 \text{ g KClO}_3}{85.0 \text{ g H}_2\text{O}} = \dfrac{35.3 \text{ g}}{100 \text{ g H}_2\text{O}}$

   This is above the saturation limit of KClO₃ at 35 °C (~12 g/100 g), so not all the KClO₃ will dissolve.

   b) $\dfrac{65.0 \text{ g NaNO}_3}{125.0 \text{ g H}_2\text{O}} = \dfrac{52.0 \text{ g}}{100 \text{ g H}_2\text{O}}$

   This is below the saturation limit of NaNO₃ at 15 °C (~84 g/100 g), so all the NaNO₃ will dissolve.

   c) $\dfrac{32.0 \text{ g KCl}}{70.0 \text{ g H}_2\text{O}} = \dfrac{45.7 \text{ g}}{100 \text{ g H}_2\text{O}}$

   This is below the saturation limit of KCl at 82 °C (~52 g/100 g), so all the KCl will dissolve.

36. a) $\dfrac{45.0 \text{ g CaCl}_2}{105.0 \text{ g H}_2\text{O}} = \dfrac{42.8 \text{ g}}{100 \text{ g H}_2\text{O}}$

   Below the saturation limit of CaCl₂ at 5 °C (~60 g/100 g), all of it dissolves.

   b) $\dfrac{15.0 \text{ g KClO}_3}{115 \text{ g H}_2\text{O}} = \dfrac{13.0 \text{ g}}{100 \text{ g H}_2\text{O}}$

   Above the saturation limit of KClO₃ at 5 °C (~8 g/100 g), not all of it dissolves.

   c) $\dfrac{50.0 \text{ g KClO}_3}{95.0 \text{ g H}_2\text{O}} = \dfrac{52.6 \text{ g}}{100 \text{ g H}_2\text{O}}$

   Above the saturation limit of Pb(NO₃)₂ at 10 °C (~45 g/100 g), not all of it dissolves.

Gases Dissolved in Water

37. The solubility of gases decreases as temperature increases. When water is boiled, the dissolved oxygen is completely removed.

38. The solubility of gases decreases as temperature increases. When water is boiled, the dissolved oxygen is completely removed, which is what fish need to breathe.

39. The solubility of gases (nitrogen) increases with increasing pressure. The diver could prevent this effect either by not diving as deep or by using a helium–oxygen, mixture as discussed in Chapter 11.

40. The amount of oxygen in the bloodstream increases at elevated pressures. This effect can be reversed either by not diving as deep or by using a helium–oxygen mixture as discussed in Chapter 11.

Mass Percent

41. a) mass % = $\dfrac{41.2 \text{ g } C_{12}H_{22}O_{11}}{(41.2 + 498) \text{ g solution}} \times 100\% = 7.64\%$

b) mass % = $\dfrac{178 \text{ mg } C_6H_{12}O_6 \times \dfrac{1 \text{ g}}{1000 \text{ mg}}}{\left(178 \text{ mg} \times \dfrac{1 \text{ g}}{1000 \text{ mg}} + 4.91\right) \text{ g solution}} \times 100\% = 3.50\%$

c) mass % = $\dfrac{7.55 \text{ g NaCl}}{(7.55 + 155) \text{ g solution}} \times 100\% = 4.64\%$

42. a) mass % = $\dfrac{132 \text{ g KCl}}{(132 + 598) \text{ g solution}} \times 100\% = 18.1\%$

b) mass % = $\dfrac{22.3 \text{ mg } KNO_3 \times \dfrac{1 \text{ g}}{1000 \text{ mg}}}{\left(22.3 \text{ mg} \times \dfrac{1 \text{ g}}{1000 \text{ mg}} + 2.84\right) \text{ g solution}} \times 100\% = 0.780\%$

c) mass % = $\dfrac{8.72 \text{ g } C_2H_6O}{(8.72 + 76.1) \text{ g solution}} \times 100\% = 10.3\%$

43. mass % = $\dfrac{42 \text{ g sugar}}{(42 + 311) \text{ g solution}} \times 100\% = 12\%$

44. mass % = $\dfrac{32 \text{ mg sodium} \times \dfrac{1 \text{ g}}{1000 \text{ mg}}}{\left(32 \text{ mg} \times \dfrac{1 \text{ g}}{1000 \text{ mg}} + 309\right) \text{ g solution}} \times 100\% = 0.010\%$

45.

Mass Solute	Mass Solvent	Mass Solution	Mass%
15.5	238.1	253.6	6.11%
22.8	167.2	190.0	12.0%
28.8	183.3	212.1	13.6%
56.9	315.2	372.1	15.3%

46.
Mass Solute	Mass Solvent	Mass Solution	Mass%
2.55	25.0	27.6	9.24%
1.81	45.8	47.6	3.8%
1.38	25.8	27.2	5.07%
23.7	3.8×10^2	4.1×10^2	5.8%

47. $\dfrac{3.5 \text{ g NaCl}}{100 \text{ g solution}} \times 254 \text{ g solution} = 8.9 \text{ g NaCl}$

48. $\dfrac{1.1 \text{ g NaCl}}{100 \text{ g solution}} \times 96.3 \text{ g solution} = 1.1 \text{ g NaCl}$

49. a) $\dfrac{3.7 \text{ g sucrose}}{100 \text{ g solution}} \times 48 \text{ g solution} = 1.8 \text{ g sucrose}$

   b) $\dfrac{10.2 \text{ mg sucrose}}{100 \text{ mg solution}} \times 103 \text{ mg solution} = 10.5 \text{ mg sucrose}$

   c) $\dfrac{14.3 \text{ kg sucrose}}{100 \text{ kg solution}} \times 3.2 \text{ kg solution} = 0.46 \text{ kg sucrose}$

50. a) $\dfrac{1.08 \text{ g KCl}}{100 \text{ g solution}} \times 19.7 \text{ g solution} = 0.213 \text{ g KCl}$

   b) $\dfrac{18.7 \text{ kg KCl}}{100 \text{ kg solution}} \times 23.2 \text{ kg solution} = 4.34 \text{ kg KCl}$

   c) $\dfrac{12 \text{ mg KCl}}{100 \text{ mg solution}} \times 38 \text{ mg solution} = 4.6 \text{ mg KCl}$

51. a) $1.5 \text{ g NaCl} \times \dfrac{100 \text{ g solution}}{0.058 \text{ g NaCl}} = 2.6 \times 10^3 \text{ g solution}$

   b) $1.5 \text{ g NaCl} \times \dfrac{100 \text{ g solution}}{1.46 \text{ g NaCl}} = 1.0 \times 10^2 \text{ g solution}$

   c) $1.5 \text{ g NaCl} \times \dfrac{100 \text{ g solution}}{8.44 \text{ g NaCl}} = 18 \text{ g solution}$

52. a) $12 \text{ g sucrose} \times \dfrac{100 \text{ g solution}}{4.1 \text{ g sucrose}} = 2.9 \times 10^2 \text{ g solution}$

   b) $12 \text{ g sucrose} \times \dfrac{100 \text{ g solution}}{3.2 \text{ g sucrose}} = 3.8 \times 10^2 \text{ g solution}$

   c) $12 \text{ g sucrose} \times \dfrac{100 \text{ g solution}}{12.5 \text{ g sucrose}} = 96 \text{ g solution}$

53. mass of AgCl solution: $4.8 \text{ L} \times \dfrac{1000 \text{ mL}}{1 \text{ L}} \times \dfrac{1.01 \text{ g}}{1 \text{ mL}} = 4.8 \times 10^3 \text{ g}$

$4.8 \times 10^3 \text{ g solution} \times \dfrac{3.4 \text{ g Ag}}{100 \text{ g solution}} = 1.6 \times 10^2 \text{ g Ag}$

54. mass of solution: $2.5 \text{ L} \times \dfrac{1000 \text{ mL}}{1 \text{ L}} \times \dfrac{1.01 \text{ g}}{1 \text{ mL}} = 2.5 \times 10^3 \text{ g}$

$2.5 \times 10^3 \text{ g solution} \times \dfrac{0.085 \text{ g dioxin}}{100 \text{ g solution}} = 2.1 \text{ g dioxin}$

55. $45.8 \text{ g NaCl} \times \dfrac{100 \text{ g solution}}{3.5 \text{ g NaCl}} = 1.3 \times 10^3 \text{ g solution}$

56. $1.2 \text{ g Ca} \times \dfrac{100 \text{ g solution}}{0.0085 \text{ g NaCl}} = 1.4 \times 10^4 \text{ g solution}$

57. $115 \text{ mg Pb} \times \dfrac{1 \text{ g}}{1000 \text{ mg}} \times \dfrac{100 \text{ g solution}}{0.0011 \text{ g Pb}} \times \dfrac{1 \text{ mL}}{1.0 \text{ g}} = 1.0 \times 10^4 \text{ mL solution}$

58. $175 \text{ mg benzene} \times \dfrac{1 \text{ g}}{1000 \text{ mg}} \times \dfrac{100 \text{ g solution}}{0.000037 \text{ g benzene}} \times \dfrac{1 \text{ mL}}{1.0 \text{ g}} = 4.7 \times 10^5 \text{ mL soln}$

Molarity

59. a) $M = \dfrac{0.127 \text{ mol sucrose}}{655 \text{ mL} \times \dfrac{1 \text{ L}}{1000 \text{ mL}}} = \dfrac{0.127 \text{ mol sucrose}}{0.655 \text{ L}} = 0.194 \text{ M sucrose}$

b) $M = \dfrac{0.205 \text{ mol KNO}_3}{0.875 \text{ L}} = 0.234 \text{ M KNO}_3$

c) $M = \dfrac{1.1 \text{ mol KCl}}{2.7 \text{ L}} = 0.41 \text{ M KCl}$

60. a) $M = \dfrac{1.54 \text{ mol LiCl}}{22.2 \text{ L}} = 0.0694 \text{ M LiCl}$

b) $M = \dfrac{0.101 \text{ mol LiNO}_3}{6.4 \text{ L}} = 0.016 \text{ M LiNO}_3$

c) $M = \dfrac{0.0323 \text{ mol glucose}}{76.2 \text{ mL} \times \dfrac{1 \text{ L}}{1000 \text{ mL}}} = \dfrac{0.0323 \text{ mol glucose}}{0.0762 \text{ L}} = 0.424 \text{ M glucose}$

61. a) moles: $22.6 \text{ g C}_{12}\text{H}_{22}\text{O}_{11} \times \dfrac{1 \text{ mol C}_{12}\text{H}_{22}\text{O}_6}{342.34 \text{ g}} = 0.0660 \text{ mol C}_{12}\text{H}_{22}\text{O}_6$

$M = \dfrac{0.0660 \text{ mol C}_{12}\text{H}_{22}\text{O}_{11}}{0.442 \text{ L}} = 0.149 \text{ M C}_{12}\text{H}_{22}\text{O}_{11}$

b) moles: $42.6 \text{ g NaCl} \times \dfrac{1 \text{ mol NaCl}}{58.44 \text{ g}} = 0.729 \text{ mol NaCl}$

$M = \dfrac{0.729 \text{ mol NaCl}}{1.58 \text{ L}} = 0.461 \text{ M NaCl}$

c) moles: $315 \text{ mg C}_6\text{H}_{12}\text{O}_6 \times \dfrac{1 \text{ g}}{1 \times 10^3 \text{ mg}} \times \dfrac{1 \text{ mol C}_6\text{H}_{12}\text{O}_6}{180.18 \text{ g}} = 1.75 \times 10^{-3} \text{ mol C}_6\text{H}_{12}\text{O}_6$

liters: $58.2 \text{ mL} \times \dfrac{1 \text{ L}}{1000 \text{ mL}} = 0.0582 \text{ L}$

$M = \dfrac{1.75 \times 10^{-3} \text{ mol C}_6\text{H}_{12}\text{O}_6}{0.0582 \text{ L}} = 0.0300 \text{ M C}_6\text{H}_{12}\text{O}_6$

62. a) moles: $33.2 \text{ g KCl} \times \dfrac{1 \text{ mol KCl}}{74.55 \text{ g}} = 0.445 \text{ mol KCl}$

$M = \dfrac{0.445 \text{ mol KCl}}{0.895 \text{ L}} = 0.498 \text{ M KCl}$

b) moles: $61.3 \text{ g C}_2\text{H}_6\text{O} \times \dfrac{1 \text{ mol C}_2\text{H}_6\text{O}}{46.08 \text{ g}} = 1.33 \text{ mol C}_2\text{H}_6\text{O}$

$M = \dfrac{1.33 \text{ mol C}_2\text{H}_6\text{O}}{3.4 \text{ L}} = 0.39 \text{ M C}_2\text{H}_6\text{O}$

c) moles: $38.2 \text{ mg KI} \times \dfrac{1 \text{ g}}{1 \times 10^3 \text{ mg}} \times \dfrac{1 \text{ mol KI}}{166.00 \text{ g}} = 2.30 \times 10^{-4} \text{ mol KI}$

liters: $112 \text{ mL} \times \dfrac{1 \text{ L}}{1000 \text{ mL}} = 0.112 \text{ L}$

$M = \dfrac{2.30 \times 10^{-4} \text{ mol KI}}{0.112 \text{ L}} = 2.05 \times 10^{-3} \text{ M KI}$

63. $6.8 \text{ g NaCl} \times \dfrac{1 \text{ mol NaCl}}{58.44 \text{ g}} \times \dfrac{1}{205 \text{ mL}} \times \dfrac{1000 \text{ mL}}{1 \text{ L}} = 0.57 \text{ M NaCl}$

64. $41 \text{ g C}_{12}\text{H}_{22}\text{O}_{11} \times \dfrac{1 \text{ mol C}_{12}\text{H}_{22}\text{O}_{11}}{342.34 \text{ g}} \times \dfrac{1}{355 \text{ mL}} \times \dfrac{1000 \text{ mL}}{1 \text{ L}} = 0.34 \text{ M C}_{12}\text{H}_{22}\text{O}_{11}$

65. a) $\dfrac{1.2 \text{ mol NaCl}}{\text{L}} \times 1.5 \text{ L} = 1.8 \text{ mol NaCl}$

b) $\dfrac{0.85 \text{ mol}}{\text{L}} \times 0.448 \text{ L} = 0.38 \text{ mol NaCl}$

c) $\dfrac{1.65 \text{ mol NaCl}}{\text{L}} \times 144 \text{ mL} \times \dfrac{1 \text{ L}}{1000 \text{ mL}} = 0.238 \text{ mol NaCl}$

66. a) $\dfrac{0.100 \text{ mol } C_{12}H_{22}O_{11}}{\text{L}} \times 3.4 \text{ L} = 0.34 \text{ mol } C_{12}H_{22}O_{11}$

b) $\dfrac{1.88 \text{ mol } C_{12}H_{22}O_{11}}{\text{L}} \times 0.952 \text{ L} = 1.79 \text{ mol } C_{12}H_{22}O_{11}$

c) $\dfrac{0.528 \text{ mol } C_{12}H_{22}O_{11}}{\text{L}} \times 21.5 \text{ mL} \times \dfrac{1 \text{ L}}{1000 \text{ mL}} = 0.0114 \text{ mol } C_{12}H_{22}O_{11}$

67. a) $0.15 \text{ mol KCl} \times \dfrac{1 \text{ L}}{0.255 \text{ mol KCl}} = 0.59 \text{ L}$

b) $0.15 \text{ mol KCl} \times \dfrac{1 \text{ L}}{1.8 \text{ mol KCl}} = 0.083 \text{ L}$

c) $0.15 \text{ mol KCl} \times \dfrac{1 \text{ L}}{0.995 \text{ mol KCl}} = 0.15 \text{ L}$

68. a) $0.325 \text{ mol NaI} \times \dfrac{1 \text{ L}}{0.152 \text{ mol NaI}} = 2.14 \text{ L}$

b) $0.325 \text{ mol} \times \dfrac{1 \text{ L}}{0.982 \text{ mol NaI}} = 0.331 \text{ L}$

c) $0.325 \text{ mol} \times \dfrac{1 \text{ L}}{1.76 \text{ mol NaI}} = 0.185 \text{ L}$

69.

Solute	Mass Solute	Mol Solute	Vol. Soln.	Molarity
KNO$_3$	22.5 g	0.223 mol	125 mL	1.78 M
NaHCO$_3$	2.10 g	0.0250 mol	250.0 mL	0.100 M
C$_{12}$H$_{22}$O$_{11}$	55.38 g	0.162 mol	1.08 L	0.150 M

70.

Solute	Mass Solute	Mol Solute	Vol. Soln.	Molarity
MgSO$_4$	0.588 g	0.00488 mol	25.0 mL	0.195 M
NaOH	7.00 g	0.175 mol	100.0 mL	1.75 M
CH$_3$OH	12.5 g	0.390 mol	0.780 L	0.500 M

71. $35 \text{ mL} \times \dfrac{1 \text{ L}}{1000 \text{ mL}} \times \dfrac{1.3 \text{ mol NaCl}}{1 \text{ L}} \times \dfrac{58.44 \text{ g}}{1 \text{ mol NaCl}} = 2.7 \text{ g NaCl}$

72. $105 \text{ mL} \times \dfrac{1 \text{ L}}{1000 \text{ mL}} \times \dfrac{1.02 \text{ mol } C_6H_{12}O_6}{1 \text{ L}} \times \dfrac{180.18 \text{ g}}{1 \text{ mol } C_6H_{12}O_6} = 19.3 \text{ g } C_6H_{12}O_6$

73. $2.5 \text{ L} \times \dfrac{0.100 \text{ mol KCl}}{1 \text{ L}} \times \dfrac{74.55 \text{ g}}{1 \text{ mol KCl}} = 19 \text{ g KCl}$

74. $500.0 \text{ mL} \times \dfrac{1 \text{ L}}{1000 \text{ mL}} \times \dfrac{1.4 \text{ mol } KNO_3}{1 \text{ L}} \times \dfrac{101.11 \text{ g}}{1 \text{ mol } KNO_3} = 71 \text{ g } KNO_3$

75. $1.5 \text{ kg } C_{12}H_{22}O_{11} \times \dfrac{1000 \text{ g}}{1 \text{ kg}} \times \dfrac{1 \text{ mol } C_{12}H_{22}O_{11}}{342.34 \text{ g}} \times \dfrac{1 \text{ L}}{0.500 \text{ mol}} = 8.8 \text{ L}$

76. $87 \text{ g } Mg(NO_3)_2 \times \dfrac{1 \text{ mol } Mg(NO_3)_2}{148.33 \text{ g}} \times \dfrac{1 \text{ L}}{0.35 \text{ mol } Mg(NO_3)_2} = 1.7 \text{ L}$

77. a) 1 mole Cl^- per 1 mole NaCl, $[Cl^-] = 0.15$ M
    b) 2 moles Cl^- per 1 mole $CuCl_2$, $[Cl^-] = 0.30$ M
    c) 3 moles Cl^- per 1 mole $AlCl_3$, $[Cl^-] = 0.45$ M

78. a) 1 mole NO_3^- per 1 mole KNO_3, $[NO_3^-] = 0.10$ M
    b) 2 moles NO_3^- per 1 mole $Ca(NO_3)_2$, $[NO_3^-] = 0.20$ M
    c) 3 moles NO_3^- per 1 mole $Cr(NO_3)_3$, $[NO_3^-] = 0.30$ M

79. a) 0.12 M $Na_2SO_4 \Rightarrow$ 0.24 M Na^+, 0.12 M SO_4^{2-}
    b) 0.25 M $K_2CO_3 \Rightarrow$ 0.50 M K^+, 0.25 M CO_3^{2-}
    c) 0.11 M RbBr $\Rightarrow$ 0.11 M Rb^+, 0.11 M Br^-

80. a) 0.20 M $SrSO_4 \Rightarrow$ 0.20 M Sr^{2+}, 0.20 M SO_4^{2-}
    b) 0.15 M $Cr_2(SO_4)_3 \Rightarrow$ 0.30 M Cr^{3+}, 0.45 M SO_4^{2-}
    c) 0.12 M $SrI_2 \Rightarrow$ 0.12 M Sr^{2+}, 0.24 M I^-

Solution Dilution

81. $M_1V_1 = M_2V_2 \Rightarrow (1.2 \text{ M})(122 \text{ mL}) = (M_2)(500.0 \text{ mL})$
    $M_2 = \dfrac{(1.2 \text{ M})(122 \text{ mL})}{(500.0 \text{ mL})} = 0.29 \text{ M}$

82. $M_1V_1 = M_2V_2 \Rightarrow (5.8 \text{ M})(3.5 \text{ L}) = (M_2)(55 \text{ L})$
    $M_2 = \dfrac{(5.8 \text{ M})(3.5 \text{ L})}{(55 \text{ L})} = 0.37 \text{ M}$

83. $M_1V_1 = M_2V_2 \Rightarrow (5.5\ M)(V_1) = (0.100\ M)(2.5\ L)$

$V_1 = \dfrac{(0.100\ M)(2.5\ L)}{(5.5\ M)} = 0.045\ L$

You would dilute 45 mL (0.045 L) of the 5.5 M stock solution to a final volume of 2.5 L.

84. $M_1V_1 = M_2V_2 \Rightarrow (15.0\ M)(V_1) = (0.200\ M)(500.0\ mL)$

$V_1 = \dfrac{(0.200\ M)(500.0\ mL)}{(15.0\ M)} = 6.67\ mL$

Dilute 6.67 mL of the 15.0 M stock solution to a final volume of 500.0 mL.

85. $M_1V_1 = M_2V_2 \Rightarrow (12\ M)(25\ mL) = (0.500\ M)(V_2)$

$V_2 = \dfrac{(12\ M)(25\ mL)}{(0.500\ M)} = 6.0 \times 10^2\ mL$

86. $M_1V_1 = M_2V_2 \Rightarrow (10.0\ M)(75\ mL) = (1.75\ M)(V_2)$

$V_2 = \dfrac{(10.0\ M)(75\ mL)}{(1.75\ M)} = 4.3 \times 10^2\ mL$

87. $M_1V_1 = M_2V_2 \Rightarrow (12.0\ M)(V_1) = (0.250\ M)(850.0\ mL)$

$V_1 = \dfrac{(0.250\ M)(850.0\ mL)}{(12.0\ M)} = 17.7\ mL$

88. $M_1V_1 = M_2V_2 \Rightarrow (5.0\ M)(V_1) = (0.040\ M)(85.0\ mL)$

$V_1 = \dfrac{(0.040\ M)(85.0\ mL)}{(5.0\ M)} = 0.68\ mL$

Solution Stoichiometry

89. a) $\dfrac{0.150\ mol\ HCl}{1\ L} \times 25\ mL \times \dfrac{1\ L}{1000\ mL} \times \dfrac{1\ mol\ NaOH}{1\ mol\ HCl} \times \dfrac{1\ L}{0.150\ mol\ NaOH}$

$= 0.025\ L$

b) $\dfrac{0.055\ mol\ HCl}{1\ L} \times 55\ mL \times \dfrac{1\ L}{1000\ mL} \times \dfrac{1\ mol\ NaOH}{1\ mol\ HCl} \times \dfrac{1\ L}{0.150\ mol\ NaOH}$

$= 0.020\ L$

c) $\dfrac{0.885\ mol\ HCl}{1\ L} \times 175\ mL \times \dfrac{1\ L}{1000\ mL} \times \dfrac{1\ mol\ NaOH}{1\ mol\ HCl} \times \dfrac{1\ L}{0.150\ mol\ NaOH}$

$= 1.03\ L$

90. a) $\dfrac{0.225 \text{ mol H}_2\text{SO}_4}{1 \text{ L}} \times 45 \text{ mL} \times \dfrac{1 \text{ L}}{1 \times 10^3 \text{ mL}} \times \dfrac{2 \text{ mol KOH}}{1 \text{ mol H}_2\text{SO}_4} \times \dfrac{1 \text{ L}}{0.225 \text{ mol KOH}}$

   $= 0.090 \text{ L}$

   b) $\dfrac{0.125 \text{ mol H}_2\text{SO}_4}{1 \text{ L}} \times 185 \text{ mL} \times \dfrac{1 \text{ L}}{1000 \text{ mL}} \times \dfrac{2 \text{ mol KOH}}{1 \text{ mol H}_2\text{SO}_4} \times \dfrac{1 \text{ L}}{0.225 \text{ mol KOH}}$

   $= 0.206 \text{ L}$

   c) $\dfrac{0.100 \text{ mol H}_2\text{SO}_4}{1 \text{ L}} \times 75 \text{ mL} \times \dfrac{1 \text{ L}}{1000 \text{ mL}} \times \dfrac{2 \text{ mol KOH}}{1 \text{ mol H}_2\text{SO}_4} \times \dfrac{1 \text{ L}}{0.225 \text{ mol KOH}}$

   $= 0.067 \text{ L}$

91. $\dfrac{0.0112 \text{ mol NiCl}_2}{1 \text{ L}} \times 134 \text{ mL} \times \dfrac{1 \text{ L}}{1000 \text{ mL}} \times \dfrac{2 \text{ mol K}_3\text{PO}_4}{3 \text{ mol NiCl}_2} \times \dfrac{1 \text{ L}}{0.225 \text{ mol K}_3\text{PO}_4}$

   $= 0.00445 \text{ L}$

92. $\dfrac{0.115 \text{ mol Co(NO}_3)_2}{1 \text{ L}} \times 175 \text{ mL} \times \dfrac{1 \text{ L}}{1000 \text{ mL}} \times \dfrac{1 \text{ mol K}_2\text{S}}{1 \text{ mol Co(NO}_3)_2} \times \dfrac{1 \text{ L}}{0.225 \text{ mol K}_2\text{S}}$

   $= 0.0894 \text{ L}$

93. $\dfrac{0.100 \text{ mol KOH}}{1 \text{ L}} \times 112 \text{ mL} \times \dfrac{1 \text{ L}}{1000 \text{ mL}} \times \dfrac{1 \text{ mol H}_3\text{PO}_4}{3 \text{ mol KOH}} \times \dfrac{1}{10.0 \text{ mL}} \times \dfrac{1000 \text{ mL}}{1 \text{ L}}$

   $= 0.373 \text{ M H}_3\text{PO}_4$

94. $\dfrac{0.101 \text{ mol NaOH}}{1 \text{ L}} \times 45.3 \text{ mL} \times \dfrac{1 \text{ L}}{1000 \text{ mL}} \times \dfrac{1 \text{ mol HClO}_4}{1 \text{ mol NaOH}} \times \dfrac{1}{25.0 \text{ mL}} \times \dfrac{1000 \text{ mL}}{1 \text{ L}}$

   $= 0.183 \text{ M HClO}_4$

95. $15.0 \text{ g H}_2 \times \dfrac{1 \text{ mol H}_2}{2.02 \text{ g}} \times \dfrac{3 \text{ mol H}_2\text{SO}_4}{3 \text{ mol H}_2} \times \dfrac{1 \text{ L}}{6.0 \text{ mol H}_2\text{SO}_4} = 1.2 \text{ L}$

96. $15.0 \text{ g Zn} \times \dfrac{1 \text{ mol Zn}}{65.39 \text{ g}} \times \dfrac{1 \text{ mol ZnCl}_2}{1 \text{ mol Zn}} \times \dfrac{1}{175 \text{ mL}} \times \dfrac{1000 \text{ mL}}{1 \text{ L}} = 1.31 \text{ M ZnCl}_2$

Molality, Freezing Point Depression, and Boiling Point Elevation

97. a) $m = \dfrac{0.25 \text{ mol}}{0.250 \text{ kg}} = 1.0 \text{ m}$

   b) $m = \dfrac{0.882 \text{ mol}}{0.225 \text{ kg}} = 3.92 \text{ m}$

   c) $m = \dfrac{0.012 \text{ mol}}{23.1 \text{ g}} \times \dfrac{1000 \text{ g}}{1 \text{ kg}} = 0.52 \text{ m}$

98. a) $m = \dfrac{0.455 \text{ mol}}{1.97 \text{ kg}} = 0.231 \text{ m}$

b) $m = \dfrac{0.559 \text{ mol}}{1.44 \text{ kg}} = 0.388 \text{ m}$

c) $m = \dfrac{0.119 \text{ mol}}{488 \text{ g}} \times \dfrac{1000 \text{ g}}{1 \text{ kg}} = 0.244 \text{ m}$

99. $12.5 \text{ g } C_2H_6O_2 \times \dfrac{1 \text{ mol } C_2H_6O_2}{62.08 \text{ g}} \times \dfrac{1}{135 \text{ g } H_2O} \times \dfrac{1000 \text{ g}}{1 \text{ kg}} = 1.49 \text{ m } C_2H_6O_2$

100. $257 \text{ g } C_6H_{12}O_6 \times \dfrac{1 \text{ mol } C_6H_{12}O_6}{180.18 \text{ g}} \times \dfrac{1}{1.62 \text{ kg } H_2O} = 0.880 \text{ m } C_6H_{12}O_6$

101. a) $\Delta T_f = 0.85 \dfrac{\text{mol}}{\text{kg}} \times 1.86 \dfrac{°C \cdot \text{kg}}{\text{mol}} = 1.6\ °C \Rightarrow$ Freezing Point $= -1.6\ °C$

b) $\Delta T_f = 1.45 \dfrac{\text{mol}}{\text{kg}} \times 1.86 \dfrac{°C \cdot \text{kg}}{\text{mol}} = 2.70\ °C \Rightarrow$ Freezing Point $= -2.70\ °C$

c) $\Delta T_f = 4.8 \dfrac{\text{mol}}{\text{kg}} \times 1.86 \dfrac{°C \cdot \text{kg}}{\text{mol}} = 8.9\ °C \Rightarrow$ Freezing Point $= -8.9\ °C$

d) $\Delta T_f = 2.35 \dfrac{\text{mol}}{\text{kg}} \times 1.86 \dfrac{°C \cdot \text{kg}}{\text{mol}} = 4.37\ °C \Rightarrow$ Freezing Point $= -4.37\ °C$

102. a) $\Delta T_f = 0.100 \dfrac{\text{mol}}{\text{kg}} \times 1.86 \dfrac{°C \cdot \text{kg}}{\text{mol}} = 0.186\ °C \Rightarrow$ Freezing Point $= -0.186\ °C$

b) $\Delta T_f = 0.469 \dfrac{\text{mol}}{\text{kg}} \times 1.86 \dfrac{°C \cdot \text{kg}}{\text{mol}} = 0.872\ °C \Rightarrow$ Freezing Point $= -0.872\ °C$

c) $\Delta T_f = 1.44 \dfrac{\text{mol}}{\text{kg}} \times 1.86 \dfrac{°C \cdot \text{kg}}{\text{mol}} = 2.68\ °C \Rightarrow$ Freezing Point $= -2.68\ °C$

d) $\Delta T_f = 5.89 \dfrac{\text{mol}}{\text{kg}} \times 1.86 \dfrac{°C \cdot \text{kg}}{\text{mol}} = 11.0\ °C \Rightarrow$ Freezing Point $= -11.0\ °C$

103. a) $\Delta T_b = 0.118 \dfrac{\text{mol}}{\text{kg}} \times 0.512 \dfrac{°C \cdot \text{kg}}{\text{mol}} = 0.0604\ °C \Rightarrow$ Boiling Point $= 100.060\ °C$

b) $\Delta T_b = 1.94 \dfrac{\text{mol}}{\text{kg}} \times 0.512 \dfrac{°C \cdot \text{kg}}{\text{mol}} = 0.993\ °C \Rightarrow$ Boiling Point $= 100.993\ °C$

c) $\Delta T_b = 3.88 \dfrac{\text{mol}}{\text{kg}} \times 0.512 \dfrac{°C \cdot \text{kg}}{\text{mol}} = 1.99\ °C \Rightarrow$ Boiling Point $= 101.99\ °C$

d) $\Delta T_b = 2.16 \dfrac{\text{mol}}{\text{kg}} \times 0.512 \dfrac{°C \cdot \text{kg}}{\text{mol}} = 1.11\ °C \Rightarrow$ Boiling Point $= 101.11\ °C$

104. a) $\Delta T_b = 0.225 \dfrac{\text{mol}}{\text{kg}} \times 0.512 \dfrac{°C \cdot \text{kg}}{\text{mol}} = 0.115\ °C \Rightarrow$ Boiling Point $= 100.115\ °C$

b) $\Delta T_b = 2.58 \dfrac{\text{mol}}{\text{kg}} \times 0.512 \dfrac{°C \cdot \text{kg}}{\text{mol}} = 1.32\ °C \Rightarrow$ Boiling Point $= 101.32\ °C$

c) $\Delta T_b = 4.33 \dfrac{\text{mol}}{\text{kg}} \times 0.512 \dfrac{°C \cdot \text{kg}}{\text{mol}} = 2.22\ °C \Rightarrow$ Boiling Point $= 102.22\ °C$

d) $\Delta T_b = 6.77 \dfrac{\text{mol}}{\text{kg}} \times 0.512 \dfrac{°C \cdot \text{kg}}{\text{mol}} = 3.47\ °C \Rightarrow$ Boiling Point $= 103.47\ °C$

105. molality: $55.8\ \text{g}\ C_6H_{12}O_6 \times \dfrac{1\ \text{mol}\ C_6H_{12}O_6}{180.2\ \text{g}} \times \dfrac{1}{455\ \text{g}} \times \dfrac{1000\ \text{g}}{1\ \text{kg}} = 0.681\ m$

$\Delta T_f = 0.681 \dfrac{\text{mol}}{\text{kg}} \times 1.86 \dfrac{°C \cdot \text{kg}}{\text{mol}} = 1.27\ °C$

Freezing Point $= 0.00 - 1.27 = -1.27\ °C$

$\Delta T_b = 0.681 \dfrac{\text{mol}}{\text{kg}} \times 0.512 \dfrac{°C \cdot \text{kg}}{\text{mol}} = 0.349\ °C$

Boiling Point $= 100.000 + 0.349 = 100.349\ °C$

106. molality: $21.2\ \text{g}\ C_2H_6O_2 \times \dfrac{1\ \text{mol}\ C_2H_6O_2}{62.08\ \text{g}} \times \dfrac{1}{85.4\ \text{g}} \times \dfrac{1000\ \text{g}}{1\ \text{kg}} = 4.00\ m$

$\Delta T_f = 4.00 \dfrac{\text{mol}}{\text{kg}} \times 1.86 \dfrac{°C \cdot \text{kg}}{\text{mol}} = 7.44\ °C$

Freezing Point $= 0.00 - 7.44 = -7.44\ °C$

$\Delta T_b = 4.00 \dfrac{\text{mol}}{\text{kg}} \times 0.512 \dfrac{°C \cdot \text{kg}}{\text{mol}} = 2.05\ °C$

Boiling Point $= 100.00 + 2.05 = 102.05\ °C$

Cumulative Problems

107. Molarity: $133\ \text{g NaCl} \times \dfrac{1\ \text{mol NaCl}}{58.44\ \text{g}} \times \dfrac{1}{1.00\ \text{L}} = 2.28\ M$

Mass Percent: $\dfrac{133\ \text{g}}{1.00\ \text{L} \times \dfrac{1000\ \text{mL}}{1\ \text{L}} \times \dfrac{1.08\ \text{g}}{1\ \text{mL}}} \times 100\% = 12.3\%$

108. Molarity: $88.4 \text{ g KNO}_3 \times \dfrac{1 \text{ mol KNO}_3}{101.11 \text{ g}} \times \dfrac{1}{1.50 \text{ L}} = 0.583 \text{ M}$

Mass Percent: $\dfrac{88.4 \text{ g}}{1.50 \text{ L} \times \dfrac{1000 \text{ mL}}{1 \text{ L}} \times \dfrac{1.05 \text{ g}}{1 \text{ mL}}} \times 100\% = 5.61\%$

109. $(8.5 \text{ M})(0.125 \text{ L}) = (M_2)(2.5 \text{ L}) \Rightarrow M_2 = \dfrac{(8.5 \text{ M})(0.125 \text{ L})}{(2.5 \text{ L})} = 0.43 \text{ M}$

$10.8 \text{ g NaCl} \times \dfrac{1 \text{ mol NaCl}}{58.44 \text{ g}} \times \dfrac{1 \text{ L}}{0.43 \text{ mol NaCl}} = 0.43 \text{ L}$

110. $(5.8 \text{ M})(0.0458 \text{ L}) = (M_2)(1.00 \text{ L}) \Rightarrow M_2 = \dfrac{(5.8 \text{ M})(0.0458 \text{ L})}{(1.00 \text{ L})} = 0.27 \text{ M}$

$15.0 \text{ g KNO}_3 \times \dfrac{1 \text{ mol KNO}_3}{101.11 \text{ g}} \times \dfrac{1 \text{ L}}{0.27 \text{ mol NaCl}} = 0.55 \text{ L}$

111. $3.25 \text{ g KI} \times \dfrac{1 \text{ mol KI}}{166.0 \text{ g}} \times \dfrac{1}{0.0250 \text{ L}} = 0.783 \text{ M KI}$

$(5.00 \text{ M})(50.00 \text{ mL}) = (0.783 \text{ M})(V_2) \Rightarrow V_2 = \dfrac{(5.00 \text{ M})(50.00 \text{ mL})}{(0.783 \text{ M})} = 319 \text{ mL}$

112. $5.9 \text{ g CuCl}_2 \times \dfrac{1 \text{ mol CuCl}_2}{134.45 \text{ g}} \times \dfrac{1}{0.0500 \text{ L}} = 0.88 \text{ M CuCl}_2$

$(8.00 \text{ M})(125 \text{ mL}) = (0.88 \text{ M})(V_2) \Rightarrow V_2 = \dfrac{(8.00 \text{ M})(125 \text{ mL})}{(0.88 \text{ M})} = 1.1 \times 10^3 \text{ mL}$

113. $\dfrac{5.88 \text{ g NaCl}}{100 \text{ g Soln}} \times \dfrac{1.02 \text{ g Soln}}{1 \text{ mL}} \times \dfrac{1000 \text{ mL}}{1 \text{ L}} \times \dfrac{1 \text{ mol NaCl}}{58.44 \text{ g}} = 1.03 \text{ M NaCl}$

114. $\dfrac{6.75 \text{ g C}_6\text{H}_{12}\text{O}_6}{100 \text{ g Soln}} \times \dfrac{1.03 \text{ g Soln}}{1 \text{ mL}} \times \dfrac{1000 \text{ mL}}{1 \text{ L}} \times \dfrac{1 \text{ mol C}_6\text{H}_{12}\text{O}_6}{180.18 \text{ g}} = 0.386 \text{ M C}_6\text{H}_{12}\text{O}_6$

115. $15.0 \text{ L H}_2 \times \dfrac{1 \text{ mol H}_2}{22.4 \text{ L}} \times \dfrac{3 \text{ mol H}_2\text{SO}_4}{3 \text{ mol H}_2} \times \dfrac{1 \text{ L}}{4.0 \text{ mol}} = 0.17 \text{ L}$

116. $28.5 \text{ L H}_2 \times \dfrac{1 \text{ mol H}_2}{22.4 \text{ L}} \times \dfrac{2 \text{ mol HCl}}{1 \text{ mol H}_2} \times \dfrac{1 \text{ L}}{1.85 \text{ mol}} = 1.38 \text{ L}$

117. NaCl(aq) + AgNO₃(aq) → AgCl(s) + NaNO₃(aq)

$$\frac{0.45 \text{ mol AgNO}_3}{1 \text{ L}} \times 0.025 \text{ L} \times \frac{1 \text{ mol NaCl}}{1 \text{ mol AgNO}_3} \times \frac{1 \text{ L}}{1.25 \text{ mol NaCl}} \times \frac{1000 \text{ mL}}{1 \text{ L}} = 9.0 \text{ mL}$$

118. Na₂SO₄(aq) + Ba(NO₃)₂(aq) → BaSO₄(s) + 2 NaNO₃(aq)

$$\frac{0.250 \text{ mol Ba(NO}_3)_2}{1 \text{ L}} \times 0.150 \text{ L} \times \frac{1 \text{ mol Na}_2\text{SO}_4}{1 \text{ mol Ba(NO}_3)_2} \times \frac{1 \text{ L}}{1.50 \text{ mol Na}_2\text{SO}_4} \times \frac{1000 \text{ mL}}{1 \text{ mL}}$$

$$= 25.0 \text{ mL}$$

119. $\frac{70.3 \text{ g HNO}_3}{100 \text{ g Soln}} \times \frac{1.41 \text{ g Soln}}{1 \text{ mL}} \times \frac{1000 \text{ mL}}{1 \text{ L}} \times \frac{1 \text{ mol HNO}_3}{63.02 \text{ g}} = 15.7 \text{ M HNO}_3$

$M_1V_1 = M_2V_2 \Rightarrow (15.7 \text{ M})(V_1) = (0.500 \text{ M})(2.5 \text{ L})$

$V_1 = \frac{(0.500 \text{ M})(2.5 \text{ L})}{(15.7 \text{ M})} = 0.080 \text{ L}$

$0.080 \text{ L} \times \frac{1000 \text{ mL}}{1 \text{ L}} = 8.0 \times 10^1 \text{ L}$

120. $\frac{37.0 \text{ g HCl}}{100 \text{ g Soln}} \times \frac{1.20 \text{ g Soln}}{1 \text{ mL}} \times \frac{1000 \text{ mL}}{1 \text{ L}} \times \frac{1 \text{ mol HCl}}{36.46 \text{ g}} = 12.2 \text{ M HCl}$

$M_1V_1 = M_2V_2 \Rightarrow (12.2 \text{ M})(V_1) = (0.500 \text{ M})(2.5 \text{ L})$

$V_1 = \frac{(0.500 \text{ M})(2.5 \text{ L})}{(12.2 \text{ M})} = 0.10 \text{ L}$

$0.10 \text{ L} \times \frac{1000 \text{ mL}}{1 \text{ L}} = 1.0 \times 10^2 \text{ mL}$

121. moles solute: $58.5 \text{ g C}_2\text{H}_6\text{O}_2 \times \frac{1 \text{ mol C}_2\text{H}_6\text{O}_2}{62.08 \text{ g}} = 0.942 \text{ mol C}_2\text{H}_6\text{O}_2$

mass solution: $500.0 \text{ mL} \times \frac{1.09 \text{ g}}{\text{mL}} = 545 \text{ g solution}$

mass solvent: $545 \text{ g solution} - 58.5 \text{ g solute} = 486.5 \text{ g} = 0.487 \text{ kg solvent}$

molality: $m = \frac{0.942 \text{ mol C}_2\text{H}_6\text{O}_2}{0.487 \text{ kg solvent}} = 1.93 \text{ m}$

$\Delta T_f = (1.93 \text{ m}) \times \left(\frac{1.86 \text{ °C}}{\text{m}}\right) = 3.59 \text{ °C} \Rightarrow T_f = 0.00 - 3.59 = -3.59 \text{ °C}$

$\Delta T_b = (1.93 \text{ m}) \times \left(\frac{0.512 \text{ °C}}{\text{m}}\right) = 0.988 \text{ °C} \Rightarrow T_b = 100.000 + 0.988 = 100.988 \text{ °C}$

122. moles solute: $144 \text{ g } C_{12}H_{22}O_{11} \times \dfrac{1 \text{ mol } C_{12}H_{22}O_{11}}{342.34 \text{ g}} = 0.421 \text{ mol } C_{12}H_{22}O_{11}$

mass solution: $1000 \text{ mL} \times \dfrac{1.06 \text{ g}}{\text{mL}} = 1060 \text{ g solution}$

mass solvent: $1060 \text{ g solution} - 144 \text{ g solute} = 916 \text{ g} = 0.916 \text{ kg solvent}$

molality: $m = \dfrac{0.421 \text{ mol } C_2H_6O_2}{0.916 \text{ kg solvent}} = 0.460 \text{ m}$

$\Delta T_f = (0.460 \text{ m}) \times \left(\dfrac{1.86 \text{ °C}}{\text{m}}\right) = 0.856 \text{ °C} \Rightarrow T_f = 0.000 - 0.856 = -0.856 \text{ °C}$

$\Delta T_b = (0.460 \text{ m}) \times \left(\dfrac{0.512 \text{ °C}}{\text{m}}\right) = 0.236 \text{ °C} \Rightarrow T_b = 100.000 + 0.236 = 100.236 \text{ °C}$

123. moles solute: $0.2500 \text{ L} \times \dfrac{5.00 \text{ mol}}{\text{L}} = 1.25 \text{ mol } C_6H_{12}O_6$

mass solute: $1.25 \text{ moles} \times \dfrac{180.18 \text{ g}}{\text{mol}} = 225.23 \text{ g}$

mass solution: $1.40 \text{ L} \times \dfrac{1.06 \text{ g}}{\text{mL}} \times \dfrac{1000 \text{ mL}}{\text{L}} = 1484 \text{ g solution}$

mass solvent: $1484 \text{ g solution} - 225 \text{ g solute} = 1259 \text{ g} = 1.259 \text{ kg solvent}$

molality: $m = \dfrac{1.25 \text{ mol } C_6H_{12}O_6}{1.259 \text{ kg solvent}} = 0.993 \text{ m}$

$\Delta T_f = (0.993 \text{ m}) \times \left(\dfrac{1.86 \text{ °C}}{\text{m}}\right) = 1.85 \text{ °C} \Rightarrow T_f = 0.00 - 1.85 = -1.85 \text{ °C}$

$\Delta T_b = (0.993 \text{ m}) \times \left(\dfrac{0.512 \text{ °C}}{\text{m}}\right) = 0.508 \text{ °C} \Rightarrow T_b = 100.000 + 0.508 = 100.508 \text{ °C}$

124. moles solute: $0.135 \text{ L} \times \dfrac{10.00 \text{ mol}}{\text{L}} = 1.35 \text{ mol } C_2H_6O_2$

mass solute: $1.35 \text{ moles} \times \dfrac{62.08 \text{ g}}{\text{mol}} = 83.81 \text{ g}$

mass solution: $1.50 \text{ L} \times \dfrac{1.05 \text{ g}}{\text{mL}} \times \dfrac{1000 \text{ mL}}{\text{L}} = 1575 \text{ g solution}$

mass solvent: $1575 \text{ g solution} - 83.8 \text{ g solute} = 1491.2 \text{ g} = 1.49 \text{ kg solvent}$

molality: $m = \dfrac{1.35 \text{ mol } C_2H_6O_2}{1.49 \text{ kg solvent}} = 0.906 \text{ m}$

$\Delta T_f = (0.906 \text{ m}) \times \left(\dfrac{1.86 \text{ °C}}{\text{m}}\right) = 1.69 \text{ °C} \Rightarrow T_f = 0.00 - 1.69 = -1.69 \text{ °C}$

$\Delta T_b = (0.906 \text{ m}) \times \left(\dfrac{0.512 \text{ °C}}{\text{m}}\right) = 0.464 \text{ °C} \Rightarrow T_b = 100.000 + 0.464 = 100.464 \text{ °C}$

125. $\dfrac{17.5 \text{ g/MW}}{0.100 \text{ kg}} \times 1.86 \dfrac{°\text{C} \cdot \text{kg}}{\text{mol}} = 1.8 \text{ °C} \Rightarrow \dfrac{17.5 \text{ g}}{\text{MW}} = 0.0968 \text{ mol} \Rightarrow$

$\text{MW} = \dfrac{17.5 \text{ g}}{0.0968 \text{ mol}} = 1.80 \times 10^2 \text{ g/mol}$

126. $\dfrac{35.9 \text{ g/MW}}{0.150 \text{ kg}} \times 1.86 \dfrac{°\text{C} \cdot \text{kg}}{\text{mol}} = 1.3 \text{ °C} \Rightarrow \dfrac{35.9 \text{ g}}{\text{MW}} = 0.105 \text{ mol} \Rightarrow$

$\text{MW} = \dfrac{35.9 \text{ g}}{0.105 \text{ mol}} = 3.4 \times 10^2 \text{ g/mol}$

127. $\Delta T_f = m \times 1.86 \dfrac{°\text{C} \cdot \text{kg}}{\text{mol}} = 6.7 \text{ °C} \Rightarrow m = \dfrac{1.86 \text{ °C} \cdot \text{kg/mol}}{6.7 \text{ °C}} = 3.60 \, m$

$\Delta T_b = 3.60 \, m \times 0.512 \dfrac{°\text{C} \cdot \text{kg}}{\text{mol}} = 1.84 \text{ °C} \Rightarrow \text{Boiling Point} = 101.84 \text{ °C}$

128. $\Delta T_b = m \times 0.512 \dfrac{°\text{C} \cdot \text{kg}}{\text{mol}} = 2.01 \text{°C} \Rightarrow m = 3.93$

$\Delta T_f = 3.93 \, m \times 1.86 \dfrac{°\text{C} \cdot \text{kg}}{\text{mol}} = 7.31 \text{ °C} \Rightarrow \text{Freezing point} = -7.31 \text{ °C}$

129. X = mass glucose, molar mass = 180.18 g/mol

125 − X = mass of sucrose, molar mass = 342.34 g/mol

$1.75 \text{ °C} = \dfrac{\overbrace{\dfrac{X}{180.18} + \dfrac{125-X}{342.34}}^{\text{Total Moles}}}{0.500 \text{ kg}} \times 1.86 \dfrac{°\text{C} \cdot \text{kg}}{\text{mol}} \Rightarrow$

$\dfrac{1.75 \text{ °C} (0.500 \text{ kg})}{1.86 \dfrac{°\text{C} \cdot \text{kg}}{\text{mol}}} = \dfrac{X}{180.18} + \dfrac{125}{342.34} - \dfrac{X}{342.34} \Rightarrow$

$\dfrac{1.75 \text{ °C} (0.500 \text{ kg})}{1.86 \dfrac{°\text{C} \cdot \text{kg}}{\text{mol}}} - \dfrac{125}{342.34} = \dfrac{X}{180.18} - \dfrac{X}{342.34} \Rightarrow$

0.1053 = 0.002629 X

Mass of Glucose = X = 40.1 g

Mass of Sucrose = 125 − X = 85 g

130. X = mass ethylene glycol, molar mass = 62.08 g/mol
13.03 − X = mass of propylene glycol, molar mass = 76.11 g/mol

$$3.50\ °C = \frac{\overbrace{\dfrac{X\ g}{62.08} + \dfrac{13.03 - X}{76.11}}^{\text{Total Moles}}}{0.1000\ \text{kg}} \times 1.86\ \frac{°C \cdot \text{kg}}{\text{mol}} \Rightarrow$$

$$\frac{3.50\ °C\ (0.1000\ \text{kg})}{1.86\ \dfrac{°C \cdot \text{kg}}{\text{mol}}} = \frac{X}{62.08} + \frac{13.03}{76.11} - \frac{X}{76.11} \Rightarrow$$

$$\frac{3.50\ °C\ (0.1000\ \text{kg})}{1.86\ \dfrac{°C \cdot \text{kg}}{\text{mol}}} - \frac{13.03}{76.11} = \frac{X}{62.08} - \frac{X}{76.11} \Rightarrow$$

0.0170 = 0.00297 X

Mass of Ethylene Glycol = $\dfrac{0.0170}{0.00297}$ = 5.72 g

Mass of Propylene Glycol = 13.03 − 5.72 = 7.31 g

Highlight Problems

131. a) left to right
 b) right to left
 c) no movement

132. Aqueous sodium chloride is made up of ions that are dissociated from one another. The picture should show sodium ions separate from chloride ions.

133. $0.100\ \text{g Hg} \times \dfrac{1\ \text{L}}{0.004\ \text{mg}} \times \dfrac{1000\ \text{mg}}{1\ \text{g}} = 3 \times 10^4\ \text{L}$

134. $2.4\ \text{g Na} \times \dfrac{100\ \text{g solution}}{0.050\ \text{g Na}} \times \dfrac{1\ \text{mL}}{1.0\ \text{g}} \times \dfrac{1\ \text{L}}{1000\ \text{mL}} = 4.8\ \text{L}$

135. mass percent: $\dfrac{13.62\ \text{g sucrose}}{13.62\ \text{g sucrose} + \left(241.5\ \text{mL} \times \dfrac{0.997\ \text{g}}{1\ \text{mL}}\right)} \times 100\% = 5.35\%$

molarity: $\dfrac{13.62\ \text{g}\ C_{12}H_{22}O_{11} \times \dfrac{1\ \text{mol}\ C_{12}H_{22}O_{11}}{342.3\ \text{g}}}{0.250\ \text{L}} = 0.159\ M\ C_{12}H_{22}O_{11}$

molality: $\dfrac{13.62\ \text{g}\ C_{12}H_{22}O_{11} \times \dfrac{1\ \text{mol}\ C_{12}H_{22}O_{11}}{342.3\ \text{g}}}{241.5\ \text{mL} \times \dfrac{0.997\ \text{g}}{1\ \text{mL}} \times \dfrac{1\ \text{kg}}{1000\ \text{g}}} = 0.165\ m\ C_{12}H_{22}O_{11}$

136. $\Delta T_b = 0.165 \; m \times 0.512 \dfrac{°C \cdot kg}{mol} = 0.0845 \; °C$, bp = 100.0845 °C

$\Delta T_f = 0.165 \; m \times 1.86 \dfrac{°C \cdot kg}{mol} = 0.307 \; °C \Rightarrow$ Freezing point = –0.307 °C

Boiling point is elevated, so it would take longer to reach the new boiling point as compared to pure water. The syrup would freeze in a typical freezer as the freezing point is easily within the range of a typical freezer at –18 °C.

137. $M_1 V_1 = M_2 V_2$

(4 M) (1 L) = (M_2)(4 L)

$M_2 = \dfrac{(4 \; M)(1 \; L)}{(4 \; L)} = 1 \; M$

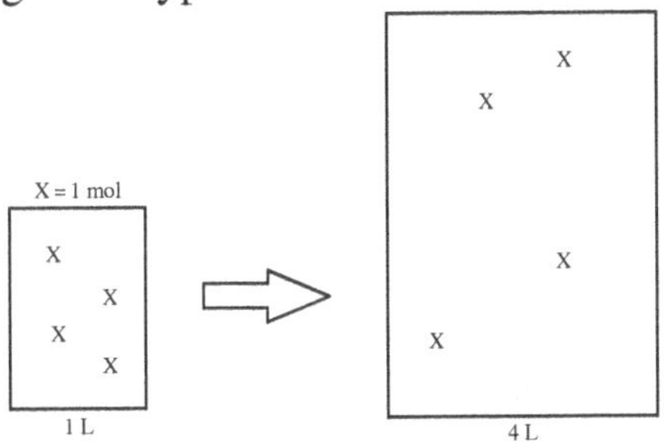

138. $1 \; lb \; U \times \dfrac{453.6 \; g \; U}{1 \; lb} \times \dfrac{1 \; L}{3.2 \; \mu g \; U} \times \dfrac{1 \; \mu g}{1 \times 10^{-6} g} = 1.4 \times 10^8 \; L$

Data Interpretation and Analysis

139. a) First Draw: $\dfrac{124.265}{15} = 8.28$ ppb Pb; 2-min Flush: $\dfrac{25.329}{15} = 1.69$ ppb Pb

b) Yes, the average level of lead drops from 8.28 ppb to 1.69 ppb. The lead came from the pipes. The longer the water is in contact with the pipes, the more lead was dissolved. By running the water for 2 minutes, freshwater was being introduced.

c) Percent of first draw samples exceeding 15 ppb: $\dfrac{3}{15} \times 100 = 20\%$

Percent of 2-min flush samples exceeding 15 ppb: $\dfrac{0}{15} \times 100 = 0\%$

Most people fill their containers with the water that first comes out of a tap, so the first draw samples are more accurate.

d) $\dfrac{2 \; L \; H_2O}{1 \; day} \times \dfrac{365 \; days}{1 \; yr} \times \dfrac{1000 \; g \; H_2O}{1 \; L \; H_2O} \times \dfrac{40.63 \; g \; Pb}{10^9 \; g \; H_2O} = 0.02966 \; g \; Pb$

Acids and Bases

14

Questions

1. The sour taste is due to the presence of acids in the candy, specifically citric acid and tartaric acid.

2. The properties of acids are:
   a) Sour taste
   b) The ability to dissolve many metals
   c) The ability to turn litmus paper red
   Some foods that contain acids are lemons, limes, sourdough bread, and tomatoes.

3. The main component of stomach acid is hydrochloric acid. The role of acid in our stomach is to kill bacteria and to start breaking food down.

4. Organic acids contain the carboxylic acid group of atoms. Two examples of organic acids are citric acid and acetic acid.

5. The properties of bases are:
   a) Bitter taste
   b) Slippery feel
   c) The ability to turn litmus paper blue
   Some products that contain bases include drain clog removers such as Drano, baking soda, and antacids for acid indigestion.

6. Alkaloids are organic bases found in plants that are often poisonous.

7. An Arrhenius acid produces H+ in aqueous solution. $HCl(aq) \rightarrow H^+(aq) + Cl^-(aq)$

8. An Arrhenius base produces OH⁻ ions in aqueous solutions.
   $NaOH(aq) \rightarrow Na^+(aq) + OH^-(aq)$

9. A Brønsted-Lowry acid is a proton (H⁺) donor.
   $HCl(aq) + H_2O(l) \rightarrow H_3O^+(aq) + Cl^-(aq)$
   A Brønsted-Lowry base is a proton acceptor.
   $NH_3(aq) + H_2O(l) \rightarrow NH_4^+(aq) + OH^-(aq)$

10. A conjugate acid–base pair is when two substances differ only by a single proton (H⁺). An example of a conjugate acid–base pair is H_2O and H_3O^+.

11. An acid–base neutralization reaction occurs when the hydrogen ion from the acid reacts with the hydroxide ion from the base to form water.
    $HCl(aq) + NaOH(aq) \rightarrow H_2O(l) + NaOH(aq)$

12. $2HCl(aq) + Fe(s) \rightarrow H_2(g) + FeCl_2(aq)$

13. $2HCl(aq) + K_2O(s) \rightarrow H_2O(l) + 2KCl(aq)$

14. Aluminum will react with a base:
    $2NaOH(aq) + 2Al(s) + 6H_2O(l) \rightarrow 2NaAl(OH)_4(aq) + 3H_2(g)$

15. A titration is a laboratory procedure in which a reactant of known concentration reacts with another of unknown concentration. The volumes of each solution are carefully monitored until the equivalence point is reached, usually when an indicator causes the reaction mixture to change colors. The equivalence point is an experimental point when an exact stoichiometric amount of each reactant has been added.

16. It would take 0.85 mol of H^+ to reach the equivalence point.

17. A strong acid will completely dissociate into component ions in solution to form a strong electrolyte. A weak acid will partially dissociate into component ions in solution to form a weak electrolyte.

18. The stronger the acid, the weaker the conjugate base is, and vice versa.

19. A monoprotic acid contains only 1 hydrogen ion that will dissociate in solution. A diprotic acid contains 2 hydrogen ions that will dissociate in solution.

20. A strong base will completely dissociate into component ions in solution to form a strong electrolyte. A weak base will react with water to a small extent to form a weak electrolyte solution of the hydroxide ion.

21. Yes, pure water contains H_3O^+ because of the process of self-ionization.

22. The $[OH^-]$ decreases when $[H_3O^+]$ increases.

23. a) $[H_3O^+] > 1.0 \times 10^{-7}$ and $[OH^-] < 1.0 \times 10^{-7}$
    b) $[H_3O^+] < 1.0 \times 10^{-7}$ and $[OH^-] > 1.0 \times 10^{-7}$
    c) $[H_3O^+] = 1.0 \times 10^{-7}$ and $[OH^-] = 1.0 \times 10^{-7}$

24. $pH = -\log [H_3O^+]$; A one-unit change in pH corresponds to a 10× change in $[H_3O^+]$.

25. $pOH = -\log [OH^-]$; A change of 2 pOH units corresponds to a 100× change in $[OH^-]$.

26. The equation $pH + pOH = 14$ is derived using K_w, as shown here:
    $[H_3O^+][OH^-] = 1.0 \times 10^{-14}$
    $\log \{[H_3O^+][OH^-]\} = \log (1.0 \times 10^{-14})$
    $\log [H_3O^+] + \log [OH^-] = -14.00$
    $-\log [H_3O^+] - \log [OH^-] = 14.00$
    $pH + pOH = 14.00$

27. A buffer is a solution that will resist a change in pH, as the weak acid can react with added base and the conjugate base can react with added acid.

28. The main components of a buffer are a weak acid and its conjugate base (or a weak base and its conjugate acid).

Problems

Acid and Base Definitions

29. a) acid: $H_2SO_4(aq) \rightarrow H^+(aq) + HSO_4^-(aq)$
    b) base: $Sr(OH)_2(aq) \rightarrow Sr^{2+}(aq) + 2OH^-(aq)$
    c) acid: $HBr(aq) \rightarrow H^+(aq) + Br^-(aq)$
    d) base: $NaOH(aq) \rightarrow Na^+(aq) + OH^-(aq)$

30. a) base: $Ca(OH)_2(aq) \rightarrow Ca^{2+}(aq) + 2OH^-(aq)$
    b) acid: $HC_2H_3O_2(aq) \rightarrow H^+(aq) + C_2H_3O_2^-(aq)$
    c) base: $KOH(aq) \rightarrow K^+(aq) + OH^-(aq)$
    d) acid: $HNO_3(aq) \rightarrow H^+(aq) + NO_3^-(aq)$

31. | B-L Acid | B-L Base | Conj. Acid | Conj. Base |
    |---|---|---|---|
    | a) HBr | H_2O | H_3O^+ | Br^- |
    | b) H_2O | NH_3 | NH_4^+ | OH^- |
    | c) HNO_3 | H_2O | H_3O^+ | NO_3^- |
    | d) H_2O | C_5H_5N | $C_5H_5NH^+$ | OH^- |

32. | B-L Acid | B-L Base | Conj. Acid | Conj. Base |
    |---|---|---|---|
    | a) HI | H_2O | H_3O^+ | I^- |
    | b) H_2O | CH_3NH_2 | $CH_3NH_3^+$ | OH^- |
    | c) H_2O | CO_3^{2-} | HCO_3^- | OH^- |
    | d) H_2CO_3 | H_2O | H_3O^+ | HCO_3^- |

33. a) conjugate acid–base pairs
    b) not conjugate acid–base pairs
    c) conjugate acid–base pairs
    d) not conjugate acid–base pairs

34. a) conjugate acid–base pairs
    b) not conjugate acid–base pairs
    c) conjugate acid–base pairs
    d) not conjugate acid–base pairs

35. a) Cl^-
    b) HSO_3^-
    c) CHO_2^-
    d) F^-

36. a) Br⁻
    b) HCO₃⁻
    c) ClO₄⁻
    d) C₂H₃O₂⁻

37. a) NH₄⁺
    b) HClO₄
    c) H₂SO₄
    d) HCO₃⁻

38. a) CH₃NH₃⁺
    b) HC₅H₅N⁺
    c) HCl
    d) HF

Acid–Base Reactions

39. a) HI(aq) + NaOH(aq) → H₂O(l) + NaI(aq)
    b) HBr(aq) + KOH(aq) → H₂O(l) + KBr(aq)
    c) 2HNO₃(aq) + Ba(OH)₂(aq) → 2H₂O(l) + Ba(NO₃)₂(aq)
    d) 2HClO₄(aq) + Sr(OH)₂(aq) → 2H₂O(l) + Sr(ClO₄)₂(aq)

40. a) 2HF(aq) + Ba(OH)₂(aq) → 2H₂O(l) + BaF₂(aq)
    b) HClO₄(aq) + NaOH(aq) → H₂O(l) + NaClO₄(aq)
    c) 2HBr(aq) + Ca(OH)₂(aq) → 2H₂O(l) + CaBr₂(aq)
    d) HCl(aq) + KOH(aq) → H₂O(l) + KCl(aq)

41. a) 2Rb(s) + 2HBr(aq) → H₂(g) + 2RbBr(aq)
    b) Mg(s) + 2HBr(aq) → H₂(g) + MgBr₂(aq)
    c) Ba(s) + 2HBr(aq) → H₂(g) + BaBr₂(aq)
    d) 2Al(s) + 6HBr(aq) → 3H₂(g) + 2AlCl₃(aq)

42. a) 2K(s) + 2HCl(aq) → H₂(g) + 2KCl(aq)
    b) Ca(s) + 2HCl(aq) → H₂(g) + CaCl₂(aq)
    c) 2Na(s) + 2HCl(aq) → H₂(g) + 2NaCl(aq)
    d) Sr(s) + 2HCl(aq) → H₂(g) + SrCl₂(aq)

43. a) MgO(s) + 2HI(aq) → H₂O(l) + MgI₂(aq)
    b) K₂O(s) + 2HI(aq) → H₂O(l) + 2KI(aq)
    c) Rb₂O(s) + 2HI(aq) → H₂O(l) + 2RbI(aq)
    d) CaO(s) + 2HI(aq) → H₂O(l) + CaI₂(aq)

44. a) SrO(s) + 2HCl(aq) → H₂O(l) + SrCl₂(aq)
    b) Na₂O(s) + 2HCl(aq) → H₂O(l) + 2NaCl(aq)
    c) Li₂O(s) + 2HCl(aq) → H₂O(l) + 2LiCl(aq)
    d) BaO(s) + 2HCl(aq) → H₂O(l) + BaCl₂(aq)

45. a) $6HClO_4(aq) + Fe_2O_3(s) \rightarrow 2Fe(ClO_4)_3(aq) + 3H_2O(l)$
    b) $H_2SO_4(aq) + Sr(s) \rightarrow SrSO_4(aq) + H_2(g)$
    c) $H_3PO_4(aq) + 3KOH(aq) \rightarrow 3H_2O(l) + K_3PO_4(aq)$

46. a) $6HI(aq) + 2Al(s) \rightarrow 3H_2(g) + 2AlI_3(aq)$
    b) $2H_2SO_4(aq) + TiO_2(s) \rightarrow 2H_2O(l) + Ti(SO_4)_2(aq)$
    c) $H_2CO_3(aq) + 2LiOH(aq) \rightarrow 2H_2O(l) + Li_2CO_3(aq)$

Acid–Base Titrations

47. $HCl(aq) + NaOH(aq) \rightarrow H_2O(l) + NaCl(aq)$

    a) $0.02844 \text{ L NaOH} \times \dfrac{0.1231 \text{ mol NaOH}}{1 \text{ L}} \times \dfrac{1 \text{ mol HCl}}{1 \text{ mol NaOH}} \times \dfrac{1}{0.02500 \text{ L HCl}} =$
    0.1400 M HCl

    b) $0.02122 \text{ L NaOH} \times \dfrac{0.0972 \text{ mol NaOH}}{1 \text{ L}} \times \dfrac{1 \text{ mol HCl}}{1 \text{ mol NaOH}} \times \dfrac{1}{0.01500 \text{ L HCl}} =$
    0.138 M HCl

    c) $0.01488 \text{ L NaOH} \times \dfrac{0.1178 \text{ mol NaOH}}{1 \text{ L}} \times \dfrac{1 \text{ mol HCl}}{1 \text{ mol NaOH}} \times \dfrac{1}{0.02000 \text{ L HCl}} =$
    0.08764 M HCl

    d) $0.00688 \text{ L NaOH} \times \dfrac{0.1325 \text{ mol NaOH}}{1 \text{ L}} \times \dfrac{1 \text{ mol HCl}}{1 \text{ mol NaOH}} \times \dfrac{1}{0.00500 \text{ L HCl}} =$
    0.182 M HCl

48. $HCl(aq) + NaOH(aq) \rightarrow H_2O(l) + NaCl(aq)$

    a) $0.00977 \text{ L HCl} \times \dfrac{0.1599 \text{ mol HCl}}{1 \text{ L}} \times \dfrac{1 \text{ mol NaOH}}{1 \text{ mol HCl}} \times \dfrac{1}{0.00500 \text{ L HCl}} =$
    0.312 M NaOH

    b) $0.01134 \text{ L HCl} \times \dfrac{0.1311 \text{ mol HCl}}{1 \text{ L}} \times \dfrac{1 \text{ mol NaOH}}{1 \text{ mol HCl}} \times \dfrac{1}{0.01500 \text{ L HCl}} =$
    0.09911 M NaOH

    c) $0.01055 \text{ L HCl} \times \dfrac{0.0889 \text{ mol HCl}}{1 \text{ L}} \times \dfrac{1 \text{ mol NaOH}}{1 \text{ mol HCl}} \times \dfrac{1}{0.01000 \text{ L HCl}} =$
    0.0938 M NaOH

    d) $0.03618 \text{ L HCl} \times \dfrac{0.1021 \text{ mol HCl}}{1 \text{ L}} \times \dfrac{1 \text{ mol NaOH}}{1 \text{ mol HCl}} \times \dfrac{1}{0.03000 \text{ L HCl}} =$
    0.1231 M NaOH

49. $H_2SO_4(aq) + 2KOH(aq) \rightarrow 2H_2O(l) + K_2SO_4(aq)$

$0.04122 \text{ L KOH} \times \dfrac{0.1322 \text{ mol KOH}}{1 \text{ L}} \times \dfrac{1 \text{ mol } H_2SO_4}{2 \text{ mol KOH}} \times \dfrac{1}{0.02500 \text{ L } H_2SO_4} =$

$0.1090 \text{ M } H_2SO_4$

50. $H_3PO_4(aq) + 3NaOH(aq) \rightarrow 3H_2O(l) + Na_3PO_4(aq)$

$0.00712 \text{ L NaOH} \times \dfrac{0.1090 \text{ mol NaOH}}{1 \text{ L}} \times \dfrac{1 \text{ mol } H_3PO_4}{3 \text{ mol NaOH}} \times \dfrac{1}{0.00500 \text{ L } H_3PO_4}$

$= 0.0517 \text{ M } H_3PO_4$

51. $H_2SO_4(aq) + 2NaOH(aq) \rightarrow 2H_2O(l) + Na_2SO_4(aq)$

$0.0100 \text{ L } H_2SO_4 \times \dfrac{0.102 \text{ mol } H_2SO_4}{1 \text{ L}} \times \dfrac{2 \text{ mol NaOH}}{1 \text{ mol } H_2SO_4} \times \dfrac{1 \text{ L}}{0.121 \text{ mol NaOH}} \times \dfrac{1000 \text{ mL}}{1 \text{ L}}$

$= 16.9 \text{ mL}$

52. $H_3PO_4(aq) + 3NaOH(aq) \rightarrow 3H_2O(l) + Na_3PO_4(aq)$

$0.0150 \text{ L } H_3PO_4 \times \dfrac{0.124 \text{ mol } H_3PO_4}{1 \text{ L}} \times \dfrac{3 \text{ mol NaOH}}{1 \text{ mol } H_3PO_4} \times \dfrac{1 \text{ L}}{0.0985 \text{ mol NaOH}} \times \dfrac{1000 \text{ mL}}{1 \text{ L}}$

$= 56.6 \text{ mL}$

Strong and Weak Acids and Bases

53. a) strong
    b) weak
    c) strong
    d) weak

54. a) weak
    b) strong
    c) strong
    d) weak

55. a) $[H_3O^+] = 1.7$ M
    b) $[H_3O^+] = 1.5$ M
    c) $[H_3O^+] < 0.38$ M
    d) $[H_3O^+] < 1.75$ M

56. a) $[H_3O^+] < 0.125$ M
    b) $[H_3O^+] < 1.25$ M
    c) $[H_3O^+] = 2.77$ M
    d) $[H_3O^+] < 0.95$ M

57. a) strong
    b) weak
    c) strong
    d) weak

58. a) weak
    b) strong
    c) strong
    d) strong

59. a) [OH⁻] = 0.25 M
    b) [OH⁻] < 0.25 M
    c) [OH⁻] = 0.50 M
    d) [OH⁻] = 1.25 M

60. a) [OH⁻] = 2.5 M
    b) [OH⁻] < 1.95 M
    c) [OH⁻] = 0.450 M
    d) [OH⁻] < 1.8 M

Acidity, Basicity, and K_w

61. a) acidic
    b) acidic
    c) neutral
    d) basic

62. a) basic
    b) basic
    c) acidic
    d) basic

63. a) $K_w = [H_3O^+][OH^-] \Rightarrow [OH^-] = \dfrac{K_w}{[H_3O^+]} = \dfrac{1.0 \times 10^{-14}}{1.5 \times 10^{-9}} = 6.7 \times 10^{-6}$ M, Basic

   b) $K_w = [H_3O^+][OH^-] \Rightarrow [OH^-] = \dfrac{K_w}{[H_3O^+]} = \dfrac{1.0 \times 10^{-14}}{9.3 \times 10^{-9}} = 1.1 \times 10^{-6}$ M, Basic

   c) $K_w = [H_3O^+][OH^-] \Rightarrow [OH^-] = \dfrac{K_w}{[H_3O^+]} = \dfrac{1.0 \times 10^{-14}}{2.2 \times 10^{-6}} = 4.5 \times 10^{-9}$ M, Acidic

   d) $K_w = [H_3O^+][OH^-] \Rightarrow [OH^-] = \dfrac{K_w}{[H_3O^+]} = \dfrac{1.0 \times 10^{-14}}{7.4 \times 10^{-4}} = 1.4 \times 10^{-11}$ M, Acidic

64. a) $K_w = [H_3O^+][OH^-] \Rightarrow [OH^-] = \dfrac{K_w}{[H_3O^+]} = \dfrac{1.0 \times 10^{-14}}{1.3 \times 10^{-3}} = 7.7 \times 10^{-12}$ M, Acidic

b) $K_w = [H_3O^+][OH^-] \Rightarrow [OH^-] = \dfrac{K_w}{[H_3O^+]} = \dfrac{1.0 \times 10^{-14}}{9.1 \times 10^{-12}} = 1.1 \times 10^{-3}$ M, Basic

c) $K_w = [H_3O^+][OH^-] \Rightarrow [OH^-] = \dfrac{K_w}{[H_3O^+]} = \dfrac{1.0 \times 10^{-14}}{5.2 \times 10^{-4}} = 1.9 \times 10^{-11}$ M, Acidic

d) $K_w = [H_3O^+][OH^-] \Rightarrow [OH^-] = \dfrac{K_w}{[H_3O^+]} = \dfrac{1.0 \times 10^{-14}}{6.1 \times 10^{-9}} = 1.6 \times 10^{-6}$ M, Basic

65. a) $K_w = [H_3O^+][OH^-] \Rightarrow [H_3O^+] = \dfrac{K_w}{[OH^-]} = \dfrac{1.0 \times 10^{-14}}{2.7 \times 10^{-12}} = 3.7 \times 10^{-3}$ M, Acidic

b) $K_w = [H_3O^+][OH^-] \Rightarrow [H_3O^+] = \dfrac{K_w}{[OH^-]} = \dfrac{1.0 \times 10^{-14}}{2.5 \times 10^{-2}} = 4.0 \times 10^{-13}$ M, Basic

c) $K_w = [H_3O^+][OH^-] \Rightarrow [H_3O^+] = \dfrac{K_w}{[OH^-]} = \dfrac{1.0 \times 10^{-14}}{1.1 \times 10^{-10}} = 9.1 \times 10^{-5}$ M, Acidic

d) $K_w = [H_3O^+][OH^-] \Rightarrow [H_3O^+] = \dfrac{K_w}{[OH^-]} = \dfrac{1.0 \times 10^{-14}}{3.3 \times 10^{-4}} = 3.0 \times 10^{-11}$ M, Basic

66. a) $K_w = [H_3O^+][OH^-] \Rightarrow [H_3O^+] = \dfrac{K_w}{[OH^-]} = \dfrac{1.0 \times 10^{-14}}{2.1 \times 10^{-11}} = 4.8 \times 10^{-4}$ M, Acidic

b) $K_w = [H_3O^+][OH^-] \Rightarrow [H_3O^+] = \dfrac{K_w}{[OH^-]} = \dfrac{1.0 \times 10^{-14}}{7.5 \times 10^{-9}} = 1.3 \times 10^{-6}$ M, Acidic

c) $K_w = [H_3O^+][OH^-] \Rightarrow [H_3O^+] = \dfrac{K_w}{[OH^-]} = \dfrac{1.0 \times 10^{-14}}{2.1 \times 10^{-4}} = 4.8 \times 10^{-11}$ M, Basic

d) $K_w = [H_3O^+][OH^-] \Rightarrow [H_3O^+] = \dfrac{K_w}{[OH^-]} = \dfrac{1.0 \times 10^{-14}}{1.0 \times 10^{-2}} = 1.0 \times 10^{-12}$ M, Basic

pH

67. a) basic
    b) neutral
    c) acidic
    d) acidic

68. a) acidic
    b) acidic
    c) basic
    d) acidic

69.
a) $pH = -\log[H^+] = -\log[1.7 \times 10^{-8}] = 7.77$
b) $pH = -\log[H^+] = -\log[1.0 \times 10^{-7}] = 7.00$
c) $pH = -\log[H^+] = -\log[2.2 \times 10^{-6}] = 5.66$
d) $pH = -\log[H^+] = -\log[7.4 \times 10^{-4}] = 3.13$

70.
a) $pH = -\log[H^+] = -\log[2.4 \times 10^{-10}] = 9.62$
b) $pH = -\log[H^+] = -\log[7.6 \times 10^{-2}] = 1.12$
c) $pH = -\log[H^+] = -\log[9.2 \times 10^{-13}] = 12.04$
d) $pH = -\log[H^+] = -\log[3.4 \times 10^{-5}] = 4.47$

71.
a) $[H_3O^+] = 10^{-pH} = 10^{-8.55} = 2.8 \times 10^{-9} M$
b) $[H_3O^+] = 10^{-pH} = 10^{-11.23} = 5.9 \times 10^{-12} M$
c) $[H_3O^+] = 10^{-pH} = 10^{-2.87} = 1.3 \times 10^{-3} M$
d) $[H_3O^+] = 10^{-pH} = 10^{-1.22} = 6.0 \times 10^{-2} M$

72.
a) $[H_3O^+] = 10^{-pH} = 10^{-1.76} = 1.7 \times 10^{-2} M$
b) $[H_3O^+] = 10^{-pH} = 10^{-3.88} = 1.3 \times 10^{-4} M$
c) $[H_3O^+] = 10^{-pH} = 10^{-8.43} = 3.7 \times 10^{-9} M$
d) $[H_3O^+] = 10^{-pH} = 10^{-12.32} = 4.8 \times 10^{-13} M$

73.
a) $[H_3O^+] = \dfrac{K_w}{[OH^-]} = \dfrac{1.0 \times 10^{-14}}{1.9 \times 10^{-7}} = 5.3 \times 10^{-8} M$, $pH = -\log 5.3 \times 10^{-8} = 7.28$
b) $[H_3O^+] = \dfrac{K_w}{[OH^-]} = \dfrac{1.0 \times 10^{-14}}{2.6 \times 10^{-8}} = 3.8 \times 10^{-7} M$, $pH = -\log 3.8 \times 10^{-7} = 6.41$
c) $[H_3O^+] = \dfrac{K_w}{[OH^-]} = \dfrac{1.0 \times 10^{-14}}{7.2 \times 10^{-11}} = 1.4 \times 10^{-4} M$, $pH = -\log 1.4 \times 10^{-4} = 3.86$
d) $[H_3O^+] = \dfrac{K_w}{[OH^-]} = \dfrac{1.0 \times 10^{-14}}{9.5 \times 10^{-2}} = 1.05 \times 10^{-13} M$, $pH = -\log 1.05 \times 10^{-13} = 12.98$

74.
a) $[H_3O^+] = \dfrac{K_w}{[OH^-]} = \dfrac{1.0 \times 10^{-14}}{2.8 \times 10^{-11}} = 3.57 \times 10^{-4} M$, $pH = -\log 3.57 \times 10^{-4} = 3.45$
b) $[H_3O^+] = \dfrac{K_w}{[OH^-]} = \dfrac{1.0 \times 10^{-14}}{9.6 \times 10^{-3}} = 1.04 \times 10^{-12} M$, $pH = -\log 1.04 \times 10^{-12} = 11.98$
c) $[H_3O^+] = \dfrac{K_w}{[OH^-]} = \dfrac{1.0 \times 10^{-14}}{3.8 \times 10^{-12}} = 2.63 \times 10^{-3} M$, $pH = -\log 2.63 \times 10^{-3} = 2.58$
d) $[H_3O^+] = \dfrac{K_w}{[OH^-]} = \dfrac{1.0 \times 10^{-14}}{6.4 \times 10^{-4}} = 1.56 \times 10^{-11} M$, $pH = -\log 1.56 \times 10^{-11} = 10.81$

75. a) $[H_3O^+] = 10^{-pH} = 10^{-4.25} = 5.62 \times 10^{-5} M$, $[OH^-] = \dfrac{K_w}{[H_3O^+]} = \dfrac{1.0 \times 10^{-14}}{5.62 \times 10^{-5}}$

$= 1.8 \times 10^{-10} M$

b) $[H_3O^+] = 10^{-pH} = 10^{-12.53} = 2.95 \times 10^{-13} M$, $[OH^-] = \dfrac{K_w}{[H_3O^+]} = \dfrac{1.0 \times 10^{-14}}{2.95 \times 10^{-13}}$

$= 3.4 \times 10^{-2} M$

c) $[H_3O^+] = 10^{-pH} = 10^{-1.50} = 3.16 \times 10^{-2} M$, $[OH^-] = \dfrac{K_w}{[H_3O^+]} = \dfrac{1.0 \times 10^{-14}}{3.16 \times 10^{-2}}$

$= 3.2 \times 10^{-13} M$

d) $[H_3O^+] = 10^{-pH} = 10^{-8.25} = 5.62 \times 10^{-9} M$, $[OH^-] = \dfrac{K_w}{[H_3O^+]} = \dfrac{1.0 \times 10^{-14}}{5.62 \times 10^{-9}}$

$= 1.8 \times 10^{-6} M$

76. a) $[H_3O^+] = 10^{-pH} = 10^{-1.82} = 1.51 \times 10^{-2} M$, $[OH^-] = \dfrac{K_w}{[H_3O^+]} = \dfrac{1.0 \times 10^{-14}}{1.51 \times 10^{-2}}$

$= 6.6 \times 10^{-13} M$

b) $[H_3O^+] = 10^{-pH} = 10^{-13.28} = 5.24 \times 10^{-14} M$, $[OH^-] = \dfrac{K_w}{[H_3O^+]} = \dfrac{1.0 \times 10^{-14}}{5.24 \times 10^{-14}}$

$= 1.9 \times 10^{-1} M$

c) $[H_3O^+] = 10^{-pH} = 10^{-8.29} = 5.13 \times 10^{-9} M$, $[OH^-] = \dfrac{K_w}{[H_3O^+]} = \dfrac{1.0 \times 10^{-14}}{5.13 \times 10^{-9}}$

$= 1.9 \times 10^{-6} M$

d) $[H_3O^+] = 10^{-pH} = 10^{-2.32} = 4.78 \times 10^{-3} M$, $[OH^-] = \dfrac{K_w}{[H_3O^+]} = \dfrac{1.0 \times 10^{-14}}{4.78 \times 10^{-3}}$

$= 2.1 \times 10^{-12} M$

77. a) pH = −log 0.0155 = 1.810
b) pOH = −log 1.28 × 10^{-3} = 2.893; pH = 14.000 − 2.893 = 11.107
c) pH = −log 1.89 × 10^{-3} = 2.724
d) pOH = −log 2(1.54 × 10^{-4}) = 3.511; pH = 14.000 − 3.511 = 10.489

78. a) pH = −log 1.34 × 10^{-3} = 2.873
b) pOH = −log 0.0211 = 1.676; pH = 14.000 − 1.676 = 12.324
c) pH = −log 0.0109 = 1.963
d) pOH = −log 2(7.02 × 10^{-5}) = 3.853; pH = 14.000 − 3.853 = 10.147

pOH

79. a) pOH = −log [1.5 × 10⁻⁹] = 8.82, acidic solution (pH = 5.18)
    b) pOH = −log [7.0 × 10⁻⁵] = 4.15, basic solution (pH = 9.85)
    c) pOH = −log [1.0 × 10⁻⁷] = 7.00, neutral solution (pH = 7.00)
    d) pOH = −log [8.8 × 10⁻³] = 2.06, basic solution (pH = 11.94)

80. a) pOH = −log [4.5 × 10⁻²] = 1.35, basic solution (pH = 12.65)
    b) pOH = −log [3.1 × 10⁻¹²] = 11.51, acidic solution (pH = 2.49)
    c) pOH = −log [5.4 × 10⁻⁵] = 4.27, basic solution (pH = 9.73)
    d) pOH = −log [1.2 × 10⁻²] = 1.92, basic solution (pH = 12.08)

81. a) pOH = 14 − pH; pOH = 14 − −log [1.2 × 10⁻⁸] = 6.08
    b) pOH = 14 − pH; pOH = 14 − −log [5.5 × 10⁻²] = 12.74
    c) pOH = 14 − pH; pOH = 14 − −log [3.9 × 10⁻⁹] = 5.59
    d) pOH = −log [OH⁻] = −log [1.88 × 10⁻¹³] = 12.726

82. a) pOH = 14 − pH; pOH = 14 − −log [8.3 × 10⁻¹⁰] = 4.92
    b) pOH = 14 − pH; pOH = 14 − −log [1.6 × 10⁻⁷] = 7.20
    c) pOH = 14 − pH; pOH = 14 − −log [7.3 × 10⁻²] = 12.86
    d) pOH = −log [OH⁻] = −log [4.32 × 10⁻⁴] = 3.365

83. a) pH = 14 − pOH; pH = 14 − 8.5 = 5.5, acidic
    b) pH = 14 − pOH; pH = 14 − 4.2 = 9.8, basic
    c) pH = 14 − pOH; pH = 14 − 1.7 = 12.3, basic
    d) pH = 14 − pOH; pH = 14 − 7.0 = 7.0, neutral

84. a) pH = 14 − pOH; pH = 14 − 12.5 = 1.5, acidic
    b) pH = 14 − pOH; pH = 14 − 5.5 = 8.5, basic
    c) pH = 14 − pOH; pH = 14 − 0.55 = 13.45, basic
    d) pH = 14 − pOH; pH = 14 − 7.98 = 6.02, acidic

Buffers and Acid Rain

85. a) not a buffer
    b) not a buffer
    c) buffer
    d) buffer

86. a) not a buffer
    b) buffer
    c) not a buffer
    d) not a buffer

87. c) NaF(*aq*) + HCl(*aq*) → HF(*aq*) + NaCl(*aq*)
    d) KC₂H₃O₂(*aq*) + HCl(*aq*) → HC₂H₃O₂(*aq*) + KCl(*aq*)

88. b) $HCHO_2(aq) + NaOH(aq) \rightarrow H_2O(l) + NaCHO_2(aq)$

89. a) $HC_2H_3O_2$
    b) NaH_2PO_4
    c) $NaCHO_2$

90. a) H_2SO_3
    b) NaF
    c) $HCHO_2$

Cumulative Exercises

91. $HCl(aq) + NaOH(aq) \rightarrow H_2O(l) + NaCl(aq)$

$$\frac{0.250 \text{ mol NaOH}}{1 \text{ L}} \times 0.0200 \text{ L} \times \frac{1 \text{ mol HCl}}{1 \text{ mol NaOH}} \times \frac{1 \text{ L}}{0.100 \text{ mol HCl}} = 0.0500 \text{ L}$$

92. $HClO_4(aq) + KOH(aq) \rightarrow H_2O(l) + KClO_4(aq)$

$$\frac{0.150 \text{ mol HClO}_4}{1 \text{ L}} \times 0.0250 \text{ L} \times \frac{1 \text{ mol KOH}}{1 \text{ mol HClO}_4} \times \frac{1 \text{ L}}{0.200 \text{ mol KOH}} = 0.0188 \text{ L}$$

93. $Mg(s) + 2HCl(aq) \rightarrow H_2(g) + MgCl_2(aq)$

$$10.0 \text{ g Mg} \times \frac{1 \text{ mol Mg}}{24.31 \text{ g}} \times \frac{2 \text{ mol HCl}}{1 \text{ mol Mg}} \times \frac{1 \text{ L}}{5.0 \text{ mol HCl}} = 0.16 \text{ L}$$

94. $2K(s) + 2HBr(aq) \rightarrow H_2(g) + 2KBr(aq)$

$$15.0 \text{ g K} \times \frac{1 \text{ mol K}}{39.10 \text{ g}} \times \frac{2 \text{ mol HBr}}{2 \text{ mol K}} \times \frac{1 \text{ L}}{3.0 \text{ mol HBr}} = 0.13 \text{ L}$$

95. $K_2O(s) + 2HI(aq) \rightarrow H_2O(l) + 2KI(aq)$

$$18.5 \text{ g K}_2\text{O} \times \frac{1 \text{ mol K}_2\text{O}}{94.20 \text{ g}} \times \frac{2 \text{ mol KI}}{1 \text{ mol K}_2\text{O}} \times \frac{166.0 \text{ g}}{1 \text{ mol KI}} = 65.2 \text{ g KI}$$

96. $CaO(s) + 2HBr(aq) \rightarrow H_2O(l) + CaBr_2(aq)$

$$5.88 \text{ g CaO} \times \frac{1 \text{ mol CaO}}{56.08 \text{ g}} \times \frac{1 \text{ mol CaBr}_2}{1 \text{ mol CaO}} \times \frac{199.9 \text{ g}}{1 \text{ mol CaBr}_2} = 21.0 \text{ g CaBr}_2$$

97. $HX(aq) + NaOH(aq) \rightarrow H_2O(l) + NaX(aq)$

$$\frac{0.1003 \text{ mol NaOH}}{1 \text{ L}} \times 0.02077 \text{ L} \times \frac{1 \text{ mol HX}}{1 \text{ mol NaOH}} = 2.083 \times 10^{-3} \text{ mol}$$

$$\text{molar mass} = \frac{0.125 \text{ g}}{2.083 \times 10^{-3} \text{ mol}} = 60.0 \text{ g/mol}$$

98. $H_2X(aq) + 2NaOH(aq) \rightarrow 2H_2O(l) + Na_2X(aq)$

$$\frac{0.1288 \text{ mol NaOH}}{1 \text{ L}} \times 0.0152 \text{ L} \times \frac{1 \text{ mol } H_2X}{2 \text{ mol NaOH}} = 9.79 \times 10^{-4} \text{ mol}$$

$$\text{molar mass} = \frac{0.105 \text{ g}}{9.79 \times 10^{-4} \text{ mol}} = 107 \text{ g/mol}$$

99. $2HCl(aq) + Mg(OH)_2(aq) \rightarrow 2H_2O(l) + MgCl_2(aq)$

$pH = 1.1 \Rightarrow [H^+] = 10^{-1.1} = 0.079 \text{ M}$

$$0.400 \text{ g } Mg(OH)_2 \times \frac{1 \text{ mol } Mg(OH)_2}{58.33 \text{ g}} \times \frac{2 \text{ mol HCl}}{1 \text{ mol } Mg(OH)_2} \times \frac{1 \text{ L}}{0.079 \text{ mol}} = 0.2 \text{ L}$$

100. $\dfrac{0.200 \text{ mol HCl}}{\text{L}} \times 0.02582 \text{ L} = 0.00516 \text{ mol HCl/tablet}$

$pH = 1.1 \Rightarrow [H^+] = 10^{-1.1} = 0.079 \text{ M}$

$$\frac{0.0052 \text{ mol HCl}}{\text{tablet}} \times \frac{\text{L}}{0.079 \text{ mol HCl}} \times \frac{1000 \text{ mL}}{1 \text{ L}} = 7 \times 10^1 \text{ mL stomach acid}$$

101. a) $pH = -\log 0.0025 = 2.60$; acidic
 b) $pH = -\log 1.8 \times 10^{-12} = 11.74$; basic
 c) $pH = -\log 9.6 \times 10^{-9} = 8.02$; basic
 d) $pH = -\log 0.0195 = 1.710$; acidic

102. a) $[H_3O^+] = \dfrac{K_w}{[OH^-]} = \dfrac{1.0 \times 10^{-14}}{1.8 \times 10^{-5}} = 5.56 \times 10^{-10} \text{ M}$, $pH = -\log 5.56 \times 10^{-10}$
  $= 9.26$, basic

 b) $[H_3O^+] = \dfrac{K_w}{[OH^-]} = \dfrac{1.0 \times 10^{-14}}{8.9 \times 10^{-12}} = 1.12 \times 10^{-3} \text{ M}$, $pH = -\log 1.12 \times 10^{-3}$
  $= 2.95$, acidic

 c) $[H_3O^+] = \dfrac{K_w}{[OH^-]} = \dfrac{1.0 \times 10^{-14}}{3.1 \times 10^{-2}} = 3.22 \times 10^{-13} \text{ M}$, $pH = -\log 3.22 \times 10^{-13}$
  $= 12.49$, basic

 d) $[H_3O^+] = \dfrac{K_w}{[OH^-]} = \dfrac{1.0 \times 10^{-14}}{1.96 \times 10^{-9}} = 5.10 \times 10^{-6} \text{ M}$, $pH = -\log 5.10 \times 10^{-6}$
  $= 5.29$, acidic

103.

$[H_3O^+]$	$[OH^-]$	pH	pOH	acidic or basic
1.0×10^{-4}	1.0×10^{-10}	4.00	10.00	acidic
5.5×10^{-3}	1.8×10^{-12}	2.26	11.74	acidic
3.1×10^{-9}	3.2×10^{-6}	8.51	5.49	basic
4.8×10^{-9}	2.1×10^{-6}	8.32	5.68	basic
2.8×10^{-8}	3.5×10^{-7}	7.55	6.45	basic

104.

[H_3O^+]	[OH^-]	pH	pOH	acidic or basic
1.0×10^{-8}	1.0×10^{-6}	8.00	6.00	basic
2.8×10^{-4}	3.6×10^{-11}	3.55	10.45	acidic
1.7×10^{-9}	5.9×10^{-6}	8.77	5.23	basic
3×10^{-14}	3×10^{-1}	13.5	0.5	basic
1.2×10^{-4}	8.6×10^{-11}	3.93	10.07	acidic

105.

	[H_3O^+]	[OH^-]	pH
a)	[0.0088]	$\dfrac{1.00 \times 10^{-14}}{0.0088} = 1.1 \times 10^{-12}$	2.06
b)	[1.5×10^{-3}]	$\dfrac{1.00 \times 10^{-14}}{1.5 \times 10^{-3}} = 6.7 \times 10^{-12}$	2.82
c)	[9.77×10^{-4}]	$\dfrac{1.00 \times 10^{-14}}{9.77 \times 10^{-4}} = 1.02 \times 10^{-11}$	3.010
d)	[0.0878]	$\dfrac{1.00 \times 10^{-14}}{0.0878} = 1.14 \times 10^{-13}$	1.057

106.

	[H_3O^+]	[OH^-]	pH
a)	[0.0150]	$\dfrac{1.000 \times 10^{-14}}{0.0150} = 6.67 \times 10^{-13}$	1.824
b)	[1.9×10^{-4}]	$\dfrac{1.00 \times 10^{-14}}{1.9 \times 10^{-4}} = 5.3 \times 10^{-11}$	3.72
c)	[0.0226]	$\dfrac{1.00 \times 10^{-14}}{0.0266} = 4.42 \times 10^{-13}$	1.646
d)	[1.7×10^{-3}]	$\dfrac{1.00 \times 10^{-14}}{1.7 \times 10^{-5}} = 5.9 \times 10^{-12}$	2.77

107.

	[H_3O^+]	[OH^-]	pOH	pH
a)	$\dfrac{1.00 \times 10^{-14}}{0.15} = 6.7 \times 10^{-14}$	[0.15]	0.82	13.18
b)	$\dfrac{1.00 \times 10^{-14}}{3.0 \times 10^{-3}} = 3.3 \times 10^{-12}$	[3.0×10^{-3}]	2.52	11.48
c)	$\dfrac{1.00 \times 10^{-14}}{9.6 \times 10^{-4}} = 1.0 \times 10^{-11}$	[9.6×10^{-4}]	3.02	10.98
d)	$\dfrac{1.00 \times 10^{-14}}{8.7 \times 10^{-5}} = 1.1 \times 10^{-10}$	[8.7×10^{-5}]	4.06	9.94

108.

	$[H_3O^+]$	$[OH^-]$	pOH	pH
a)	$\dfrac{1.00\times10^{-14}}{8.77\times10^{-3}} = 1.14\times10^{-12}$	$[8.77\times10^{-3}]$	2.057	11.943
b)	$\dfrac{1.00\times10^{-14}}{0.0224} = 4.46\times10^{-13}$	$[0.0224]$	1.650	12.350
c)	$\dfrac{1.00\times10^{-14}}{1.9\times10^{-4}} = 5.3\times10^{-11}$	$[1.9\times10^{-4}]$	3.72	10.28
d)	$\dfrac{1.00\times10^{-14}}{1.0\times10^{-3}} = 1.0\times10^{-11}$	$[1.0\times10^{-3}]$	3.00	11.00

109. $2HCl(aq) + Fe(s) \rightarrow H_2(g) + FeCl_2(aq)$

$500.0 \text{ g Fe} \times \dfrac{1 \text{ mol Fe}}{55.85 \text{ g}} \times \dfrac{2 \text{ mol HCl}}{1 \text{ mol Fe}} \times \dfrac{1 \text{ L}}{12.0 \text{ mol}} = 1.49 \text{ L}$

No, a pen cannot contain this amount of acid.

110. $2HCl(aq) + Zn(s) \rightarrow H_2(g) + ZnCl_2(aq)$

$2.5 \text{ g Zn} \times \dfrac{1 \text{ mol Zn}}{65.39 \text{ g}} \times \dfrac{2 \text{ mol HCl}}{1 \text{ mol Zn}} = 0.076 \text{ mol HCl}$

$0.020 \text{ L} \times \dfrac{6.0 \text{ mol HCl}}{1 \text{ L}} = 0.12 \text{ mol}$

$\dfrac{0.12 - 0.076 \text{ mol}}{0.020 \text{ L}} = 2.2 \text{ M HCl}$

111. $HCl(aq) + NaOH(aq) \rightarrow NaCl(aq) + H_2O(l)$

$0.125 \text{ L} \times \dfrac{0.0250 \text{ mol HCl}}{1 \text{ L}} = 0.00313 \text{ mol HCl}$

$0.0750 \text{ L} \times \dfrac{0.0500 \text{ mol NaOH}}{1 \text{ L}} = 0.00375 \text{ mol NaOH}$

$0.00375 - 0.00313 = 6.25\times10^{-4}$ mol NaOH excess

$[OH^-] = \dfrac{6.25\times10^{-4} \text{ mol NaOH}}{(0.125 \text{ L} + 0.075 \text{ L})} = 3.13\times10^{-3} \text{ M}$

$pOH = -\log 3.13\times10^{-3} = 2.51; \quad pH = 14 - 2.51 = 11.49$

112. HI(aq) + KOH(aq) → KI(aq) + H$_2$O(l)

$0.175 \text{ L} \times \dfrac{0.0880 \text{ mol HI}}{1 \text{ L}} = 0.0154 \text{ mol HI}$

$0.125 \text{ L} \times \dfrac{0.0570 \text{ mol KOH}}{1 \text{ L}} = 0.007125 \text{ mol NaOH}$

$0.0154 - 0.007125 = 8.28 \times 10^{-3} \text{ mol HI excess}$

$[\text{H}^+] = \dfrac{8.28 \times 10^{-3} \text{ mol H}^+}{(0.175 \text{ L} + 0.125 \text{ L})} = 2.76 \times 10^{-3} \text{ M}$

$\text{pH} = -\log 2.76 \times 10^{-3} = 2.56$

113. $0.050 \text{ mL} \times \dfrac{1 \text{ L}}{1000 \text{ mL}} \times \dfrac{1.00 \times 10^{-7} \text{ mol H}^+}{1 \text{ L}} \times \dfrac{6.022 \times 10^{23} \text{ H}^+}{1 \text{ mol H}^+} = 3.0 \times 10^{12} \text{ H}^+ \text{ ions}$

114. $0.0010 \text{ L} \times \dfrac{0.100 \text{ mol H}^+}{1 \text{ L}} \times \dfrac{6.022 \times 10^{23} \text{ H}^+}{1 \text{ mol H}^+} = 6.0 \times 10^{19} \text{ H}^+ \text{ ions}$

$[\text{OH}^-] = \dfrac{1.00 \times 10^{-14}}{0.100} = 1.00 \times 10^{-13} \text{ M}$

$0.0010 \text{ L} \times \dfrac{1.00 \times 10^{-13} \text{ mol OH}^-}{1 \text{ L}} \times \dfrac{6.022 \times 10^{23} \text{ OH}^-}{1 \text{ mol OH}^-} = 6.0 \times 10^7 \text{ OH}^- \text{ ions}$

115. Total $[\text{OH}^-] = 10^{-1.51} = 0.031$ M
    moles OH$^-$ = 0.031 mol OH$^-$/L × 4.00 L = 0.124 mol OH$^-$
    *moles OH$^-$ = 1 × moles NaOH + 2 × moles Sr(OH)$_2$ = 0.124 mol OH$^-$
    mol NaOH + mol Sr(OH)$_2$ = 0.100 mol ⇒ *mol NaOH = 0.100 – mol Sr(OH)$_2$
    Using the two * equations:
    1 × (0.100 – mol Sr(OH)$_2$) + 2 × moles Sr(OH)$_2$ = 0.124 mol OH$^-$
    mol Sr(OH)$_2$ = (0.124 – 0.100) mol Sr(OH)$_2$ = 0.024 mol Sr(OH)$_2$
    and 0.076 mol NaOH.

116. moles HCl + moles HBr = moles H$^+$ ⇒

$\text{g HCl} \times \dfrac{1 \text{ mol HCl}}{36.46 \text{ g HCl}} + \text{g HBr} \times \dfrac{1 \text{ mol HBr}}{80.91 \text{ g HBr}} = 10^{-2.40} \text{ M H}^+ \times 1.50 \text{ L} \Rightarrow$

g HCl × 0.02743 + g HBr × 0.01236 = 0.0060

given: g HCl + g HBr = 0.35 g ⇒ g HCl = 0.35 – g HBr

(0.35 – g HBr) × 0.02743 + g HBr × 0.01236 = 0.0060

0.0096 – 0.02743 g HBr + 0.01236 g HBr = 0.0060

–0.01507 g HBr = –0.0036 ⇒ g HBr = 0.24

g HCl = 0.35 – g HBr ⇒ 0.35 – 0.24 = 0.11 g HCl

Highlight Problems

117. a) weak
  b) strong
  c) weak
  d) strong

118. $2H^+(aq) + CaCO_3(s) \rightarrow Ca^{2+}(aq) + H_2O(l) + CO_2(g)$
  $pH = 5.5 \Rightarrow [H^+] = 10^{-5.5} = 3.2 \times 10^{-6}$ M, $V = 3.8 \times 10^9$ L
  $$\frac{3.2 \times 10^{-6} \text{ mol H}^+}{1 \text{ L}} \times 3.8 \times 10^9 \text{ L} \times \frac{1 \text{ mol CaCO}_3}{2 \text{ mol H}^+} \times \frac{100.09 \text{ g}}{1 \text{ mol CaCO}_3} \times \frac{1 \text{ kg}}{1000 \text{ g}} =$$
  6×10^2 kg CaCO$_3$

119. Great Lakes: $pH = 4.5 \Rightarrow [H^+] = 10^{-4.5} = 3.2 \times 10^{-5}$ M
  West Coast: $pH = 5.4 \Rightarrow [H^+] = 10^{-5.4} = 4.0 \times 10^{-6}$ M
  Ratio: $\dfrac{3.2 \times 10^{-5}}{4.0 \times 10^{-6}}$ = 8 times more concentrated

120. Answers will vary.

121. Answers will vary.

122. Answers will vary based on the calculator model.

123. Answers will vary.

Data Interpretation and Analysis

124. (a) $HX + NaOH \rightarrow NaX + H_2O$
  (b) The pH increases because the acid is being neutralized and then excess sodium hydroxide is added.
  (c) The pH at the equivalence point is approximately 8.0, and the pH halfway to the equivalence point is 4.0.
  (d) $\dfrac{0.100 \text{ mol NaOH}}{\text{L}} \times 0.025 \text{ L} \times \dfrac{1 \text{ mol HX}}{1 \text{ mol NaOH}} = 4.0$ mol HX

Chemical Equilibrium   15

Questions

1. The two general concepts involved in equilibrium are sameness and changelessness. A reaction with a fast rate proceeds quickly; a large amount of reactant is converted to product in a certain period of time. A reaction with a slow rate proceeds slowly; only a small amount of reactant is converted to product in the same period of time.

2. The rate of a chemical reaction is the amount of reactant that changes to product in a given period of time. A chemical reaction with a fast rate proceeds quickly, with a large amount of reactant being converted to product. A reaction with a slow rate proceeds slowly, with only a small amount of reactant being converted to product.

3. Chemists seek to control reaction rates so that a desired result is obtained. The goal may be to prevent a reaction from going too fast and becoming dangerous, or it may be to speed it up so that it proceeds at a fast enough rate to be used in industrial settings.

4. Most chemical reactions occur as a result of collisions between reactant atoms or molecules, which results in the breaking of chemical bonds in the reactant molecules and allows for the formation of new bonds to form products.

5. The two factors that influence reaction rates are concentration and temperature. The rate of a reaction increases with increasing concentration. The rate of a reaction increases with increasing temperature.

6. As the reaction proceeds, the rate of the forward reaction will decrease due to a decrease in the concentration of the reactants.

7. Dynamic equilibrium exists when the rate of the forward reaction is equal to the rate of the reverse reaction.

8. The rate of the forward reaction is the same as the rate of the reverse reaction (sameness). The concentrations of the reactants and products no longer change (changelessness).

9. Because the rates of the forward and reverse reactions are the same at equilibrium, the relative concentrations of reactants and products become constant. It doesn't matter if it is a high or a low concentration; it simply remains constant.

10. Correct answers may vary.

11. The equilibrium constant is a measure of how far a reaction goes toward completion. It is significant because it is used to quantify the concentration of all compounds in a reaction at equilibrium.

12. $K_{eq} = \dfrac{[C]^c[D]^d}{[A]^a[B]^b}$

13. A small equilibrium constant shows that a reverse reaction is favored, and when equilibrium is reached, there will be more reactants than products. A large equilibrium constant shows that a forward reaction is favored, and when equilibrium is reached, there will be more products than reactants.

14. The concentrations of pure solids and liquids are omitted because they remain constant and are incorporated into the value of the equilibrium constant itself.

15. No, the concentrations of reactants and products will not always be the same in every equilibrium mixture of a particular reaction at a given temperature. The final concentrations will depend on the initial concentrations and will adjust accordingly so that the K_{eq} value is achieved.

16. Le Châtelier's Principle: When a chemical system at equilibrium is disturbed, the system shifts in a direction that minimizes the disturbance.

17. Correct answers may vary.

18. The reaction will proceed in the forward (right) direction.

19. The reaction will proceed in the reverse (left) direction.

20. The reaction will proceed in the reverse (left) direction.

21. The reaction will proceed in the forward (right) direction.

22. The reaction will proceed in the reverse (left) direction.

23. The reaction will proceed in the forward (right) direction.

24. The reaction will proceed in the forward (right) direction.

25. The reaction will proceed in the reverse (left) direction.

26. The reaction will proceed in the forward (right) direction if you increase the temperature of an endothermic reaction at equilibrium. The reaction will proceed in the reverse (left) direction if you decrease the temperature of an endothermic reaction at equilibrium.

27. The reaction will proceed in the reverse (left) direction if you increase the temperature of an exothermic reaction at equilibrium. The reaction will proceed in the forward (right) direction if you decrease the temperature of an exothermic reaction at equilibrium.

28. The solubility product constant is the equilibrium constant for a reaction that represents the dissolving of a slightly soluble salt.

29. $K_{sp} = [A^{2+}][B^-]^2$

30. $K_{sp} = [A^+]^2[B^{2-}]$

31. Solubility is the amount of a compound that dissolves in a specified amount of liquid. The molar solubility is just the solubility expressed in molarity.

32. The activation energy represents an energy barrier between the reactants and the formation of products. In other words, collisions between reactant molecules must have enough energy to overcome this energy barrier in order for the reaction to proceed to form products.

33. Two reactants with a large K_{eq} may not react immediately when combined because there may be a large activation energy barrier that the reactants must overcome in order for the reaction to occur.

34. The effect of a catalyst is to lower the activation energy for a reaction that will increase the rate of a reaction. Catalysts are important in chemistry because many reactions would be too slow to be useful without catalysts.

35. The catalyst does NOT affect the value of the equilibrium constant; it merely allows the reaction to reach equilibrium faster than without a catalyst.

36. Enzymes are biochemical catalysts; they speed up the rate of important biochemical or metabolic reactions to produce adequate amounts of products needed by the organism at normal body temperature.

Problems

The Rate of Reaction

37. The rate would decrease because the effective concentration of the reactants has been decreased, which lowers the rate of a reaction.

38. The second experiment at 0.30 atm would have a greater reaction rate as it corresponds to a higher reactant concentration.

39. Reaction rates tend to decrease with decreasing temperature, so all life processes (chemical reactions) would have decreased rates.

40. As the temperature increases, the rate of collisions and the energy of the collisions will increase. Both of these factors will increase the number of collisions per unit time that have enough energy to get over the activation energy barrier and, therefore, will increase the rate of the reaction.

41. The reaction rate at the second reading would be slower than the initial rate because the reactants are being consumed, thus lowering their concentrations and the resulting reaction rate.

42. As the reactants are being consumed, the reaction rate decreases due to the lower concentrations of reactants.

The Equilibrium Constant

43. a) $K_{eq} = \dfrac{[N_2O_4]}{[NO_2]^2}$

　b) $K_{eq} = \dfrac{[NO]^2[Br_2]}{[BrNO]^2}$

　c) $K_{eq} = \dfrac{[H_2][CO_2]}{[H_2O][CO]}$

　d) $K_{eq} = \dfrac{[CS_2][H_2]^4}{[CH_4][H_2S]^2}$

44. a) $K_{eq} = \dfrac{[CO_2]^2}{[CO]^2[O_2]}$

　b) $K_{eq} = \dfrac{[NO]^2}{[N_2][O_2]}$

　c) $K_{eq} = \dfrac{[SbCl_3][Cl_2]}{[SbCl_5]}$

　d) $K_{eq} = \dfrac{[COCl_2]}{[CO][Cl_2]}$

45. a) $K_{eq} = \dfrac{[Cl_2]}{[PCl_5]}$

　b) $K_{eq} = [O_2]^3$

　c) $K_{eq} = \dfrac{[H_3O^+][F^-]}{[HF]}$

　d) $K_{eq} = \dfrac{[NH_4^+][OH^-]}{[NH_3]}$

46. a) $K_{eq} = \dfrac{[CHO_2^-][H_3O^+]}{[HCHO_2]}$

b) $K_{eq} = \dfrac{[OH^-][HCO_3^-]}{[CO_3^{-2}]}$

c) $K_{eq} = \dfrac{[CO]^2}{[O_2]}$

d) $K_{eq} = \dfrac{[CO]^2}{[CO_2]}$

47. $K_{eq} = \dfrac{[H_2]^2[S_2]}{[H_2S]^2}$

48. $K_{eq} = \dfrac{[COCl_2]}{[Cl_2][CO]}$

49. a) $K_{eq} \gg 1$, products dominate equilibrium
b) $K_{eq} \approx 1$, significant amounts of reactants and products
c) $K_{eq} \ll 1$, reactants dominate equilibrium
d) $K_{eq} \approx 1$, significant amounts of reactants and products

50. a) $K_{eq} \approx 1$, significant amounts of reactants and products
b) $K_{eq} \ll 1$, reactants dominate equilibrium
c) $K_{eq} \gg 1$, products dominate equilibrium
d) $K_{eq} \ll 1$, reactants dominate equilibrium

Calculating and Using Equilibrium Constants

51. $K_{eq} = \dfrac{[0.105][0.0844]}{[0.225]} = 0.0394$

52. $K_{eq} = \dfrac{[0.185]}{[0.105][0.114]^2} = 136$

53. $K_{eq} = \dfrac{[2.74 \times 10^{-2}]^2[7.54 \times 10^{-3}]}{[0.562]^2} = 1.79 \times 10^{-5}$

54. $K_{eq} = \dfrac{[0.175][0.0274]}{[0.0233][0.0115]} = 17.9$

55. $K_{eq} = [0.278][0.355] = 0.0987$

56. $K_{eq} = [0.548] = 0.548$

57. $K_{eq} = \dfrac{[0.0255][0.135]}{[SbCl_5]} = 4.9 \times 10^{-4} \Rightarrow [SbCl_5] = \dfrac{[0.0255][0.135]}{4.9 \times 10^{-4}} = 7.0$ M

58. $K_{eq} = \dfrac{[I]^2}{[0.0205]} = 1.1 \times 10^{-2} \Rightarrow [I] = \sqrt{(1.1 \times 10^{-2})[0.0205]} = 0.015$ M

59. $K_{eq} = \dfrac{[ICl]^2}{[0.0112][0.0155]} = 81.9 \Rightarrow [ICl] = \sqrt{(81.9)[0.0112][0.0155]} = 0.119$

60. $K_{eq} = \dfrac{[0.391]^2}{[SO_2]^2[0.125]} = 4.34 \Rightarrow [SO_2] = \sqrt{\dfrac{[0.391]^2}{(4.34)[0.125]}} = 0.531$

61.

T(K)	$[N_2]$	$[H_2]$	$[NH_3]$	K_{eq}
500	0.115	0.105	0.439	1.45×10^3
575	0.110	0.25	0.128	9.6
775	0.120	0.140	4.39×10^{-3}	0.0584

62.

T(K)	$[H_2]$	$[I_2]$	$[HI]$	K_{eq}
25	0.355	0.388	0.0922	0.0617
340	0.34	0.0455	0.387	9.6
445	0.0485	0.0468	0.338	50.2

Le Châtelier's Principle

63. a) shift right
    b) shift left
    c) shift right

64. a) shift right
    b) shift left
    c) shift left

65. a) unchanged
    b) shift left
    c) shift left
    d) shift right

66. a) unchanged
    b) unchanged
    c) shift left
    d) shift right

67. a) shift right
    b) shift left

68. a) shift right
    b) shift left

69. a) no effect
    b) no effect

70. a) no effect
    b) no effect

71. a) shift right
    b) shift left

72. a) shift right
    b) shift left

73. a) shift left
    b) shift right

74. a) shift left
    b) shift right

75. a) no effect
    b) shift right
    c) shift left
    d) shift right
    e) no effect

76. a) no effect
    b) shift right
    c) shift right
    d) shift right
    e) no effect

The Solubility-Product Constant

77. a) $CaSO_4(s) \rightleftarrows Ca^{2+}(aq) + SO_4^{2-}(aq)$; $K_{sp} = [Ca^{2+}][SO_4^{2-}]$

b) $AgCl(s) \rightleftarrows Ag^+(aq) + Cl^-(aq)$; $K_{sp} = [Ag^+][Cl^-]$

c) $CuS(s) \rightleftarrows Cu^{2+}(aq) + S^{2-}(aq)$; $K_{sp} = [Cu^{2+}][S^{2-}]$

d) $FeCO_3(s) \rightleftarrows Fe^{2+}(aq) + CO_3^{2-}(aq)$; $K_{sp} = [Fe^{2+}][CO_3^{2-}]$

78. a) $Mg(OH)_2(s) \rightleftarrows Mg^{2+}(aq) + 2OH^-(aq)$; $K_{sp} = [Mg^{2+}][OH^-]^2$

b) $FeCO_3(s) \rightleftarrows Fe^{2+}(aq) + CO_3^{2-}(aq)$; $K_{sp} = [Fe^{2+}][CO_3^{2-}]$

c) $PbS(s) \rightleftarrows Pb^{2+}(aq) + S^{2-}(aq)$; $K_{sp} = [Pb^{2+}][S^{2-}]$

d) $PbSO_4(s) \rightleftarrows Pb^{2+}(aq) + SO_4^{2-}(aq)$; $K_{sp} = [Pb^{2+}][SO_4^{2-}]$

79. $K_{sp} = [Fe^{2+}][OH^-]^2$

80. $K_{sp} = [Ba^{2+}][OH^-]^2$

81. $K_{sp} = [2.6 \times 10^{-4}][5.2 \times 10^{-4}]^2 = 7.0 \times 10^{-11}$

82. $K_{sp} = [9.2 \times 10^{-9}][9.2 \times 10^{-9}] = 8.5 \times 10^{-17}$

83. $K_{sp} = [Pb^{2+}][SO_4^{2-}] \Rightarrow [SO_4^{2-}] = \dfrac{K_{sp}}{[Pb^{2+}]} = \dfrac{1.82 \times 10^{-8}}{1.35 \times 10^{-4}} = 1.35 \times 10^{-4}$ M

84. $K_{sp} = [Pb^{2+}][Cl^-]^2 \Rightarrow [Pb^{2+}] = \dfrac{K_{sp}}{[Cl^-]^2} = \dfrac{1.17 \times 10^{-5}}{(2.86 \times 10^{-2})^2} = 1.43 \times 10^{-2}$ M

85. $[Ca^{2+}] = [CO_3^{2-}] = S \Rightarrow K_{sp} = S^2 \Rightarrow S = \sqrt{4.96 \times 10^{-9}} = 7.04 \times 10^{-5}$ M

86. $[Pb^{2+}] = [S^{2-}] = S \Rightarrow K_{sp} = S^2 \Rightarrow S = \sqrt{9.04 \times 10^{-29}} = 9.51 \times 10^{-15}$ M

87. $[Mg^{2+}] = [CO_3^{2-}] = S \Rightarrow K_{sp} = S^2 \Rightarrow S = \sqrt{6.82 \times 10^{-6}} = 2.61 \times 10^{-3}$ M

88. $[Cu^+] = [I^-] = S \Rightarrow K_{sp} = S^2 \Rightarrow S = \sqrt{1.27 \times 10^{-12}} = 1.13 \times 10^{-6}$ M

89.
Compound	[Cation]	[Anion]	K_{sp}
$SrCO_3$	2.4×10^{-5}	2.4×10^{-5}	$\underline{5.8 \times 10^{-10}}$
SrF_2	1.0×10^{-3}	$\underline{2.0 \times 10^{-3}}$	4.0×10^{-9}
Ag_2CO_3	$\underline{2.6 \times 10^{-4}}$	1.3×10^{-4}	8.8×10^{-12}

90.
Compound	[Cation]	[Anion]	K_{sp}
CdS	3.7×10^{-15}	3.7×10^{-15}	$\underline{1.4 \times 10^{-29}}$
BaF_2	$\underline{3.7 \times 10^{-3}}$	7.2×10^{-3}	1.9×10^{-7}
Ag_2SO_4	2.8×10^{-2}	$\underline{1.4 \times 10^{-2}}$	1.1×10^{-5}

Cumulative Problems

91. $Fe^{3+}_{equilibrium} = Fe^{3+}_{initial} - Fe^{3+}_{reacted}$

$$\frac{1.7 \times 10^{-4} \text{ mol FeSCN}^{2+}}{L} \times \frac{1 \text{ mol Fe}^{3+}}{1 \text{ mol FeSCN}^{2+}} = 1.7 \times 10^{-4} \text{ M Fe}^{2+}$$

$[Fe^{3+}] = 1.0 \times 10^{-3} - 1.7 \times 10^{-4} = 8.3 \times 10^{-4}$

$SCN^-_{equilibrium} = SCN^-_{initial} - SCN^-_{reacted}$

$$\frac{1.7 \times 10^{-4} \text{ mol FeSCN}^{2+}}{L} \times \frac{1 \text{ mol SCN}^-}{1 \text{ mol FeSCN}^{2+}} = 1.7 \times 10^{-4} \text{ M SCN}^-$$

$[SCN^-] = 8.0 \times 10^{-4} - 1.7 \times 10^{-4} = 6.3 \times 10^{-4}$

$$K_{eq} = \frac{[FeSCN^{2+}]}{[Fe^{3+}][SCN^-]} = \frac{[1.7 \times 10^{-4}]}{[8.3 \times 10^{-4}][6.3 \times 10^{-4}]} = 3.3 \times 10^2$$

92. $SO_2Cl_{2\ equilibrium} = SO_2Cl_{2\ initial} - SO_2Cl_{2\ reacted}$

$[SO_2Cl_2] = 2.0 \times 10^{-2} - 1.2 \times 10^{-2} = 8.0 \times 10^{-3}$ M

$SO_{2\ equilibrium} = Cl_{2\ equilibrium} = 1.2 \times 10^{-2}$ M

$$K_{eq} = \frac{[SO_2][Cl_2]}{[SO_2Cl_2]} = \frac{[1.2 \times 10^{-2}][1.2 \times 10^{-2}]}{[8.0 \times 10^{-3}]} = 1.8 \times 10^{-2} \text{ M}$$

93. $K_{eq} = \frac{[HI]^2}{[H_2][I_2]} \Rightarrow 6.17 \times 10^{-2} = \frac{[HI]^2}{[0.104][0.0202]} \Rightarrow$

$[HI] = \sqrt{(6.17 \times 10^{-2})[0.104][0.0202]} = 0.0114$ M

$$\frac{0.0114 \text{ mol HI}}{1 \text{ L}} \times 3.67 \text{ L} \times \frac{127.91 \text{ g}}{1 \text{ mol HI}} = 5.34 \text{ g HI}$$

94. $K_{eq} = \dfrac{[COCl_2]}{[CO][Cl_2]} \Rightarrow 2.9 \times 10^{10} = \dfrac{[COCl_2]}{[1.8 \times 10^{-6}][7.3 \times 10^{-7}]} \Rightarrow$

$[COCl_2] = (2.9 \times 10^{10})[1.8 \times 10^{-6}][7.3 \times 10^{-7}] = 0.038$ M

$\dfrac{0.038 \text{ mol } COCl_2}{1 \text{ L}} \times 5.19 \text{ L} \times \dfrac{98.91 \text{ g}}{1 \text{ mol } COCl_2} = 2.0 \times 10^1 \text{ g } COCl_2$

95. a) no
    b) yes
    c) yes
    d) yes

96. a) yes
    b) no
    c) yes
    d) yes

97. $[Cu^{2+}] = [S^{2-}] = S \Rightarrow K_{sp} = S^2 \Rightarrow S = \sqrt{1.27 \times 10^{-36}} = 1.13 \times 10^{-18}$ M

$\dfrac{1.13 \times 10^{-18} \text{ mol CuS}}{1 \text{ L}} \times 15.0 \text{ L} \times \dfrac{95.62 \text{ g}}{1 \text{ mol CuS}} = 1.62 \times 10^{-15}$ g

98. $[Fe^{2+}] = [CO_3^{2-}] = S \Rightarrow K_{sp} = S^2 \Rightarrow S = \sqrt{3.07 \times 10^{-11}} = 5.54 \times 10^{-6}$ M

$\dfrac{5.54 \times 10^{-6} \text{ mol FeCO}_3}{1 \text{ L}} \times 15.0 \text{ L} \times \dfrac{115.86 \text{ g}}{1 \text{ mol FeCO}_3} = 9.63 \times 10^{-3}$ g

99. $\dfrac{0.105 \text{ g Na}_2SO_4}{0.100 \text{ L}} \times \dfrac{1 \text{ mol Na}_2SO_4}{142.0 \text{ g}} = 7.39 \times 10^{-3}$ M

$Q = (0.025)(7.39 \times 10^{-3}) = 1.85 \times 10^{-4}$; $K_{sp} = 7.10 \times 10^{-5}$

$Q > K_{sp}$ ∴ Yes, a precipitate will form.

100. $\dfrac{0.0500 \text{ g Na}_2CO_3}{0.1500 \text{ L}} \times \dfrac{1 \text{ mol Na}_2CO_3}{105.99 \text{ g}} = 3.14 \times 10^{-3}$ M

$Q = (1.5 \times 10^{-3})(3.14 \times 10^{-3}) = 4.72 \times 10^{-6}$; $K_{sp} = 6.82 \times 10^{-6}$

$Q < K_{sp}$ ∴ No, a precipitate will not form.

101. $\dfrac{4.15 \text{ g CaCrO}_4}{1 \text{ L}} \times \dfrac{1 \text{ mol CaCrO}_4}{156.1 \text{ g}} = 0.0266$ M

$K_{sp} = [Ca^{2+}][CrO_4^{2-}] = (0.0266)^2 = 7.08 \times 10^{-4}$

102. $\dfrac{0.042 \text{ g NiCO}_3}{1 \text{ L}} \times \dfrac{1 \text{ mol NiCO}_3}{118.7 \text{ g}} = 3.54 \times 10^{-4} \text{ M}$

$K_{sp} = [\text{Ni}^{2+}][\text{CO}_3^{2-}] = (3.54 \times 10^{-4})^2 = 1.3 \times 10^{-7}$

103. $K = [\text{CO}_2] = 4.1 \times 10^{-4}$ mol/L

mol $CO_2 = 4.1 \times 10^{-4}$ mol/L $\times 0.500$ L $= 0.00021$ mol $CO_2 \Rightarrow$

$0.00021 \text{ mol CO}_2 \times \dfrac{1 \text{ mol CaO}}{1 \text{ mol CO}_2} \times \dfrac{56.08 \text{ g}}{1 \text{ mol CaO}} = 0.012 \text{ g CaO}$

104. $K = [\text{NH}_3][\text{H}_2\text{S}] = 3.5 \times 10^{-3}$

$[\text{NH}_3] = [\text{H}_2\text{S}]$ at equilibrium due to stoichiometry

$[\text{NH}_3]^2 = 3.5 \times 10^{-3} \Rightarrow [\text{NH}_3] = \sqrt{3.5 \times 10^{-3}} = 0.059$ mol/L

$\dfrac{0.059 \text{ mol NH}_3}{\text{L}} \times 2.0 \text{ L} \times \dfrac{17.03 \text{ g}}{1 \text{ mol NH}_3} = 2.0 \text{ g NH}_3$

105. $\text{MgCO}_3(s) \rightleftarrows \text{Mg}^{2+}(aq) + \text{CO}_3^{2-}(aq) \quad K_{sp} = [\text{Mg}^{2+}][\text{CO}_3^{2-}] = 6.82 \times 10^{-6}$

$[0.115][\text{CO}_3^{2-}] = 6.82 \times 10^{-6} \Rightarrow [\text{CO}_3^{2-}] = \dfrac{6.82 \times 10^{-6}}{[0.115]} = 5.93 \times 10^{-5}$ M

$\dfrac{5.93 \times 10^{-5} \text{ mol K}_2\text{CO}_3}{\text{L}} \times 2.55 \text{ L} \times \dfrac{138.21 \text{ g}}{1 \text{ mol K}_2\text{CO}_3} = 0.0209 \text{ g K}_2\text{CO}_3$

106. $\text{CaSO}_4(s) \rightleftarrows \text{Ca}^{2+}(aq) + \text{SO}_4^{2-}(aq) \quad K_{sp} = [\text{Ca}^{2+}][\text{SO}_4^{2-}] = 7.10 \times 10^{-5}$

$[0.0251][\text{SO}_4^{2-}] = 7.10 \times 10^{-5} \Rightarrow [\text{SO}_4^{2-}] = \dfrac{7.10 \times 10^{-5}}{[0.0251]} = 2.83 \times 10^{-3}$ M

$\dfrac{2.83 \times 10^{-3} \text{ mol Na}_2\text{SO}_4}{\text{L}} \times 75.0 \text{ L} \times \dfrac{142.05 \text{ g}}{1 \text{ mol Na}_2\text{SO}_4} = 30.2 \text{ g Na}_2\text{SO}_4$

Highlight Problems

107. Equilibrium was reached at figure e.

108. a > b > c

109. $[\text{Ca}^{2+}] = [\text{CO}_3^{2-}] = S \Rightarrow K_{sp} = S^2 \Rightarrow S = \sqrt{4.96 \times 10^{-9}} = 7.04 \times 10^{-5}$ M

$0.250 \text{ g CaCO}_3 \times \dfrac{1 \text{ mol CaCO}_3}{100.09 \text{ g}} \times \dfrac{1 \text{ L}}{7.04 \times 10^{-5} \text{ mol}} = 35.5 \text{ L}$

110. Diagram A: [A] = 24 spheres/ 2.666 cm² = 9.0 spheres/cm²
Diagram B: [A] = 18 spheres/ 2 cm² = 9 spheres/cm²
[B] = 6 spheres/ 47 cm² = 0.128 spheres/cm²
Diagram C: [A] = 9 spheres/ 1 cm² = 9 spheres/cm²
[B] = 15 spheres/ 48 cm² = 0.313 spheres/cm²

The concentration of A remains constant throughout the representations; however, the concentration of B increases. The concentration of a solid remains constant as it does not expand to fill its container; therefore, K= [B] = 0.313 spheres/cm².

111. (a) $K_{sp} = 4.96 \times 10^{-9}$

(b) $CaCO_3(s) \rightleftarrows Ca^{2+}(aq) + CO_3^{2-}(aq)$

$x^2 = 4.96 \times 10^{-9}$

$x = 7.04 \times 10^{-5} M$

(c) $7.04 \times 10^{-5} \dfrac{mol\ CaCO_3}{L} \times 0.200\ L \times \dfrac{100.1\ g\ CaCO_3}{1\ mol} = 1.41 \times 10^{-3} g\ CaCO_3$

112. (a) product C favored when $K > 0$.

(b) Before (c) At equilibrium

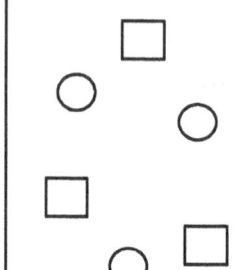
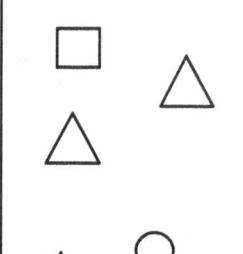

(d) The solution will produce more C in order to restore the equilibrium.

113. (a) Change the concentration of a reactant
Increasing concentration—shifts reaction equilibrium right
Decreasing concentration—shifts reaction equilibrium left
   (b) Change the concentration of a product
Increasing concentration—shifts reaction equilibrium left
Decreasing concentration—shifts reaction equilibrium right
   (c) Change the pressure on a reaction involving at least one gaseous compound.
Increase volume = decrease in pressure—reaction shifts in the direction of more moles of gaseous compounds.
Decrease volume = increase in pressure—reaction shifts in the direction of fewer moles of gaseous compounds.
   (d) Add or remove heat to a reaction.
Add heat: if the reaction is exothermic, shifts left; if endothermic, shifts right
Remove heat: if the reaction is exothermic, shifts right; if endothermic, shifts left

Data Interpretation and Analysis

114. (a) $CaCO_3(s) \rightleftarrows CaO(s) + CO_2(g)$
   (b) $K = [CO_2]$
   (c) At 950 K, $K = 0.0003$; at 1050 K, $K = 0.0030$
   (d) At 1000 K, $K = 0.0012 = [CO_2]$;

   $$\frac{0.0012 \text{ mol CO}_2}{1 \text{ L}} \times 1.00 \text{ L} \times \frac{44.01 \text{ g}}{1 \text{ mol CO}_2} = 0.053 \text{ g CO}_2$$

   (e) $0.053 \text{ g CO}_2 \times \frac{1 \text{ mol CO}_2}{44.01 \text{ g}} \times \frac{1 \text{ mol CaCO}_3}{1 \text{ mol CO}_2} \times \frac{100.1}{1 \text{ mol CaCO}_3} = 0.12 \text{ g CaCO}_3$

Oxidation and Reduction 16

Questions

1. The fuel cells use the electron-gaining tendency of oxygen and the electron-losing tendency of hydrogen to force electrons to move through a wire, creating the electricity that powers the car.

2. Redox reactions involve the transfer of electrons between reactants.

3. a) Oxidation occurs when a substance gains oxygen atoms in the reaction. Reduction occurs when oxygen is lost from a substance.
   b) Oxidation occurs when a substance loses electrons. Reduction occurs when a substance gains electrons.
   c) Oxidation is an increase in the oxidation state of a substance. Reduction is a decrease in the oxidation state of a substance.

4. An oxidizing agent is a reactant that causes another substance to be oxidized by taking electrons from that substance. The oxidizing agent is reduced in the process of oxidizing another substance. A reducing agent is the substance being oxidized that causes the reduction of the other substance.

5. Good oxidizing agents have a strong tendency to <u>gain</u> electrons in reactions.

6. Good reducing agents have a strong tendency to <u>lose</u> electrons in reactions.

7. The oxidation state of a free element in its naturally occurring state is zero. The oxidation state of a monatomic ion is equal to the charge of the ion.

8. For a neutral molecule, the sum of the oxidation states of the individual atoms must add up to <u>zero</u>.

9. For an ion, the sum of the oxidation states of the individual atoms must sum to <u>the ion charge</u>.

10. In their compounds
    -Group I metals have oxidation states equal to +1.
    -Group II metals have oxidation states equal to +2.
    -Nonmetals according to F = –1, H = +1, O = –2, Group 7A = –1, Group 6A = –2, Group 5A = –3

11. In a redox reaction, an atom that undergoes an increase in oxidation state is <u>oxidized</u>. An atom that undergoes a decrease in oxidation state is <u>reduced</u>.

12. Hydrogen peroxide oxidizes melanin, the dark pigment that gives hair color. After being oxidized, melanin no longer exhibits its typical dark color, and the hair looks lighter or bleached.

13. When balancing redox equations, the number of electrons lost in the oxidation half-reaction must <u>equal</u> the number of electrons gained in the reduction half-reaction.

14. When balancing aqueous redox reactions, oxygen is balanced using H_2O and hydrogen is balanced using H^+.

15. When balancing aqueous redox reactions, charge is balanced using electrons (e^-).

16. When balancing aqueous redox reactions in basic media, hydrogen ions are neutralized using OH^-.

17. The metals at the top of the activity series are the most reactive.

18. The metals at the top of the activity series are the easiest to oxidize.

19. The metals at the bottom of the activity series are least likely to lose electrons.

20. Any half-reaction in the activity series will be spontaneous when paired with the reverse of any half-reaction <u>below</u> it.

21. Metals above H_2 on the activity series will dissolve in acids, while metals below H_2 will not dissolve in acids.

22. Electrical current is the flow of electronic charge. A battery creates electrical current by taking advantage of the transfer of electrons between the oxidizing agent and the reducing agent during a spontaneous redox reaction. The electron transfer flows through an external wire, creating the current.

23. Oxidation occurs at the <u>anode</u> of an electrochemical cell.

24. Reduction occurs at the <u>cathode</u> of an electrochemical cell.

25. The role of a salt bridge is to allow for the flow of ions between two halves of an electrochemical cell, which completes the electrical circuit.

26. A high voltage in an electrochemical cell is analogous to a steep landscape for a river.

27. The common dry cell battery contains only a small amount of liquid water within it—hence the name. The cell voltage is approximately 1.5 volts.
    The anode reaction: $Zn(s) \rightarrow Zn^{2+}(aq) + 2e^-$.
    The cathode reaction: $2MnO_2(s) + 2NH_4^+(aq) + 2e^- \rightarrow Mn_2O_3(s) + 2NH_3(g) + H_2O(l)$.

28. The lead–acid storage battery actually contains six individual electrochemical cells that each produce 2 volts. Arranged in series, they produce an overall voltage of 12 volts. This type of battery is used in automobiles.
The anode reaction is: $Pb(s) + SO_4^{2-}(aq) \rightarrow PbSO_4(s) + 2e^-$.
The cathode reaction is: $PbO_2(s) + 4H^+(aq) + SO_4^{2-} aq) + 2e^- \rightarrow PbSO_4(s) + 2H_2O(l)$.

29. A fuel cell operates much like a battery; however, the reactants in a fuel cell are continually replenished from an external supply as the reaction proceeds. The most common fuel cell is the hydrogen–oxygen fuel cell. The reaction at the anode is the oxidation of hydrogen gas to form water according to the half-reaction:
$H_2(g) + 4OH^-(aq) \rightarrow 4H_2O(l) + 4e^-$.
The reaction at the cathode corresponds to the reduction of oxygen gas to form hydroxide ion according to the half-reaction: $O_2(g) + 2H_2O(l) + 4e^- \rightarrow 4OH^-$.

30. Electrolysis is when an electrical current is used to force a nonspontaneous redox reaction to occur. Most metals undergo a spontaneous reaction to form metal oxides, which is not a desirable state for their use in manufacturing. Electrolysis can be used to force the nonspontaneous back reaction to occur, so the pure metal can be recovered.

31. The oxidation of metals to form a metal oxide is called corrosion. For the corrosion of iron: The oxidation half-reaction is $2Fe(s) \rightarrow 2Fe^{2+}(aq) + 4e^-$. The reduction half-reaction is $O_2(g) + 2H_2O(l) + 4e^- \rightarrow 4OH^- (aq)$.

32. Rust can be prevented by both physical and chemical means. An example of physical prevention would be to keep the metal dry, since the redox reaction cannot occur without water. Chemical prevention involves either galvanizing the metal (coating it with a more active metal) or connecting a sacrificial electrode to the metal. The sacrificial electrode consists of a more active metal that will preferentially oxidize instead of the iron (i.e., it will act as the anode), thereby protecting the iron metal from corrosion.

Problems

Oxidation and Reduction

33. a) H_2
    b) Al
    c) Al

34. a) Zn
    b) C in CH_4
    c) Sr

35. | | Oxidized | Reduced |
    |---|---|---|
    | a) | Sr | O_2 |
    | b) | Ca | Cl_2 |
    | c) | Mg | Ni^{2+} |

36. | | Oxidized | Reduced |
    |---|---|---|
    | a) | Mg | Br_2 |
    | b) | Mn | Cr^{3+} |
    | c) | Ni | H^+ |

37. | | Oxidizing Agent | Reducing Agent |
    |---|---|---|
    | a) | O_2 | Sr |
    | b) | Cl_2 | Ca |
    | c) | Ni^{2+} | Mg |

38. | | Oxidizing Agent | Reducing Agent |
    |---|---|---|
    | a) | Br_2 | Mg |
    | b) | Cr^{3+} | Mn |
    | c) | H^+ | Ni |

39. A good oxidizing agent is a substance that is easily reduced (gains electrons). Those elements that form anions will serve as good oxidizing agents.
    a) No
    b) Yes
    c) No
    d) Yes

40. A good oxidizing agent is a substance that is easily reduced (gains electrons). Those elements that form anions will serve as good oxidizing agents.
    a) Yes
    b) Yes
    c) No
    d) No

41. A good reducing agent is a substance that is easily oxidized (loses electrons). Those elements that form cations will serve as good reducing agents.
    a) Yes
    b) No
    c) Yes
    d) No

42. A good reducing agent is a substance that is easily oxidized (loses electrons). Those elements that form cations will serve as good reducing agents.
   a) No
   b) No
   c) Yes
   d) Yes

43. | | Oxidized | Reduced | Oxidizing Agent | Reducing Agent |
     |---|---|---|---|---|
     | a) | N_2 | O_2 | O_2 | N_2 |
     | b) | C in CO | O_2 | O_2 | C in CO |
     | c) | Sb in $SbCl_3$ | Cl in Cl_2 | Cl_2 | Sb in $SbCl_3$ |

44. | | Oxidized | Reduced | Oxidizing Agent | Reducing Agent |
     |---|---|---|---|---|
     | a) | H_2 | I_2 | I_2 | H_2 |
     | b) | H_2 | C in CO | C in CO | H_2 |
     | c) | Al | H^+ | H^+ | Al |

Oxidation States

45. a) 0
    b) +2
    c) +3
    d) 0

46. a) 0
    b) 0
    c) +1
    d) +3

47. a) Na = +1, Cl = −1
    b) Ca = +2, F = −1
    c) S = +4, O = −2
    d) H = +1, S = −2

48. a) C = −4, H = +1
    b) C = 0, H = +1, Cl = −1
    c) Cu = +2, Cl = −1
    d) H = +1, I = −1

49. a) +2
    b) +4
    c) +1

50. a) +2
    b) +6
    c) +3

51. a) C = +4, O = −2
    b) O = −2, H = +1
    c) N = +5, O = −2
    d) N = +3, O = −2

52. a) Cr = +6, O = −2
    b) Cr = +6, O = −2
    c) P = +5, O = −2
    d) Mn = +7, O = −2

53. a) +1
    b) +3
    c) +5
    d) +7

54. a) +6
    b) +4
    c) +4
    d) +6

55. a) Cu = +2, N = +5, O = −2
    b) Sr = +2, O = −2, H = +1
    c) K = +1, O = −2, Cr = +6
    d) Na = +1, H = +1, O = −2, C = +4

56. a) Na = +1, O = −2, P = +5
    b) S = −2, Hg = +1
    c) Fe = +3, N = −3, C = +2
    d) H = +1, N = −3, Cl = −1

57. a) Sb: +5 → +3, reduced; Cl: −1 → 0, oxidized
    b) C: +2 → +4, oxidized; Cl: 0 → −1, reduced; O: −2 → −2, neither
    c) N: +2 → +3, oxidized; O: −2 → −2, neither; Br: 0 → −1, reduced
    d) H: 0 → +1, oxidized; C: +4 → +2, reduced; O: −2 → −2, neither

58. a) C: −4 → +4, oxidized; H: +1 → 0, reduced; S: −2 → −2, neither
    b) H: +1 → 0, reduced; S: −2 → 0, oxidized
    c) C: 0 → +4, oxidized; O: 0 → −2, reduced; H: +1 → +1, neither
    d) C: −2 → −1, oxidized; H: +1 → +1, neither; Cl: 0 → −1, reduction

59. Na: 0 → +1, reducing agent; H: +1 → 0, oxidizing agent; O: −2 → −2, neither

60. N: 0 → −3, oxidizing agent; H: 0 → +1, reducing agent

Balancing Redox Reactions

Refer to the following guide when examining the solutions for questions 63–72 to determine what is being done in each line.

1. Assign oxidation states.
2. Separate into half-reactions.
3. Balance half-reactions;
   - balance all atoms other than O and H.
   - balance O by adding H_2O to side lacking O.
   - balance H by adding H^+ to side lacking H.
4. Balance charge by adding e^- to one side.
5. Make # of electrons gained/lost equal by multiplying half-reactions by appropriate coefficients.
6. Add half-reactions and cancel species that appear on both sides.
7. (In basic solution) add the number of OH^- equal to H^+ to both sides.
   - $OH^- + H^+ \rightarrow H_2O$
   - cancel water molecules if on both sides

61. a) 1: $\underset{0}{K(s)} + \underset{+3}{Cr^{3+}(aq)} \rightarrow \underset{0}{Cr(s)} + \underset{+1}{K^+(aq)}$

    2: $K(s) \rightarrow K^+(aq)$ $Cr^{+3}(aq) \rightarrow Cr(s)$

    3: $K(s) \rightarrow K^+(aq)$ $Cr^{+3}(aq) \rightarrow Cr(s)$

    4: $K(s) \rightarrow K^+(aq) + 1e^-$ $Cr^{+3}(aq) + 3e^- \rightarrow Cr(s)$

    5: $3K(s) \rightarrow 3K^+(aq) + 3e^-$ $Cr^{+3}(aq) + 3e^- \rightarrow Cr(s)$

    6: Overall: $3K(s) + Cr^{+3}(aq) \rightarrow 3K^+(aq) + Cr(s)$

    b) 1: $\underset{0}{Mg(s)} + \underset{+3}{Cr^{+3}(aq)} \rightarrow \underset{+2}{Mg^{2+}(aq)} + \underset{0}{Cr(s)}$

    2: $Mg(s) \rightarrow Mg^{2+}(aq)$ $Cr^{3+}(aq) \rightarrow Cr(s)$

    3: $Mg(s) \rightarrow Mg^{2+}(aq)$ $Cr^{3+}(aq) \rightarrow Cr(s)$

    4: $Mg(s) \rightarrow Mg^{2+}(aq) + 2e^-$ $Cr^{3+}(aq) + 3e^- \rightarrow Cr(s)$

    5: $3Mg(s) \rightarrow 3Mg^{2+}(aq) + 6e^-$ $2Cr^{3+}(aq) + 6e^- \rightarrow 2Cr(s)$

    6: Overall: $3Mg(s) + 2Cr^{3+}(aq) \rightarrow 3Mg^{+2}(aq) \; 2Cr(s)$

    c) 1: $\underset{0}{Al(s)} + \underset{+2}{Fe^{2+}(aq)} \rightarrow \underset{+3}{Al^{3+}(aq)} + \underset{0}{Fe(s)}$

    2: $Al(s) \rightarrow Al^{3+}(aq)$ $Fe^{2+}(aq) \rightarrow Fe(s)$

    3: $Al(s) \rightarrow Al^{3+}(aq)$ $Fe^{2+}(aq) \rightarrow Fe(s)$

    4: $Al(s) \rightarrow Al^{3+}(aq) + 3e^-$ $Fe^{2+}(aq) + 2e^- \rightarrow Fe(s)$

    5: $2Al(s) \rightarrow 2Al^{3+}(aq) + 6e^-$ $3Fe^{2+}(aq) + 6e^- \rightarrow 3Fe(s)$

    6: Overall: $2Al(s) + 3Fe^{2+}(aq) \rightarrow 2Al^{3+}(aq) + 3Fe(s)$

62. a) 1: $\underset{0}{Zn}(s) + \underset{+2}{Sn^{2+}}(aq) \rightarrow \underset{+2}{Zn^{2+}}(aq) + \underset{0}{Sn}(s)$

   2: $Zn(s) \rightarrow Zn^{2+}(aq)$ $Sn^{2+}(aq) \rightarrow Sn(s)$

   3: $Zn(s) \rightarrow Zn^{2+}(aq)$ $Sn^{2+}(aq) \rightarrow Sn(s)$

   4: $Zn(s) \rightarrow Zn^{2+}(aq) + 2e^-$ $Sn^{2+}(aq) + 2e^- \rightarrow Sn(s)$

   5: $Zn(s) \rightarrow Zn^{2+}(aq) + 2e^-$ $Sn^{2+}(aq) + 2e^- \rightarrow Sn(s)$

   6: Overall: $Zn(s) + Sn^{2+}(aq) \rightarrow Zn^{2+}(aq) + Sn(s)$

b) 1: $\underset{0}{Mg}(s) + \underset{+3}{Cr^{3+}}(aq) \rightarrow \underset{+2}{Mg^{2+}}(aq) + \underset{0}{Cr}(s)$

   2: $Mg(s) \rightarrow Mg^{2+}(aq)$ $Cr^{3+}(aq) \rightarrow Cr(s)$

   3: $Mg(s) \rightarrow Mg^{2+}(aq)$ $Cr^{3+}(aq) \rightarrow Cr(s)$

   4: $Mg(s) \rightarrow Mg^{2+}(aq) + 2e^-$ $Cr^{3+}(aq) + 3e^- \rightarrow Cr(s)$

   5: $3Mg(s) \rightarrow 3Mg^{2+}(aq) + 6e^-$ $2Cr^{3+}(aq) + 6e^- \rightarrow 2Cr(s)$

   6: Overall: $3Mg(s) + 2Cr^{3+}(aq) \rightarrow 3Mg^{+2}(aq) + 2Cr(s)$

c) 1: $\underset{0}{Al}(s) + \underset{+}{Ag^+}(aq) \rightarrow \underset{+3}{Al^{3+}}(aq) + \underset{0}{Ag}(s)$

   2: $Al(s) \rightarrow Al^{3+}(aq)$ $Ag^+(aq) \rightarrow Ag(s)$

   3: $Al(s) \rightarrow Al^{3+}(aq)$ $Ag^+(aq) \rightarrow Ag(s)$

   4: $Al(s) \rightarrow Al^{3+}(aq) + 3e^-$ $Ag^+(aq) + e^- \rightarrow Ag(s)$

   5: $Al(s) \rightarrow Al^{3+}(aq) + 3e^-$ $3Ag^+(aq) + 3e^- \rightarrow 3Ag(s)$

   6: Overall: $Al(s) + 3Ag^+(aq) \rightarrow Al^{3+}(aq) + 3Ag(s)$

63. a) 1: $\underset{+7}{MnO_4^-}(aq) \rightarrow \underset{+2}{Mn^{2+}}(aq)$ reduction   2: $MnO_4^-(aq) \rightarrow Mn^{2+}(aq)$

   3: $MnO_4^-(aq) + 8H^+(aq) \rightarrow Mn^{2+}(aq) + 4H_2O(l)$

   4: $MnO_4^-(aq) + 8H^+(aq) + 5e^- \rightarrow Mn^{2+}(aq) + 4H_2O(l)$

b) 1: $\underset{+2}{Pb^{2+}}(aq) \rightarrow \underset{+4}{PbO_2}(s)$ oxidation   2: $Pb^{2+}(aq) \rightarrow PbO_2(s)$

   3: $Pb^{2+}(aq) + 2H_2O(l) \rightarrow PbO_2(s) + 4H^+(aq)$

   4: $Pb^{+2}(aq) + 2H_2O(l) \rightarrow PbO_2(s) + 4H^+(aq) + 2e^-$

c) 1: $\underset{+5}{IO_3^-}(aq) \rightarrow \underset{0}{I_2}(s)$ reduction   2: $IO_3^-(aq) \rightarrow I_2(s)$

   3: $2IO_3^-(aq) + 12H^+(aq) \rightarrow I_2(s) + 6H_2O(l)$

   4: $2IO_3^-(aq) + 12H^+(aq) + 10e^- \rightarrow I_2(s) + 6H_2O(l)$

d) 1: $\underset{+4}{SO_2}(g) \rightarrow \underset{+6}{SO_4^{2-}}(aq)$ oxidation   2: $SO_2(g) \rightarrow SO_4^{2-}(aq)$

   3: $SO_2(g) + 2H_2O(l) \rightarrow SO_4^{2-}(aq) + 4H^+(aq)$

   4: $SO_2(g) + 2H_2O(l) \rightarrow SO_4^{2-}(aq) + 4H^+(aq) + 2e^-$

64. a) 1: $\underset{0}{S}(s) \rightarrow \underset{-2}{H_2S}(g)$ reduction   2: $S(s) \rightarrow H_2S(g)$

   3: $S(s) + 2H^+(aq) \rightarrow H_2S(g)$ 4: $S(s) + 2H^+(aq) + 2e^- \rightarrow H_2S(g)$

   b) 1: $\underset{+7}{S_2O_8^{2-}}(aq) \rightarrow 2\underset{+6}{SO_4^{2-}}(aq)$ reduction   2: $S_2O_8^{2-}(aq) \rightarrow 2SO_4^{2-}(aq)$

   3: $S_2O_8^{2-}(aq) \rightarrow 2SO_4^{2-}(aq)$ 4: $S_2O_8^{2-}(aq) + 2e^- \rightarrow 2SO_4^{2-}(aq)$

   c) 1: $\underset{+6}{Cr_2O_7^{2-}}(aq) \rightarrow \underset{+3}{Cr^{3+}}(aq)$ reduction   2: $Cr_2O_7^{2-}(aq) \rightarrow Cr^{3+}(aq)$

   3: $Cr_2O_7^{2-}(aq) + 14H^+ \rightarrow 2Cr^{3+}(aq) + 7H_2O(l)$

   4: $Cr_2O_7^{2-}(aq) + 14H^+ + 6e^- \rightarrow 2Cr^{3+}(aq) + 7H_2O(l)$

   d) 1: $\underset{+2}{NO}(g) \rightarrow \underset{+5}{NO_3^-}(aq)$ oxidation   2: $NO(g) \rightarrow NO_3^-(aq)$

   3: $NO(g) + 2H_2O(l) \rightarrow NO_3^-(aq) + 4H^+(aq)$

   4: $NO(g) + 2H_2O(l) \rightarrow NO_3^-(aq) + 4H^+(aq) + 3e^-$

65. a) 1: $\underset{+4}{PbO_2}(s) + \underset{-1}{I^-}(aq) \rightarrow \underset{+2}{Pb^{2+}}(aq) + \underset{0}{I_2}(s)$

   2: $PbO_2 \rightarrow Pb^{2+}$ $I^- \rightarrow I_2$

   3: $PbO_2 + 4H^+ \rightarrow Pb^{2+} + 2H_2O$ $2I^- \rightarrow I_2$

   4: $PbO_2 + 4H^+ + 2e^- \rightarrow Pb^{2+} + 2H_2O$ $2I^- \rightarrow I_2 + 2e^-$

   5: $PbO_2 + 4H^+ + 2e^- \rightarrow Pb^{2+} + 2H_2O$ $2I^- \rightarrow I_2 + 2e^-$

   6: $PbO_2(s) + 4H^+(aq) + 2I^-(aq) \rightarrow Pb^{2+}(aq) + 2H_2O(l) + I_2(s)$

   b) 1: $\underset{+4}{SO_3^{2-}}(aq) + \underset{+7}{MnO_4^-}(aq) \rightarrow \underset{+6}{SO_4^{2-}}(aq) + \underset{+2}{Mn^{2+}}(s)$

   2: $SO_3^{2-} \rightarrow SO_4^{2-}$ $MnO_4^- \rightarrow Mn^{2+}$

   3: $SO_3^{2-} + H_2O \rightarrow SO_4^{2-} + 2H^+$ $MnO_4^- + 8H^+ \rightarrow Mn^{2+} + 4H_2O$

   4: $SO_3^{2-} + H_2O \rightarrow SO_4^{2-} + 2H^+ + 2e^-$ $MnO_4^- + 8H^+ + 5e^- \rightarrow Mn^{2+} + 4H_2O$

   5: $5SO_3^{2-} + 5H_2O \rightarrow 5SO_4^{2-} + 10H^+ + 10e^-$ $2MnO_4^- + 16H^+ + 10e^- \rightarrow 2Mn^{2+} + 8H_2O$

   6: $5SO_3^{2-}(aq) + 2MnO_4^-(aq) + 6H^+(aq) \rightarrow 5SO_4^{2-}(aq) + 2Mn^{2+}(aq) + 3H_2O(l)$

   c) 1: $\underset{+2}{S_2O_3^{2-}}(aq) + \underset{0}{Cl_2}(g) \rightarrow \underset{+6}{SO_4^{2-}}(aq) + \underset{-1}{Cl^-}(aq)$

   2: $S_2O_3^{2-} \rightarrow SO_4^{2-}$ $Cl_2 \rightarrow Cl^-$

   3: $S_2O_3^{2-} + 5H_2O \rightarrow 2SO_4^{2-} + 10H^+$ $Cl_2 \rightarrow 2Cl^-$

   4: $S_2O_3^{2-} + 5H_2O \rightarrow 2SO_4^{2-} + 10H^+ + 8e^-$ $Cl_2 + 2e^- \rightarrow 2Cl^-$

   5: $S_2O_3^{2-} + 5H_2O \rightarrow 2SO_4^{2-} + 10H^+ + 8e^-$ $4Cl_2 + 8e^- \rightarrow 8Cl^-$

   6: $S_2O_3^{2-}(aq) + 5H_2O(l) + 4Cl_2(g) \rightarrow 2SO_4^{2-}(aq) + 10H^+(aq) + 8Cl^-(aq)$

66. a) 1: $\underset{-1}{I^-}(aq) + \underset{+3}{NO_2^-}(s) \rightarrow \underset{0}{I_2}(s) + \underset{+2}{NO}(g)$

   2: $NO_2^- \rightarrow NO$ $\qquad\qquad\qquad$ $I^- \rightarrow I_2$

   3: $NO_2^- + 2H^+ \rightarrow NO + H_2O$ $\qquad$ $2I^- \rightarrow I_2$

   4: $NO_2^- + 2H^+ + 1e^- \rightarrow NO + H_2O$ $\qquad$ $2I^- \rightarrow I_2 + 2e^-$

   5: $2NO_2^- + 4H^+ + 2e^- \rightarrow 2NO + 2H_2O$ $\qquad$ $2I^- \rightarrow I_2 + 2e^-$

   6: $2NO_2^-(aq) + 4H^+(aq) + 2I^-(aq) \rightarrow 2NO(g) + 2H_2O(l) + I_2(s)$

b) 1: $\underset{+5}{BrO_3^-}(aq) + \underset{-2}{N_2H_4}(g) \rightarrow \underset{-1}{Br^-}(aq) + \underset{0}{N_2}(aq)$

   2: $BrO_3^- \rightarrow Br^-$ $\qquad\qquad\qquad$ $N_2H_4 \rightarrow N_2$

   3: $BrO_3^- + 6H^+ \rightarrow Br^- + 3H_2O$ $\qquad$ $N_2H_4 \rightarrow N_2 + 4H^+$

   4: $BrO_3^- + 6H^+ + 6e^- \rightarrow Br^- + 3H_2O$ $\qquad$ $N_2H_4 \rightarrow N_2 + 4H^+ + 4e^-$

   5: $2BrO_3^- + 12H^+ + 12e^- \rightarrow Br^- + 6H_2O$ $\quad$ $3N_2H_4 \rightarrow 3N_2 + 12H^+ + 12e^-$

   6: $2BrO_3^-(aq) + 3N_2H_4(g) \rightarrow 2Br^-(aq) + 6H_2O(l) + 3N_2(g)$

c) 1: $\underset{+5}{NO_3^-}(aq) + \underset{+2}{Sn^{2+}}(s) \rightarrow \underset{+4}{Sn^{4+}}(aq) + \underset{+2}{NO}(aq)$

   2: $NO_3^- \rightarrow NO$ $\qquad\qquad\qquad$ $Sn^{2+} \rightarrow Sn^{4+}$

   3: $NO_3^- + 4H^+ \rightarrow NO + 2H_2O$ $\qquad$ $Sn^{2+} \rightarrow Sn^{4+}$

   4: $NO_3^- + 4H^+ + 3e^- \rightarrow NO + 2H_2O$ $\qquad$ $Sn^{2+} \rightarrow Sn^{4+} + 2e^-$

   5: $2NO_3^- + 8H^+ + 6e^- \rightarrow 2NO + 4H_2O$ $\qquad$ $3Sn^{2+} \rightarrow 3Sn^{4+} + 6e^-$

   6: $2NO_3^-(aq) + 8H^+(aq) + 3Sn^{2+}(aq) \rightarrow 2NO(g) + 4H_2O(l) + 3Sn^{4+}(aq)$

67. a) 1: $\underset{+7}{ClO_4^-}(aq) + \underset{-1}{Cl^-}(aq) \rightarrow \underset{+5}{ClO_3^-}(aq) + \underset{0}{Cl_2}(g)$

   2: $ClO_4^- \rightarrow ClO_3^-$ $\qquad\qquad\qquad$ $Cl^- \rightarrow Cl_2$

   3: $ClO_4^- + 2H^+ \rightarrow ClO_3^- + H_2O$ $\qquad$ $2Cl^- \rightarrow Cl_2$

   4: $ClO_4^- + 2H^+ + 2e^- \rightarrow ClO_3^- + H_2O$ $\qquad$ $2Cl^- \rightarrow Cl_2 + 2e^-$

   5: $ClO_4^- + 2H^+ + 2e^- \rightarrow ClO_3^- + H_2O$ $\qquad$ $2Cl^- \rightarrow Cl_2 + 2e^-$

   6: $ClO_4^-(aq) + 2H^+(aq) + 2Cl^-(aq) \rightarrow ClO_3^-(aq) + H_2O(l) + Cl_2(g)$

b) 1: $\underset{+7}{MnO_4^-}(aq) + \underset{0}{Al}(s) \rightarrow \underset{+2}{Mn^{2+}}(aq) + \underset{+3}{Al^{3+}}(aq)$

   2: $MnO_4^- \rightarrow Mn^{2+}$ $\qquad\qquad\qquad$ $Al \rightarrow Al^{3+}$

   3: $MnO_4^- + 8H^+ \rightarrow Mn^{2+} + 4H_2O$ $\qquad$ $Al \rightarrow Al^{3+}$

   4: $MnO_4^- + 8H^+ + 5e^- \rightarrow Mn^{2+} + 4H_2O$ $\qquad$ $Al \rightarrow Al^{3+} + 3e^-$

   5: $3MnO_4^- + 24H^+ + 15e^- \rightarrow 3Mn^{+2} + 12H_2O$ $\quad$ $5Al \rightarrow 5Al^{3+} + 15e^-$

   6: $3MnO_4^-(aq) + 24H^+(aq) + 5Al(s) \rightarrow 3Mn^{2+}(aq) + 12H_2O(l) + 5Al^{3+}(aq)$

c) 1: $\underset{0}{Br_2}(aq) + \underset{0}{Sn}(s) \rightarrow \underset{+2}{Sn^{2+}}(aq) + \underset{-1}{Br^-}(aq)$

   2: $Br_2 \rightarrow Br^-$ $\qquad\qquad\qquad$ $Sn \rightarrow Sn^{2+}$

   3: $Br_2 \rightarrow 2Br^-$ $\qquad\qquad\qquad$ $Sn \rightarrow Sn^{2+}$

   4: $Br_2 + 2e^- \rightarrow 2Br^-$ $\qquad\qquad$ $Sn \rightarrow Sn^{2+} + 2e^-$

   5: $Br_2 + 2e^- \rightarrow 2Br^-$ $\qquad\qquad$ $Sn \rightarrow Sn^{2+} + 2e^-$

   6: $Br_2(aq) + Sn(s) \rightarrow Sn^{2+}(aq) + 2Br^-(aq)$

68. a) 1: $\underset{+5}{I}O_3^-(aq) + \underset{+4}{S}O_2(g) \rightarrow \underset{0}{I_2}(s) + \underset{+6}{S}O_4^{2-}(aq)$

    2: $IO_3^- \rightarrow I_2$ $SO_2 \rightarrow SO_4^{2-}$

    3: $2IO_3^- + 12H^+ \rightarrow I_2 + 6H_2O$ $SO_2 + 2H_2O \rightarrow SO_4^{2-} + 4H^+$

    4: $2IO_3^- + 12H^+ + 10e^- \rightarrow I_2 + 6H_2O$ $SO_2 + 2H_2O \rightarrow SO_4^{2-} + 4H^+ + 2e^-$

    5: $2IO_3^- + 12H^+ + 10e^- \rightarrow I_2 + 6H_2O$ $5SO_2 + 10H_2O \rightarrow 5SO_4^{2-} + 20H^+ + 10e^-$

    6: $2IO_3^-(aq) + 5SO_2(g) + 4H_2O(l) \rightarrow I_2(s) + 5SO_4^{2-}(aq) + 8H^+(aq)$

b) 1: $\underset{+4}{Sn}^{4+}(aq) + H_2(g) \rightarrow \underset{+2}{Sn}^{2+}(aq) + \underset{+1}{H}^+(aq)$

    2: $Sn^{4+} \rightarrow Sn^{2+}$ $H_2 \rightarrow H^+$

    3: $Sn^{4+} \rightarrow Sn^{2+}$ $H_2 \rightarrow 2H^+$

    4: $Sn^{4+} + 2e^- \rightarrow Sn^{2+}$ $H_2 \rightarrow 2H^+ + 2e^-$

    5: $Sn^{4+} + 2e^- \rightarrow Sn^{2+}$ $H_2 \rightarrow 2H^+ + 2e^-$

    6: $Sn^{4+}(aq) + H_2(g) \rightarrow Sn^{2+}(aq) + 2H^+(aq)$

c) 1: $\underset{+6}{Cr_2}O_7^{2-}(aq) + Br^-(aq) \rightarrow \underset{+3}{Cr}^{3+}(aq) + Br_2(l)$

    2: $Cr_2O_7^{2-} \rightarrow Cr^{3+}$ $Br^- \rightarrow Br_2$

    3: $Cr_2O_7^{2-} + 14H^+ \rightarrow 2Cr^{3+} + 7H_2O$ $2Br^- \rightarrow Br_2$

    4: $Cr_2O_7^{2-} + 14H^+ + 6e^- \rightarrow 2Cr^{3+} + 7H_2O$ $2Br^- \rightarrow Br_2 + 2e^-$

    5: $Cr_2O_7^{2-} + 14H^+ + 6e^- \rightarrow 2Cr^{3+} + 7H_2O$ $6Br^- \rightarrow 3Br_2 + 6e^-$

    6: $Cr_2O_7^{2-}(aq) + 14H^+(aq) + 6Br^-(aq) \rightarrow 2Cr^{3+}(aq) + 7H_2O(l) + 3Br_2(l)$

69. a) 1. $ClO^-(aq) + Cr(OH)_4^-(aq) \rightarrow CrO_4^{2-}(aq) + Cl^-(aq)$
    $\;\;+1\;-2\;+3\;-2\;+1\;+6\;-2-1$

   2. $ClO^- \rightarrow Cl^-$ $Cr(OH)_4^- \rightarrow CrO_4^{2-}$

   3. $ClO^- + 2H^+ \rightarrow Cl^- + H_2O$ $Cr(OH)_4^- \rightarrow CrO_4^{2-} + 4H^+$

   4. $ClO^- + 2H^+ + 2e^- \rightarrow Cl^- + H_2O$ $Cr(OH)_4^- \rightarrow CrO_4^{2-} + 4H^+ + 3e^-$

   5. $3ClO^- + 6H^+ + 6e^- \rightarrow 3Cl^- + 3H_2O$ $\;\;2Cr(OH)_4^- \rightarrow 2CrO_4^{2-} + 8H^+ + 6e^-$

   6. $3ClO^- + 2Cr(OH)_4^- \rightarrow 2CrO_4^{2-} + 3Cl^- + 2H^+ + 3H_2O$

   7. Overall: $3ClO^- + 2Cr(OH)_4^- + 2OH^- \rightarrow 2CrO_4^{2-} + 3Cl^- + 5H_2O$

   b) 1. $MnO_4^-(aq) + Br^-(aq) \rightarrow MnO_2(s) + BrO_3^-(aq)$
      $\;\;+7\;-2-1\;+4\;-2\;+5\;-2$

   2. $MnO_4^- \rightarrow MnO_2$ $Br^- \rightarrow BrO_3^-$

   3. $MnO_4^- + 4H^+ \rightarrow MnO_2 + 2H_2O$ $Br^- + 3H_2O \rightarrow BrO_3^- + 6H^+$

   4. $MnO_4^- + 4H^+ + 3e^- \rightarrow MnO_2 + 2H_2O$ $Br^- + 3H_2O \rightarrow BrO_3^- + 6H^+ + 6e^-$

   5. $2MnO_4^- + 8H^+ + 6e^- \rightarrow 2MnO_2 + 4H_2O$ $\;\;Br^- + 3H_2O \rightarrow BrO_3^- + 6H^+ + 6e^-$

   6. $2MnO_4^- + Br^- + 2H^+ \rightarrow 2MnO_2 + BrO_3^- + H_2O$

   7. Overall: $2MnO_4^- + Br^- + H_2O \rightarrow 2MnO_2 + BrO_3^- + 2OH^-$

70. a) 1. $NO_2^-(aq) + Al(s) \rightarrow NH_3(g) + AlO_2^-(aq)$
    $\;\;+3\;-2\;0-3\;+1\;+3\;-2$

   2. $NO_2^- \rightarrow NH_3$ $Al \rightarrow AlO_2^-$

   3. $NO_2^- + 7H^+ \rightarrow NH_3 + 2H_2O$ $Al + 2H_2O \rightarrow AlO_2^- + 4H^+$

   4. $NO_2^- + 7H^+ + 6e^- \rightarrow NH_3 + 2H_2O$ $Al + 2H_2O \rightarrow AlO_2^- + 4H^+ + 3e^-$

   5. $NO_2^- + 7H^+ + 6e^- \rightarrow NH_3 + 2H_2O$ $2Al + 4H_2O \rightarrow 2AlO_2^- + 8H^+ + 6e^-$

   6. $NO_2^- + 2Al + 2H_2O \rightarrow NH_3 + 2AlO_2^- + H^+$

   7. Overall: $NO_2^- + 2Al + H_2O + OH^- \rightarrow NH_3 + 2AlO_2^-$

   b) 1. $Al(s) + MnO_4^-(aq) \rightarrow MnO_2(s) + Al(OH)_4^-(aq)$
      $\;\;0\;+7\;-2\;+4\;-2\;+3\;-2\;+1$

   2. $MnO_4^- \rightarrow MnO_2$ $Al \rightarrow Al(OH)_4^-$

   3. $MnO_4^- + 4H^+ \rightarrow MnO_2 + 2H_2O$ $Al + 4H_2O \rightarrow Al(OH)_4^- + 4H^+$

   4. $MnO_4^- + 4H^+ + 3e^- \rightarrow MnO_2 + 2H_2O$ $Al + 4H_2O \rightarrow Al(OH)_4^- + 4H^+ + 3e^-$

   5. $MnO_4^- + 4H^+ + 3e^- \rightarrow MnO_2 + 2H_2O$ $Al + 4H_2O \rightarrow Al(OH)_4^- + 4H^+ + 3e^-$

   6. $MnO_4^- + Al + 2H_2O \rightarrow MnO_2 + Al(OH)_4^-$

   7. Overall: $MnO_4^- + Al + 2H_2O \rightarrow MnO_2 + Al(OH)_4^-$

The Activity Series

71. a) Ag —For the elements listed, it is lowest on the activity series of metals.

72. b) Cu —For the elements listed, it is lowest on the activity series of metals.

73. b) Cu^{2+}

74. a) Pb^{2+}

75. b) Al

76. b) Mg

77. a) no reaction
    c) spontaneous
    b) spontaneous
    d) no reaction

78. a) no reaction
    c) no reaction
    b) spontaneous
    d) spontaneous

79. You could use any of the following metals: Fe, Cr, Zn, Mn, Al, Mg, Na, Ca, K, Li.

80. You could use any of the following metals: Sn, Ni, Fe, Cr, Zn, Mn, Al, Mg, Na, Ca, K, Li.

81. Mg will reduce Al^{3+} but not Na^+.

82. Fe will be oxidized by Ni^{2+} but not Cr^{3+}.

83. a) no reaction
    c) no reaction
    b) $Fe + 2HCl \rightarrow H_2 + Fe^{2+} + 2Cl^-$
    d) $2Al + 6HCl \rightarrow 3H_2 + 2Al^{3+} + 6Cl^-$

84. a) $2Cr + 6HCl \rightarrow 3H_2 + 2Cr^{3+} + 6Cl^-$
    b) $Pb + 2HCl \rightarrow H_2 + Pb^{2+} + 2Cl^-$
    c) no reaction
    d) $Zn + 2HCl \rightarrow H_2 + Zn^{2+} + 2Cl^-$

Batteries, Electrochemical Cells, and Electrolysis

85. The electrochemical cell:

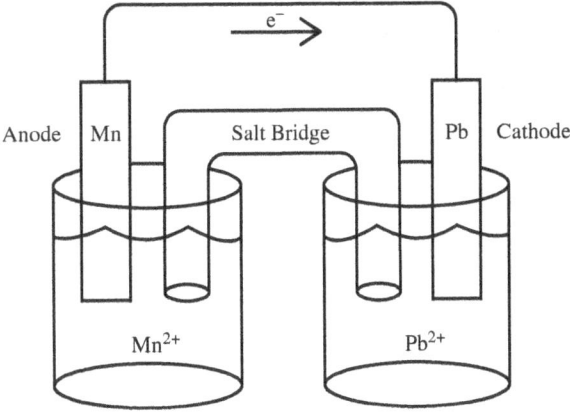

86. The electrochemical cell:

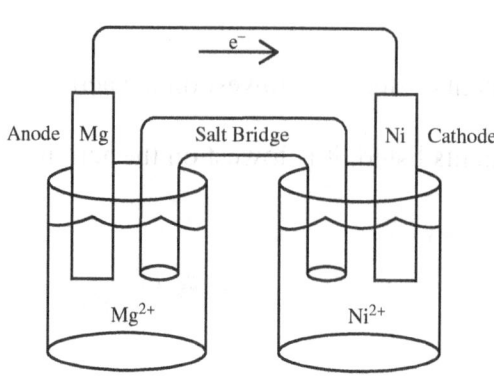

87. Choice d would produce the highest voltage.

88. Choice b would produce the highest voltage.

89. The overall reaction for an alkaline battery is:
cathode: $2MnO_2(s) + 2H_2O(l) + 2e^- \rightarrow 2MnO(OH)(s) + 2OH^-(aq)$
+anode: $Zn(s) + 2OH^-(aq) \rightarrow Zn(OH)_2(s) + 2e^-$
Overall: $2MnO_2(s) + 2H_2O(l) + Zn(s) \rightarrow 2MnO(OH)(s) + Zn(OH)_2(s)$

90. The overall reaction for a lead–acid storage battery is:
cathode: $PbO_2(s) + 4H^+(aq) + SO_4^{2-}(aq) + 2e^- \rightarrow PbSO_4(s) + 2H_2O(l)$
+anode: $Pb(s) + SO_4^{2-}(aq) \rightarrow PbSO_4(s) + 2e^-$
Overall: $PbO_2(s) + 4H^+(aq) + 2SO_4^{2-}(aq) + Pb(s) \rightarrow 2PbSO_4(s) + 2H_2O(l)$

91. The electrolysis cell for electroplating copper:
anode: $Cu(s) \rightarrow Cu^{2+}(aq) + 2e^-$
cathode: $Cu^{2+}(aq) + 2e^- \rightarrow Cu(s)$

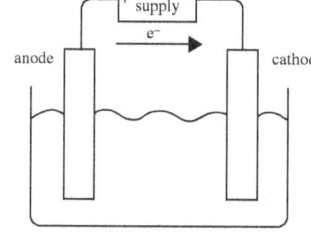

92. The electrolysis cell for electroplating nickel:
anode: $Ni(s) \rightarrow Ni^{2+}(aq) + 2e^-$
cathode: $Ni^{2+}(aq) + 2e^- \rightarrow Ni(s)$

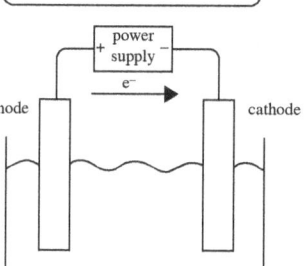

Corrosion

93. a) yes
    b) no
    c) yes

94. a) yes
    b) yes
    c) no

Cumulative Problems

95. a) Redox reaction, with Zn being oxidized and Co being reduced.
    b) Not a redox reaction.
    c) Not a redox reaction.
    d) Redox reaction, with K being oxidized and Br₂ being reduced.

96. a) Not a redox reaction.
    b) Not a redox reaction.
    c) Redox reaction, with Al being oxidized and the Fe in Fe₂O₃ being reduced.
    d) Not a redox reaction.

97. 1: $\underset{+7}{MnO_4^-}(aq) + \underset{0}{Zn}(s) \rightarrow \underset{+2}{Mn^{2+}}(aq) + \underset{+2}{Zn^{2+}}(aq)$

    2: $MnO_4^- \rightarrow Mn^{2+}$ $\quad\quad\quad\quad\quad\quad\quad\quad$ $Zn \rightarrow Zn^{2+}$

    3: $MnO_4^- + 8H^+ \rightarrow Mn^{2+} + 4H_2O$ $\quad\quad$ $Zn \rightarrow Zn^{2+}$

    4: $MnO_4^- + 8H^+ + 5e^- \rightarrow Mn^{2+} + 4H_2O$ $\quad$ $Zn \rightarrow Zn^{2+} + 2e^-$

    5: $2MnO_4^- + 16H^+ + 10e^- \rightarrow 2Mn^{2+} + 8H_2O$ $\quad$ $5Zn \rightarrow 5Zn^{2+} + 10e^-$

    6: $2MnO_4^-(aq) + 16H^+(aq) + 5Zn(s) \rightarrow 2Mn^{2+}(aq) + 8H_2O(l) + 5Zn^{2+}(aq)$

    $2.85 \text{ g Zn} \times \dfrac{1 \text{ mol Zn}}{65.39 \text{ g}} \times \dfrac{2 \text{ mol KMnO}_4}{5 \text{ mol Zn}} \times \dfrac{1 \text{ L}}{0.500 \text{ mol KMnO}_4} = 0.0349 \text{ L}$

98. 1: $\underset{+6}{Cr_2O_7^{2-}}(aq) + \underset{0}{Cu}(s) \rightarrow \underset{+3}{Cr^{3+}}(aq) + \underset{+2}{Cu^{2+}}(aq)$

    2: $Cr_2O_7^{2-} \rightarrow Cr^{3+}$ $\quad\quad\quad\quad\quad\quad\quad\quad$ $Cu \rightarrow Cu^{2+}$

    3: $Cr_2O_7^{2-} + 14H^+ \rightarrow 2Cr^{3+} + 7H_2O$ $\quad\quad$ $Cu \rightarrow Cu^{2+}$

    4: $Cr_2O_7^{2-} + 14H^+ + 6e^- \rightarrow 2Cr^{3+} + 7H_2O$ $\quad$ $Cu \rightarrow Cu^{2+} + 2e^-$

    5: $Cr_2O_7^{2-} + 14H^+ + 6e^- \rightarrow 2Cr^{3+} + 7H_2O$ $\quad$ $3Cu \rightarrow 3Cu^{2+} + 6e^-$

    6: $Cr_2O_7^{2-}(aq) + 14H^+(aq) + 3Cu(s) \rightarrow 2Cr^{3+}(aq) + 7H_2O(l) + 3Cu^{2+}(aq)$

    $5.25 \text{ g Cu} \times \dfrac{1 \text{ mol Cu}}{63.55 \text{ g}} \times \dfrac{1 \text{ mol K}_2\text{Cr}_2\text{O}_7}{3 \text{ mol Cu}} \times \dfrac{1 \text{ L}}{0.850 \text{ mol K}_2\text{Cr}_2\text{O}_7} = 0.0324 \text{ L}$

99. Yes, Mg is more easily oxidized than Ag; therefore, Mg will react with Ag^+.
   1. $Mg(s) + Ag^+(aq) \rightarrow Mg^{2+}(aq) + Ag(s)$
   2. $Mg(s) \rightarrow Mg^{2+}(aq)$ $Ag^+(aq) \rightarrow Ag(s)$
   3. $Mg(s) \rightarrow Mg^{2+}(aq)$ $Ag^+(aq) \rightarrow Ag(s)$ (no change)
   4. $Mg(s) \rightarrow Mg^{2+}(aq) + 2e^-$ $Ag^+(aq) + e^- \rightarrow Ag(s)$
   5. $Mg(s) \rightarrow Mg^{2+}(aq) + 2e^-$ $2Ag^+(aq) + 2e^- \rightarrow 2Ag(s)$
   6. $Mg(s) + 2Ag^+(aq) \rightarrow Mg^{2+}(aq) + 2Ag(s)$

100. No reaction will occur, as zinc is more easily oxidized than tin. Therefore, tin metal will not spontaneously react with zinc ions.

101. $5H_2O_2(aq) + 2MnO_4^-(aq) + 6H^+(aq) \rightarrow 2Mn^{2+}(aq) + 5O_2(g) + 8H_2O(l)$

$$0.03481\,L\,MnO_4^- \times \frac{0.0998\,mol\,MnO_4^-}{1\,L} \times \frac{5\,mol\,H_2O_2}{2\,mol\,MnO_4^-} \times \frac{34.01\,g}{1\,mol\,H_2O_2} = 0.295\,g\,H_2O_2$$

$$\frac{0.295\,g\,H_2O_2}{10.00\,g} \times 100\% = 2.95\%$$

102. $5Fe^{2+}(aq) + MnO_4^-(aq) + 8H^+(aq) \rightarrow Mn^{2+}(aq) + 5Fe^{3+}(aq) + 4H_2O(l)$

$$0.02245\,L\,MnO_4^- \times \frac{0.1201\,mol\,MnO_4^-}{1\,L} \times \frac{5\,mol\,Fe^{2+}}{1\,mol\,MnO_4^-} \times \frac{55.85\,g}{1\,mol\,Fe^{2+}} = 0.7529\,g\,Fe^{2+}$$

$$\frac{0.7529\,g\,Fe^{2+}}{1.012\,g} \times 100\% = 74.40\%\,Fe^{2+}$$

103. $5.8\,g\,Ag \times \frac{1\,mol\,Ag}{107.87\,g} \times \frac{1\,mol\,e^-}{1\,mol\,Ag} = 0.054\,mol\,e^-$

104. $1.40\,g\,Au \times \frac{1\,mol\,Au}{196.97\,g} \times \frac{3\,mol\,e^-}{1\,mol\,Au} = 0.0213\,mol\,e^-$

105. a) $6HI + 2Cr \rightarrow 2Cr^{3+} + 3H_2 + 6I^-$

$$5.95\,g\,Cr \times \frac{1\,mol\,Cr}{52.00\,g} \times \frac{6\,mol\,HI}{2\,mol\,Cr} \times \frac{1\,L}{3.5\,mol\,HI} = 0.098\,L$$

b) $6HI + 2Al \rightarrow 2Al^{3+} + 3H_2 + 6I^-$

$$2.15\,g\,Al \times \frac{1\,mol\,Al}{26.98\,g} \times \frac{6\,mol\,HI}{2\,mol\,Al} \times \frac{1\,L}{3.5\,mol\,HI} = 0.068\,L$$

c) no reaction

d) no reaction

106. a) no reaction

b) $2HCl + Pb \rightarrow Pb^{2+} + H_2 + 2Cl^-$

$2.55 \text{ g Pb} \times \dfrac{1 \text{ mol Pb}}{207.2 \text{ g}} \times \dfrac{2 \text{ mol HCl}}{1 \text{ mol Pb}} \times \dfrac{1 \text{ L}}{6.0 \text{ mol HCl}} = 0.0041 \text{ L}$

c) $2HCl + Sn \rightarrow Sn^{2+} + H_2 + 2Cl^-$

$4.83 \text{ g Sn} \times \dfrac{1 \text{ mol Sn}}{118.7 \text{ g}} \times \dfrac{2 \text{ mol HCl}}{1 \text{ mol Sn}} \times \dfrac{1 \text{ L}}{6.0 \text{ mol HCl}} = 0.014 \text{ L}$

d) $2HCl + Mg \rightarrow Mg^{2+} + H_2 + 2Cl^-$

$1.25 \text{ g Mg} \times \dfrac{1 \text{ mol Mg}}{24.31 \text{ g}} \times \dfrac{2 \text{ mol HCl}}{1 \text{ mol Mg}} \times \dfrac{1 \text{ L}}{6.0 \text{ mol HCl}} = 0.017 \text{ L}$

107. $2Al(s) + 6HCl(aq) \rightarrow 2AlCl_3(aq) + 3H_2(g)$

$\dfrac{6.0 \text{ mol HCl}}{1 \text{ L}} \times 0.000050 \text{ L} \times \dfrac{2 \text{ mol Al}}{6 \text{ mol HCl}} \times \dfrac{26.98 \text{ g}}{1 \text{ mol Al}} = 0.0027 \text{ g}$

Volume Dissolved: $\dfrac{0.0027 \text{ g}}{2.7 \text{ g/cm}^3} = 1.0 \times 10^{-3} \text{ cm}^3$

Volume Cylinder $= \pi r^2 h \Rightarrow 1.0 \times 10^{-3} \text{ cm}^3 = 3.14 r^2 (0.0028 \text{ cm})$

$r = \sqrt{\dfrac{1.0 \times 10^{-3} \text{ cm}^3}{3.14(0.0028 \text{ cm})}} = 0.337 \text{ cm}$

diameter = 0.67 cm

108. $Mn(s) + 2HCl(aq) \rightarrow MnCl_2(aq) + H_2(g)$

$\dfrac{12.0 \text{ mol HCl}}{1 \text{ L}} \times 0.00100 \text{ L} \times \dfrac{1 \text{ mol Mn}}{2 \text{ mol HCl}} \times \dfrac{54.94 \text{ g}}{1 \text{ mol Mn}} = 0.330 \text{ g}$

Volume Dissolved: $\dfrac{0.330 \text{ g}}{7.47 \text{ g/cm}^3} = 0.0441 \text{ cm}^3$

Volume Cylinder $= \pi r^2 h \Rightarrow 0.0441 \text{ cm}^3 = 3.14 r^2 (0.0055 \text{ cm})$

$r = \sqrt{\dfrac{0.0441 \text{ cm}^3}{3.14(0.0055 \text{ cm})}} = 1.6 \text{ cm}$

diameter = 3.2 cm

109. $1.0 \text{ g Ag} \times \dfrac{1 \text{ mol Ag}}{107.87 \text{ g}} \times \dfrac{1 \text{ mol e}^-}{1 \text{ mol Ag}} \times \dfrac{6.022 \times 10^{23} \text{ e}^-}{1 \text{ mol e}^-} \times \dfrac{1.60 \times 10^{-19} \text{ C}}{1 \text{ e}^-} \times \dfrac{1 \text{ s}}{0.100 \text{ C}}$

$= 8.9 \times 10^3 \text{ s}$

110. $0.400 \text{ g Au} \times \dfrac{1 \text{ mol Au}}{196.97 \text{ g}} \times \dfrac{1 \text{ mol e}^-}{1 \text{ mol Au}} \times \dfrac{6.022 \times 10^{23} \text{ e}^-}{1 \text{ mol e}^-} \times \dfrac{1.60 \times 10^{-19} \text{ C}}{1 \text{ e}^-} \times \dfrac{1 \text{ s}}{0.200 \text{ C}}$

= 978 s

Highlight Problems

111. The sketch should show aluminum atoms going into solution as +3 ions, which dissolve the metal electrode. The copper ions are forming solid, elemental copper atoms on the aluminum strip.

112. $\dfrac{2.65 \times 10^4 \text{ mol e}^-}{\text{month}} \times \dfrac{1 \text{ mol H}_2}{2 \text{ mol e}^-} \times \dfrac{2.02 \text{ g}}{1 \text{ mol H}_2} \times \dfrac{1 \text{ kg}}{1000 \text{ g}} = 26.7 \text{ kg H}_2/\text{month}$

113. The sketch should show the formation of an increased number of zinc ions in solution and the loss of zinc atoms from the surface of the anode. The cathode should have an increased number of nickel atoms and a corresponding decrease in the nickel ions from solution.

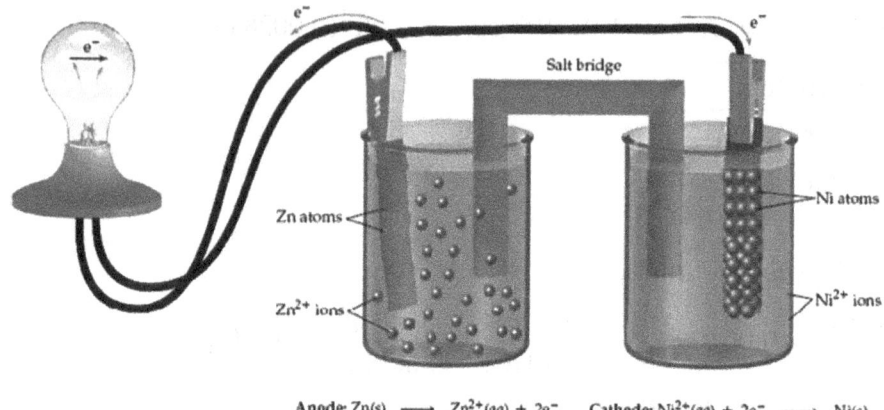

114. Answers will vary.

115. (a) $\underset{-2\ +1}{C}\underset{}{H_3}\underset{-2\ +1}{O}\underset{}{H} + \underset{0}{O_2} \rightarrow \underset{+4\ -2}{C}\underset{}{O_2} + \underset{+1\ -2}{H_2}\underset{}{O}$

(b) Carbon in methanol lost electrons to go from −2 to +4; carbon is oxidized. Oxygen in O₂ gained electrons to go from 0 to −2; oxygen gas is reduced.

(c) $CH_3OH \rightarrow CO_2$ $\quad\quad\quad\quad\quad\quad$ $O_2 \rightarrow H_2O$
$CH_3OH + H_2O \rightarrow CO_2$ $\quad\quad$ $O_2 \rightarrow 2H_2O$
$CH_3OH + H_2O \rightarrow CO_2 + 6H^+$ $\quad$ $O_2 + 4H^+ \rightarrow 2H_2O$
$CH_3OH + H_2O \rightarrow CO_2 + 6H^+ + 6e^-$ $\quad$ $O_2 + 4H^+ + 4e^- \rightarrow 2H_2O$

(d) $2CH_3OH + 2H_2O \rightarrow 2CO_2 + 12H^+ + 12e^-$ $\quad$ $3O_2 + 12H^+ + 12e^- \rightarrow 6H_2O$
$2CH_3OH + \cancel{2H_2O} + 3O_2 + \cancel{12H^+} + \cancel{12e^-} \rightarrow 2CO_2 + \cancel{12H^+} + \cancel{12e^-} + \cancel{6}4H_2O$
$2CH_3OH + 3O_2 \rightarrow 2CO_2 + 4H_2O$

(e) The hydrogen ions are produced by the methanol reaction and consumed by the oxygen reaction.

Data Interpretation and Analysis

116. (a) [Scatter plot of Voltage vs $\log [Cu^{2+}]$, with voltage ranging from 0.30 to 0.35 and $\log [Cu^{2+}]$ ranging from −1.1 to 0.1, showing a roughly linear increasing trend.]

(b) $Y = 0.0298X + 0.3399$
(c) Solution i: $Y = 0.0298(0.303) + 0.3399 = 0.349$
Solution ii: $Y = 0.0298(0.338) + 0.3399 = 0.350$

Radioactivity and Nuclear Chemistry

17

Questions

1. Radioactivity is the emission of tiny, invisible (to the human eye) particles by the nuclei of certain atoms. A radioactive atom will spontaneously emit these tiny, invisible particles.

2. Becquerel first discovered radioactivity in 1896 by placing radioactive uranium minerals on photographic plates in an attempt to look for X-rays.

3. Uranic rays were the name given to the radioactive particles being emitted from uranium minerals in Becquerel's studies.

4. Marie Sklodowska Curie discovered two new elements that were both radioactive, which is a term that she created to replace the term "uranic rays." Curie was awarded the Nobel Prize in physics for the discovery of radioactivity.

5. X = chemical symbol, used to identify element
   Z = atomic number, the number of protons in the nucleus; determines the identity of the element
   A = atomic mass, the sum of the number of protons and the number of neutrons in the nucleus

6. Radioactivity originates from the <u>nucleus</u> of radioactive atoms.

7. Alpha (α) radiation occurs when the nucleus emits a particle that contains 2 protons and 2 neutrons. Alpha particles are symbolized by ^4_2He.

8. When an atom emits an alpha particle, the atomic number decreases by 2 and the atomic mass decreases by 4.

9. Alpha particles have a high ionizing power in comparison with other radioactive particles; however, they have very low penetrating power in comparison with the other radioactive particles.

10. Beta (β) radiation occurs when the nucleus emits an electron. The beta particle is symbolized as $^{\ \ 0}_{-1}\text{e}$.

11. When an atom emits a beta particle, a neutron in the nucleus is converted to a proton. As a result, the atomic number increases by 1 while the atomic mass remains constant.

12. Beta particles have intermediate ionizing power and penetrating power in comparison with other radioactive particles.

13. Gamma radiation is a type of electromagnetic radiation that is essentially a high-energy photon. The gamma particle is symbolized by $^{0}_{0}\gamma$.

14. Since a gamma ray has no charge and no mass, it does not change the mass or atomic number of an atom. Gamma rays provide a release of energy from a nucleus and are usually emitted in conjunction with other types of radiation.

15. Gamma rays have low ionizing power and high penetrating power in comparison with other radioactive particles.

16. Positron emission occurs during the conversion of a proton in the nucleus to a neutron. The positron particle emitted has the same mass as an electron, but it has a positive charge. The positron is symbolized by $^{0}_{+1}e$.

17. When an atom emits a positron, it converts a proton to a neutron, which converts it into an atom of the next lighter element (Z decreases by 1, while A remains constant).

18. Positrons have similar ionizing power and penetrating power as beta particles, which are intermediate in comparison with other radioactive particles.

19. A nuclear equation represents nuclear processes such as radioactivity. For a nuclear equation to be balanced, the sum of the atomic numbers on both sides of the equation must be equal, and the sum of the mass numbers on both sides of the equation must be equal.

20. The parent nuclide is the reactant $^{231}_{91}Pa$; the daughter nuclides are the products $^{227}_{89}Ac$ and $^{4}_{2}He$. This equation represents alpha emission.

21. A film-badge dosimeter is a radioactivity detector used as a safety precaution for people who work with or near radioactive compounds. The dosimeter is a small piece of photographic film that is attached to clothing and regularly collected and developed in order to monitor the amount of radioactivity to which the badge has been exposed in the recent past.

22. The Geiger-Müller counter is an instant method of detecting radioactivity that is based on the radioactive particle passing through a chamber of argon gas. The radioactive particles ionize part of the argon gas, which can then conduct electricity across electrodes within the gas chamber.

23. In a scintillation counter, the radioactive particles pass through a crystal of NaI or CsI, which emits UV-Vis photons as it becomes excited by the radioactive particles. The photons are then detected and converted into an electrical signal.

24. Natural sources of radioactivity include some naturally occurring radioactive atoms in soil and minerals (uranium, radium), in the food that we eat, and radiation from space that makes its way through the atmosphere.

25. The half-life of a radioactive nuclide is the time it will take for one-half of the original parent nuclides to undergo decay. The half-life can be used to determine radioactive decay rates.

26. A radioactive decay series is a sequence of radioactive decays that occur when a very heavy radioactive atom, such as uranium, undergoes radioactive decay to produce a daughter nuclide that is also radioactive. Each radioactive daughter in the series also undergoes radioactive decay until, after many decay steps, a stable nuclide is formed.

27. Radon in the environment comes from the radioactive decay series of uranium. It presents a danger because radon is a gaseous compound and its daughter nuclides can attach to dust particles, which can be inhaled into the human body and increase a person's risk for developing lung cancer.

28. Carbon-14 comes from the reaction of atomic N_2 with neutrons in the upper atmosphere. Carbon-14 is converted into carbon dioxide and then into plant material during photosynthesis. Since animals ingest plant material, all living organisms contain a residual amount of carbon-14. The carbon-14 level is continuous, as all living organisms continually ingest new amounts of carbon-14.

29. When an organism dies, it no longer uptakes carbon-14; therefore, the amount of carbon-14 starts to decay at a rate equal to the half-life of carbon-14. By measuring how much carbon-14 remains in an organism, you can use the half-life to determine how long it took the carbon-14 to decay to the current level.

30. Carbon-14 has been proven accurate by measuring the levels of carbon-14 in objects that are of <u>known</u> age, proving it is an accurate method. For example, tree rings from trees that are hundreds of years old can be used to test accuracy. Carbon-14 dating is limited to about 50,000-year-old objects because the amount of radioactive carbon in older objects is too small to accurately measure.

31. $^{235}_{92}\text{Ur} + ^{1}_{0}\text{n} \rightarrow ^{140}_{56}\text{Ba} + ^{93}_{36}\text{Kr} + 3^{1}_{0}\text{n} + \text{energy}$

Enrico Fermi bombarded uranium-235 with neutrons in the hope that it would undergo beta decay and produce a new element with atomic number 93. Instead, it underwent nuclear fission, breaking into several smaller elements.

32. Nuclear fission is the process of splitting a nucleus into smaller elements. The first fission experiment was done by Enrico Fermi, although he did not realize that his reaction had undergone fission. Meitner, Strassmann, and Hahn were the first to report that bombarding uranium-235 with neutrons produces a fission reaction.

33. The nuclear fission reaction is triggered by a neutron colliding with a fissionable nuclide, which produces lighter elements and more neutrons. The neutrons produced in the reaction can then collide with other nuclides, producing a chain reaction. This reaction can then be used to make a bomb because a tremendous amount of energy is released during each fission reaction.

34. A critical mass is the minimum amount of uranium needed to create a chain reaction.

35. The main goal of the Manhattan Project was to build an atomic bomb before the Germans did. The project was led by J. R. Oppenheimer.

36. The heat generated by a controlled nuclear fission reaction is used to boil water; the steam produced turns turbines, which generate electricity.

37. The control rods serve to regulate the rate of the nuclear fission reaction in the reactor core so that the temperature is maintained at a desired level. If the reactor core temperature is too low, the control rods are raised, which allows more neutrons to collide with uranium, producing heat and the fission products. If the reactor core temperature is too hot, the control rods are lowered into the core, absorbing a greater number of neutrons; therefore, fewer neutrons will collide with uranium and less heat is produced.

38. The main benefits of nuclear electricity generation are that it requires very little fuel and does not generate air pollution or greenhouse gases. The problems associated with nuclear power include the danger of a nuclear accident and the disposal of radioactive waste.

39. A nuclear power plant cannot detonate like a nuclear bomb because the uranium fuel is not enriched with the quantity of uranium-235 that is needed to produce a bomb.

40. Nuclear fusion is the process of two lighter elements coming together to form a heavier element, along with the release of a large amount of energy.

41. Traditional nuclear bombs are of the fission type. However, hydrogen bombs are based on fusion. Note: In order for a hydrogen bomb to work, a small fission reaction takes place to generate sufficient heat to initiate the fusion process.

42. Fusion is not currently used for electricity generation. Fusion is an excellent candidate for electricity generation because it provides 10 times more energy per gram of fuel than fission reactions and the products of the reaction are not radioactive. However, due to the extreme temperatures needed to initiate fusion reactions, there is no known material that can withstand the temperature.

43. Radiation can ionize molecules in living organisms.

44. Acute radiation damage results from exposure to large amounts of radiation in a short period of time. These high levels of radiation kill large numbers of cells, particularly rapidly dividing cells such as those found in the immune system and the intestinal lining.

45. Radiation can increase the risk of cancer because it can damage DNA, which can cause cells to grow abnormally.

46. Genetic defects in offspring could occur if radiation damaged reproductive cells. These defects have been observed in laboratory animals exposed to high levels of radiation but have never been observed in humans.

47. The main unit of radiation is the roentgen equivalent man, also known as rem. A typical person is exposed to one-third of a rem per year.

48. <u>Radiation Exposure and Probable Outcome</u>
    20–100 rem   Decreased white blood cell count and possible increase in cancer risk.
    100–400 rem   Radiation sickness, skin lesions, increase in cancer risk
    500 rem   Death

49. Isotope scanning is a method used to diagnose different diseases based on the use of radioactive isotopes and their ability to target different organs and tissues. The radioactive isotopes are detected with either photographic film or a scintillation counter.

50. Cancer cells are known for their ability to rapidly divide and grow. Radiation is effective in killing rapidly growing cells and therefore is used to kill cancer cells. Some healthy cells are also killed, causing patients to suffer from radiation sickness.

Problems

Isotopic and Nuclear Particle Symbols

51. $^{210}_{82}Pb$

52. $^{207}_{83}Bi$

53. Protons (Z): 81
    Neutrons (A–Z): 126

54. Protons (Z): 86
    Neutrons (A–Z): 133

55. a) beta particle
    b) neutron
    c) gamma ray

56. a) proton
    b) alpha particle
    c) positron

57.

Chemical Symbol	Atomic Number (Z)	Mass Number (A)	#Protons	#Neutrons
Tc	<u>43</u>	95	<u>43</u>	<u>52</u>
Ba	56	128	<u>56</u>	<u>72</u>
Eu	<u>63</u>	<u>145</u>	<u>63</u>	82
Fr	<u>87</u>	223	<u>87</u>	136

58.

Chemical Symbol	Atomic Number (Z)	Mass Number (A)	#Protons	#Neutrons
Pd	46	<u>100</u>	<u>46</u>	54
Ce	<u>58</u>	136	<u>58</u>	<u>78</u>
Po	84	208	<u>84</u>	<u>124</u>
<u>Ra</u>	<u>88</u>	<u>226</u>	88	138

Radioactive Decay

59. a) $^{234}_{92}\text{U} \rightarrow {}^{230}_{90}\text{Th} + {}^{4}_{2}\text{He}$

    b) $^{230}_{90}\text{Th} \rightarrow {}^{226}_{88}\text{Ra} + {}^{4}_{2}\text{He}$

    c) $^{226}_{88}\text{Ra} \rightarrow {}^{222}_{86}\text{Rn} + {}^{4}_{2}\text{He}$

    d) $^{222}_{86}\text{Rn} \rightarrow {}^{218}_{84}\text{Po} + {}^{4}_{2}\text{He}$

60. a) $^{218}_{84}\text{Po} \rightarrow {}^{214}_{82}\text{Pb} + {}^{4}_{2}\text{He}$

    b) $^{214}_{84}\text{Po} \rightarrow {}^{210}_{82}\text{Pb} + {}^{4}_{2}\text{He}$

    c) $^{210}_{84}\text{Po} \rightarrow {}^{206}_{82}\text{Pb} + {}^{4}_{2}\text{He}$

    d) $^{227}_{90}\text{Th} \rightarrow {}^{223}_{88}\text{Ra} + {}^{4}_{2}\text{He}$

61. a) $^{214}_{82}\text{Pb} \rightarrow {}^{214}_{83}\text{Bi} + {}^{0}_{-1}e$
    b) $^{214}_{83}\text{Bi} \rightarrow {}^{214}_{84}\text{Po} + {}^{0}_{-1}e$
    c) $^{231}_{90}\text{Th} \rightarrow {}^{231}_{91}\text{Pa} + {}^{0}_{-1}e$
    d) $^{227}_{89}\text{Ac} \rightarrow {}^{227}_{90}\text{Th} + {}^{0}_{-1}e$

62. a) $^{211}_{82}\text{Pb} \rightarrow {}^{211}_{83}\text{Bi} + {}^{0}_{-1}e$
    b) $^{207}_{81}\text{Tl} \rightarrow {}^{207}_{82}\text{Pb} + {}^{0}_{-1}e$
    c) $^{234}_{90}\text{Th} \rightarrow {}^{234}_{91}\text{Pa} + {}^{0}_{-1}e$
    d) $^{234}_{91}\text{Pa} \rightarrow {}^{234}_{92}\text{U} + {}^{0}_{-1}e$

63. a) $^{11}_{6}\text{C} \rightarrow {}^{11}_{5}\text{B} + {}^{0}_{+1}e$
    b) $^{13}_{7}\text{N} \rightarrow {}^{13}_{6}\text{C} + {}^{0}_{+1}e$
    c) $^{15}_{8}\text{O} \rightarrow {}^{15}_{7}\text{N} + {}^{0}_{+1}e$

64. a) $^{55}_{27}\text{Co} \rightarrow {}^{55}_{26}\text{Fe} + {}^{0}_{+1}e$
    b) $^{22}_{11}\text{Na} \rightarrow {}^{22}_{10}\text{Ne} + {}^{0}_{+1}e$
    c) $^{18}_{9}\text{F} \rightarrow {}^{18}_{8}\text{O} + {}^{0}_{+1}e$

65. $^{241}_{94}\text{Pu} \rightarrow {}^{241}_{95}\text{Am} + {}^{0}_{-1}e$
    $^{241}_{95}\text{Am} \rightarrow {}^{237}_{93}\text{Np} + {}^{4}_{2}\text{He}$
    $^{237}_{93}\text{Np} \rightarrow {}^{233}_{91}\text{Pa} + {}^{4}_{2}\text{He}$
    $^{233}_{91}\text{Pa} \rightarrow {}^{233}_{92}\text{U} + {}^{0}_{-1}e$

66. $^{225}_{88}\text{Ra} \rightarrow {}^{225}_{89}\text{Ac} + {}^{0}_{-1}e$
    $^{225}_{89}\text{Ac} \rightarrow {}^{221}_{87}\text{Fr} + {}^{4}_{2}\text{He}$
    $^{221}_{87}\text{Fr} \rightarrow {}^{217}_{85}\text{At} + {}^{4}_{2}\text{He}$
    $^{217}_{85}\text{At} \rightarrow {}^{213}_{83}\text{Bi} + {}^{4}_{2}\text{He}$

67. $^{232}_{90}\text{Th} \rightarrow {}^{228}_{88}\text{Ra} + {}^{4}_{2}\text{He}$
    $^{228}_{88}\text{Ra} \rightarrow {}^{228}_{89}\text{Ac} + {}^{0}_{-1}e$
    $^{228}_{89}\text{Ac} \rightarrow {}^{228}_{90}\text{Th} + {}^{0}_{-1}e$
    $^{228}_{90}\text{Th} \rightarrow {}^{224}_{88}\text{Ra} + {}^{4}_{2}\text{He}$

68. $^{220}_{86}\text{Rn} \rightarrow {}^{216}_{84}\text{Po} + {}^{4}_{2}\text{He}$

$^{216}_{84}\text{Po} \rightarrow {}^{212}_{82}\text{Pb} + {}^{4}_{2}\text{He}$

$^{212}_{82}\text{Pb} \rightarrow {}^{212}_{83}\text{Bi} + {}^{0}_{-1}\text{e}$

$^{212}_{83}\text{Bi} \rightarrow {}^{208}_{81}\text{Tl} + {}^{4}_{2}\text{He}$

Half-Life

69. $\underbrace{100{,}000 \xrightarrow{2\text{ days}} 50{,}000 \xrightarrow{2\text{ days}} 25{,}000 \xrightarrow{2\text{ days}} 12{,}500 \xrightarrow{2\text{ days}} 6{,}250 \xrightarrow{2\text{ days}} 3{,}125}_{10\text{ days}}$

There would be 3,125 radioactive atoms remaining after 10 days.

70. $\underbrace{4.0 \times 10^{10} \xrightarrow{8\text{ days}} 2.0 \times 10^{10} \xrightarrow{8\text{ days}} 1.0 \times 10^{10} \xrightarrow{8\text{ days}} 5.0 \times 10^{9} \xrightarrow{8\text{ days}} 2.5 \times 10^{9}}_{32\text{ days} \approx 1\text{ month}}$

There would be approximately 2.5×10^9 atoms remaining after 1 month.

71. $5.0 \times 10^{-2} \xrightarrow{6\text{ hrs}} 2.5 \times 10^{-2} \xrightarrow{6\text{ hrs}} 1.25 \times 10^{-2} \xrightarrow{6\text{ hrs}} 6.25 \times 10^{-3}$

It would take 18 hours for technetium-99 to decay to 6.3×10^{-3} mg.

72. $0.240 \xrightarrow{11.4\text{ days}} 0.120 \xrightarrow{11.4\text{ days}} 0.0600 \xrightarrow{11.4\text{ days}} 0.0300 \xrightarrow{11.4\text{ days}} 0.0150$

It would take 45.6 days for radium-223 to decay to 1.50×10^{-2} moles.

73. $2.80 \xrightarrow{1\text{st}} 1.40 \xrightarrow{2\text{nd}} 0.700 \xrightarrow{3\text{rd}} 0.350 \xrightarrow{4\text{th}} 0.175 \xrightarrow{5\text{th}} 0.0875$

It would take 5 half-lives or 1.22×10^6 years.

74. $1 \xrightarrow{1\text{st}} 1/2 \xrightarrow{2\text{nd}} 1/4 \xrightarrow{3\text{rd}} 1/8$

It would take 3 half-lives or 2.11×10^9 years.

75. $2.45 \xrightarrow{3.8\text{ days}} 1.23 \xrightarrow{3.8\text{ days}} 0.613 \xrightarrow{3.8\text{ days}} 0.306$

There would be 0.306 grams of the isotope after 11.4 days.

76. $4\text{ days} \times \dfrac{24\text{ hrs}}{1\text{ day}} \times \dfrac{1\text{ half-life}}{12\text{ hrs}} = 8\text{ half-lives}$

$68 \xrightarrow{1\text{st}} 34 \xrightarrow{2\text{nd}} 17 \xrightarrow{3\text{rd}} 8.5 \xrightarrow{4\text{th}} 4.25 \xrightarrow{5\text{th}} 2.125 \xrightarrow{6\text{th}} 1.0625 \xrightarrow{7\text{th}} 0.531 \xrightarrow{8\text{th}} 0.266$

There would be 0.27 mg of tracer remaining in the patient.

77. Ga-67 > P-32 > Cr-51 > Sr-89

78. Tc-99m > Y-90 > In-111 > I-131

Radiocarbon Dating

79. The age of the boat would be equal to one half-life of C-14, or approximately 5,730 years old.

80. The last ice age would have taken place 11,460 years ago, or two half-lives of C-14.

81. The skull has undergone six half-lives of C-14, which would make it approximately 34,380 years old.

82. The age of the mammoth is equal to three half-lives of C-14, which is approximately 17,190 years old.

Fission and Fusion

83. $^{235}_{92}U + ^{1}_{0}n \rightarrow ^{144}_{54}Xe + ^{90}_{38}Sr + 2^{1}_{0}n$; 2 neutrons produced.

84. $^{235}_{92}U + ^{1}_{0}n \rightarrow ^{137}_{52}Te + ^{97}_{40}Zr + 2^{1}_{0}n$; 2 neutrons produced.

85. $^{2}_{1}H + ^{2}_{1}H \rightarrow ^{3}_{2}He + ^{1}_{0}n$

86. $^{3}_{1}H + ^{1}_{1}H \rightarrow ^{4}_{2}He$

Cumulative Problems

87. a) $^{1}_{1}p + ^{9}_{4}Be \rightarrow ^{6}_{3}Li + ^{4}_{2}He$

    b) $^{209}_{83}Bi + ^{64}_{28}Ni \rightarrow ^{272}_{111}Rg + ^{1}_{0}n$

    c) $^{179}_{74}W + ^{0}_{-1}e^- \rightarrow ^{179}_{73}Ta$

88. a) $^{27}_{13}Al + ^{4}_{2}He \rightarrow ^{30}_{15}P + ^{1}_{0}n$

    b) $^{32}_{16}S + ^{1}_{0}n \rightarrow ^{29}_{14}Si + ^{4}_{2}He$

    c) $^{241}_{95}Am \rightarrow ^{237}_{93}Np + ^{4}_{2}He$

89. $^{238}_{92}U + ^{1}_{0}n \rightarrow ^{239}_{92}U$; $^{239}_{92}U \rightarrow ^{0}_{-1}\beta + ^{239}_{93}Np$; $^{239}_{93}Np \rightarrow ^{0}_{-1}\beta + ^{239}_{94}Pu$

90. $^{27}_{13}Al + ^{1}_{0}n \rightarrow ^{28}_{13}Al$; $^{28}_{13}Al \rightarrow ^{4}_{2}He + ^{24}_{11}Na$; $^{24}_{11}Na \rightarrow ^{0}_{-1}\beta + ^{24}_{12}Mg$

91. $\dfrac{3.2 \times 10^{-11} \text{ J}}{\text{atom}} \times \dfrac{6.022 \times 10^{23} \text{ atoms}}{1 \text{ mol U}} = 1.9 \times 10^{13} \text{ J/mol}$

$\dfrac{1.9 \times 10^{13} \text{ J}}{\text{mol}} \times \dfrac{1 \text{ mol U}}{238 \text{ g}} \times \dfrac{1000 \text{ g}}{1 \text{ kg}} = 8.1 \times 10^{13} \text{ J/kg}$

92. $^{2}_{1}H + ^{3}_{1}H \rightarrow ^{4}_{2}He + ^{1}_{0}n$

$2 \text{ H atoms} \times \dfrac{2.8 \times 10^{-12} \text{ J}}{\text{H atom}} \times \dfrac{6.022 \times 10^{23} \text{ atoms}}{1 \text{ mol H}} = 3.4 \times 10^{12} \text{ J/mol}$

93. In one half-life (5 days), we will lose 0.60 g of material. Assuming that each atom produces one beta particle in the decay process, the number of beta emissions is

$0.60 \text{ g Bi} \times \dfrac{1 \text{ mol Bi}}{208.98 \text{ g}} \times \dfrac{6.022 \times 10^{23} \text{ atoms}}{1 \text{ mol Bi}} \times \dfrac{1 \beta \text{ particle}}{1 \text{ atom}} = 1.7 \times 10^{21} \beta \text{ particles}$

94. In two half-lives (6 min), 41 mg of Po-218 would have decayed. Assuming that each atom produces one alpha particle in the decay process, the number of alpha particles is:

$0.041 \text{ g } ^{218}\text{Po} \times \dfrac{1 \text{ mol } ^{218}\text{Po}}{218 \text{ g}} \times \dfrac{6.022 \times 10^{23} \text{ atoms}}{1 \text{ mol } ^{218}\text{Po}} \times \dfrac{1 \alpha \text{ particle}}{\text{atomic decay}} = 1.1 \times 10^{20} \alpha \text{ particles}$

95. $\dfrac{0.400 \text{ rem}}{0.585 \text{ rem}} \times 100\% = 68.4\%$ due to radon

96. $\dfrac{0.020 \text{ rem}}{0.36 \text{ rem}} \times 100\% = 5.6\%$ due to work exposure

97. $\dfrac{1.6 \times 10^{3} \text{ yrs}}{\text{half-life}} \times \dfrac{365 \text{ d}}{1 \text{ yr}} = 5.84 \times 10^{5} \text{ days/half-life}$

$45 \text{ days} \times \dfrac{\text{half-life}}{5.84 \times 10^{5} \text{ days}} = 7.7 \times 10^{-5} \text{ half-lives}$

$\underbrace{(0.5)^{7.7 \times 10^{-5}} = 0.999946629}_{\text{Fraction remaining}}$

$\underbrace{1.5 \text{ g Rn} \times 0.999946629 = 1.499919944}_{\text{Mass remaining}}$

$8.01 \times 10^{-5} \text{ g Ra} \times \dfrac{1 \text{ mol Ra}}{226.03 \text{ g}} \times \dfrac{1 \text{ mol Rn}}{1 \text{ mol Ra}} = 3.54 \times 10^{-7} \text{ mol Rn}$

$V = \dfrac{nRT}{P} = \dfrac{(3.54 \times 10^{-7} \text{ mol Rn})(0.08206 \text{ L} \cdot \text{atm/mol} \cdot \text{K})(298.15 \text{ K})}{1 \text{ atm}} = 8.67 \times 10^{-6} \text{ L Rn}$

98. $15 \text{ g } ^{235}\text{U} \times \dfrac{1 \text{ mol } ^{235}\text{U}}{235.04 \text{ g}} \times \dfrac{1 \text{ mol } ^{91}\text{Kr}}{1 \text{ mol } ^{235}\text{U}} \times \dfrac{92.93 \text{ g}}{1 \text{ mol } ^{91}\text{Kr}} = 5.9 \text{ g Kr}$

Highlight Problems

99. The missing nucleus contains 9 protons and 7 neutrons (fluorine-16).

100. The missing nucleus contains 10 protons and 11 neutrons (neon-21).

101. The missing nucleus contains 5 protons and 5 neutrons (boron-10).

102. There are equal amounts of uranium and lead; this indicates that the uranium has gone through one half-life. The mineral and the earth are 4.5 billion years old.

103.

Particle Name	Symbol	Mass Number	Atomic Number or Charge
alpha particle	^4_2He	4	2
beta particle	$^0_{-1}e$	0	−1
gamma ray	$^0_0\gamma$	0	0
positron	$^0_{+1}e$	0	1
neutron	^1_0n	1	0
proton	^1_1p	1	1

104.

Time (s)	Atom Count
0	16,000
55.6	8,000
111.2	4,000
166.8	2,000
222.4	1,000
278.0	500

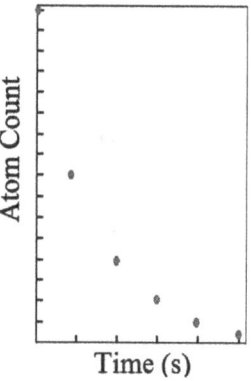

105. $^{238}_{92}U \rightarrow {}^4_2He + {}^{234}_{90}Th$

$^{234}_{90}Th \rightarrow {}^0_{-1}e + {}^{234}_{91}Pa$

$^{234}_{91}Pa \rightarrow {}^0_{-1}e + {}^{234}_{92}U$

106. Answers will vary.

Data Interpretation and Analysis

107. (a) Mass at 200 minutes equals 3.4 g; mass at 400 minutes is 2.4 g.
(b) Half-life is 360 min or 6 hrs.
(c) Twenty-four hours = 4 half-lives, 6.25% remains or 0.00313 mg.

Organic Chemistry

Questions

1. Organic molecules are responsible for most odors.

2. Organic chemistry is the study of carbon-containing compounds and their reactions.

3. At the end of the eighteenth century, it was believed that organic compounds came from living things and were easily decomposed, while inorganic compounds came from the earth and were more difficult to decompose. A final difference was that many inorganic compounds could be easily synthesized, but organic compounds could not be.

4. In the 1700s, vitalism was the belief that some mystical or supernatural power (a vital force) was needed to synthesize organic compounds. This was the accepted explanation for why chemists had been unsuccessful in their attempts to synthesize organic compounds.

5. Carbon chemistry is complex because of carbon's ability to bond with itself to form chains, branches, and ring structures. As a result, the number of molecules that can be formed from carbon is very large. It is this complexity and diversity of carbon-based compounds that makes life possible.

6. a) tetrahedral
   b) trigonal planar
   c) linear

7. Hydrocarbons are compounds that contain only carbon and hydrogen atoms. The main uses of hydrocarbons are as fuels and as raw materials in the synthesis of many products, including fabrics, soaps, dyes, cosmetics, drugs, plastic, and rubber.

8. The main categories of hydrocarbons are:
   a) alkanes C_nH_{2n+2}
   b) alkenes C_nH_{2n}
   c) alkynes C_nH_{2n-2}

9. Alkanes are considered saturated hydrocarbons because they contain only single bonds and they have the maximum number of hydrogen atoms possible (i.e., saturated). Alkenes and alkynes are considered unsaturated hydrocarbons because they contain double and triple bonds, and therefore have fewer hydrogen atoms than the similar alkane compound.

10. A molecular formula lists the type and number of each atom in a compound. The structural formula shows the type and number of each atom and a two-dimensional structure of how the atoms are bonded to each other. The condensed structural formula is a shorthand method of providing structural information in what appears to be a modified molecular formula format.

11. The n-alkane compounds have all carbon atoms bonded in a straight chain, where the branched alkanes have branches of carbon atoms coming off of a main straight chain of carbon atoms.

12. Isomers are compounds with the same exact chemical formula, but they have completely different structures. Isobutane and butane are examples of isomers.

13. Alkenes are hydrocarbons that contain at least one double bond between carbon atoms, where alkanes only contain single bonds.

14. Alkynes are hydrocarbons that contain at least one triple bond between carbon atoms, where alkanes only contain single bonds.

15. Hydrocarbon combustion reactions are the burning of hydrocarbons in the presence of oxygen. An example of a hydrocarbon combustion reaction is:
$$CH_3CH_2CH_3(g) + 5O_2(g) \rightarrow 3CO_2(g) + 4H_2O(g)$$

16. Alkane substitution reactions occur when a hydrogen atom(s) is replaced by another atom(s). An example of an alkane substitution reaction is:
$$CH_4(g) + Cl_2(g) \rightarrow CH_3Cl(g) + HCl(g)$$

17. Alkene addition reactions occur when atoms add across the double bond. An example of an alkene addition reaction is:
$$CH_2=CH_2(g) + Cl_2(g) \rightarrow CH_2ClCH_2Cl(g)$$

18. Alkyne addition reactions occur when atoms add across the triple bond. An example of an alkyne addition reaction is:
$$CH \equiv CH(g) + Cl_2(g) \rightarrow CHCl=CHCl(g)$$

19. The structure of benzene is 6 carbon atoms connected together in a ring, with a single hydrogen atom bonded to each carbon. Two common ways of showing the structure of benzene are shown here.

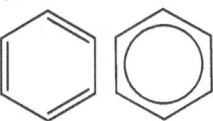

20. A functional group is a characteristic atom or group of atoms that has been incorporated into a hydrocarbon compound. Examples include alcohols (OH groups), carboxylic acids (COOH groups), and amines (NR$_3$ groups).

21. The generic structure of alcohols is ROH. Two examples of alcohols are ethanol: CH$_3$CH$_2$OH and 1-butanol: CH$_3$CH$_2$CH$_2$CH$_2$OH.

22. Alcoholic beverages contain ethanol, rubbing alcohol consists of isopropyl alcohol, and methanol is a common fuel additive.

23. The generic structure of ethers is R—O—R. Two examples of ethers are dimethyl ether: CH$_3$OCH$_3$ and diethyl ether: CH$_3$CH$_2$OCH$_2$CH$_3$.

24. Diethyl ether is a common ether that is used as a laboratory solvent, and it was once commonly used as an anesthetic.

25. | Functional Group | Generic Structure | Example |
| --- | --- | --- |
| aldehydes | R—C(=O)—H | CH$_3$—C(=O)—H (acetaldehyde) |
| ketones | R—C(=O)—R | CH$_3$—C(=O)—CH$_3$ (acetone) |

26. Two common aldehydes are formaldehyde, which is used as a preservative in formalin, and cinnamaldehyde, which is the fragrant component of cinnamon. Two common ketones are acetone, which is the main component in nail-polish remover, and 2-heptanone, which is the fragrant component of cloves.

27. | Functional Group | Generic Structure | Example |
| --- | --- | --- |
| carboxylic acid | R—C(=O)—OH | CH$_3$—C(=O)—OH (acetic acid) |
| esters | R—C(=O)—OR | CH$_3$—CH$_2$—C(=O)—O—CH$_2$—CH$_3$ (ethyl propanoate) |

28. Acetic acid is a common carboxylic acid that is the main ingredient in vinegar. Another common carboxylic acid is citric acid, which is present in citrus fruits such as limes, lemons, and oranges. Several common esters are ethyl butanoate and methyl butanoate, which are responsible for the fragrance of pineapples and apples, respectively.

29. The generic structure of amines is NR_3. Examples of specific amines are ammonia

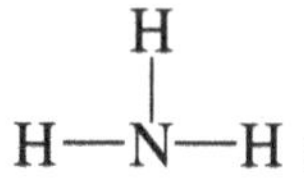

$$H-\underset{\underset{H}{|}}{N}-H$$ and ethylamine $CH_3CH_2-\underset{\underset{H}{|}}{N}-H$.

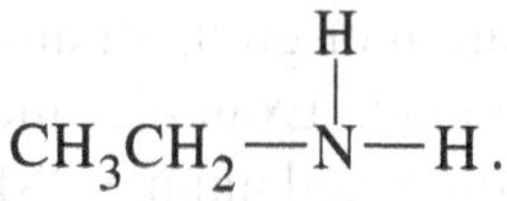

30. An example of a commonly found amine is cadaverine, (H_2N-$(CH_2)_5$-NH_2), which is responsible for the odor of decaying animal flesh.

31. A polymer is a long, chainlike molecule that is made up of small repeating units called monomers. A polymer is made up of one type of monomer, where a copolymer has two different monomers.

32. An addition polymer is made when monomers bond to each other without the loss of atoms. A condensation polymer is formed when the monomers lose an atom or group of atoms while forming the polymer. A condensation polymer often loses an H^+ and an OH^-, which forms water.

Problems

Hydrocarbons

33. Choices c and d are hydrocarbons because they contain only C and H.

34. Choices b and c are hydrocarbons because they contain only C and H.

35. a) alkyne (2n − 2)
    b) alkane (2n + 2)
    c) alkyne (2n − 2)
    d) alkene (2n)

36. a) alkene (2n)
    b) alkyne (2n − 2)
    c) alkane (2n + 2)
    d) alkene (2n)

Alkanes

37. a) CH₃CH₂CH₂CH₂CH₂CH₂CH₃

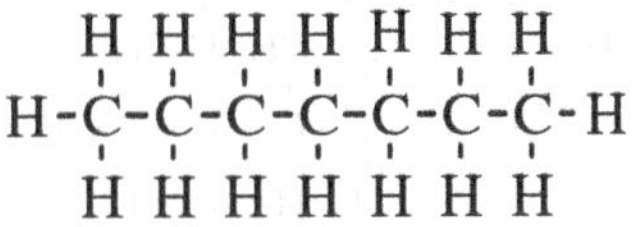

H H H H H H H
H-C-C-C-C-C-C-C-H
H H H H H H H

b) CH₃CH₂CH₂CH₂CH₂CH₂CH₂CH₃

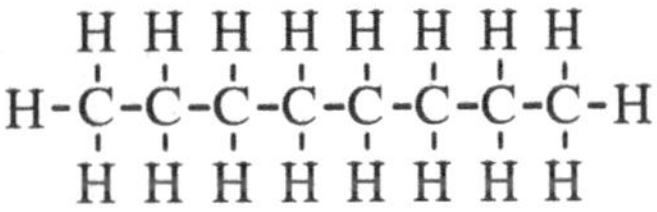

H H H H H H H H
H-C-C-C-C-C-C-C-C-H
H H H H H H H H

c) CH₃CH₂CH₂CH₂CH₂CH₃

H H H H H H
H-C-C-C-C-C-C-H
H H H H H H

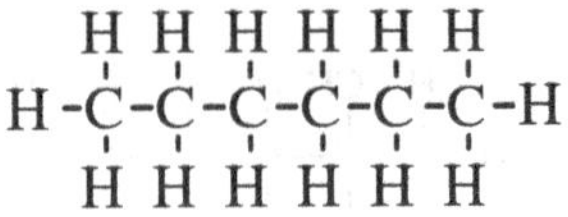

d) CH₃CH₃

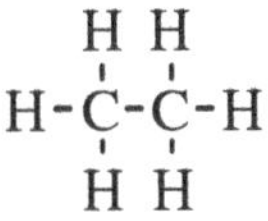

H H
H-C-C-H
H H

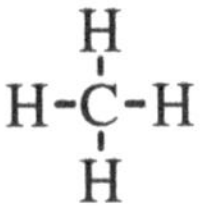

38. a) CH₄

H
H-C-H
H

b) CH₃CH₂CH₂CH₂CH₃

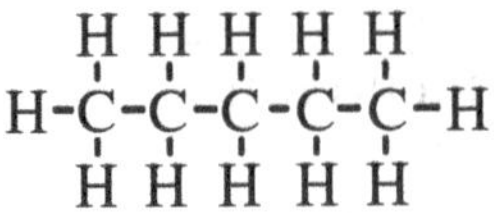

H H H H H
H-C-C-C-C-C-H
H H H H H

c) CH₃CH₂CH₂CH₃

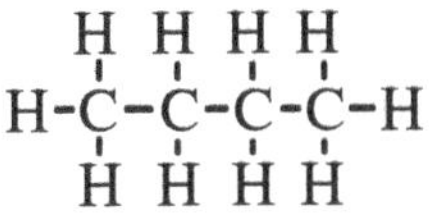

H H H H
H-C-C-C-C-H
H H H H

d) CH₃CH₂CH₃

H H H
H-C-C-C-H
H H H

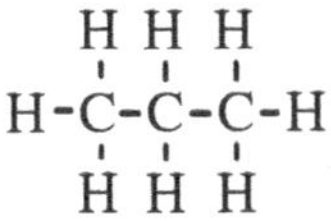

39. Two isomers of butane:

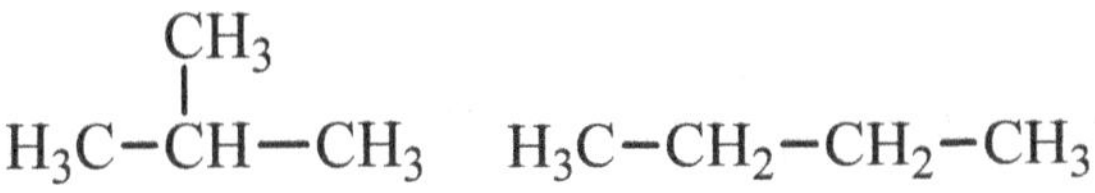

$$H_3C-\underset{\underset{CH_3}{|}}{CH}-CH_3 \quad H_3C-CH_2-CH_2-CH_3$$

40. Three isomers of pentane:

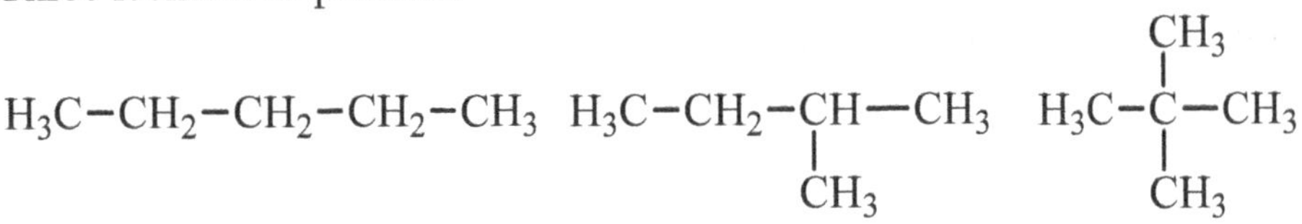

$$H_3C-CH_2-CH_2-CH_2-CH_3 \quad H_3C-CH_2-\underset{\underset{CH_3}{|}}{CH}-CH_3 \quad H_3C-\overset{\overset{CH_3}{|}}{\underset{\underset{CH_3}{|}}{C}}-CH_3$$

41. Five isomers of octane (18 total isomers):

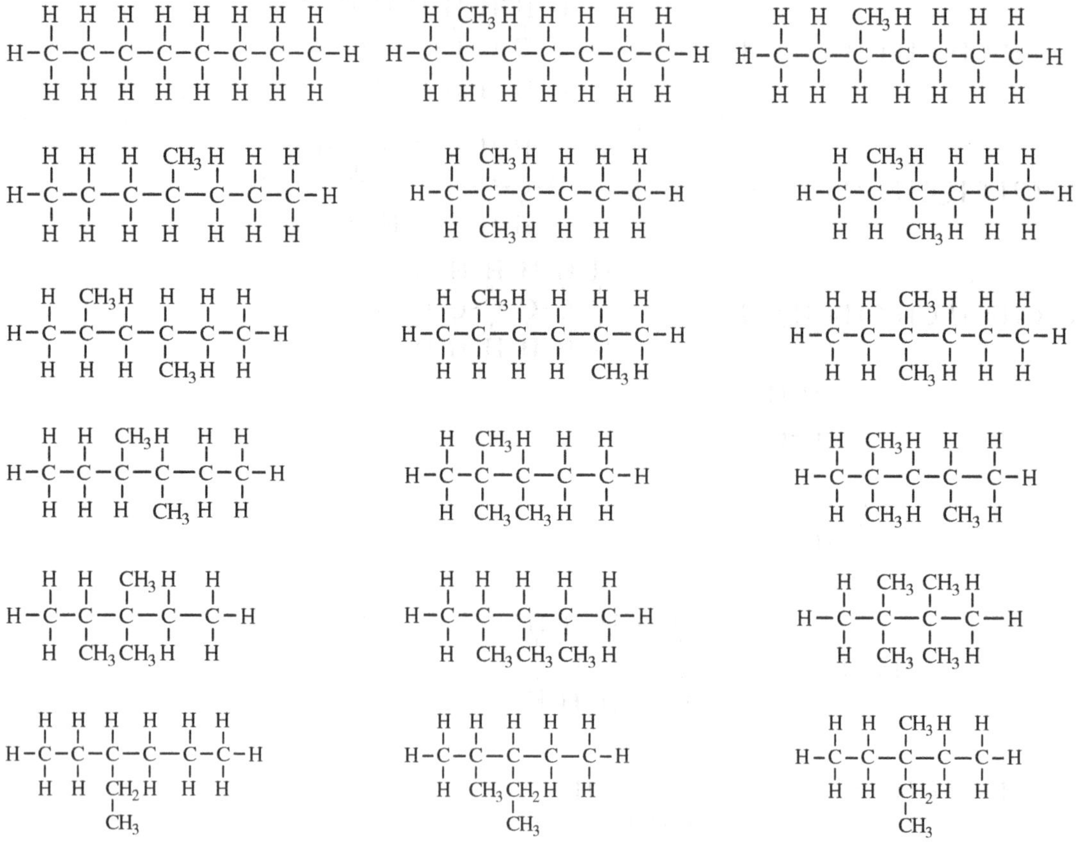

[Structural formulas of 18 octane isomers shown]

42. The nine isomers of heptane:

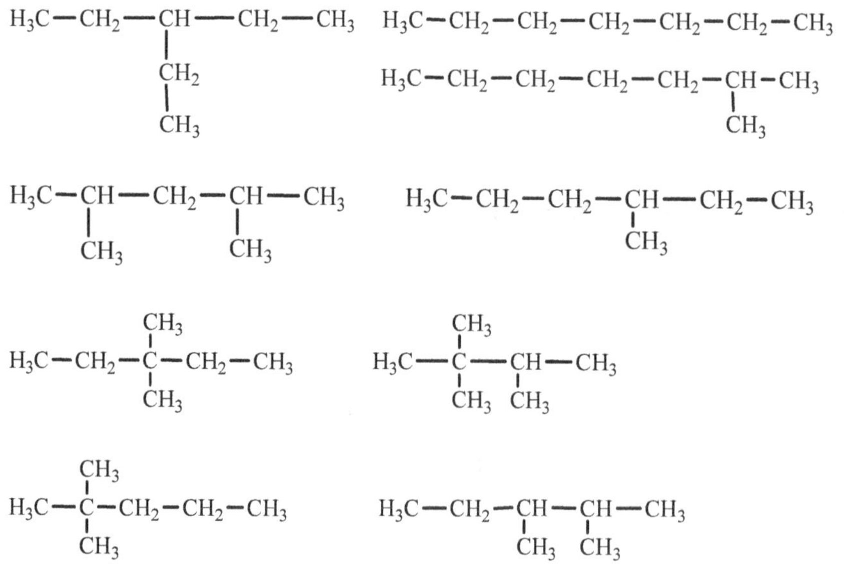

$H_3C-CH_2-CH-CH_2-CH_3$
       |
       CH_2
       |
       CH_3

$H_3C-CH_2-CH_2-CH_2-CH_2-CH_2-CH_3$

$H_3C-CH_2-CH_2-CH_2-CH_2-CH-CH_3$
                              |
                              CH_3

$H_3C-CH-CH_2-CH-CH_3$
     | |
     CH_3 CH_3

$H_3C-CH_2-CH_2-CH-CH_2-CH_3$
                 |
                 CH_3

$H_3C-CH_2-\underset{\underset{CH_3}{|}}{\overset{\overset{CH_3}{|}}{C}}-CH_2-CH_3$

$H_3C-\underset{\underset{CH_3}{|}}{\overset{\overset{CH_3}{|}}{C}}-CH-CH_3$
                                    |
                                    CH_3

$H_3C-\underset{\underset{CH_3}{|}}{\overset{\overset{CH_3}{|}}{C}}-CH_2-CH_2-CH_3$

$H_3C-CH_2-CH-CH-CH_3$
            | |
            CH_3 CH_3

43. a) n-pentane
    b) 2-methylbutane
    c) 4-ethyl-2-methylhexane
    d) 3,3-dimethylpentane

44. a) n-butane
    b) 4-propyloctane
    c) 4-ethyl-2-methylhexane
    d) 2,2,3,3-tetramethylbutane

45. Alkane structures:

   a) 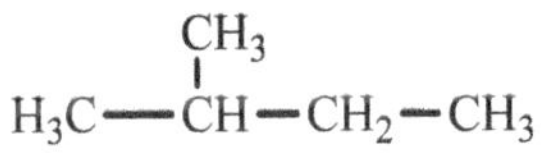 $H_3C-\underset{\underset{CH_3}{|}}{CH}-CH_2-CH_3$

   b) 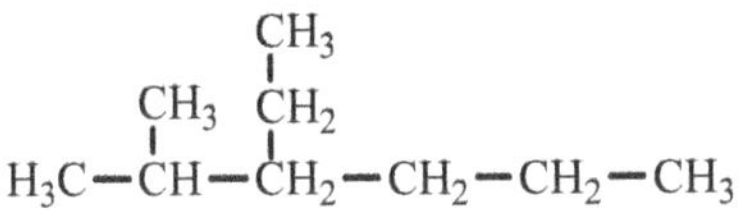 $H_3C-\underset{\underset{CH_3}{|}}{CH}-\underset{\underset{CH_2-\overset{|}{CH_3}}{|}}{CH}-CH_2-CH_2-CH_2-CH_3$

   c) 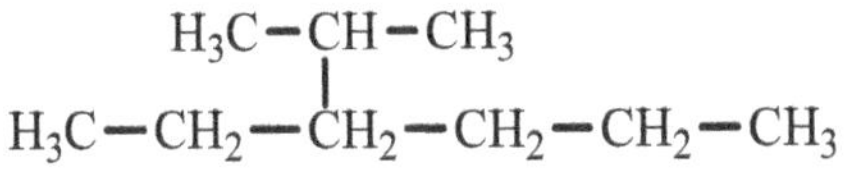 $H_3C-CH_2-\underset{\underset{CH_3-CH-CH_3}{|}}{CH_2}-CH_2-CH_2-CH_3$ (with $H_3C-CH-CH_3$ branch on top)

   d) 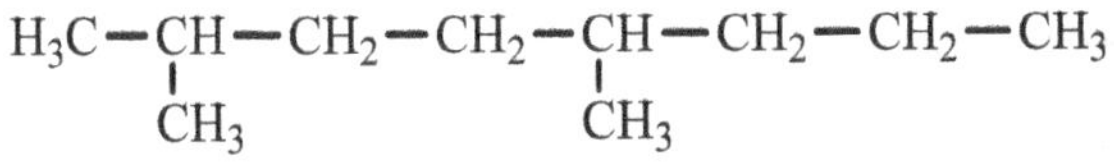 $H_3C-\underset{\underset{CH_3}{|}}{CH}-CH_2-CH_2-\underset{\underset{CH_3}{|}}{CH}-CH_2-CH_2-CH_3$

46. Alkane structures:

   a) 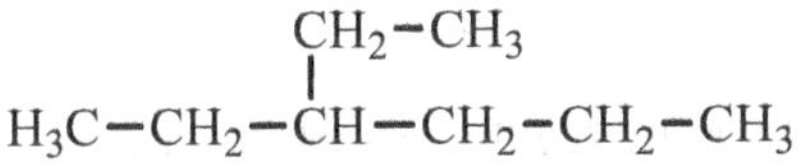 $H_3C-CH_2-\underset{\underset{CH_2-CH_3}{|}}{CH}-CH_2-CH_2-CH_3$

   b) 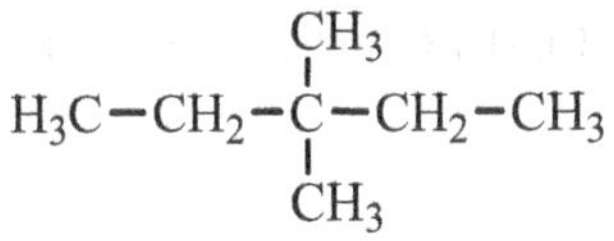 $H_3C-CH_2-\underset{\underset{CH_3}{|}}{\overset{\overset{CH_3}{|}}{C}}-CH_2-CH_3$

   c) $H_3C-CH_2-\underset{\underset{CH_3}{|}}{\overset{\overset{CH_2-CH_3}{|}}{CH}}-CH_2-CH_3$

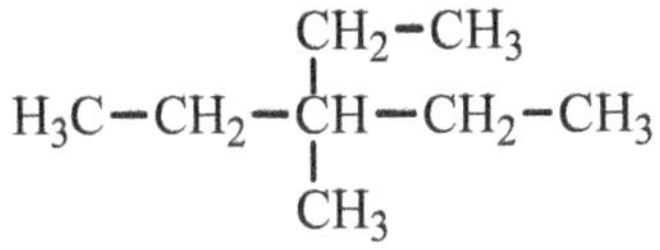

   d) 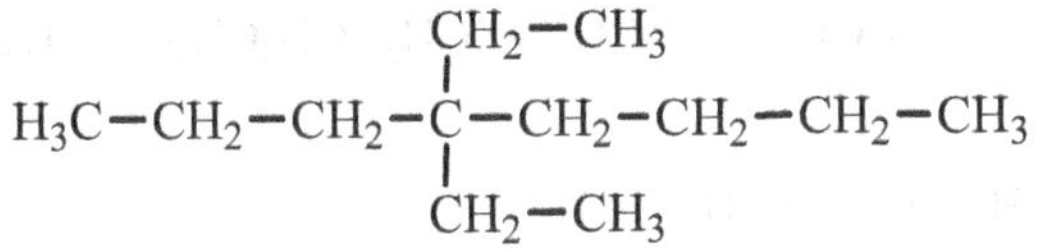 $H_3C-CH_2-CH_2-\underset{\underset{CH_2-CH_3}{|}}{\overset{\overset{CH_2-CH_3}{|}}{C}}-CH_2-CH_2-CH_2-CH_3$

47. a) n-pentane
    b) 3-methylhexane
    c) 2,3-dimethylpentane

48. a) n-hexane
    b) 3,4-dimethyloctane
    c) 2,3-dimethylpentane

49.

Name	Molecular Formula	Structural Formula	Condensed Structural Formula
2, 2, 3 trimethyl pentane	C_8H_{18}	H₃C—C(CH₃)(CH₃)—CH(CH₃)—CH₂—CH₃	$CH_3C(CH_3)_2CH(CH_3)CH_2CH_3$
2-methyl-3-propylhexane	$C_{10}H_{22}$	H₃C—CH(CH₃)—CH(CH₂CH₂CH₃)—CH₂—CH₂—CH₃	$CH_3CH(CH_3)CH(CH_2CH_2CH_3)CH_2CH_2CH_3$
2, 2, 3, 3 tetramethyl hexane	$C_{10}H_{22}$	H₃C—C(CH₃)(CH₃)—C(CH₃)(CH₃)—CH₂—CH₂—CH₃	$CH_3C(CH_3)_2C(CH_3)_2CH_2CH_2CH_3$
4,4-diethyl-2, 3-dimethylhexane	$C_{12}H_{26}$	H₃C—CH(CH₃)—CH(CH₃)—C(CH₂CH₃)(CH₂CH₂CH₃)—CH₂—CH₃	$CH_3CH(CH_3)CH(CH_3)C(CH_2CH_3)_2CH_2CH_3$

50.

Name	Molecular Formula	Structural Formula	Condensed Structural Formula
3, 3-diethyl-2, 2-dimethyloctane	$C_{14}H_{30}$	H₃C—C(CH₃)(CH₃)—C(CH₂CH₃)(CH₂CH₃)—CH₂—CH₂—CH₂—CH₂—CH₃	$CH_3C(CH_3)_2C(CH_2CH_3)_2CH_2CH_2CH_2CH_2CH_3$
3, 3-diethyl-4, 4-dimethyloctane	$C_{14}H_{30}$	H₃C—CH₂—C(CH₂CH₃)(CH₂CH₃)—C(CH₃)(CH₃)—CH₂—CH₂—CH₂—CH₃	$CH_3CH_2C(CH_2CH_3)_2C(CH_3)_2CH_2CH_2CH_2CH_3$
2, 2-dimethyl 3-ethylpentane	C_9H_{20}	CH₃—C(CH₃)(CH₃)—CH(CH₂CH₃)—CH₂—CH₃	$CH_3C(CH_3)_2CH(CH_2CH_3)CH_2CH_3$
4-propylnonane	$C_{12}H_{26}$	CH₃—CH₂—CH₂—CH(CH₂CH₂CH₃)—CH₂—CH₂—CH₂—CH₂—CH₃	$CH_3CH_2CH_2CH(CH_2CH_2CH_3)CH_2CH_2CH_2CH_3$

Alkenes and Alkynes

51. 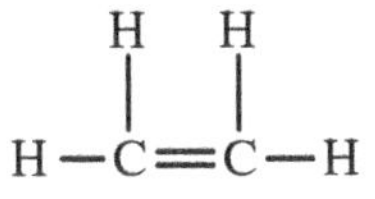 H—C(H)=C(H)—H ⟹ CH₂CH₂

    H—C(H)=C(H)—C(H)(H)—H ⟹ CH₂CHCH₃

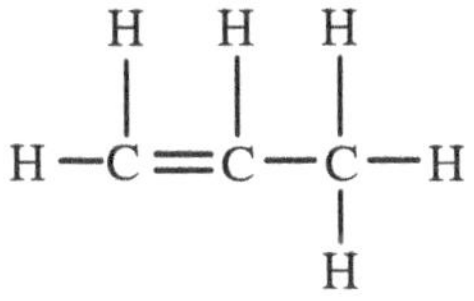

52. H—C≡C—H ⟹ CHCH

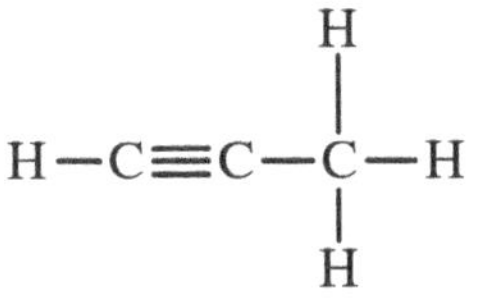

    H—C≡C—C(H)(H)—H ⟹ CHCCH₃

53. 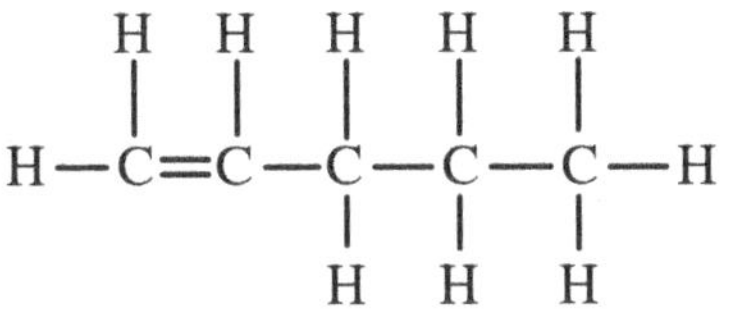 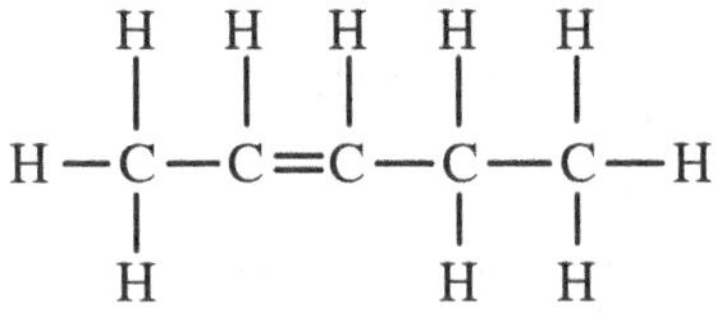 H—C(H)=C(H)—C(H)(H)—C(H)(H)—C(H)(H)—H   H—C(H)(H)—C(H)=C(H)—C(H)(H)—C(H)(H)—H

54. 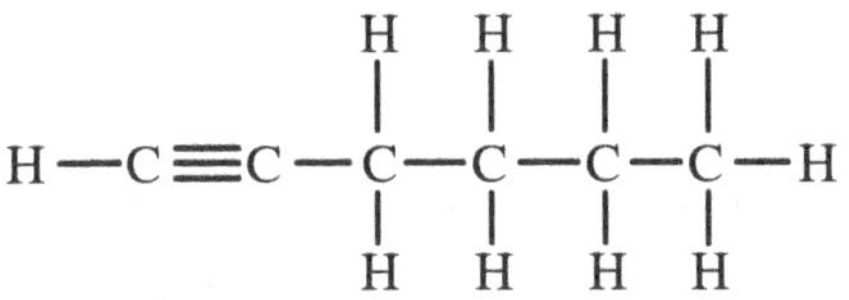 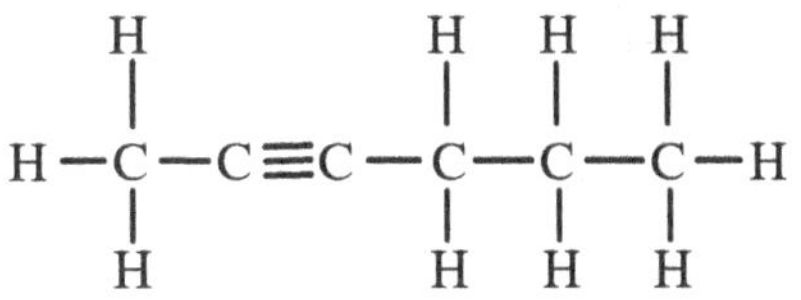 H—C≡C—C(H)(H)—C(H)(H)—C(H)(H)—C(H)(H)—H   H—C(H)(H)—C≡C—C(H)(H)—C(H)(H)—C(H)(H)—H

    H—C(H)(H)—C(H)(H)—C≡C—C(H)(H)—C(H)(H)—H

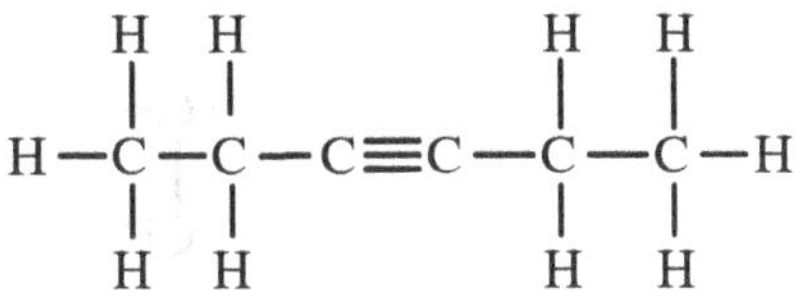

55. a) 2-pentene
    b) 4-methyl-2-pentene
    c) 3,3-dimethyl-1-butene
    d) 3,4-dimethyl-1-hexene

56. a) 1-butene
    b) 4-ethyl-2-hexene
    c) 3,4-dimethyl-1-pentene
    d) 3-isopropyl-1-hexene

57. a) 2-butyne
    b) 4-methyl-2-pentyne
    c) 4,4-dimethyl-2-hexyne
    d) 3-ethyl-3-methyl-1-pentyne

58. a) 1-pentyne
    b) 3-isopropyl-1-hexyne
    c) 2,5-dimethyl-3-hexyne
    d) 5-ethyl-1,3-heptyne

59. a) H$_3$C—CH=CH$_3$—CH$_2$—CH$_2$—CH$_3$
    b) H$_3$C—CH$_2$—C≡C—CH$_2$—CH$_2$—CH$_3$
    c) 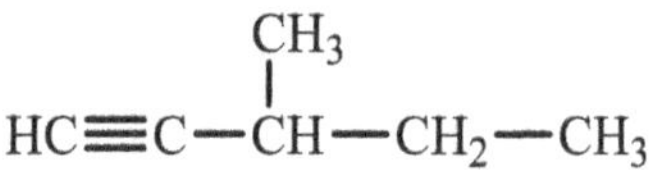 HC≡C—CH(CH$_3$)—CH$_2$—CH$_3$
    d) 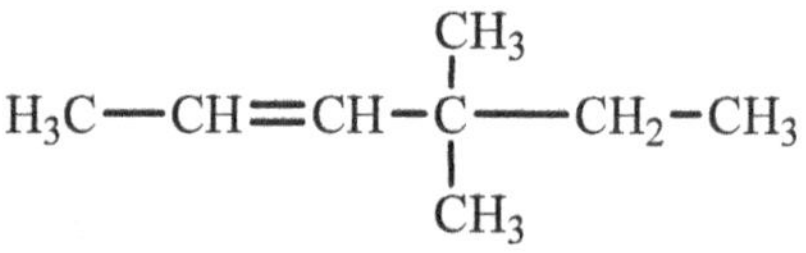 H$_3$C—CH=CH—C(CH$_3$)(CH$_3$)—CH$_2$-CH$_3$

60. a) H$_3$C—CH$_2$—C≡C—CH$_2$—CH$_2$—CH$_2$—CH$_3$
    b) H$_2$C=CH—CH$_2$—CH$_2$—CH$_3$
    c) 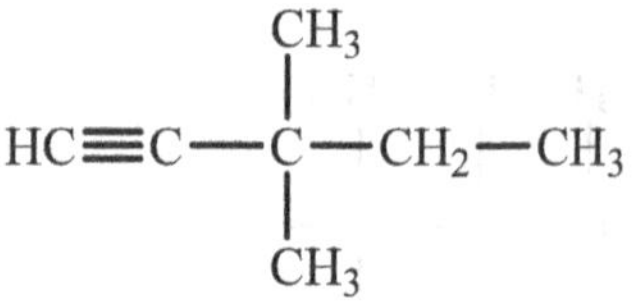 HC≡C—C(CH$_3$)(CH$_3$)—CH$_2$—CH$_3$
    d) 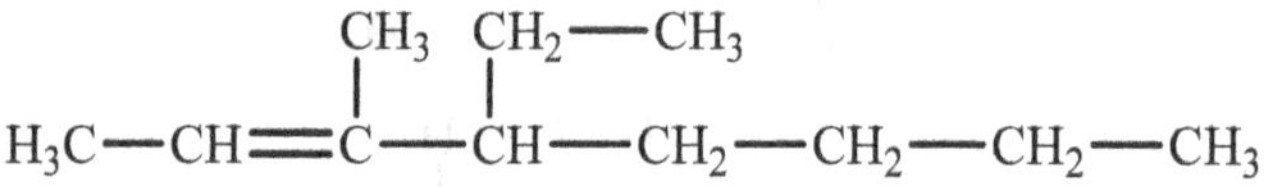 H$_3$C—CH=C(CH$_3$)—CH(CH$_2$—CH$_3$)—CH$_2$—CH$_2$—CH$_2$—CH$_3$

61. 1-pentene H$_2$C=CH—CH$_2$—CH$_2$—CH$_3$ 3-methyl-1-butene H$_2$C=CH—CH(CH$_3$)—CH$_3$

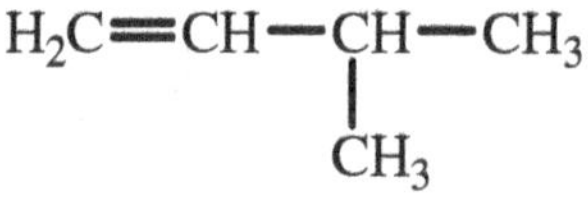

    2-pentene H$_3$C—CH=CH—CH$_2$—CH$_3$

    2-methyl-1butene H$_2$C=C(CH$_3$)—CH$_2$—CH$_3$ 2-methyl-2-butene H$_3$C—CH=C(CH$_3$)—CH$_3$

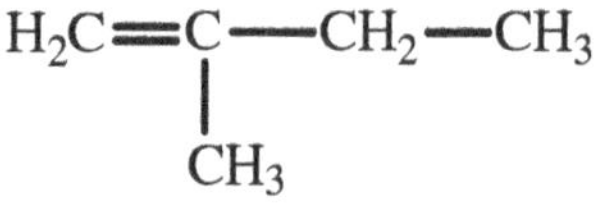

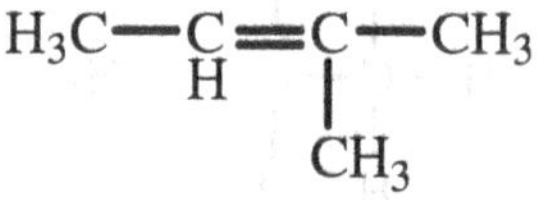

62. 1-hexyne HC≡C—CH$_2$—CH$_2$—CH$_2$—CH$_3$ 4-methyl-2-pentyne H$_3$C—C≡C—CH(CH$_3$)—CH$_3$

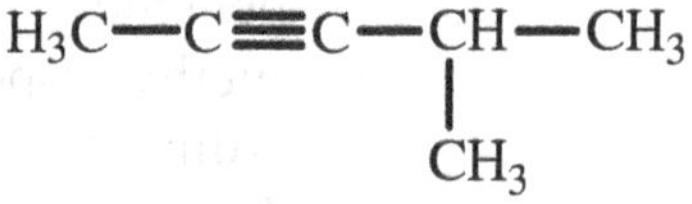

    2-hexyne H$_3$C—C≡C—CH$_2$—CH$_2$—CH$_3$

    3,3-dimethyl-1butyne HC≡C—C(CH$_3$)(CH$_3$)—CH$_3$ 4-methyl-1-pentyne HC≡C—CH$_2$—CH(CH$_3$)—CH$_3$

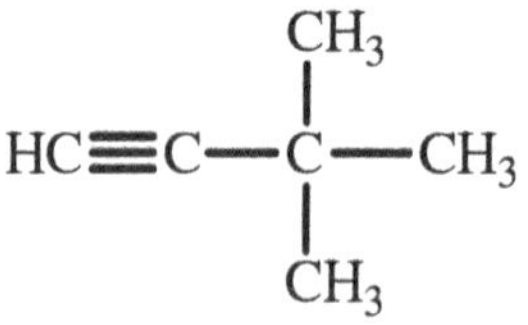

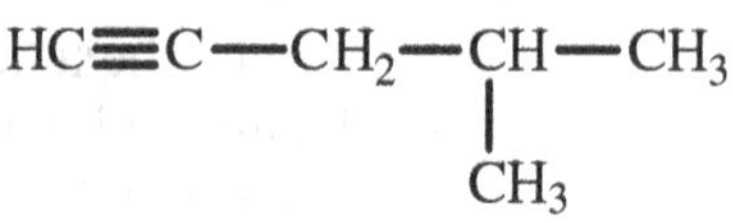

63.

Name	Molecular Formula	Structural Formula	Condensed Structural Formula
2,2,dimethyl-3-hexene	C_8H_{16}	H₃C—C(CH₃)(CH₃)—CH=CH—CH₂—CH₃	$CH_3C(CH_3)_2CH=CHCH_2CH_3$
4,4-diethyl-5,5-dimethyl-2-hexyne	$C_{12}H_{22}$	H₃C—C(CH₃)(CH₃)—C(CH₂CH₃)(CH₂CH₃)—C≡C—CH₃	$CH_3C(CH_3)_2C(CH_2CH_3)_2C≡CCH_3$
3,4-dimethyl-1-octyne	$C_{10}H_{18}$	HC≡C—CH(CH₃)—CH(CH₃)—CH₂—CH₂—CH₂—CH₃	$CH≡CCH(CH_3)CH(CH_3)CH_2CH_2CH_2CH_3$
4,4-diethyl-5,5-dimethyl-2-hexene	$C_{12}H_{24}$	H₃C—C(CH₃)(CH₃)—C(CH₂CH₃)(CH₂CH₃)—CH=CH—CH₃	$CH_3C(CH_3)_2C(CH_2CH_3)_2CH=CHCH_3$

64.

Name	Molecular Formula	Structural Formula	Condensed Structural Formula
3-ethyl-4-methyl-1-heptene	$C_{10}H_{20}$	H₂C=CH—CH—CH—CH₂—CH₂—CH₃ 　　　　　｜　　｜ 　　　　　CH₃　CH₃ 　　　　　｜ 　　　　　CH₃	$CH_2CHCH(CH_2CH_3)CH(CH_3)CH_2CH_2CH_3$
3-ethyl-2-pentene	C_7H_{14}	CH₃—CH=C—CH₂—CH₃ 　　　　　　｜ 　　　　　　CH₂ 　　　　　　｜ 　　　　　　CH₃	$CH_3CH=C(CH_2CH_3)CH_2CH_3$
3-methyl-4-propyl-1-heptyne	$C_{11}H_{20}$	HC≡C—CH—CH—CH₂—CH₂—CH₃ 　　　　｜　　｜ 　　　　CH₃　CH₂ 　　　　　　　｜ 　　　　　　　CH₂ 　　　　　　　｜ 　　　　　　　CH₃	$CHCCH(CH_3)CH(CH_2CH_2CH_3)CH_2CH_2CH_3$
3,3-dimethyl-4-ethyl-1-hexyne	$C_{10}H_{18}$	CH₃ 　　　　　｜ HC≡C—C—CH—CH₂—CH₃ 　　　　｜　｜ 　　　　CH₃ CH₂ 　　　　　　｜ 　　　　　　CH₃	$CHCC(CH_3)_2CH(CH_2CH_3)CH_2CH_3$

Hydrocarbon Reactions

65. a) $2CH_3CH_3(g) + 7O_2(g) \rightarrow 4CO_2(g) + 6H_2O(g)$
    b) $2CH_2=CHCH_3(g) + 9O_2(g) \rightarrow 6CO_2(g) + 6H_2O(g)$
    c) $2CH≡CH(g) + 5O_2(g) \rightarrow 4CO_2(g) + 2H_2O(g)$

66. a) $2CH_3CH_2CH_2CH_3(g) + 13O_2(g) \rightarrow 8CO_2(g) + 10H_2O(g)$
    b) $CH_2=CH_2(g) + 3O_2(g) \rightarrow 2CO_2(g) + 2H_2O(g)$
    c) $2CH≡CCH_2CH_3(g) + 11O_2(g) \rightarrow 8CO_2(g) + 6H_2O(g)$

67. $CH_4(g) + Br_2(g) \rightarrow CH_3Br(g) + HBr(g)$

68. $CH_3CH_3(g) + I_2(g) \rightarrow CH_3CH_2I(g) + HI(g)$

69. $CH_3CH=CHCH_3(g) + Cl_2(g) \rightarrow CH_3CHClCHClCH_3(g)$

70. $CH_3CH_2\underset{\underset{CH_3}{|}}{C}=CH_2(g) + Cl_2(g) \rightarrow CH_3CH_2\underset{\underset{CH_3}{|}}{C}ClCH_2Cl(g)$

71. $CH_2=CH_2(g) + H_2(g) \rightarrow CH_3CH_3(g)$

72. $CH_3CH_2CH_2CH=CH_2(g) + H_2(g) \rightarrow CH_3CH_2CH_2CH_2CH_3(g)$

Aromatic Hydrocarbons

73. The structural formula that represents both shorthand formulas is:

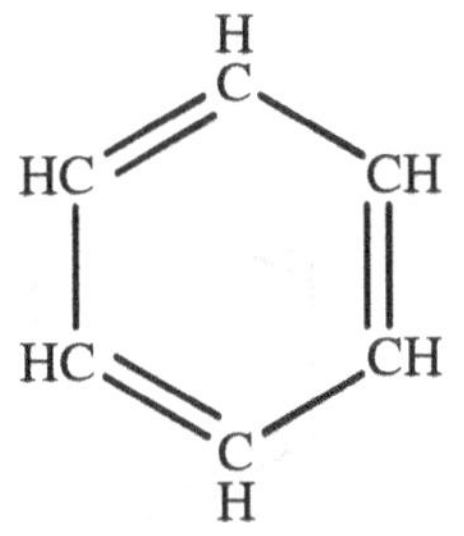

    [benzene structural formula with alternating double bonds]

74. The actual bonds in benzene are shorter than a single bond but longer than a double bond. The actual structure is between the alternating double and single bonds as shown in the resonance structures.

75. a) fluorobenzene
    b) isopropylbenzene
    c) ethylbenzene

76. a) iodobenzene
    b) toluene
    c) tert-butylbenzene

77. a) 4-phenyloctane
    b) 5-phenyl-3-heptene
    c) 7-phenyl-2-heptyne

78. a) 3-methyl-4-phenylhexane
    b) 2-phenyl-3-octene
    c) 4-ethyl-2-phenylheptane

79. a) 1-bromo-2-chlorobenzene
    b) 1,2-diethylbenzene or orthodiethylbenzene or o-diethylbenzene
    c) 1,3-difluorobenzene or metadichlorobenzene or m-dichlorobenzene

80. a) 2-chloro-1-fluorobenzene
    b) 4-ethyl-1-fluorobenzene
    c) 1,4-diiodobenzene or paradiiodobenzene or p-diiodobenzene

81. The structures are:
    a) phenyl–CH$_2$–CH$_2$–CH$_2$–CH$_3$
    b) 2-iodo-(ethyl)benzene (CH$_2$–CH$_3$ with I ortho)
    c) 1,4-dimethylbenzene (para CH$_3$, CH$_3$)

82. The structures are:
    a) H$_3$C–CH–CH$_3$ attached to phenyl
    b) 1,3-dibromobenzene (meta Br, Br)
    c) 1-bromo-4-ethylbenzene (para Br, CH$_2$CH$_3$)

Functional Groups

83.
$$R-\overset{O}{\underset{\|}{C}}-H \Rightarrow \text{aldehyde}$$

$$R-\overset{O}{\underset{\|}{C}}-R \Rightarrow \text{ketone}$$

$$R-O-R \Rightarrow \text{ether}$$

$$R-\underset{|}{\overset{R}{N}}-R \Rightarrow \text{amine}$$

84.
$$R-\overset{O}{\underset{\|}{C}}-OR \Rightarrow \text{ester}$$

$$R-\overset{O}{\underset{\|}{C}}-OH \Rightarrow \text{carboxylic acid}$$

$$R-OH \Rightarrow \text{alcohol}$$

$$R-O-R \Rightarrow \text{ether}$$

85. Functional groups and families:

a) 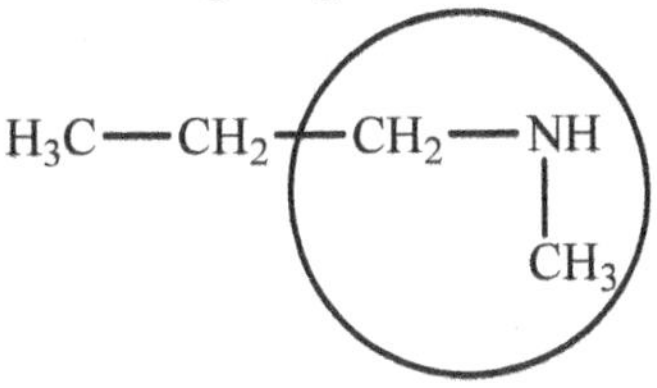 H₃C—CH₂—CH₂—NH—CH₃   amine

b) 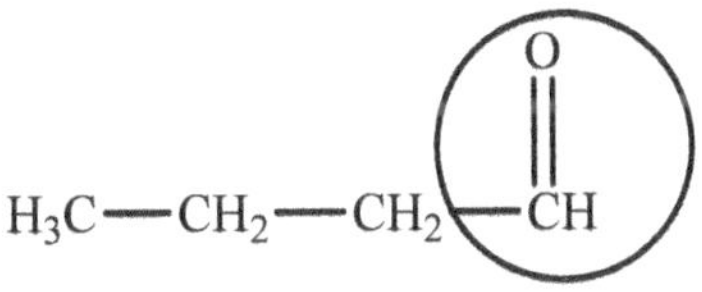 H₃C—CH₂—CH₂—CH=O   aldehyde

c) 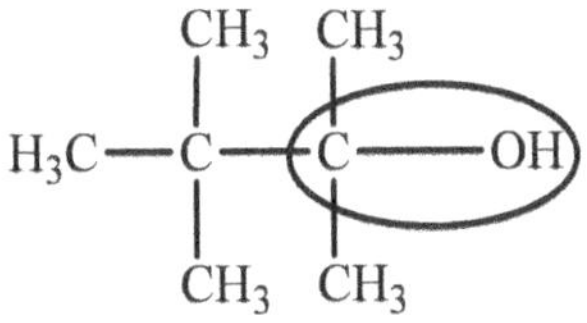 H₃C—C(CH₃)(CH₃)—C(CH₃)(CH₃)—OH   alcohol

d) 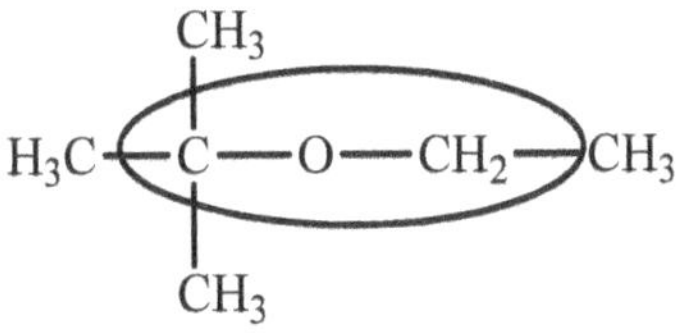 H₃C—C(CH₃)(CH₃)—O—CH₂—CH₃   ether

86. Functional groups and families:

a) 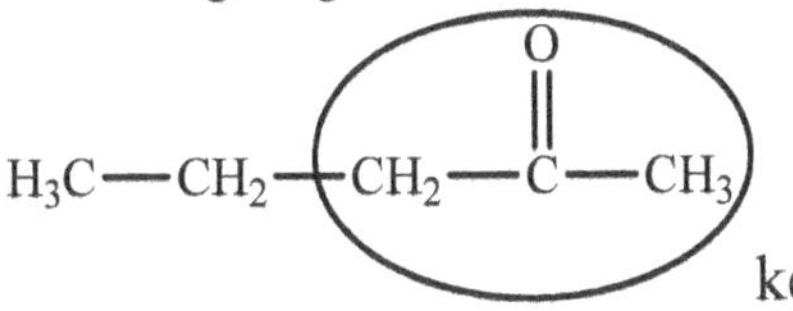 H₃C—CH₂—CH₂—C(=O)—CH₃   ketone

b) 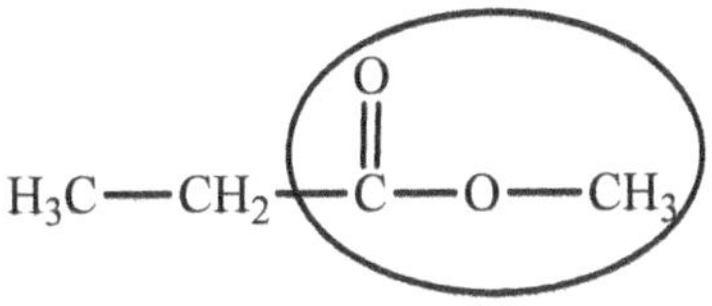 H₃C—CH₂—C(=O)—O—CH₃   ester

c) 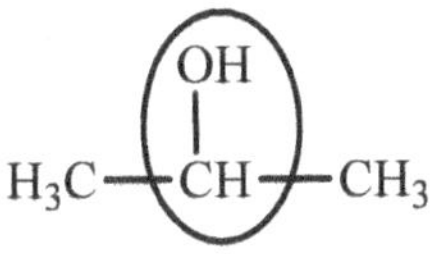 H₃C—CH(OH)—CH₃   alcohol

d) 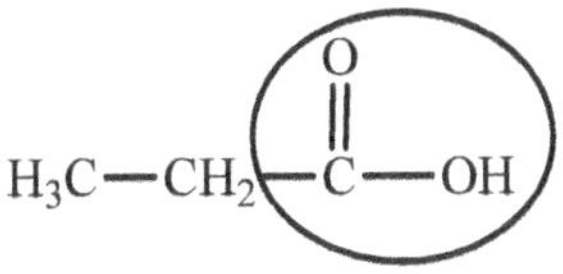 H₃C—CH₂—C(=O)—OH   carboxylic acid

Alcohols

87. a) 2-butanol
    b) 2-methyl-1-propanol
    c) 3-ethyl-1-hexanol
    d) 3-methyl-3-pentanol

88. a) 3-octanol
    b) 4-methyl-2-hexanol
    c) 2-methyl-2-pentanol
    d) 2,3-dimethyl-1-heptanol

89. a) 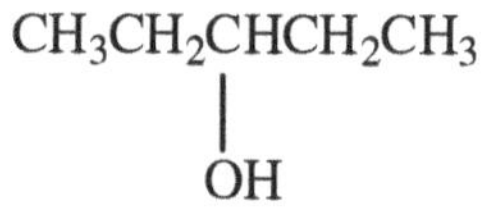 CH₃CH₂CHCH₂CH₃
              |
              OH

    b) 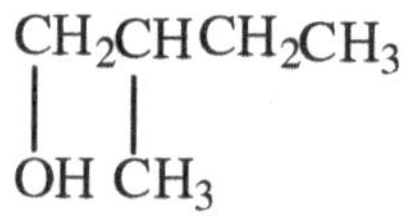 CH₂CHCH₂CH₃
       | |
       OH CH₃

    c) 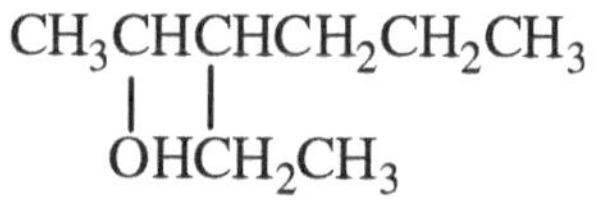 CH₃CHCHCH₂CH₂CH₃
          | |
          OH CH₂CH₃

    d) CH₃CH₂OH

90. a) HO—CH₂—CH₂—CH₂—CH₂—CH₂—CH₃

    b) 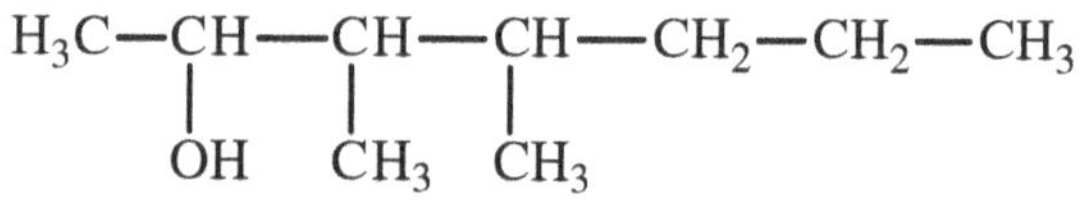 H₃C—CH—CH—CH—CH₂—CH₂—CH₃
           | | |
           OH CH₃ CH₃

    c) 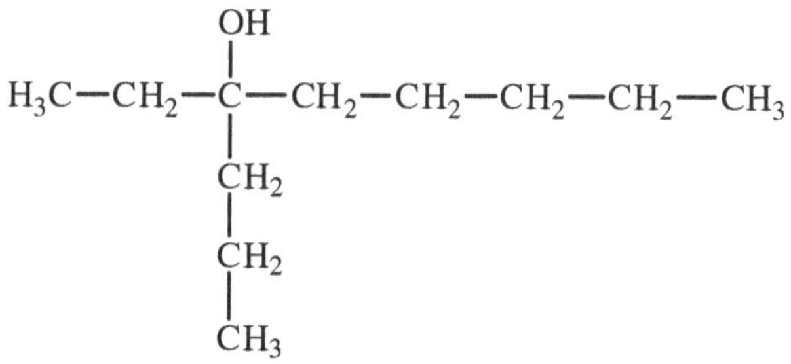 H₃C—CH₂—C(OH)—CH₂—CH₂—CH₂—CH₂—CH₃
                |
                CH₂
                |
                CH₂
                |
                CH₃

    d) 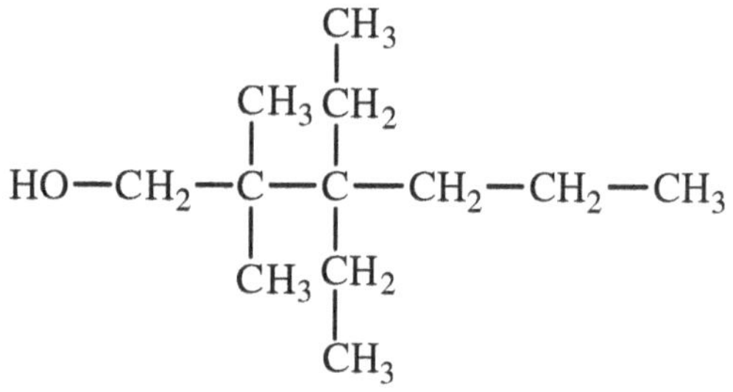 HO—CH₂—C(CH₃)(CH₃)—C(CH₂CH₃)(CH₂CH₃)—CH₂—CH₂—CH₃ with CH₃ on lower CH₂

Ethers

91. a) CH₃CH₂CH₂CH₂-O-CH₂CH₂CH₂CH₃
    b) ethyl propyl ether
    c) dipropyl ether
    d) CH₃-O-CH₂CH₂CH₂CH₂CH₃

92. a) CH₃CH₂-O-CH₂CH₂CH₂CH₃
    b) diethyl ether
    c) pentyl propyl ether
    d) CH₃CH₂-O-CH₂CH₂CH₂CH₂CH₂CH₃

93. a) H₃C—CH₂—CH₂—CH₂—CH₂—CH₂—CH₂—CH(=O)
    b) butanal
    c) 4-heptanone

94. a) 2-pentanone

    b) H₃C—CH₂—C(=O)—CH₂—CH₃

    c) H₃C—CH₂—CH(=O)

    d) heptanal

95. a) H₃C—CH₂—CH₂—CH₂—CH₂—CH₂—CH₂—C(=O)—OH

    b) methyl ethanoate

    c) H₃C—CH₂—CH₂—C(=O)—O—CH₂—CH₃

    d) heptanoic acid

96. a) H₃C—CH₂—CH₂—CH₂—CH₂—C(=O)—OH

    b) butanoic acid

    c) propyl propanoate

    d) H₃C—CH₂—C(=O)—O—CH₂—CH₂—CH₂—CH₃

Amines

97. a) H₃C—CH₂—NH—CH₂—CH₃

    b) triethylamine

    c) butylproplyamine

98. a) 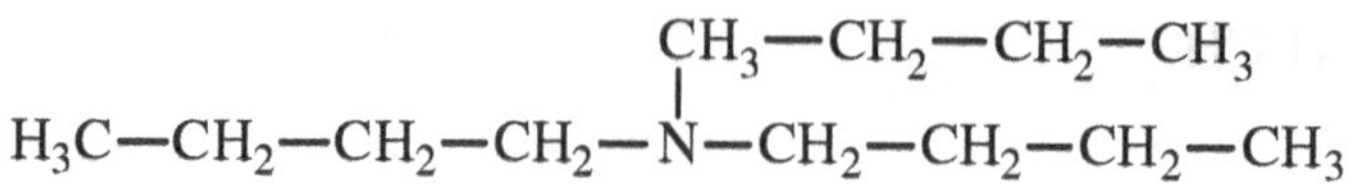

b) isopropylamine

c) $H_3C—NH—CH_2—CH_3$

Polymers

99. 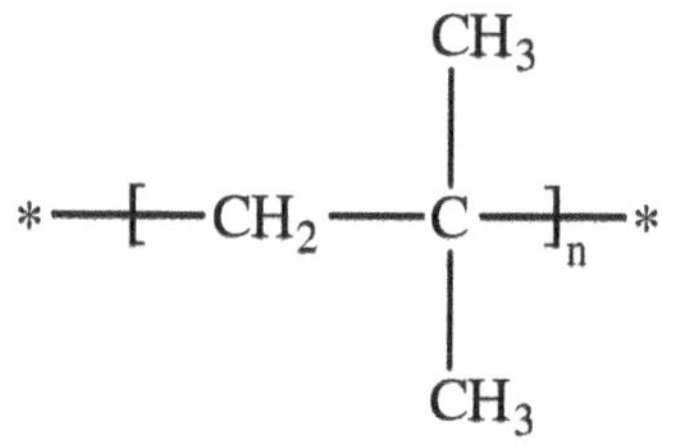

100. 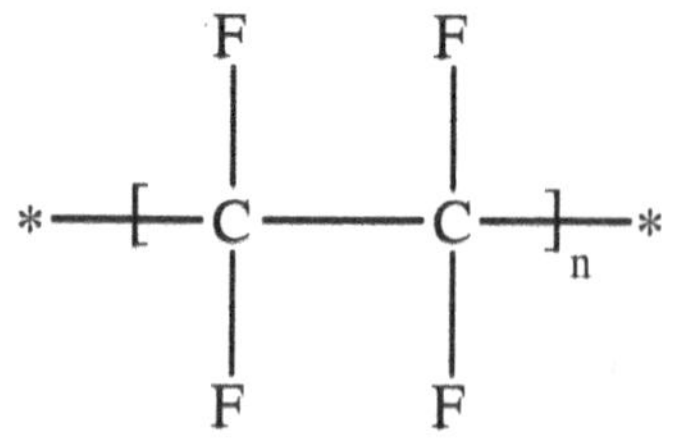

101. 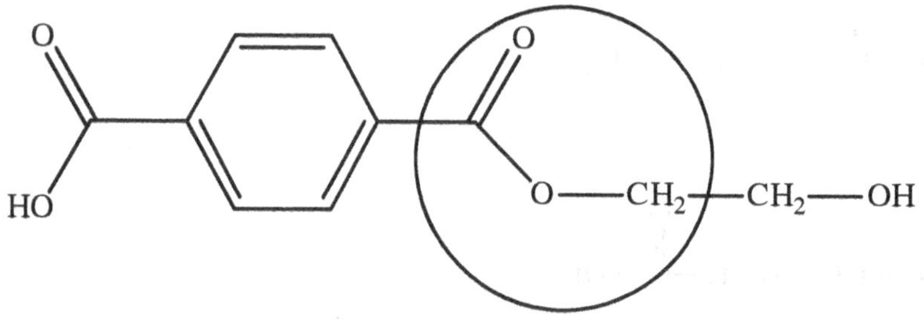

102. 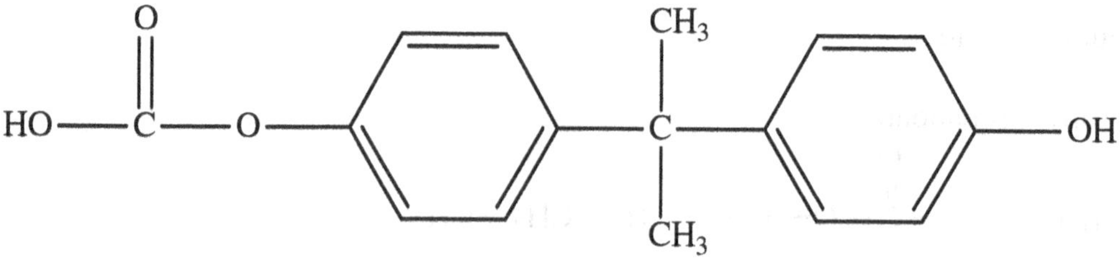

Cumulative Problems

103. a) alcohol
b) amine
c) alkane
d) carboxylic acid
e) ether
f) alkene

104. a) aromatic
b) alkyne
c) ester
d) ketone
e) aldehyde
f) alcohol

105. a) 3-methyl-4-tert-butylheptane
b) 3-methyl butanal
c) 4-isopropyl-3-methyl-2-heptene
d) propyl butanoate

106. a) 4-ethyl-4,5-dimethyl-2-heptene
b) 1,4 difluorobenzene or parafluorobenzene or p-fluorobenzene
c) 2-methyl-1-propanol
d) 3-methyl-5-sec-butylnonane

107. a) same molecule
b) isomers
c) same molecule

108. a) isomers
b) same molecule
c) same molecule

109. $CH_2 = CH_2 + HCl \rightarrow CH_3CH_2Cl$

110. $CH \equiv CH + HI \rightarrow CH_2 = CHI$

111. $CH_3CH = CHCH_3 + H_2 \rightarrow CH_3CH_2CH_2CH_3$

$15.5 \text{ kg 2-butene} \times \dfrac{1 \times 10^3 \text{ g}}{1 \text{ kg}} \times \dfrac{1 \text{ mol 2-butene}}{56.12 \text{ g}} \times \dfrac{1 \text{ mol H}_2}{1 \text{ mol 2-butene}} \times \dfrac{2.02 \text{ g}}{1 \text{ mol H}_2}$

$= 558 \text{ g H}_2$

112. $2C_8H_{18} + 25O_2 \rightarrow 16CO_2 + 18H_2O$

$3.8 \text{ kg octane} \times \dfrac{1000 \text{ g}}{1 \text{ kg}} \times \dfrac{1 \text{ mol octane}}{114.26 \text{ g}} \times \dfrac{16 \text{ mol CO}_2}{2 \text{ mol octane}} \times \dfrac{44.01 \text{ g}}{1 \text{ mol CO}_2} \times \dfrac{1 \text{ kg}}{1000 \text{ g}}$

$= 12 \text{ kg CO}_2$

113. $2C_8H_{18} + 25O_2 \rightarrow 16CO_2 + 18H_2O$

$18.9 \times 10^3 \text{ g C}_8H_{18} \times \dfrac{1 \text{ mol C}_8H_{18}}{114.2 \text{ g}} \times \dfrac{25 \text{ mol O}_2}{2 \text{ mol C}_8H_{18}} \times \dfrac{22.4 \text{ L}}{1 \text{ mol O}_2} = 4.63 \times 10^4 \text{ L O}_2$

114. $C_3H_4 + 2H_2 \rightarrow C_3H_8$

$$15.5 \times 10^3 \text{ g } C_3H_4 \times \frac{1 \text{ mol } C_3H_4}{40.06 \text{ g}} \times \frac{2 \text{ mol } H_2}{1 \text{ mol } C_3H_4} \times \frac{22.4 \text{ L}}{1 \text{ mol } H_2} = 1.73 \times 10^4 \text{ L } H_2$$

Highlights

115. a) alcohol
   b) amine
   c) carboxylic acid
   d) ester
   e) alkane
   f) ether

116. dioxin: aromatics, ethers, halogen (chlorine)
   furan: alkene, ether
   hexachlorobenzene: aromatic, halogen (chlorine)
   DDT: aromatic, halogen (chlorine)
   Structural features in common:
       1) closed rings / aromatics (all) 2) chlorine (3 out of 4) 3) ethers (2 out of 4)

117. Answers will vary.

118. • Salt and sugar: Salt is an inorganic compound composed of ions. Sugar is an organic compound composed of covalent bonds. Salt and sugar are both essential components in our diets but can become a problem when consumed in excess.

   • Methane and 3-methylheptane: Both compounds are alkanes; however, 3-methyl heptane is a branched compound containing eight total carbon atoms and 18 hydrogen atoms where methane contains 1 carbon atom and 4 hydrogen atoms.

   • Aldehydes and ketones: Aldehydes and ketones both contain a carbon atom double bonded to an oxygen atom (C=O) called a carbonyl group. The location of the carbonyl group is at a terminal carbon for aldehydes and at an internal carbon atom for ketones. Aldheydes and ketones are responsible for many fragrant odors.

   • Polystrene and polyurethane: Polystyrene and polyurethane are both polymers that can be used as insulation. Polystyrene is an addition polymer that contains a benzene ring, and polyurethane is a condensation polymer that contains an amine functional group.

119. a) 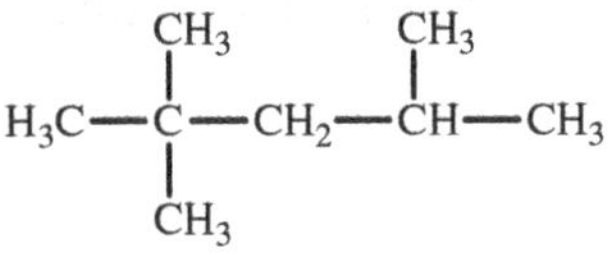 H₃C—C(CH₃)(CH₃)—CH₂—CH(CH₃)—CH₃ C_8H_{18}

b) There are 18 possible isomers of octane:

octane

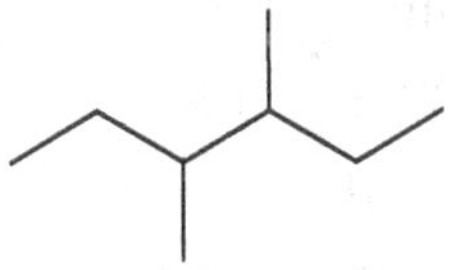

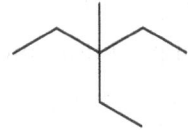
3,4-dimethylhexane

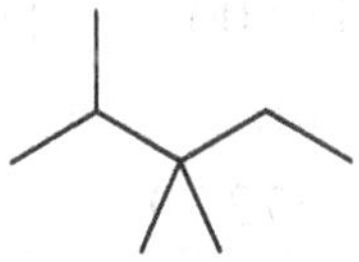

2,3,3-trimethylpentane

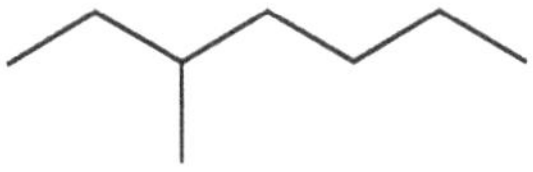

3-methylheptane

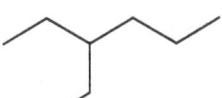
3-ethyl-3-methylpentane

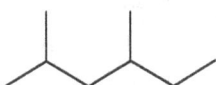
2,2,3,3-tetramethylbutane

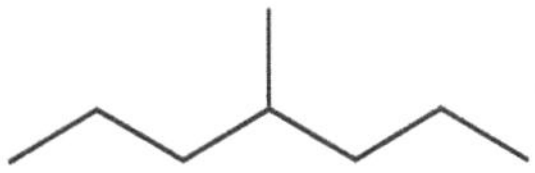

4-methylheptane

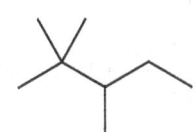
3-ethylhexane

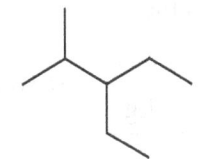
2,4-dimethtlhexane

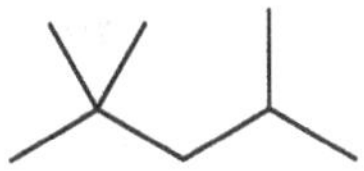

2,2,4-trimethlpentane

2,2,3-trimethlpentane

3-ethyl-2-methylpentane

3,3-dimethtlhexane

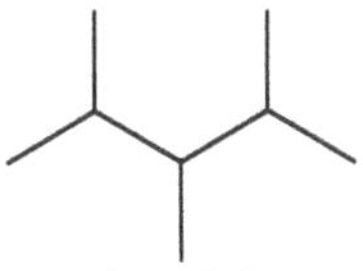

2,3,4-trimethtlpentane

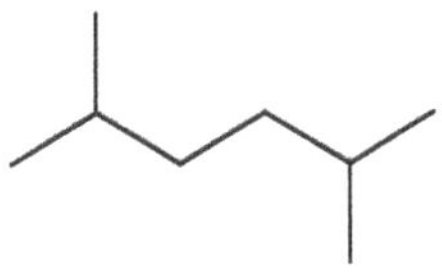

2,5-dimethtlhexane

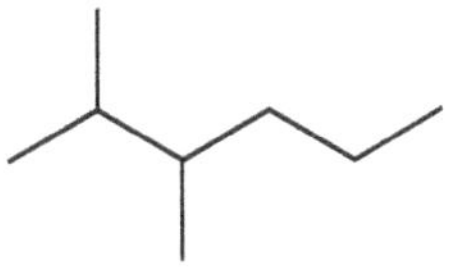

2,3-dimethtlhexane

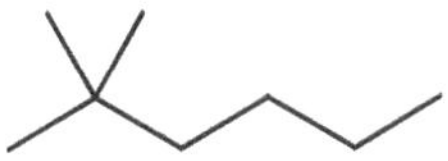

2,2-dimethtlhexane

2-methylheptane

c) Answers will vary based on which isomer is chosen.
d) The names of the 18 isomers are shown above.

120. Isomers are molecules with the same molecular formula but different structures.

Data Interpretation and Analysis

121. (a) Methanol: $\dfrac{\$46.20}{1\,L} \times \dfrac{1\,L}{1000\,mL} \times \dfrac{1\,mL}{0.79\,g} \times \dfrac{1000\,g}{1\,kg} = \$58.48\,/\,kg$

Ethanol: $\dfrac{\$112.00}{1\,L} \times \dfrac{1\,L}{1000\,mL} \times \dfrac{1\,mL}{0.79\,g} \times \dfrac{1000\,g}{1\,kg} = \$141.77\,/\,kg$

1-propanol: $\dfrac{\$72.70}{1\,L} \times \dfrac{1\,L}{1000\,mL} \times \dfrac{1\,mL}{0.80\,g} \times \dfrac{1000\,g}{1\,kg} = \$90.88\,/\,kg$

1-butanol: $\dfrac{\$72.60}{1\,L} \times \dfrac{1\,L}{1000\,mL} \times \dfrac{1\,mL}{0.81\,g} \times \dfrac{1000\,g}{1\,kg} = \$89.63\,/\,kg$

Ethanol is the most expensive alcohol per kilogram.

(b) Methanol: $\dfrac{\$46.20}{1\,L} \times \dfrac{1\,L}{1000\,mL} \times \dfrac{1\,mL}{0.79\,g} \times \dfrac{32.04\,g}{1\,mol} = \$1.87\,/\,mol$

Ethanol: $\dfrac{\$112.00}{1\,L} \times \dfrac{1\,L}{1000\,mL} \times \dfrac{1\,mL}{0.79\,g} \times \dfrac{46.07\,g}{1\,mol} = \$6.53\,/\,mol$

1-propanol: $\dfrac{\$72.70}{1\,L} \times \dfrac{1\,L}{1000\,mL} \times \dfrac{1\,mL}{0.80\,g} \times \dfrac{60.10\,g}{1\,mol} = \$5.46\,/\,mol$

1-butanol: $\dfrac{\$72.60}{1\,L} \times \dfrac{1\,L}{1000\,mL} \times \dfrac{1\,mL}{0.81\,g} \times \dfrac{74.12\,g}{1\,mol} = \$6.64\,/\,mol$

1-butanol is the most expensive alcohol per mole.

(c) $\underbrace{0.522\,kg \times \dfrac{\$58.48}{kg}}_{\text{methanol}} + \underbrace{0.184\,kg \times \dfrac{\$141.77}{kg}}_{\text{ethanol}} + \underbrace{0.225\,kg \times \dfrac{\$89.63}{kg}}_{\text{1-butanol}} = \$76.78$

Biochemistry 19

Questions

1. The goal of the Human Genome Project was to map all of the genetic material (genome) of a human being. One of the surprises from the Human Genome Project results was that humans have only 20,000–25,000 genes; scientists had always predicted that there would be many more.

2. The benefits of the Human Genome Project are expected to include the ability to identify persons who are genetically predisposed to contracting certain diseases and the ability to design new drugs to fight diseases.

3. The cell is the smallest structural unit of living things that has the properties associated with life. The main chemical components of the cell can be divided into four classes: carbohydrates, lipids, proteins, and nucleic acids.

4. Carbohydrates are organic compounds having the generic chemical formula $(CH_2O)n$. Carbohydrates are used for short-term energy storage and are the main structural components of plants.

5. Glucose is soluble in water because of the many OH groups that allow water to hydrogen-bond to glucose. This is important because glucose needs to be transported by blood to the cells and then through the cell wall into the aqueous interior of the cell.

6. A monosaccharide is a carbohydrate unit that cannot be broken down into simpler carbohydrates. A disaccharide is a carbohydrate unit that can be broken down into two simpler monosaccharides. A polysaccharide is a carbohydrate unit that is composed of many monosaccharides.

7. During digestion, the links in disaccharides and polysaccharides are broken, allowing individual monosaccharides to pass through the intestinal wall and enter the bloodstream.

8. Monosaccharides and disaccharides are collectively known as simple sugars. Polysaccharides are also referred to as complex carbohydrates.

9. Starch and cellulose are both polysaccharides, but the difference is the bond angles that form between saccharide units. Because of the difference in bond angle, humans can digest starch and use it for energy, while cellulose cannot be digested and passes directly through humans.

Copyright © 2018 Pearson Education, Inc.   Page 315

10. Lipids are chemical components of a cell that are insoluble in water but are soluble in nonpolar solvents. Lipids include fatty acids, fats, oils, phospholipids, glycolipids, and steroids. The main functions of lipids are as structural units of cells, long-term energy storage, and insulation.

11. A fatty acid is a carboxylic acid with a long hydrocarbon chain. The general structure of a fatty acid is shown below, where R is a carbon chain ranging from 3 to 19 atoms long.

$$R-\underset{\underset{\text{Generic fatty acid structure}}{}}{\overset{\overset{O}{\|}}{C}}-OH$$

12. A saturated fatty acid has no double bonds between carbon atoms in the R group. An unsaturated fatty acid has at least one (possibly more) double bond.

13. Triglycerides are tri-esters composed of glycerol and three fatty acids. Oils and fats are triglycerides.

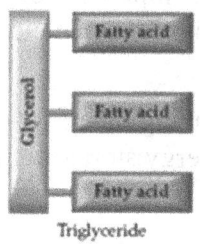

Triglyceride

14. A saturated fat is a triglyceride containing saturated fatty acids (no double- or triple-bonded carbons) and is usually a solid at room temperature. An unsaturated fat is a triglyceride containing unsaturated fatty acids (double-bonded carbons) and is usually a liquid at room temperature.

15. Phospholipids are molecules having the same generic structure as triglycerides in which one of the fatty acid esters has been replaced with a phosphate ester. Glycolipids have similar structures and properties as phospholipids, but the polar phosphate group has been replaced by a sugar molecule such as glucose.

16. The main function of phospholipids and glycolipids in the body is as components of cell walls.

17. Steroids are lipids containing a four-ring structure. Examples include cholesterol, testosterone, and estrogen. Cholesterol is a steroid that is part of cell membranes and also serves as a starting material for the body to synthesize other steroids. Also, steroids serve as male and female hormones in the body.

18. Proteins are polymers consisting of long chains of amino acids linked together by peptide bonds.

19. Proteins have many roles in the body such as catalysts, structural unit of muscle, skin and cartilage, transportation of oxygen, disease-fighting antibodies, and hormones.

20. Amino acids are compounds that contain an amine group, a carboxylic acid group, and an R group. The general structure for amino acids is:

$$H_2N-CH(R)-C(=O)-OH$$

21. Amino acids differ from each other by the nature of the R group.

22. A peptide bond is formed from the reaction between the amine end of one amino acid and the carboxylic acid end of another amino acid.

23. The formation of a peptide bond:

$$H_2N-CH(R)-C(=O)-OH + H_2N-CH(R)-C(=O)-OH \rightarrow H_2N-CH(R)-C(=O)-NH-CH(R)-C(=O)-OH + H_2O$$

24. The shape of proteins is determined by the sequence of amino acids and by the short-range and long-range interactions between the amino acids in the protein chain. The shape of the protein is important because it determines biological function.

25. The primary protein structure simply refers to the sequence of amino acids in the protein chain. The primary protein structure is maintained by the peptide bonds formed between the amino acids.

26. The short-ranged periodic or repeating patterns along a protein chain are referred to as secondary protein structure. The secondary protein structure is maintained by interactions between individual amino acid R units that are fairly close to each other in the linear sequence.

27. The tertiary protein structure refers to the large-scale bends and folds in the protein structure. The tertiary protein structure results from interactions between individual amino acid R units that are separated from each other by a large distance along the linear sequence of amino acids.

28. The quaternary protein structure is the shape that occurs when several proteins interact with one another to form a larger protein structure. The quaternary protein structure is maintained by interactions of individual amino acid R groups between different chains.

29. The α-helix structure occurs when the amino acid chain is wrapped into a tight coil, much the same as a spring, with the R groups extending outward. The β-pleated sheet structure has an extended chain that is in a zigzag pattern.

30. Nucleic acids are polymers in which the individual units are called nucleotides. Each nucleotide is made up of a phosphate, a sugar, and a base.

31. Nucleic acids contain a chemical code that specifies the correct amino acid sequences for the creation of proteins.

32. The two main types of nucleic acids are deoxyribonucleic acid (DNA) and ribonucleic acid (RNA).

33. The four bases found within DNA are adenine (A), cytosine (C), guanine (G), and thymine (T).

34. A codon is a sequence of three bases that form the code for a single amino acid in a protein.

35. The genetic code links a specific codon to an amino acid.

36. The genetic code is identical for all living organisms.

37. A gene is a sequence of codons within a DNA molecule that codes for a single protein. Genes vary in length from fifty to thousands of amino acids.

38. There are three nucleotides needed to specify a single amino acid; therefore, the gene would have to consist of $3 \times 300 = 900$ nucleotides.

39. A chromosome is the structure within a cell that contains the genes.

40. Most cells in the human body contain a complete set of genes for all the proteins needed by the human body, although all cells do not synthesize every protein specified by the DNA in the cell.

41. No, cells only produce those proteins that are critical to the cell type and function.

42. DNA replication begins with the DNA uncoiling and separating into individual strands. Enzymes then aid in the creation of new strands that are the complements of the opposite strands. The hydrogen bonding re-forms the strands, and the result is two complete copies of the original DNA.

43. a) thymine (T)
    b) adenine (A)
    c) guanine (G)
    d) cytosine (C)

44. The synthesis of a protein involves the DNA code for the protein unraveling and the creation of a complementary copy of the gene of interest using m-RNA. The m-RNA leaves the nucleus of the cell and proceeds to the ribosome, where protein synthesis occurs. The ribosome then follows along the m-RNA and reads each codon, which specifies the amino acid sequence in the new protein. Many of the amino acids used in protein synthesis come from digested protein in your diet.

Problems

Carbohydrates

45. a) carbohydrate, monosaccharide
    b) not a carbohydrate
    c) not a carbohydrate
    d) carbohydrate, disaccharide

46. a) not a carbohydrate
    b) carbohydrate, monosaccharide
    c) carbohydrate, trisaccharide
    d) not a carbohydrate

47. a) hexose
    b) tetrose
    c) pentose
    d) tetrose

48. a) pentose
    b) triose
    c) pentose
    d) pentose

49. The linear and ring structures for glucose are:

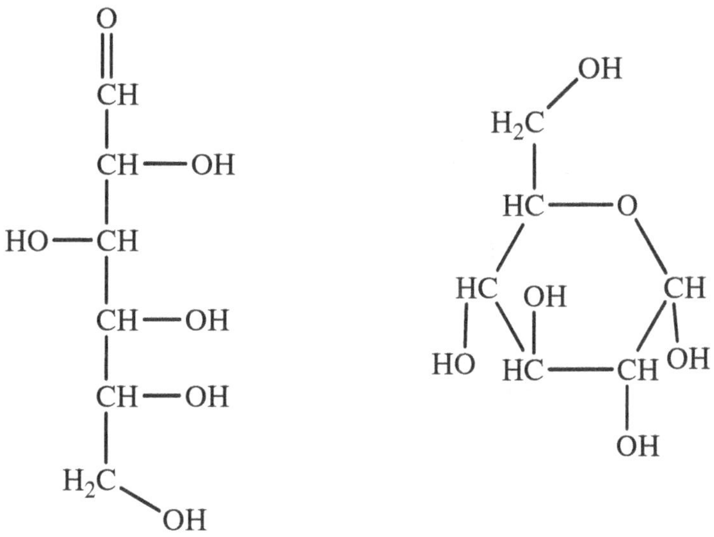

50. The ring structure of fructose is given below. Fructose and glucose are isomers.

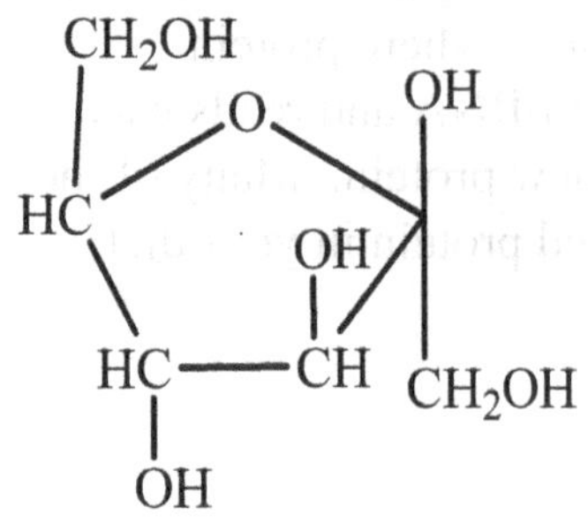

51. The structure of sucrose:

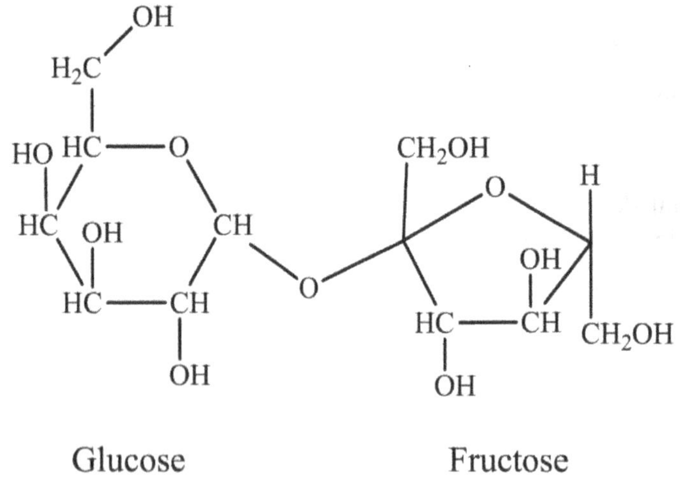

Glucose   Fructose

52. The structure of lactose:

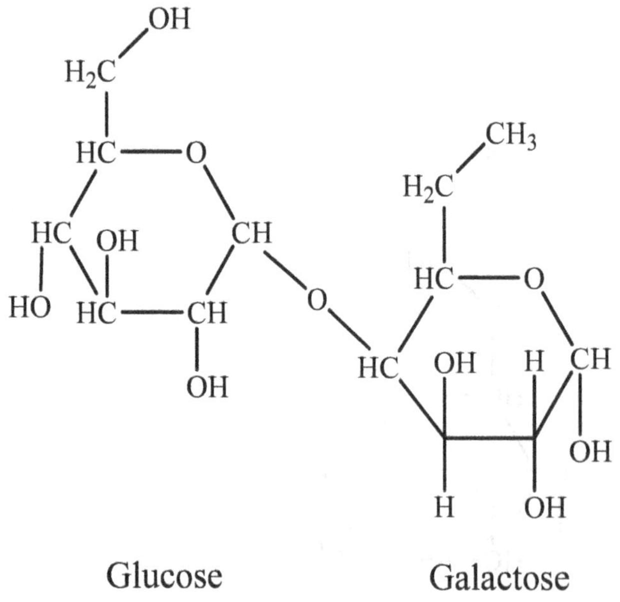

Glucose   Galactose

Lipids

53. a) lipid-fatty acid-saturated
    b) lipid-steroid
    c) lipid-triglyceride-unsaturated
    d) not a lipid

54. a) lipid-fatty acid-unsaturated
    b) not a lipid
    c) lipid-steroid
    d) lipid-triglyceride-saturated

55. The block diagram of a triglyceride is:

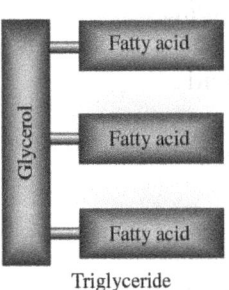

56. Phospholipids differ from triglycerides in that one of the fatty acid ester groups (nonpolar) has been replaced with a phosphate ester (polar).

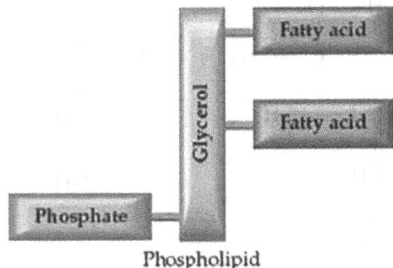

57. The structure of the triglyceride is as follows Because it is a saturated fat, you would expect it to be a solid.

$$\begin{array}{l} H_2C-O-\overset{O}{\underset{\|}{C}}-(CH_2)_{12}CH_3 \\ | \\ CH-O-\overset{O}{\underset{\|}{C}}-(CH_2)_{12}CH_3 \\ | \\ CH_2-O-\overset{O}{\underset{\|}{C}}-(CH_2)_{12}CH_3 \end{array}$$

58. The structure of the triglyceride is as follows. Because it is an unsaturated fat, you would expect it to be a liquid.

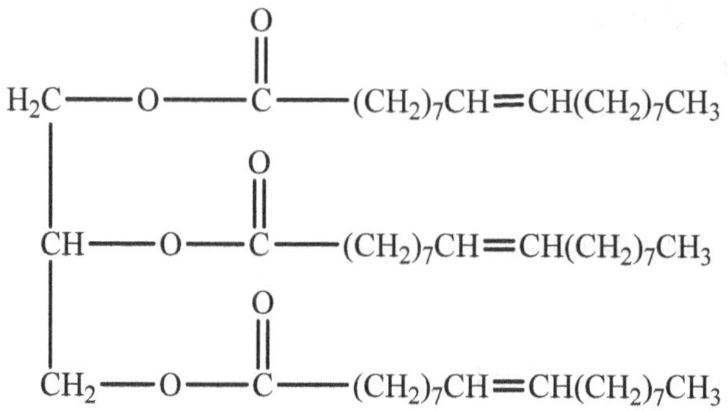

$$\begin{array}{l} H_2C-O-\overset{O}{\underset{\|}{C}}-(CH_2)_7CH=CH(CH_2)_7CH_3 \\ | \\ CH-O-\overset{O}{\underset{\|}{C}}-(CH_2)_7CH=CH(CH_2)_7CH_3 \\ | \\ CH_2-O-\overset{O}{\underset{\|}{C}}-(CH_2)_7CH=CH(CH_2)_7CH_3 \end{array}$$

Amino Acids and Proteins

59. a) not an amino acid
    b) amino acid
    c) not an amino acid
    d) amino acid

60. a) amino acid
    b) not an amino acid
    c) not an amino acid
    d) amino acid

61. Isoleucine(Ile) + Serine(Ser) → dipeptide + H₂O

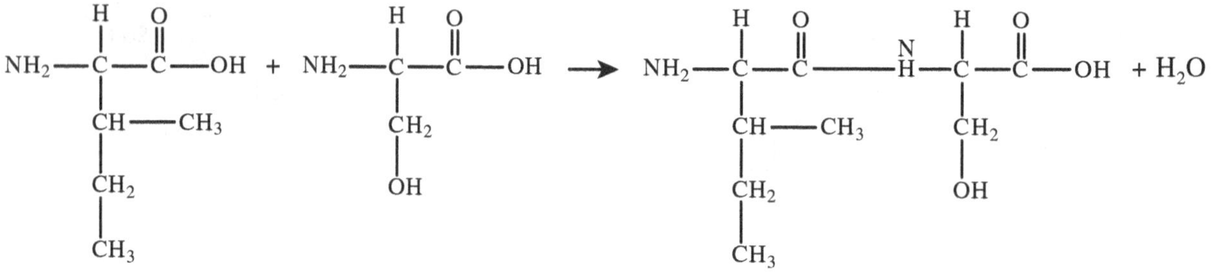

62. Valine(Val) + Lysine(Lys) → dipeptide + H₂O

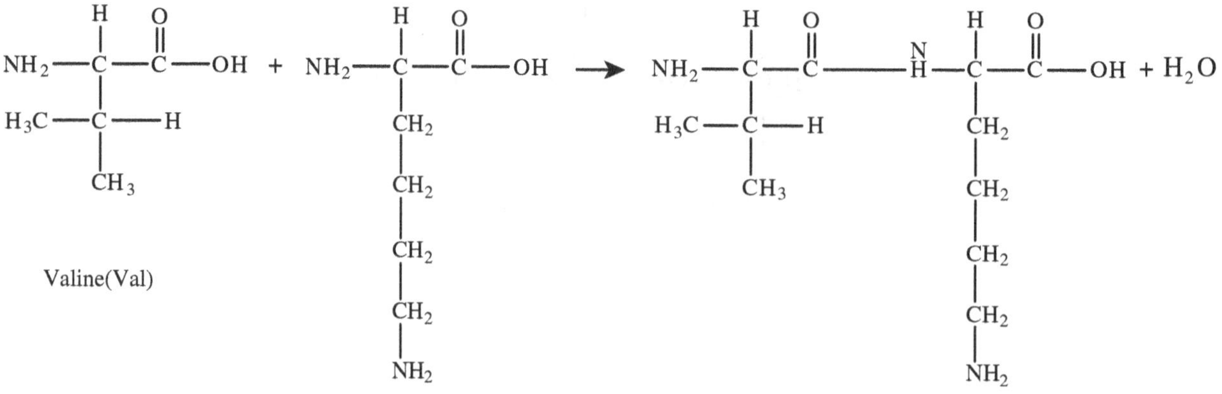

Copyright © 2018 Pearson Education, Inc.

63. a)

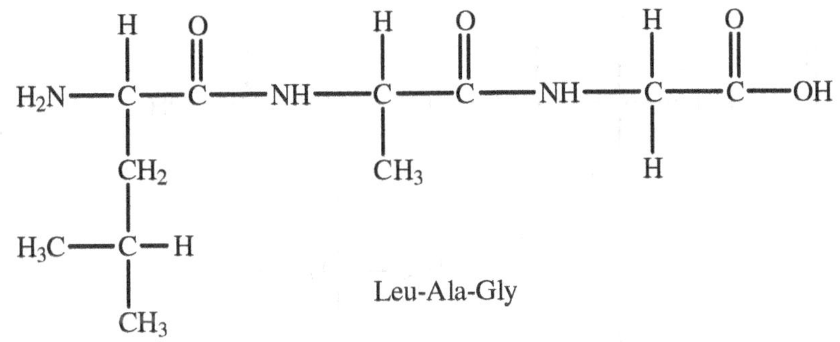

```
         H    O           H    O           H    O
         |    ||          |    ||          |    ||
H₂N—C—C—NH—C—C—NH—C—C—OH
         |                |                |
        CH₂              CH₃               H
         |
 H₃C—C—H
         |
        CH₃
```

Leu-Ala-Gly

b)

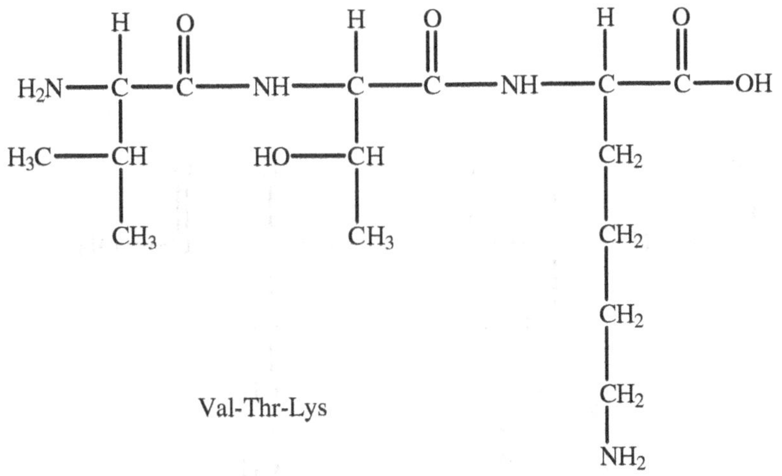

```
         H    O           H    O           H    O
         |    ||          |    ||          |    ||
H₂N—C—C—NH—C—C—NH—C—C—OH
         |                |                |
 H₃C—CH           HO—CH              CH₂
         |                |                |
        CH₃              CH₃              CH₂
                                            |
                                           CH₂
                                            |
                                           CH₂
                                            |
                                           NH₂
```

Val-Thr-Lys

c)

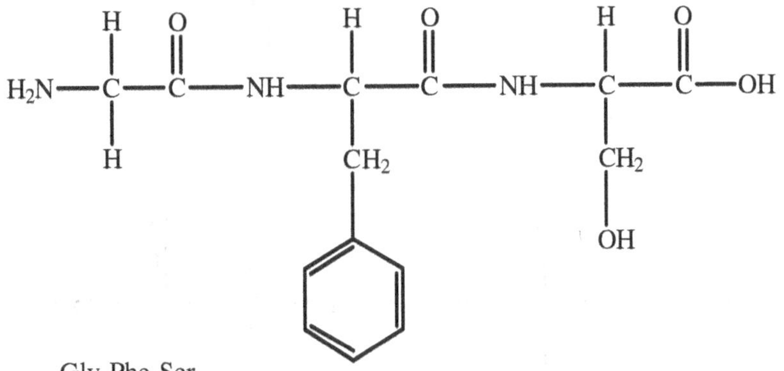

```
         H    O           H    O           H    O
         |    ||          |    ||          |    ||
H₂N—C—C—NH—C—C—NH—C—C—OH
         |                |                |
         H               CH₂              CH₂
                          |                |
                        (phenyl)          OH
```

Gly-Phe-Ser

64. a)

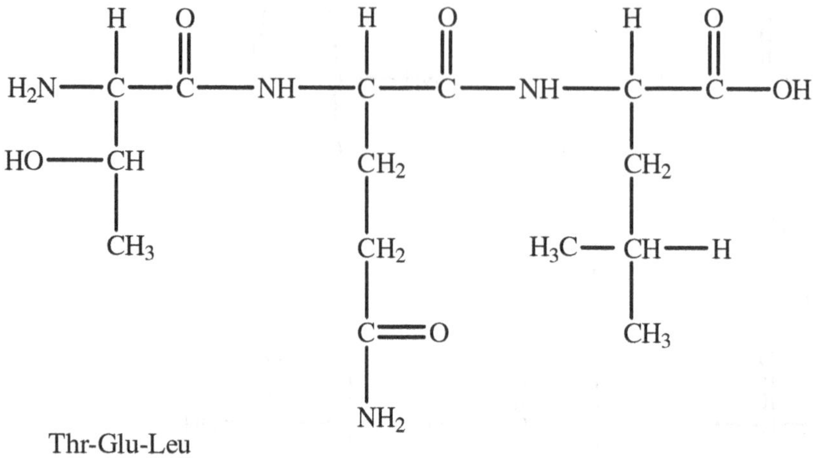

Structure of tripeptide Thr-Glu-Leu:
$H_2N-CH(CH(OH)CH_3)-CO-NH-CH(CH_2CH_2CONH_2)-CO-NH-CH(CH_2CH(CH_3)_2)-COOH$

Thr-Glu-Leu

b) 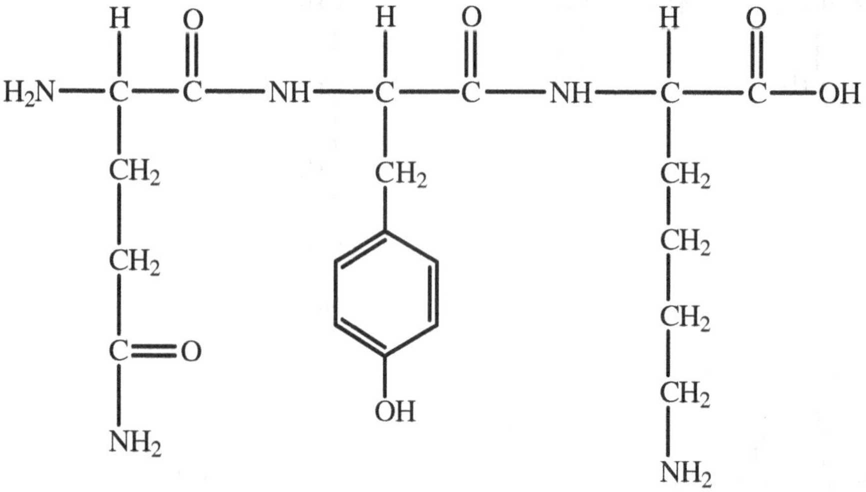 Structure of tripeptide Glu-Tyr-Lys:
$H_2N-CH(CH_2CH_2CONH_2)-CO-NH-CH(CH_2-C_6H_4-OH)-CO-NH-CH(CH_2CH_2CH_2CH_2NH_2)-COOH$

Glu-Tyr-Lys

c) 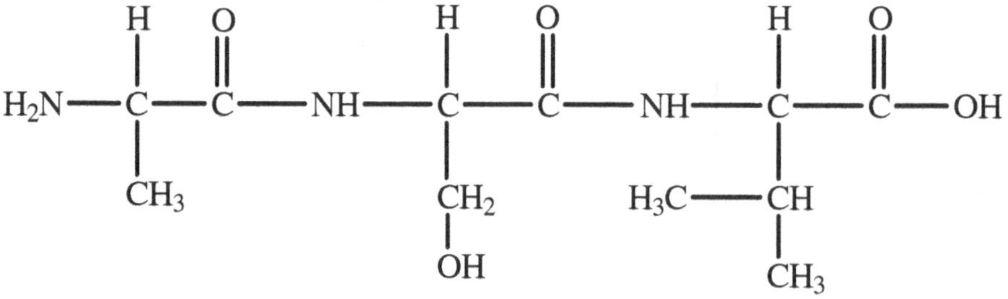 Structure of tripeptide Ala-Ser-Val:
$H_2N-CH(CH_3)-CO-NH-CH(CH_2OH)-CO-NH-CH(CH(CH_3)_2)-COOH$

Ala-Ser-Val

65. This interaction is an example of a tertiary structure because the amino acids involved are a long distance apart in terms of their placement in the chain.

66. The interactions between amino acids that are located a short distance apart, in terms of their placement in the chain, determine the secondary structure of the protein.

67. A listing of the amino acids in order of their appearance in the chain is a primary structure.

68. The interaction of multiple amino acid chains is the quaternary structure.

Nucleic Acids

69. a) nucleotide, G
    b) not a nucleotide
    c) not a nucleotide
    d) not a nucleotide

70. a) not a nucleotide
    b) nucleotide, A
    c) nucleotide, C
    d) not a nucleotide

71. The complementary strand of DNA is: T T A C G C G

72. The complementary strand of DNA is: A T A G C C A

73. DNA replicates as follows:

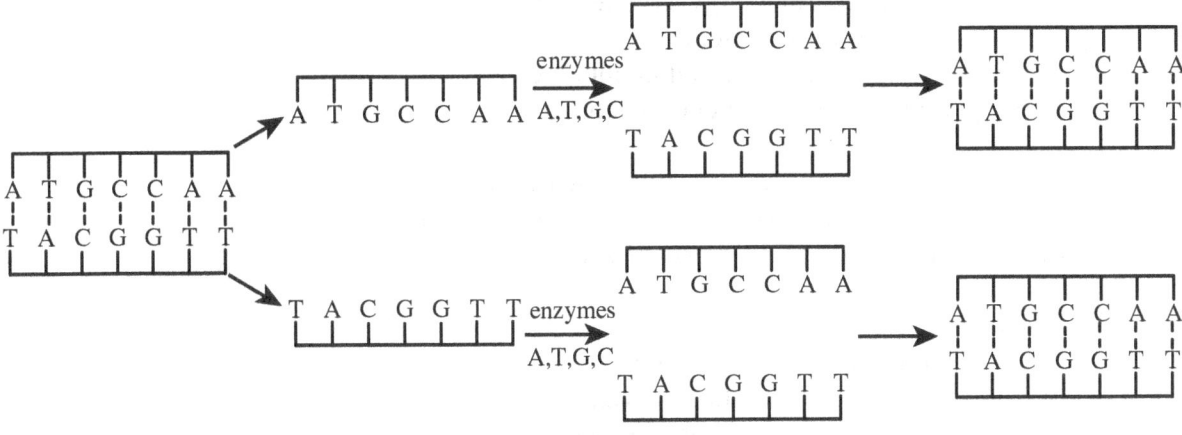

1. Hydrogen bonds break
2. Complementary bases match to each strand
3. Hydrogen bonds reform

74. DNA replicates as follows:

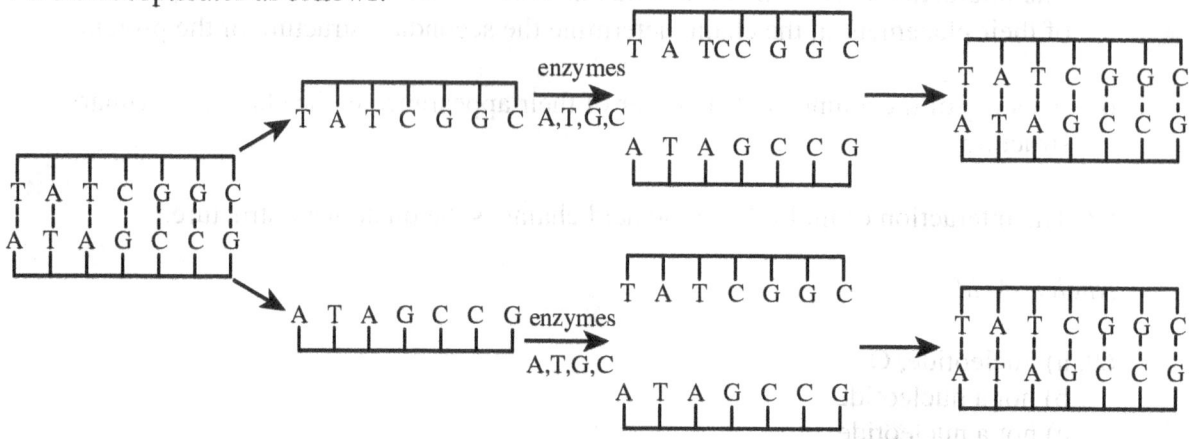

1. Hydrogen bonds break 2. Complementary bases match to each strand 3. Hydrogen bonds reform

Cumulative Problems

75. a) glycoside linkage—carbohydrates
    b) peptide bonds—proteins
    c) ester linkage—triglycerides

76. a) nucleotide—DNA
    b) saccharide—starch
    c) amino acid—protein

77. a) glucose—short-term energy storage
    b) DNA—blueprint for proteins
    c) phospholipids—compose cell membranes
    d) triglycerides—long-term energy storage

78. a) proteins—act as enzymes (among other things)
    b) cellulose—compose structural components of plants
    c) RNA—involved in protein synthesis

79. a) codon—codes for a single amino acid
    b) gene—codes for a single protein
    c) genome—all of the genetic material of an organism
    d) chromosome—structure that contains genes

80. a) pentose—a five-carbon sugar
    b) dipeptide—two amino acids joined by a peptide bond
    c) diglyceride—a glycerol molecule with two fatty acids attached
    d) fatty acid—a carboxylic acid with a long hydrocarbon R group

81. Nitrogen (1): Trigonal Pyramidal
    Carbon (2): Tetrahedral
    Carbon (3): Trigonal Planar
    Oxygen (4): Bent

82. Oxygen (1): Bent
    Carbon (2): Tetrahedral
    Carbon (3): Tetrahedral
    Nitrogen (4): Trigonal Pyramidal
    Carbon (5): Trigonal Planar
    Oxygen (6): Bent

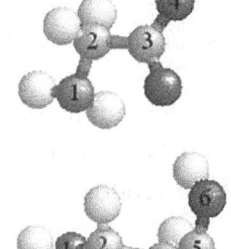

83.

Gly-Val

Val-Gly

84. thr-gly-ala
    thr-ala-gly
    gly-thr-ala
    gly-ala-thr
    ala-thr-gly
    ala-gly-thr

85. Lining up fragment pieces that overlap patterns:
           ala-ser-phe-gly-asn-lys
        gly-arg-ala-ser-phe
        gly-arg
              gly-asn-lys-trp
                trp-glu-val
                  glu-val

    Protein: gly-arg-ala-ser-phe-gly-asn-lys-trp-glu-val

86. Lining up fragment pieces that overlap patterns:
             asp-thr-ala-trp
          ser-lys-trp-arg-asp
        gly-glu-ser-lys
            trp-arg
               thr-ala-trp
      gly-glu

    Protein: gly-glu-ser-lys-trp-arg-asp-thr-ala-trp

87. Because each amino acid in the protein requires a codon consisting of three base pairs for synthesis, the number of base pairs is:
   51 amino acids × (3 bases/codon) = 153 bases

88. Since each amino acid in the protein requires a codon consisting of three base pairs for synthesis, the number of base pairs is:
   146 amino acids × (3 bases/codon) = 438 bases

89. $3.66 \text{ torr} \times \dfrac{1 \text{ atm}}{760 \text{ torr}} = n\left(0.0821 \dfrac{\text{L} \cdot \text{atm}}{\text{mol} \cdot \text{K}}\right) 298 \text{ K} \Rightarrow$

   $n = 1.97 \times 10^{-4}$

   $n = \dfrac{(\text{mass/molar mass})}{\text{L}} \Rightarrow 1.97 \times 10^{-4} = \dfrac{0.02388/\text{Molar Mass}}{0.0200} \Rightarrow$

   Molar Mass $= \dfrac{0.02388}{(1.97 \times 10^{-4}) 0.0200} = 6.06 \times 10^3$ g/mol

90. $4.55 \text{ torr} \times \dfrac{1 \text{ atm}}{760 \text{ torr}} = n\left(0.0821 \dfrac{\text{L} \cdot \text{atm}}{\text{mol} \cdot \text{K}}\right) 298 \text{ K} \Rightarrow$

   $n = 2.45 \times 10^{-4}$

   $n = \dfrac{(\text{mass/Molar Mass})}{\text{L}} \Rightarrow 2.45 \times 10^{-4} = \dfrac{0.02865/\text{Molar Mass}}{0.0250} \Rightarrow$

   Molar Mass $= \dfrac{0.02865}{(2.45 \times 10^{-4}) 0.0250} = 4.68 \times 10^3$ g/mol

Highlight Problems

91. The actual thymine-containing nucleotide uses the -OH end to bond and replicate; however, with the fake nucleotide having a nitrogen-based end instead, the possibility of replication is halted.

92. The side chain on valine is an isopropyl group, which is nonpolar. The side chain on glutamic acid contains a carboxylic acid functional group, which is capable of forming hydrogen bonds with water. The hydrogen bond promotes solubility in water and, when this is replaced by an isopropyl group, solubility decreases.

93. Cornstarch is a white powder, while cotton balls are made up of many individual strands or fibers. Both cornstarch and cotton balls are polysaccharides, which are complex carbohydrates composed of glucose units. The difference between starch and cellulose is the link between the glucose units. In starch, the oxygen atom that joins glucose units is an alpha linkage. In cellulose, the oxygen atom that joins glucose units is in a beta linkage configuration.

94. Answers will vary.

95. Answers will vary.

96. CUA GCG GAC AUU ACU GUC AAC GCG CGC AGU UGG UAG
    B   i   o   c   h   e   m   i   s   t   r   y

    GCG CGC UUC AAC UUC GGU GCG AAG CCG!
    i   s   a   m   a   z   i   n   g

Data Interpretation and Analysis

97. (a) pH = 5.9
    (b) pH = 4.7 and 7.2
    (c) As lactase is found in the small intestine, the pH should be near the optimum value of 5.9. If the pH differed too much, everyone would suffer from the symptoms of lactose intolerance.

Instructor Resource Manual Table of Contents

Chapter 1	The Chemical World	1
Chapter 2	Measurement and Problem Solving	3
Chapter 3	Matter and Energy	8
Chapter 4	Atoms and Elements	13
Chapter 5	Molecules and Compounds	17
Chapter 6	Chemical Composition	22
Chapter 7	Chemical Reactions	26
Chapter 8	Quantities in Chemical Reactions	32
Chapter 9	Electrons in Atoms and the Periodic Table	36
Chapter 10	Chemical Bonding	41
Chapter 11	Gases	46
Chapter 12	Liquids, Solids, and Intermolecular Forces	52
Chapter 13	Solutions	58
Chapter 14	Acids and Bases	63
Chapter 15	Chemical Equilibrium	68
Chapter 16	Oxidation and Reduction	73
Chapter 17	Radioactivity and Nuclear Chemistry	79
Chapter 18	Organic Chemistry	83
Chapter 19	Biochemistry	89

Copyright © 2018 Pearson Education, Inc.

The Chemical World

Chapter Overview

    This chapter presents an understanding of the history of chemical investigation. It is hoped that this will help the student understand the history of experimentation and scientific inquiry so that he or she feels a real-world association with the material to be covered later in the course. A few examples that the student may be familiar with are presented.

Lecture Outline

1.1 Sand and Water
    A. All things are made of atoms
    B. How atoms are arranged dictates properties of substances
    C. What is chemistry?
1.2 Chemicals Compose Ordinary Things
    Learning Objective: Recognize that chemicals make up virtually everything we come into contact with in our world.
    A. Everything we can hold or touch is made of chemicals
    B. Molecules interact all around you all the time
    C. Understanding how the universe works means understanding how molecules interact.
1.3 The Scientific Method: How Chemists Think
    Learning Objective: Identify and understand the key characteristics of the scientific method: observation, the formulation of hypotheses, the testing of hypotheses by experiment, and the formulation of laws and theories.
    A. Observation – hypothesis – law – theory – experiment
    B. Scientific law (e.g., law of conservation of mass)
    C. Dalton's atomic theory
1.4 Analyzing and Interpreting Data
    Learning Objective: Identify patterns in data and interpret graphs.
    A. Sets of measurements constitute data
    B. Identifying patterns in data
    C. Interpreting graphs
1.5 A Beginning Chemist: How to Succeed
    A. Curiosity and imagination
    B. Calculation
    C. Commitment

Chemical Principle Teaching Ideas

Matter and Molecules
     Go around the room and point out how everything around the students, including the room and their notebooks, are made of matter. Emphasizing the real-world association with what is covered in lecture is always a good idea.

The Scientific Method
     Using the scientific method to cover a simple concept such as putting together a children's puzzle or baking a cake will help them understand the method, which is most important here.

Success as a Beginning Chemist
     If the students are to do well in this course, they must be willing to expand their horizons outside the classroom and try to use everyday interactions with their world to help understand the concepts to be covered.

Skill Builder Solutions

1.1. According to the graph, the concentration of carbon dioxide in 1880 was 290 ppm. In 1920, the concentration of carbon dioxide was 304 ppm. The change in concentration was 304 ppm – 290 ppm = 14 ppm, over a change in time of 40 years. The average rate of increase was then 14 ppm / 40 years = 0.35 ppm/year. The increase is likely due to increased burning of fossil fuels.

Suggested Demonstrations

Open a can of soda in the students' presence and talk about the myriad of reactions and interactions taking place. The more often students associate chemical principles with real-life events, the better.

Burning of Magnesium, *Chemical Demonstrations* 1:38, Shakhashiri, B. Z. University of Wisconsin Press, 1983.

Use the scientific method to analyze an everyday occurrence in students' lives. Suggestions are a jigsaw puzzle, the burning of a combustible material, or some aspect of body chemistry.

Guided Inquiry Ideas

Below are a few example questions that students answer in the guided inquiry activities provided in the Guided Activity Workbook.

In a grammatically correct sentence, describe the relationship between laws and theories.

Considering the way the terms are used in science, are laws more certain than theories? Explain.

In a grammatically correct sentence, describe the difference between atoms and molecules.

How many hydrogen atoms are there in one molecule of water? What do all the atoms have in common?

What famous law might be explained with the kinetic theory of gases?

Measurement and Problem Solving

2

Chapter Overview

Chapter 2 introduces the student to a cornerstone of the chemical sciences, the manipulation of numbers and their associated units. These concepts are very important for the rest of the course, and in order to be successful in this course, students must understand them well. Simple and complex unit conversions as well as problem-solving strategies will be covered and explained in detail.

Lecture Outline

2.1 The Metric Mix-up: $125 Million Unit Error
2.2 Scientific Notation: Writing Large and Small Numbers
  <u>Learning Objective:</u> Express very large and very small numbers using scientific notation.
  A. Shorthand notation for numbers
  B. Two main pieces: decimal and power-of-10 exponent
  C. Measured value does not change, just how you report it
2.3 Scientific Figures: Writing Numbers to Reflect Precision
  <u>Learning Objective:</u> Report measured quantities to the right number of digits.
  <u>Learning Objective:</u> Determine which digits in a number are significant.
  A. How many digits can I report? How many should I report?
  B. Certain digits and estimated digits
  C. Counting significant figures
    1. All nonzero digits are significant
    2. Interior zeros are significant
    3. Trailing zeros after a decimal point are significant
    4. Trailing zeros before a decimal point are significant
    5. Leading zeros are not significant
    6. Zeros at the end of a number, but to the left of a decimal point, are ambiguous
  D. Exact numbers
2.4 Significant Figures in Calculations
  <u>Learning Objective:</u> Round numbers to the correct number of significant figures.
  <u>Learning Objective:</u> Determine the correct number of significant figures in the results of multiplication and division calculations.
  <u>Learning Objective:</u> Determine the correct number of significant figures in the results of addition and subtraction calculations.
  <u>Learning Objective:</u> Determine the correct number of significant figures in the results of calculations involving both addition/subtraction and multiplication/division.

A. Multiplication and Division
    1. Result carries as many significant digits as the factor with the fewest significant digits
B. Rounding
    1. If leftmost dropped digit is 4 or less, round down
    2. If leftmost dropped digit is 5 or higher, round up
C. Addition and Subtraction
    1. Result carries as many decimal places as the quantity with the fewest decimal places
D. Calculations Involving Both Multiplication/Division and Addition/Subtraction
    1. Do steps in parentheses first
    2. Determine the number of significant figures in intermediate answer
    3. Do remaining steps

2.5 The Basic Units of Measurement
    <u>Learning Objective:</u> Recognize and work with the SI base units of measurement, prefix multipliers, and derived units.
    A. English, metric, SI
    B. SI Units
        1. Length – m
        2. Mass – kg
        3. Time – s
    C. Prefix Multipliers
        1. milli (m) 0.001
        2. centi (c) 0.01
        3. kilo (k) 1000
        4. Mega (M) 1,000,000
    D. Derived Units
        1. Area – cm^2
        2. Volume – cm^3 or L

2.6 Problem Solving and Unit Conversions
    <u>Learning Objective:</u> Convert between units.
    A. Units are important, most numbers have one
    B. Include units in all calculations
    C. Conversion factors change one unit to another, the value is unchanged
    D. General problem-solving strategy
        1. Sort
        2. Strategize
        3. Solve
        4. Check

2.7 Solving Multistep Conversion Problems
    <u>Learning Objective:</u> Convert between units.
    A. Understand where you are going first
    B. Not all calculations can be done in one step

2.8 Unit Conversion in Both the Numerator and Denominator
    <u>Learning Objective:</u> Convert units in a quantity that has units in the numerator and the denominator.

2.9 Units Raised to a Power
    Learning Objective: Convert units raised to a power.
    A. 1 inch = 2.54 cm, so 1 inch^3 = (2.54)^3 cm^3 = 16.4 cm^3

2.10 Density
    Learning Objective: Calculate the density of a substance.
    Learning Objective: Use density as a conversion factor.
    A. Mass per unit volume
    B. Derived unit
    C. Can be used as a conversion factor between mass and volume

2.11 Numerical Problem-Solving Strategies and the Solution Map
    A. Come up with a plan before you use your calculator
    B. Use the units to guide your plan

Chemical Principle Teaching Ideas

Uncertainty
    Students generally have a hard time understanding this concept. One method is to refer to everyday objects that they recognize. For example, you can talk about a coffee cup containing about 200 mL of coffee. You then ask the students what the new volume would be if you were to add a drop of water with a volume of 0.05 mL.

Units
    Units are very important, and should always be used. Consider giving the students a measured value in many different units and having them guess what the unit is. Report the volume of your mug in barrels. What is the volume of the room measured in teaspoons?

Density
    Most students understand the concept of density or how much stuff is packed into a particular volume. What they have a harder time recognizing is the fact that it is a conversion factor between mass and volume. This is the easiest example that is discussed and should be emphasized as this concept is used frequently throughout the course.

Skill Builder Solutions

2.1.   Assuming all the trailing zeros are not significant, the decimal moves over 13 spaces to give $\$1.8416 \times 10^{13}$.

2.2.   Not all the leading zeros are significant, so we move the decimal over 5 places to give 3.8×10^{-5}.

2.3.   Each of the markings on the thermometer represents 1 degree Fahrenheit. We can therefore estimate one digit past the decimal place for a temperature of 103.4 degrees Fahrenheit.

2.4.   a. 4
b. 3, as leading zeros do not count, but trailing zeros after the decimal do
c. 2
d. Unlimited significant figures
e. 3
f. Ambiguous, since you do not know if the last two zeros are significant

2.5.   a. $\dfrac{1.10 \times 0.512 \times 1.301 \times 0.005}{3.4} = 0.001$ or 1×10^{-3}. There is only one significant digit in the final answer, as the 0.005 has only one significant digit in the numerator.

b. $\dfrac{4.562 \times 3.99870}{89.5} = 0.204$. The number 89.5 has the smallest number of significant digits, 3, so that is how many are quoted in the final answer.

2.6.   a. $2.18 + 5.621 + 1.5870 - 1.8 = 7.6$. Only one digit past the decimal place is quoted because the least accurately known number (1.8) has one digit past the decimal.
b. $7.876 - 0.56 + 123.792 = 131.11$. Two digits past the decimal are quoted because 0.56 has two past the decimal and is the number with the fewest digits past the decimal.

2.7.   a. $3.897 \times (782.3 - 451.88) = 3.897 \times 330.42 = 1288$. Four digits are quoted because the number in the second (multiplication) step with the fewest significant digits has four of them.
b. $\dfrac{4.58}{1.239} - 0.578 = 3.70 - 0.578 = 3.12$. Two digits past the decimal are quoted because the first part of the subtraction (3.70) has two digits past the decimal place.

2.8. $56.0 \text{ cm} \times \dfrac{1 \text{ inch}}{2.54 \text{ cm}} = 22.0 \text{ inch}$

2.9. $5{,}678 \text{ m} \times \dfrac{1 \text{ km}}{1000 \text{ m}} = 5.678 \text{ km}$

2.10. $1.2 \text{ qu} \times \dfrac{1 \text{ qt}}{4 \text{ qu}} \times \dfrac{1 \text{ L}}{1.057 \text{ qt}} = 0.28 \text{ L}$

2.11. $15.0 \text{ km} \times \dfrac{0.6214 \text{ mi}}{1 \text{ km}} \times \dfrac{5280 \text{ ft}}{1 \text{ mi}} \times \dfrac{1 \text{ lap}}{1056 \text{ ft}} = 46.6 \text{ laps}$

Plus. $5.72 \text{ naut mi} \times \dfrac{1.151 \text{ mi}}{1 \text{ naut mi}} \times \dfrac{1 \text{ km}}{0.6214 \text{ mi}} \times \dfrac{1000 \text{ m}}{1 \text{ km}} = 1.06 \times 10^4 \text{ m}$

2.12 $\dfrac{65 \text{ km}}{1 \text{ hr}} \times \dfrac{1 \text{ hr}}{60 \text{ min}} \times \dfrac{1 \text{ min}}{60 \text{ s}} \times \dfrac{1000 \text{ m}}{1 \text{ km}} = 18 \text{ m/s}$

2.13. $289.7 \text{ in}^3 \times \dfrac{(2.54)^3 \text{ cm}^3}{1 \text{ in}^3} = 4747 \text{ cm}^3$

2.14. $3.25 \text{ yd}^3 \times \dfrac{(36)^3 \text{ inch}^3}{1 \text{ yd}^3} = 1.52 \times 10^5 \text{ inch}^3$

2.15. $\dfrac{9.67 \text{ g}}{0.452 \text{ cm}^3} = 21.4 \text{ g/cm}^3$; Therefore, the ring is genuine platinum.

2.16. $35 \text{ mg} \times \dfrac{1 \text{ g}}{1000 \text{ mg}} \times \dfrac{1 \text{ cm}^3}{0.788 \text{ g}} = 4.4 \times 10^{-2} \text{ cm}^3$

Plus. $246 \text{ cm}^3 \times \dfrac{7.93 \text{ g}}{1 \text{ cm}^3} \times \dfrac{1 \text{ kg}}{1000 \text{ g}} = 1.95 \text{ kg}$

2.17. $0.82 \text{ L} \times \dfrac{1000 \text{ mL}}{1 \text{ L}} \times \dfrac{19.3 \text{ g}}{1 \text{ mL}} \times \dfrac{1 \text{ kg}}{1000 \text{ g}} = 16 \text{ kg}$

2.18. $\dfrac{23.2 \text{ mg} \times \dfrac{1 \text{ g}}{1000 \text{ mg}}}{1.20 \text{ mm}^3 \times \dfrac{1 \text{ cm}^3}{(10)^3 \text{ mm}^3}} = \dfrac{2.32 \times 10^{-2} \text{ g}}{1.20 \times 10^{-3} \text{ cm}^3} = 19.3 \text{ g/cm}^3$

Yes, it is consistent with the density of gold.

Suggested Demonstrations

Density and Miscibility of Liquids, *Chemical Demonstrations* 3:233, Shakhashiri, B. Z. University of Wisconsin Press, 1989.

Guided Inquiry Ideas

Below are a few example questions that students answer in the guided inquiry activities provided in the Guided Activity Workbook.

How many significant figures are there in the number 0.0051? Underline it/them.

How many significant figures are there in the number 5.00? Underline it/them.

In a complete sentence or two, describe when you know a "trailing zero" is significant.

In a complete sentence, describe the significance of "leading zeros."

Which of the following is a correct conversion factor from cm^3 to in^3? Circle all that apply.

$\left(\dfrac{1 \text{ in}^3}{2.54 \text{ cm}^3}\right)$ $\left(\dfrac{1 \text{ in}}{2.54 \text{ cm}}\right)^3$ $\left(\dfrac{1 \text{ in}^3}{16.4 \text{ cm}^3}\right)$

Matter and Energy

3

Chapter Overview

The universe is made up of a finite amount of matter and energy. It is these two concepts that tie most chemical principles together. Therefore, we must have a very good understanding of these ideas to develop the more detailed topics later in the course. Temperature, states and properties of matter, and the associated calculations will be covered.

Lecture Outline

3.1 In Your Room
  A. What you see, touch, smell is all matter
3.2 What Is Matter?
  <u>Learning Objective:</u> Define matter, atoms, and molecules.
  A. Occupies space and has mass
  B. Atom – smallest unit of matter
  C. Molecule – atoms joined together
3.3 Classifying Matter According to Its State: Solid, Liquid, and Gas
  <u>Learning Objective:</u> Classify matter as solid, liquid, or gas.
  A. Solid
    1. Crystalline
    2. Amorphous
  B. Liquid
    1. Fixed volume
    2. Fluid
  C. Gas
    1. Compressible
    2. Fluid
3.4 Classifying Matter According to Its Composition: Elements, Compounds, and Mixtures
  <u>Learning Objective:</u> Classify matter as element, compound, or mixture.
  A. Pure substance
    1. Composed of one type of atom or molecule
    2. Elements or compounds
  B. Mixture
    1. Composed of two or more types of atoms or molecules
    2. Homogeneous or heterogeneous
3.5 Differences in Matter: Physical and Chemical Properties
  <u>Learning Objective:</u> Distinguish between physical and chemical properties.
  A. Physical property
    1. Observable without changing the identity
    2. Examples: melting point, odor, color

B. Chemical property
  1. Observable only by changing the identity
  2. Example: flammability
3.6 Changes in Matter: Physical and Chemical Changes
  Learning Objective: Distinguish between physical and chemical changes.
  A. Physical change
    1. Appearance and properties can change
    2. Composition does not change
  B. Chemical change
    1. Appearance and properties can change
    2. Composition changes
  C. Separation of mixtures through physical changes
    1. Decanting
    2. Distillation
    3. Filtration
3.7 Conservation of Mass: There Is No New Matter
  Learning Objective: Apply the law of conservation of mass.
  A. Matter is neither created nor destroyed in a chemical reaction
  B. Mass sum of chemicals before reaction is the same as after
3.8 Energy
  Learning Objective: Recognize the different forms of energy.
  Learning Objective: Identify and convert among energy units.
  A. Energy cannot be created or destroyed
  B. Forms of energy
    1. Kinetic – energy of motion
    2. Potential – energy associated with position
    3. Electrical – energy from the flow of electron charge
    4. Thermal – energy from random motions of atoms and molecules
    5. Chemical – energy associated with potential chemical change
  C. Units of energy
    1. Joule (J)
    2. calorie (cal)
    3. Calorie (Cal)
    4. Kilowatt-hour (kWh)
3.9 Energy and Chemical and Physical Change
  Learning Objective: Distinguish between exothermic and endothermic reactions.
  A. Exothermic reaction: molecules release energy to surroundings
  B. Endothermic reaction: molecules absorb energy from surroundings
3.10 Temperature: Random Motion of Molecules and Atoms
  Learning Objective: Convert between Fahrenheit, Celsius, and Kelvin temperature scales.
  A. Fahrenheit (°F)
  B. Celsius (°C)
  C. Kelvin (K)
3.11 Temperature Changes: Heat Capacity
  Learning Objective: Relate energy, temperature change, and heat capacity.
  A. How much energy is required to change the temperature of a substance
  B. Varies from one substance to another
  C. Metals have low heat capacity, water has a high heat capacity
  D. Units: J/g°C

3.12 Energy and Heat Capacity Calculations
   Learning Objective: Perform calculations involving transfer of heat and changes in temperature.
   A. Heat = mass × heat capacity × temperature change
   B. Heat capacity is a conversion factor between temperature change, heat, and mass

Chemical Principle Teaching Ideas

Matter
   Try to have the students come up with something that is NOT matter, and they will quickly realize that everything around them is matter.

Classification of Matter
   A good exercise in classifying matter into certain categories is to point around the room, asking students what each object is and what group it falls under. There are plenty of examples of each grouping in a classroom; even the human body has some of each category.

Properties and Changes of Matter
   Try listing various changes and ask the students if the changes are physical or chemical. List some measurable properties and ask students to classify them as physical or chemical changes.

Conservation of Mass
   Some students will say that this is not true, as in a nuclear reaction where mass is converted into energy. Explain to them that this law is for chemical reactions, where no nuclear changes take place.

Energy
   The total energy in the universe is constant, but the universe is expanding at a furious pace. You can use this opportunity to talk about the universe, how it was born, and its history.

Temperature
   To help students understand why we have different temperature scales, you can discuss the origin of each scale. The Fahrenheit scale is based on the average temperature of the human body being at 100 °F (it was off a little bit) and the freezing point of a salt-ice bath of 0 °F. The Celsius scale is based on the freezing point and boiling point of pure water as 0° and 100° Celsius, respectively. The Kelvin scale is based on the coldest temperature that is possible, the temperature at which molecular motion ceases, and is set as 0 K.

Heat Capacity
   One easy way for students to remember the concept of heat capacity and its relative magnitude is for them to look at water heating on a stove. Most of them will agree that if you put an empty metal pot on a stove the pot will get hot very quickly. However, if we fill the pot with water, it will take a very long time for the water to get to the same temperature. If they remember this scenario, they should remember that metals have a low heat capacity and water has quite a high heat capacity.

Skill Builder Solutions

3.1.
a. Mercury is a pure substance and is also an element.
b. Exhaled air is a mixture of nitrogen, carbon dioxide, oxygen, and some other gases. The mixture is the same composition throughout, so it is a homogeneous mixture.
c. Soup is also a mixture, but each bite can taste different, so the mixture is not the same throughout. It is therefore a heterogeneous mixture.
d. Sugar is a pure substance and also a compound, as sugar is a molecule.

3.2.
a. Hydrogen burns in oxygen to give water as a product. This is a chemical change, so it is a chemical property.
b. Observing the color of copper does not require that changes be made to the copper, so it is a physical property.
c. Luster (shininess) is a physical property associated with the smoothness of the metal surface.
d. Phase changes do not alter composition, so this is a physical property.

3.3.
a. This is a chemical reaction involving copper solid becoming copper ion.
b. Since the penny is just getting a new shape, that is a physical change.
c. Phase changes like this solid-to-liquid change are physical changes.
d. Fireworks involve combustion, which is a chemical change.

3.4. 12 g gas + 48 g oxygen = 60 g of reactants. 33 g of carbon dioxide formed, so 60 g – 33 g = 27 g of water is produced.

3.5. $512 \text{ cal} \times \dfrac{4.184 \text{ J}}{1 \text{ cal}} \times \dfrac{1 \text{ kJ}}{1000 \text{ J}} = 2.14 \text{ kJ}$

Plus. $2.75 \times 10^4 \text{ kJ} \times \dfrac{1000 \text{ J}}{1 \text{ kJ}} \times \dfrac{1 \text{ cal}}{4.184 \text{ J}} = 6.57 \times 10^6 \text{ cal}$

3.6.
a. Potential energy must be removed from liquid water to turn it into ice, so it is an exothermic reaction.
b. All combustion reactions are exothermic, releasing energy to the surroundings.

3.7. $358 - 273 = 85 \,°C$

3.8. $[139 \times 1.8] + 32 = 282 \,°F$

3.9. $\dfrac{(-321 - 32)}{1.8} + 273 = 77 \text{ K}$

3.10. $(3.10 \text{ g})(42.0 \,°C)\left(\dfrac{0.385 \text{ J}}{\text{g} \,°C}\right) = 50.1 \text{ J}$

Plus. $\dfrac{11.3 \text{ J}}{(12 \,°C)\left(\dfrac{0.128 \text{ J}}{\text{g} \,°C}\right)} = 7.4 \text{ g}$

3.11. $\Delta T = \dfrac{5.78 \times 10^3 \, \text{J}}{(328 \, \text{g})\left(\dfrac{4.184 \, \text{J}}{\text{g} \, °\text{C}}\right)} = 4.21 \, °\text{C} \quad T_f = 25.0 + \Delta T = 29.2 \, °\text{C}.$

Suggested Demonstrations

Heat and Dilution of Sulfuric Acid, *Chemical Demonstrations* 1:17, Shakhashiri, B. Z. University of Wisconsin Press, 1983.

Heat of Solution of Lithium Chloride, *Chemical Demonstrations* 1:21, Shakhashiri, B. Z. University of Wisconsin Press, 1983.

Burning of Magnesium, *Chemical Demonstrations* 1:38, Shakhashiri, B. Z. University of Wisconsin Press, 1983.

Dehydration of Sugar by Sulfuric Acid, *Chemical Demonstrations* 1:77, Shakhashiri, B. Z. University of Wisconsin Press, 1983.

Collapsing Can, *Chemical Demonstrations* 2:6, Shakhashiri, B. Z. University of Wisconsin Press, 1985.

Liquid Vapor Equilibrium, *Chemical Demonstrations* 2:75, Shakhashiri, B. Z. University of Wisconsin Press, 1985.

Guided Inquiry Ideas

Below are a few example questions that students answer in the guided inquiry activities provided in the Guided Activity Workbook.

How much energy do you think it will take to raise the temperature of 1 g of water from room temperature (about 20 °C) to body temperature (about 37 °C)? Explain your answer.

If you drink 1 cup (about 250 g) of ice water that is initially at 0 °C, how much energy will your body have to use to warm it up to body temperature?

An experiment was done in which 58 g of butane (enough for about a dozen lighters) was found to consume 208 g of oxygen when it burned. The products were 176 g of carbon dioxide and 90 g of water.
   During the reaction, what changed?
   During the reaction what did not change?

Atoms and Elements 4

Chapter Overview

Chapter 4 brings us from the earliest recorded ideas on atomic theory over 2000 years ago up to the present day. This chapter serves as a historical timeline, giving the students a better understanding of the origins of the facts that we assert. The origin of current atomic theory is presented, and the student's vocabulary is again extended.

Lecture Outline

4.1 Experiencing Atoms at Tiburon
    A. Atoms are very small
    B. Atoms make up all matter
    C. ~91 different naturally occurring elements
4.2 Indivisible: The Atomic Theory
    <u>Learning Objective:</u> Recognize that all matter is composed of atoms.
    A. Democritus (~400 BC)
    B. Dalton (early 1800s)
        1. Each element is composed of tiny indestructible particles called atoms
        2. All atoms of a given element have the same mass and other properties that distinguish them from atoms of other elements
        3. Atoms combine in simple, whole-number ratios to form compounds
4.3 The Nuclear Atom
    <u>Learning Objective:</u> Explain how the experiments of Thomson and Rutherford led to the development of the nuclear theory of the atom.
    A. Positive and negative charges (Thomson)
    B. Mostly empty space (Rutherford)
    C. Current nuclear theory of the atom
        1. Most of the atom's mass and all of its positive charge are contained in the nucleus
        2. Most of the atom is empty space, with small fast-moving electrons
        3. Number of protons = number of electrons in neutral atoms
4.4 The Properties of Protons, Neutrons, and Electrons
    <u>Learning Objective:</u> Describe the respective properties and charges of electrons, neutrons, and protons.
    A. Masses expressed in atomic mass units (amu)
    B. Protons and neutrons are about 2000 times more massive than electrons
    C. Nature of electrical charge
        1. Electrical charge is a fundamental property of protons and electrons
        2. Species of opposite electric charge are attracted to each other

3. Species of the same electric charge repel each other
4. Species of opposite electric charge cancel to make a neutral atom

4.5 Elements: Defined by Their Number of Protons
Learning Objective: Determine an element's atomic symbol and atomic number using the periodic table.
A. Identity of atom comes from the number of protons
B. Atomic number (Z) = number of protons
C. Chemical symbol is a shorthand notation of chemical name
1. C = carbon
2. O = oxygen
3. Co = cobalt

4.6 Looking for Patterns: The Periodic Law and the Periodic Table
Learning Objective: Use the periodic table to classify elements by group.
A. Mendeleev (1834–1907)
B. Periodic law
C. Metals
D. Nonmetals
E. Metalloids, also known as semiconductors
F. Individual group names
1. Group 1 – alkali metals
2. Group 2 – alkaline earth metals
3. Group 7 – halogens
4. Group 8 – noble gases

4.7 Ions: Losing and Gaining Electrons
Learning Objective: Determine ion charge from numbers of protons and electrons.
Learning Objective: Determine the number of protons and electrons in an ion.
A. If an atom has different numbers of protons and electrons, it is an ion
B. Cation = positive ion
C. Anion = negative ion
D. Periodic trends
1. Group 1: +1 ion only
2. Group 2: +2 ion only
3. Group 7: usually −1 ion

4.8 Isotopes: When the Number of Neutrons Varies
Learning Objective: Determine atomic numbers, mass numbers, and isotope symbols for an isotope.
Learning Objective: Determine the number of protons and neutrons from isotope symbols.
A. Isotopes have the same chemical properties but different masses
B. Some isotopes are more prevalent than others
C. Chemical symbol with mass number indicates which isotope
D. Some elements have many isotopes, some very few

4.9 Atomic Mass: The Average Mass of an Element's Atoms
Learning Objective: Calculate atomic mass from percent natural abundances and isotopic masses.
A. Value on periodic table represents average mass of all isotopes
B. No chlorine atom weighs exactly 35.453 amu

Chemical Principle Teaching Ideas

The Atomic Theory
Many students believe that we have completely understood current atomic theory for centuries. They do not understand that the concepts we now hold as truth became "mainstream" only 200 years ago, with many changes occurring since. Providing a brief history of atomic theory, starting with Plato and Aristotle's belief that the four elements—air, water, earth, and fire—were the ingredients of all matter gives students a historical perspective.

Discovery of the Atom's Nucleus
Demonstrations of the rudimentary experiments that revolutionized atomic theory will go a long way toward making students understand that chemistry is an experimental science and that most of what we learn was discovered through pure experimentation.

Charge
Most students understand the concept of electrical charge, but a simple demonstration of static electricity will reinforce the idea.

The Periodic Table
The periodic table can be the best exam "cheat sheet" a student could ever use. By understanding the layout of the table, many trends and quite a lot of information can be extracted. Mentioning briefly the things that have periodic trends will also reinforce the periodic law in their minds.

Atomic Number, Ions, and Isotopes
Emphasize that all atoms having the same number of protons behave the same chemically and only differ in their mass, and then only if they are different isotopes. A real world example is a pile of paper clips of various shapes, sizes, and colors. They differ in observable properties, but all behave the same when they "react" as they all hold paper. A similar example can be made with cork stoppers of different sizes.

Skill Builder Solutions

4.1.   a. Located in Group 1A, sodium is atom #11.
       b. Located in Group 8B, nickel is element #28.
       c. Located in Group 5A, phosphorus is element #15.
       d. Located in Group 5B, tantalum is element #73.

4.2.   a. Sulfur is in Group 6A and is a nonmetal.
       b. Chlorine is in Group 7A and is a nonmetal.
       c. Titanium is in Group 4B and is a metal.
       d. Antimony is considered a metalloid in Group 5A.

4.3.   a. Group 1A is called the alkali metal group.
       b. Group 3A has no special name.
       c. Iodine is in Group 7A, also known as the halogen group.
       d. Argon is in Group 8A, also known as the noble gases.

4.4. a. Nickel has 28 protons, so the net charge is 28 − 26 = +2.
b. Bromine has 35 protons, so the ion charge = 35 − 36 = −1.
c. Phosphorus has 15 protons, so the ion charge = 15 − 18 = −3.

4.5. All sulfur atoms have 16 protons. If the net charge is −2, that means there are two more electrons than protons in the atom, so there are 16 + 2 = 18 electrons.

4.6. Potassium tends to lose one electron to form a +1 ion, and selenium tends to gain two electrons to form a −2 ion.

4.7. Chlorine atoms all have 17 protons, so Z = 17, and since there are 18 neutrons, the mass number is 35. Therefore, the chemical symbol is $^{35}_{17}\text{Cl}$.

4.8. All potassium atoms have 19 protons, so there are 39 − 19 = 20 neutrons.

4.9. $\dfrac{78.99}{100}(23.99 \text{ amu}) + \dfrac{10.00}{100}(24.99 \text{ amu}) + \dfrac{11.01}{100}(25.98 \text{ amu}) = 24.31 \text{ amu}$

Guided Inquiry Ideas

Below are a few example questions that students answer in the guided inquiry activities provided in the Guided Activity Workbook.

If an atom has a net charge of zero, what could you conclude about the relative number of electrons and protons? Explain.

Rutherford's gold foil experiment showed that practically all the mass is concentrated in a very small location called the **nucleus**. Which particles must be in the nucleus?

A mad scientist proposes that Type 2 carbon has more protons. How would you respond?

How many electrons would you need to weigh the same as one proton?

Describe the relationship between the name of an element and its symbol.

Molecules and Compounds   5

Chapter Overview

Chapter 5 introduces the students to predicting and writing chemical formulas. They will also learn how to derive the name of a molecule from the formula and the type of species. The naming of ionic compounds, molecular compounds, and acids will be explained.

Lecture Outline

5.1 Sugar and Salt
5.2 Compounds Display Constant Composition
    <u>Learning Objective:</u> Restate and apply the law of constant composition.
    A. Every sample of a compound has the same proportions of constituent elements
5.3 Chemical Formulas: How to Represent Compounds
    <u>Learning Objective:</u> Write chemical formulas.
    <u>Learning Objective:</u> Determine the total number of each type of atom in a chemical formula.
    A. Subscripts in formula represent the relative number of each type of atom in the molecule
    B. Order for listing nonmetal atoms: C, P, N, H, S, I, Br, Cl, O, F
    C. Groups of atoms are set off by parentheses, for example, $Mg(NO_3)_2$
    D. Types of Chemical Formulas
        1. Empirical
        2. Molecular
        3. Structural
5.4 A Molecular View of Elements and Compounds
    <u>Learning Objective:</u> Classify elements as atomic or molecular.
    <u>Learning Objective:</u> Classify compounds as ionic or molecular.
    A. Atomic elements exist in nature as single atoms
    B. Seven molecular elements exist as diatomic molecules: H_2, N_2, O_2, F_2, Cl_2, Br_2, I_2
    C. Molecular compounds are formed between two or more nonmetals
    D. Ionic compounds are formed between a metal and one or more nonmetals
5.5 Writing Formulas for Ionic Compounds
    <u>Learning Objective:</u> Write formulas for ionic compounds.
    A. Ionic compounds always contain positive and negative ions
    B. Formula units form neutral compounds
5.6 Nomenclature: Naming Compounds
    <u>Learning Objective:</u> Distinguish between common and systematic names for compounds.
    A. Systematic name; for example, sodium chloride
    B. Common name; for example, table salt

5.7 Naming Ionic Compounds
   Learning Objective: Name binary ionic compounds containing a metal that forms only one type of ion.
   Learning Objective: Name binary ionic compounds containing a metal that forms more than one type of ion.
   Learning Objective: Name ionic compounds containing a polyatomic ion.
   A. Type I compounds
      1. Metal in compound forms only one type of ion
      2. Most main-group metals form type I compounds
   B. Naming type I binary ionic compounds
      1. Name of cation (metal) + (base name of anion + ide)
      2. Example, NaCl is sodium chloride
   C. Type II compounds
      1. Metal in compound forms more than one type of ion
      2. Transition metals usually, but not exclusively, form type II compounds
   D. Naming type II binary ionic compounds
      1. Name of cation + (charge of cation) + (base name of anion + ide)
      2. Charge of cation given in roman numerals
      3. Example, $FeCl_3$ is iron (III) chloride
   E. Naming ionic compounds containing a polyatomic ion
      1. Use the same procedure as ionic compounds
      2. Use name of polyatomic ion, not constituent atoms
      3. Example, $NaNO_3$ is sodium nitrate
5.8 Naming Molecular Compounds
   Learning Objective: Name molecular compounds.
   A. Formed between two or more nonmetals
   B. (Prefix) name of first element + (prefix) base name of second element + ide
   C. 1 = mono; 2 = di; 3 = tri; 4 = tetra…
   D. Example, N_2O is dinitrogen monoxide
5.9 Naming Acids
   Learning Objective: Name binary acids.
   Learning Objective: Name oxyacids containing an oxyanion ending in –ate.
   Learning Objective: Name oxyacids containing an oxyanion ending in –ite.
   A. Acids are molecular compounds that dissolve in water to form H^+ ions
   B. Binary acids
      1. Hydrogen and nonmetal
      2. (Hydro + base name of nonmetal + ic) + acid
      3. Example, HCl is hydrochloric acid
   C. Oxyacids
      1. Hydrogen and polyatomic oxyanion
      2. Oxyanions ending with –ate
         a. (Base name of oxyanion + ic) + acid
         b. Example, HNO_3 is nitric acid
      3. Oxyanions ending with –ite
         a. (Base name of oxyanion + ous) + acid
         b. Example, HNO_2 is nitrous acid

5.10 Nomenclature Summary
   Learning Objective: Recognize and name chemical compounds.
   A. Ionic compounds
   B. Molecular compounds
   C. Acids
5.11 Formula Mass: The Mass of a Molecule or Formula Unit
   Learning Objective: Calculate formula mass.
   A. Formula mass = mass of atom 1 × number of first element in molecule + mass of atom 2 × number of second element + …
   B. Example, formula mass of CCl_4 = 1 × (12.01 amu) + 4 × (35.45 amu) = 153.81 amu

Chemical Principle Teaching Ideas

Compounds
   To illustrate the universal makeup of a compound, show examples of different forms of the same molecule. Showing a sample of solid, liquid, and gaseous water and explaining that the ratio of hydrogen atoms to oxygen atoms is the same will help teach the concept.

Chemical Formulas
   Emphasize that all molecules are composed of atoms in small, whole-number ratios.

Chemical Nomenclature
   Students have trouble remembering which rules fit for which kind of molecule. Give the students one easy example of each to memorize, and tell them that if they remember one example of each, they can derive the rules. Suggested molecules are CO_2, NaCl, and $FeCl_2$.

Formula Mass
   Doing several examples of formula mass of molecules is the best way for students to become comfortable with this concept. Be sure to include at least one case of multiple polyatomic ions; for example, $(NH_4)_2SO_4$.

Skill Builder Solutions

5.1. $\frac{4.3 \text{ g O}}{3.2 \text{ g C}} = 1.3$, and $\frac{7.5 \text{ g O}}{5.6 \text{ g C}} = 1.3$ Yes, these data are consistent with the law of constant composition.

5.2. a. Ag_2S shows 2 silvers to every 1 sulfur.
   b. N_2O shows 2 nitrogen atoms to every oxygen.
   c. TiO_2 shows 2 oxygen atoms to every 1 titanium.

5.3 The subscript 2 on K means there are 2 potassium atoms. No subscript on the S implies 1 sulfur, and the subscript 4 behind the O represents 4 oxygen atoms.

Plus. The subscript 2 behind the Al indicates 2 aluminum atoms for each molecule. There are 3 sulfate groups, each having 1 sulfur and 4 oxygen atoms; therefore, there are 3 sulfur atoms and $3 \times 4 = 12$ oxygen atoms in $Al_2(SO_4)_3$.

5.4. a. Molecular element; chlorine is one of the seven species that exist as diatomic species.
 b. Molecular compound; this species contains two different nonmetals, so it is a compound.
 c. Atomic element; gold is a monatomic element.
 d. Ionic compound; a metal and a nonmetal form an ionic species.
 e. Ionic compound; a metal and a nonmetal form an ionic species.

5.5. Since the strontium cation has a +2 charge and chlorine anion has a −1 charge, we need 2 chlorine atoms for each strontium atom. Thus, the formula is $SrCl_2$.

5.6. The aluminum cation has a +3 charge and nitrogen anion has a −3 charge, so we need only one of each to make a neutral species: AlN.

5.7 The aluminum cation has a +3 charge and the phosphate anion has a −3 charge, so we need only one of each to make a neutral species: $AlPO_4$.

Plus. The sodium cation has a +1 charge and the sulfite anion has a −2 charge, so we need 2 sodium cations for every 1 sulfite anion to make a neutral species: Na_2SO_3.

5.8. The cation is potassium and the anion is bromine, which becomes bromide, so the correct name is potassium bromide.

Plus. The cation is zinc, and the anion is nitrogen, which becomes nitride, giving the correct name zinc nitride.

5.9. Since the oxygen has a −2 charge and there is just 1 lead atom, it must be in a +2 state. Therefore, the name is lead (II) oxide.

5.10. The nitrate oxyanion has a charge of −1, and there are two of them for a total charge of −2. Since there is only one manganese, it must have a +2 charge, giving a name of manganese (II) nitrate.

5.11. Two nitrogen atoms get a prefix of di, and 4 oxygen atoms get a prefix of tetra for the name dinitrogen tetroxide.

5.12. The base name of fluorine is fluor, so HF is called hydrofluoric acid.

5.13. Since NO_2 (nitrite) ends in *-ite*, the acid name ends in *-ous* giving nitrous acid.

5.14. 2 × the mass of nitrogen + 1 × the mass of oxygen = 2(14.01 amu) + 1(16.00 amu) = 44.02 amu.

Suggested Demonstrations

Bring in a set of molecular models and make a few of the oxyanion groups. Then, bond them with different species illustrating their combinatory nature.

Reaction of Sodium and Chlorine, *Chemical Demonstrations* 1:61, Shakhashiri, B. Z. University of Wisconsin Press, 1983.

Thermite Reaction, *Chemical Demonstrations* 1:85, Shakhashiri, B. Z. University of Wisconsin Press, 1983.

Explosive Reaction of Nitric Oxide and Carbon Disulfide, *Chemical Demonstrations* 1:117, Shakhashiri, B. Z. University of Wisconsin Press, 1983.

Guided Inquiry Ideas

Below are a few example questions that students answer in the guided inquiry activities provided in the Guided Activity Workbook.

How do you know that carbon monoxide and carbon dioxide are molecular compounds and not ionic?

In any stable ionic compound, what is the relationship between the total positive and total negative charge?

How does the name of the second atom in a formula for a molecular compound compare to the name of that atom?

Describe as completely as you can the steps in naming a molecular compound.

List three quantities that vary from one sample of water to the next. What doesn't change from one sample of water to the next?

Chapter 6

Chemical Composition

Chapter Overview

In Chapter 6, we begin to talk about measurable quantities and what they represent in the atomic reference frame. A discussion of how mass can be converted into numbers of species, expressed in moles, is presented. Determining empirical and molecular formulae via laboratory experiments is also explained.

Lecture Outline

6.1 How Much Sodium?
6.2 Counting Nails by the Pound
    <u>Learning Objective:</u> Recognize that we use the mass of atoms to count them because they are too small and numerous to count individually.
    A. Not useful if we want to know how many nails we have
    B. Need a conversion factor between mass and number of items
6.3 Counting Atoms by the Gram
    <u>Learning Objective:</u> Convert between moles and number of atoms.
    <u>Learning Objective:</u> Convert between grams and moles.
    <u>Learning Objective:</u> Convert between grams and number of atoms.
    A. Pair = 2; dozen = 12; mole = 6.02×10^{23}
    B. 1 mole of Au atoms = 6.02×10^{23} Au atoms
    C. The periodic table gives the mass of 1 mole of that atom in grams
        1. 1 mole of S atoms = 32.07 g S
        2. 1 mole of C atoms = 12.01 g C
    D. Unit conversions between grams and moles of atoms
6.4 Counting Molecules by the Gram
    <u>Learning Objective:</u> Convert between grams and moles of a compound.
    <u>Learning Objective:</u> Convert between mass of a compound and number of molecules.
    A. Mass of 1 mole H_2O = mass of 2 moles of H + mass of 1 mole O
    B. Calculations performed in a similar fashion
6.5 Chemical Formulas as Conversion Factors
    <u>Learning Objective:</u> Convert between moles of a compound and moles of a constituent element.
    <u>Learning Objective:</u> Convert between grams of a compound and grams of a constituent element.
    A. Molecular formulas give ratios of atoms in molecule
    B. Converting moles of atoms to moles of molecules
    C. Converting grams of atoms to grams of molecules
6.6 Mass Percent Composition of Compounds
    <u>Learning Objective:</u> Use mass percent composition as a conversion factor.
    A. Percentage of molecule's total mass that is due to element X
    B. Convert between mass of element and mass of compound

6.7 Mass Percent Composition from a Chemical Formula
    Learning Objective: Determine mass percent composition from a chemical formula.
6.8 Calculating Empirical Formulas for Compounds
    Learning Objective: Determine an empirical formula from experimental data.
    Learning Objective: Calculate an empirical formula from reaction data.
    A. Empirical formula is the smallest whole-number ratio of atoms
    B. Can be different from molecular formula
    C. Many different compounds may have the same empirical formula
6.9 Calculating Molecular Formulas for Compounds
    Learning Objective: Calculate a molecular formula from an empirical formula and molar mass.
    A. Empirical formula molar mass
    B. Compare molar mass to empirical formula mass

Chemical Principle Teaching Ideas

The Mole Concept
    To reinforce the concept of a mole, it helps to use several real-world examples involving moles. Calculating the mass of a pair, a dozen, and a mole of M&Ms can help a student understand how a mole is just a number, nothing else.

Chemical Formulas and Chemical Composition
    Explain that when we are talking about making large things, such as cars, bicycles, and houses, we are concerned not with the mass of each component, but with the actual number of each reactant. In the chemistry lab, we cannot count the actual number of atoms in a beaker, but we can measure the mass of the reactants. With the molar mass, we can then easily convert between the mass and the number of molecules (moles).

Empirical and Molecular Formulas from Laboratory Data
    Any time you can reinforce a concept with a simple demonstration, it will help the students with the material. Here is a valuable opportunity to show them exactly how it can be done in the laboratory.

Skill Builder Solutions

6.1. 8.83×10^{-2} mol Au $\times \dfrac{6.02 \times 10^{23} \text{ Au atoms}}{1 \text{ mol Au}} = 5.32 \times 10^{22}$ Au atoms

6.2. 2.78 mol S $\times \dfrac{32.07 \text{ g S}}{1 \text{ mol S}} = 89.1$ g S

6.3. 1.23×10^{24} He atoms $\times \dfrac{1 \text{ mol He}}{6.02 \times 10^{23} \text{ He atoms}} \times \dfrac{4.00 \text{ g He}}{1 \text{ mol He}} = 8.17$ g He

6.4 1.18 g $NO_2 \times \dfrac{1 \text{ mol } NO_2}{46.01 \text{ g } NO_2} = 2.56 \times 10^{-2}$ mol NO_2

6.5. 3.64 g $H_2O \times \dfrac{1 \text{ mol } H_2O}{18.01 \text{ g } H_2O} \times \dfrac{6.02 \times 10^{23} \text{ H}_2\text{O molecules}}{1 \text{ mol } H_2O} = 1.22 \times 10^{23}$ H_2O molecules

6.6. $1.4 \text{ mol H}_2\text{SO}_4 \times \dfrac{4 \text{ mol O}}{1 \text{ mol H}_2\text{SO}_4} = 5.6 \text{ mol O}$

6.7. $5.8 \text{ g NaHCO}_3 \times \dfrac{1 \text{ mol NaHCO}_3}{84.01 \text{ g NaHCO}_3} \times \dfrac{3 \text{ mol O}}{1 \text{ mol NaHCO}_3} \times \dfrac{16.00 \text{ g O}}{1 \text{ mol O}} = 3.3 \text{ g O}$

Plus. $7.20 \text{ g Al}_2(\text{SO}_4)_3 \times \dfrac{1 \text{ mol Al}_2(\text{SO}_4)_3}{342.17 \text{ g Al}_2(\text{SO}_4)_3} \times \dfrac{12 \text{ mol O}}{1 \text{ mol Al}_2(\text{SO}_4)_3} \times \dfrac{16.00 \text{ g O}}{1 \text{ mol O}} = 4.04 \text{ g O}$

6.8. $22 \text{ g NaCl} \times \dfrac{39 \text{ g Na}}{100 \text{ g NaCl}} = 8.6 \text{ g Na}$

6.9. mass % O in acetic acid $= \dfrac{2 \times 16.00 \text{ g O}}{\text{mass of 1 mol HC}_2\text{H}_3\text{O}_2} \times 100\% = \dfrac{32.00 \text{ g}}{60.06 \text{ g}} \times 100\% = 53.28\%$

6.10. $165 \text{ g C} \times \dfrac{1 \text{ mol C}}{12.01 \text{ g C}} = 13.7 \text{ mol C}$

$27.8 \text{ g H} \times \dfrac{1 \text{ mol H}}{1.01 \text{ g H}} = 27.5 \text{ mol H}$

$220.2 \text{ g O} \times \dfrac{1 \text{ mol O}}{16.00 \text{ g O}} = 13.76 \text{ mol O}$

$C_{13.7}H_{27.6}O_{13.8} \rightarrow C_{\frac{13.7}{13.7}}H_{\frac{27.6}{13.7}}O_{\frac{13.76}{13.7}} \rightarrow CH_2O$

6.11. $75.69 \text{ g C} \times \dfrac{1 \text{ mol C}}{12.01 \text{ g C}} = 6.302 \text{ mol C}$

$8.80 \text{ g H} \times \dfrac{1 \text{ mol H}}{1.01 \text{ g H}} = 8.71 \text{ mol H}$

$15.51 \text{ g O} \times \dfrac{1 \text{ mol O}}{16.00 \text{ g O}} = 0.9694 \text{ mol O}$

$C_{6.302}H_{8.73}O_{0.9694} \rightarrow C_{\frac{6.302}{0.9694}}H_{\frac{8.73}{0.9694}}O_{\frac{0.9694}{0.9694}} \rightarrow C_{6.5}H_9O \rightarrow C_{13}H_{18}O_2$

6.12. Mass O = mass oxide − mass Cu = 1.95 g − 1.56 g = 0.39 g O

$1.56 \text{ g Cu} \times \dfrac{1 \text{ mol Cu}}{63.55 \text{ g Cu}} = 0.0245 \text{ mol Cu}$

$0.39 \text{ g O} \times \dfrac{1 \text{ mol O}}{16.00 \text{ g O}} = 0.0244 \text{ mol O}$

$Cu_{0.0245}O_{0.0244} \rightarrow Cu_{\frac{0.0245}{0.0244}}O_{\frac{0.0245}{0.0244}} \rightarrow CuO$

6.13. C_2H_5 weighs $2 \times 12.01 + 5 \times 1.01 = 29.07$ g.

$\dfrac{58.12}{29.07} \approx 2$, so the molecular formula is $2(C_2H_5)$ or C_4H_{10}.

Plus. $39.97 \text{ g C} \times \dfrac{1 \text{ mol C}}{12.01 \text{ g C}} = 3.328$ mol C

$13.41 \text{ g H} \times \dfrac{1 \text{ mol H}}{1.01 \text{ g H}} = 13.3$ mol H

$46.62 \text{ g N} \times \dfrac{1 \text{ mol N}}{14.01 \text{ g N}} = 3.328$ mol N

$C_{3.328}H_{13.3}N_{3.328} \rightarrow C_{\frac{3.328}{3.328}}H_{\frac{13.3}{3.328}}N_{\frac{3.328}{3.328}} \rightarrow CH_4N$

CH_4N has a mass of ≈ 30, $\dfrac{60.10}{30} \approx 2$, so the molecular formula is $2(CH_4N) = C_2H_8N_2$

Suggested Demonstrations

Burning of Magnesium, *Chemical Demonstrations* 1:38, Shakhashiri, B. Z. University of Wisconsin Press, 1983.

Reaction of Sodium and Chlorine, *Chemical Demonstrations* 1:61, Shakhashiri, B. Z. University of Wisconsin Press, 1983.

Relative Velocity of Sound Propagation: Musical Molecular Weights, *Chemical Demonstrations* 2:88, Shakhashiri, B. Z. University of Wisconsin Press, 1985.

Guided Inquiry Ideas

Below are a few example questions that students answer in the guided inquiry activities provided in the Guided Activity Workbook.

How many nails are in 3.5 doz nails? Show all your work using conversion factors correctly.

How many He atoms are in 3.5 mol He atoms? Show all your work using conversion factors correctly.

Using grammatically correct English sentences, describe how a mole is similar to a dozen. How is it different?

A certain mysterious compound was found to contain 3.0 g hydrogen and 24 g oxygen.
    What is the mass percent of hydrogen in the compound? Of oxygen?

    Calculate the moles of hydrogen and oxygen in the sample. Are there more oxygen atoms or hydrogen atoms in the compound?

Chemical Reactions

Chapter Overview

This chapter covers many aspects of chemical reactions, including how to recognize and classify them. Several examples of the main categories of reaction are presented and discussed. The concept of the balanced chemical equation is introduced and manipulated.

Lecture Outline

7.1 Grade School Volcanoes, Automobiles, and Laundry Detergents
7.2 Evidence of a Chemical Reaction
    Learning Objective: Identify evidence of a chemical reaction.
    A. Color change
    B. Formation of solid or gas
    C. Heat absorption or emission
    D. Light emission
7.3 The Chemical Equation
    Learning Objective: Identify balanced chemical equations.
    A. Reactants go to products
    B. Cannot create or destroy atoms in a chemical equation
    C. Reaction must be balanced
7.4 How to Write Balanced Chemical Equations
    Learning Objective: Write balanced chemical equations.
    A. Write reactant and product species, including physical states
    B. If an element occurs only once on each side of the reaction, balance that first
    C. Balance free elements last
    D. Clear non integer coefficients
7.5 Aqueous Solutions and Solubility: Compounds Dissolved in Water
    Learning Objective: Determine whether a compound is soluble.
    A. Aqueous: homogeneous mixture of substance in water
    B. Soluble substances dissolve in water
    C. Insoluble substances do not dissolve in water
    D. Solubility rules
7.6 Precipitation Reactions: Reactions in Aqueous Solution that Form a Solid
    Learning Objective: Predict and write equations for precipitation reactions.
    A. Aqueous ions combine to form insoluble products
    B. Predicting precipitation reactions
        1. Predict potential solid products
        2. Check solubility rules
        3. If all potential solid products are soluble, there is no reaction

7.7 Writing Chemical Equations for Reactions in Solutions: Molecular, Complete Ionic, and Net Ionic Equations
    Learning Objective: Write molecular, complete ionic, and net ionic equations.
    A. Molecular equation shows complete formulas for every compound
    B. Complete ionic equation shows species as they appear in solution
    C. Net ionic equation shows only the species that actually change during reaction

7.8 Acid–Base and Gas-Evolution Reactions
    Learning Objective: Identify and write equations for acid–base reactions.
    Learning Objective: Identify and write equations for gas-evolution reactions.
    A. Acid + base → neutral salt + water
    B. Gas-evolution reactions have different possible products

7.9 Oxidation–Reduction Reactions
    Learning Objective: Identify redox reactions.
    Learning Objective: Identify and write equations for combustion reactions.
    A. Also known as redox reactions
    B. Reaction in which
        1. A substance reacts with elemental oxygen
        2. A metal reacts with a nonmetal
        3. One substance transfers electrons to another substance
    C. Combustion reactions
        1. An important type of oxidation–reduction reaction
        2. Substance reacts with O_2 to form oxygen-containing species
        3. Emits heat energy

7.10 Classifying Chemical Reactions
    Learning Objective: Classify chemical reactions.
    A. Synthesis (aka combination) reactions
    B. Decomposition reactions
    C. Displacement reactions
    D. Double-displacement reactions

Chemical Principle Teaching Ideas

Chemical Equations
    Start with a very simple example, such as 2 wheels + 1 bike → 1 bicycle. Emphasize that each species should have a coefficient in front of it, letting the reader know that the equations represent nothing more than a recipe. It says nothing about the actual number of species in the system. Remind students that you can multiply all the coefficients by the same value and the equation will still be balanced.

Aqueous Solutions and Solubility
    All solutions that a student encounters have dissolved solids, even tap water, soda, and fruit juice. Use these and other common examples to help students understand the solubility of various species.

Types of Reactions and How to Classify Them
There are many examples of each kind of reaction either that the students are familiar with or that can be done safely in the lecture room. Showing students an example of each to memorize will help them when they are trying to classify new reactions. Make sure to mention that a chemical reaction can be of more than one type.

Skill Builder Solutions

7.1. a. During combustion, butane turns into carbon dioxide and water, giving off heat and light. Since the identity of the substances is changing, it is a chemical reaction.
b. During evaporation, butane exhibits a phase change, from liquid to gas, so it is not a chemical reaction.
c. The combustion of wood involves long organic molecules combining with oxygen to yield water and carbon dioxide. The substances are changing identity, so it is a chemical reaction.
d. Sublimation is only a phase change and does not involve the rearrangement of atoms; therefore, it is not a chemical reaction.

7.2. $Cr_2O_3(s) + C(s) \rightarrow Cr(s) + CO_2(g)$
Balance the chromium atoms first, so we need a 1 in front of the $Cr_2O_3(s)$ and a 2 in front of the $Cr(s)$:
$Cr_2O_3(s) + C(s) \rightarrow 2\ Cr(s) + CO_2(g)$
There are 3 oxygen atoms on the left, so we need a 3/2 in front of the $CO_2(g)$:
$Cr_2O_3(s) + C(s) \rightarrow 2\ Cr(s) + 3/2\ CO_2(g)$
There are 3/2 carbons on the right, so we need a 3/2 in front of $C(s)$:
$Cr_2O_3(s) + 3/2\ C(s) \rightarrow 2\ Cr(s) + 3/2\ CO_2(g)$
To clear the fractions, multiply everything by 2:
$2\ Cr_2O_3(s) + 3\ C(s) \rightarrow 4\ Cr(s) + 3\ CO_2(g)$

7.3. $C_4H_{10}(g) + O_2(g) \rightarrow CO_2(g) + H_2O(g)$
Balance the carbon atoms first by putting a 1 in front of C_4H_{10} and a 4 in front of CO_2:
$C_4H_{10}(g) + O_2(g) \rightarrow 4\ CO_2(g) + H_2O(g)$
There are 10 hydrogen atoms on the left, so we need 5 H_2O molecules on the right:
$C_4H_{10}(g) + O_2(g) \rightarrow 4\ CO_2(g) + 5\ H_2O(g)$
There are 13 oxygen atoms on the right, so to balance the equation, we need a 13/2 in front of $O_2(g)$:
$C_4H_{10}(g) + 13/2\ O_2(g) \rightarrow 4\ CO_2(g) + 5\ H_2O(g)$
To remove the fractional coefficients, multiply everything by 2:
$2\ C_4H_{10}(g) + 13\ O_2(g) \rightarrow 8\ CO_2(g) + 10\ H_2O(g)$

7.4. Skeletal equation found by switching partners and making neutral products:
$Pb(C_2H_3O_2)_2(aq) + KI(aq) \rightarrow PbI_2(s) + KC_2H_3O_2(aq)$
Pb is balanced as it now appears.
$Pb(C_2H_3O_2)_2(aq) + KI(aq) \rightarrow PbI_2(s) + KC_2H_3O_2(aq)$

There are 2 acetate ions "fixed" on the left side, so we will balance those next by adding a 2 in front of the potassium acetate:
$Pb(C_2H_3O_2)_2(aq) + KI(aq) \rightarrow PbI_2(s) + 2\ KC_2H_3O_2(aq)$
There are 2 iodine atoms on the right, so we need two on the left for the balanced reaction:
$Pb(C_2H_3O_2)_2(aq) + 2\ KI(aq) \rightarrow PbI_2(s) + 2\ KC_2H_3O_2(aq)$

7.5. $HCl(g) + O_2(g) \rightarrow H_2O(l) + Cl_2(g)$
Cl only appears in one species on each side, so we will balance that first. There are two on the right hand side and one on the left, so we need to add a 2 to the HCl:
$2\ HCl(g) + O_2(g) \rightarrow H_2O(l) + Cl_2(g)$
There is 1 oxygen on the right, so we need a 1/2 in front of O_2 on the left:
$2\ HCl(g) + 1/2\ O_2(g) \rightarrow H_2O(l) + Cl_2(g)$
Remove the fraction by multiplying everything by 2:
$4\ HCl(g) + O_2(g) \rightarrow 2\ H_2O(l) + 2\ Cl_2(g)$

7.6. a. Insoluble; most sulfides are insoluble, and Cu salts are not an exception (Table 7.2).
b. Soluble; iron (II) sulfate is soluble because sulfates are soluble, and iron is not an exception.
c. Insoluble; lead(II) carbonate is insoluble, as are most lead salts.
d. Soluble; ammonium chloride is soluble, as are all other ammonium salts.

7.7. The two potential products are $Ni(OH)_2$ and KBr. KBr is soluble, but $Ni(OH)_2$ is not, giving- $NiBr_2(aq) + KOH(aq) \rightarrow Ni(OH)_2(s) + KBr(aq)$, which, when balanced, yield $NiBr_2(aq) + 2\ KOH(aq) \rightarrow Ni(OH)_2(s) + 2\ KBr(aq)$.

7.8. The two potential products are NH_4NO_3 and $FeCl_3$, both of which are soluble, so no precipitation occurs.

7.9. Potassium sulfate: K_2SO_4 ; Strontium nitrate: $Sr(NO_3)_2$. Potential products are KNO_3 and $SrSO_4$. The $SrSO_4$ is an insoluble product (exception to sulfate rule), so the reaction is
$K_2SO_4(aq) + Sr(NO_3)_2(aq) \rightarrow KNO_3(aq) + SrSO_4(s)$.
Balancing gives
$K_2SO_4(aq) + Sr(NO_3)_2(aq) \rightarrow 2\ KNO_3(aq) + SrSO_4(s)$.

7.10. We must first break up all the aqueous species, giving the overall ionic equation
$2\ H^+(aq) + 2\ Br^-(aq) + Ca^{2+}(aq) + 2\ OH^-(aq) \rightarrow 2\ H_2O(l) + Ca^{2+}(aq) + 2\ Br^-(aq)$.
In this reaction, bromine and calcium are spectators, so the net ionic equation is
$2\ H^+(aq) + 2\ OH^-(aq) \rightarrow 2\ H_2O(l)$.
Now divide all coefficients by 2 to get the final balanced net ionic equation.
$H^+(aq) + OH^-(aq) \rightarrow H_2O(l)$.

7.11. This is a reaction between an acid and a base, so the products are a salt and water. The initial reaction is thus
$H_2SO_4(aq) + 2\ KOH(aq) \rightarrow 2\ H_2O(l) + K_2SO_4(aq)$.

We break up all aqueous species to get the overall ionic equation:
2 H$^+$(aq) + SO$_4^{2-}$(aq) + 2 K$^+$(aq) + 2 OH$^-$(aq) → 2 H$_2$O(l) + 2 K$^+$(aq) + SO$_4^{2-}$(aq)
Removing the spectator ions K$^+$ and SO$_4^{2-}$, we get
2 H$^+$(aq) + 2 OH$^-$(aq) → 2 H$_2$O(l).
Now divide all coefficients by 2 to get the final balanced net ionic equation.
H$^+$(aq) + OH$^-$(aq) → H$_2$O(l).

7.12. By simply switching partners, we get the following reaction:
2 HBr(aq) + K$_2$SO$_3$(aq) → H$_2$SO$_3$(aq) + 2 KBr(aq).
The H$_2$SO$_3$(aq) will decompose into H$_2$O(l) and SO$_2$(g), giving
2 HBr(aq) + K$_2$SO$_3$(aq) → H$_2$O(l) + SO$_2$(g) + 2 KBr(aq).

Plus. We must first write the complete ionic equation:
2 H$^+$(aq) + 2 Br$^-$(aq) + 2 K$^+$(aq) + SO$_3^{2-}$(aq) → H$_2$O(l) + SO$_2$(g) + 2 K$^+$(aq) + 2 Br$^-$(aq).
Removing the spectator ions, we have the following for the net ionic equation:
2 H$^+$(aq) + SO$_3^{2-}$(aq) → H$_2$O(l) + SO$_2$(g).

7.13. a. Lithium goes from an oxidation state of 0 to +1, losing 1 electron, while chlorine goes from an oxidation state of 0 to −1, gaining an electron, so it is a redox reaction.
b. The oxidation state of Al goes from 0 to +3, losing 3 electrons, while each Sn^{2+} gains 2 electrons, so it is a redox reaction.
c. This is a precipitation reaction, and the oxidation states of all species remain constant. Thus, it is not a redox reaction.
d. Carbon goes from an oxidation state of 0 to +4, while carbon goes from 0 to −2, so the reaction is a redox type.

7.14. The skeletal equation of pentane combining with oxygen to form carbon dioxide and water is
C$_5$H$_{12}$(l) + O$_2$(g) → CO$_2$(g) + H$_2$O(g)
Balancing this equation gives
C$_5$H$_{12}$(l) + 8 O$_2$(g) → 5 CO$_2$(g) + 6 H$_2$O(g)

Plus. The skeletal equation of liquid propanol burning in the presence of oxygen to form carbon dioxide and water is
C$_3$H$_7$OH(l) + O$_2$(g) → CO$_2$(g) + H$_2$O(g)
Balancing this equation gives
2 C$_3$H$_7$OH(l) + 9 O$_2$(g) → 6 CO$_2$(g) + 8 H$_2$O(g)

7.15. a. Al changes position with the hydrogen that is bonded with the phosphate, so the reaction is a single-displacement type.
b. Both partners are switched in this reaction, so it is a double-displacement type.
c. Two species combine to form one species, so this is a synthesis reaction.
d. One species becomes two separate species, so this is a decomposition reaction.

Suggested Demonstrations

In a clear plastic soda bottle, place a small amount of vinegar. Put some baking soda into a balloon, attach the balloon over the neck of the bottle, and then invert the balloon to combine the reactants. Explain all of the reactions going on and classify them accordingly.

Read the ingredient label of various liquids with which the students are familiar, such as soda, and talk about the dissolved chemicals, where they come from, and how they get into the solution.

Fill a test tube with a very dilute solution of Cu^{2+} ion and place a small amount of solid Zn inside and shake. Explain why the blue color is leaving the solution as the electrons are transferred from the Zn to the Cu.

Burning of Magnesium, *Chemical Demonstrations* 1:38, Shakhashiri, B. Z. University of Wisconsin Press, 1983.

Combustion of Magnesium in Carbon Dioxide, *Chemical Demonstrations* 1:90, Shakhashiri, B. Z. University of Wisconsin Press, 1983.

Guided Inquiry Ideas

Below are a few example questions that students answer in the guided inquiry activities provided in the Guided Activity Workbook.

Some antacids are used by dropping a tablet into a glass of water, which results in significant fizzing, and then drinking the water. Is there any evidence to suggest there might be a chemical reaction involved? What is the evidence?

Can you name any process that results in a gas being produced that is not a chemical reaction?

Suggest one or two phenomena from your daily experience that you believe to be chemical reactions. What is your evidence that a chemical reaction is occurring?

The reaction of zinc with iron (II) chloride is an example of an oxidation–reduction reaction:
$$Zn(s) + FeCl_2(aq) \rightarrow ZnCl_2(aq) + Fe(s)$$
What is the spectator ion in the above reaction?

In order for the positive charge to switch from iron to zinc, which particle must be moving: protons, electrons, or neutrons? Explain your answer.

Quantities in Chemical Reactions

8

Chapter Overview

Chapter 8 incorporates more in-depth calculations involved in the study of chemistry. Conversions between amounts of different species using a chemical equation will be discussed. A simple example involving pancakes is used to introduce a traditionally difficult concept. As this concept is one that students will use again later in the text, the content here is very important.

Lecture Outline

8.1 Climate Change: Too Much Carbon Dioxide
    A. Greenhouse gases
    B. Global warming

8.2 Making Pancakes: Relationships between Ingredients
    <u>Learning Objective:</u> Recognize the numerical relationship between chemical quantities in a balanced chemical equation.
    A. Stoichiometry is the relationship between chemical quantities in a balanced equation
    B. Similar to a cooking recipe that can be doubled or halved

8.3 Making Molecules: Mole-to-Mole Conversions
    <u>Learning Objective:</u> Carry out mole-to-mole conversions between reactants and products based on the numerical relationship between chemical quantities in a balanced chemical equation.
    A. Knowing the amount of any one ingredient allows calculation of any other
    B. Need balanced chemical equation (recipe) to do conversions

8.4 Making Molecules: Mass-to-Mass Conversions
    <u>Learning Objective:</u> Carry out mass-to-mass conversions between reactants and products in a balanced chemical equation and molar masses.
    A. Chemical equation gives ratios of moles of species, not mass
    B. Mass A → moles A → moles B → mass B

8.5 More Pancakes: Limiting Reactant, Theoretical Yield, and Percent Yield
    <u>Learning Objective:</u> Calculate limiting reactant, theoretical yield, and percent yield in a balanced chemical equation.
    A. The limiting reactant is the reactant that is completely consumed
    B. The theoretical yield is the maximum amount of product that can be made
    C. Actual yield is the amount of product actually produced
    D. Percent yield is a measure of the actual yield compared to the theoretical yield

8.6 Limiting Reactant, Theoretical Yield, and Percent Yield from Initial Masses of Reactants
   Learning Objective: Calculate limiting reactant, theoretical yield, and percent yield in a balanced chemical equation.
   A. Convert masses to number of moles
   B. Compare the possible number of product moles to determine limiting reactant
   C. Use limiting reactant amount to determine theoretical yield
8.7 Enthalpy: A Measure of Heat Evolved or Absorbed in a Reaction
   Learning Objective: Calculate the amount of thermal energy emitted or absorbed by a chemical reaction.
   A. $\Delta H_{rxn} < 0$ if exothermic
   B. $\Delta H_{rxn} > 0$ if endothermic
   C. Magnitude of ΔH_{rxn} dependent on amount of reactants

Chemical Principle Teaching Ideas

Stoichiometry
   The use of a real world example here allows for further solidification of the concept in the students' minds. Any number of examples can be used, including a bicycle shop, cooking, or the building of a house.

Limiting Reactant, Theoretical Yield, and Percent Yield
   Use a real-world example as a sample calculation of limiting reactant, theoretical yield, and percent yield. For example, given 5 slices of ham and 4 slices of bread, how many ham sandwiches can be made? Then, do a very simple chemical example; for example, given 1.0 g of H_2 and 1.0 g of O_2, what are the limiting reactant and theoretical yield?

Skill Builder Solutions

8.1. $24.6 \text{ mol } O_2 \times \dfrac{2 \text{ mol } H_2O}{1 \text{ mol } O_2} = 49.2 \text{ mol } H_2O$

8.2. $5.50 \text{ g Mg(OH)}_2 \times \dfrac{1 \text{ mol Mg(OH)}_2}{58.33 \text{ g Mg(OH)}_2} \times \dfrac{2 \text{ mol HCl}}{1 \text{ mole Mg(OH)}_2} \times \dfrac{36.46 \text{ g HCl}}{1 \text{ mol HCl}} = 6.88 \text{ g HCl}$

8.3. $2.6 \times 10^3 \text{ kg SO}_2 \times \dfrac{1000 \text{ g SO}_2}{1 \text{ kg SO}_2} \times \dfrac{1 \text{ mol SO}_2}{64.07 \text{ g SO}_2} \times \dfrac{2 \text{ mol H}_2SO_4}{2 \text{ mol SO}_2} \times \dfrac{98.09 \text{ g H}_2SO_4}{1 \text{ mol H}_2SO_4} \times \dfrac{1 \text{ kg H}_2SO_4}{1000 \text{ g H}_2SO_4} = 4.0 \times 10^3 \text{ kg H}_2SO_4$

8.4. $4.8 \text{ mol Na} \times \dfrac{2 \text{ mol NaF}}{2 \text{ mol Na}} = 4.8 \text{ mol NaF}$

$2.6 \text{ mol F}_2 \times \dfrac{2 \text{ mol NaF}}{1 \text{ mol F}_2} = 5.2 \text{ mol NaF}$

Therefore, Na is the limiting reagent, and our theoretical yield is 4.8 mol of NaF.

8.5. $25.2 \text{ g N}_2 \times \dfrac{1 \text{ mol N}_2}{28.02 \text{ g N}_2} \times \dfrac{2 \text{ mol NH}_3}{1 \text{ mol N}_2} \times \dfrac{17.04 \text{ g NH}_3}{1 \text{ mol NH}_3} = 30.7 \text{ g NH}_3$

$8.42 \text{ g H}_2 \times \dfrac{1 \text{ mol H}_2}{2.02 \text{ g H}_2} \times \dfrac{2 \text{ mol NH}_3}{3 \text{ mol H}_2} \times \dfrac{17.04 \text{ g NH}_3}{1 \text{ mol NH}_3} = 47.4 \text{ g NH}_3$

Therefore, N_2 is the limiting reagent and the theoretical yield is 30.7 g of NH_3.

Plus. $31.5 \text{ kg N}_2 \times \dfrac{1000 \text{ g N}_2}{1 \text{ kg N}_2} \times \dfrac{1 \text{ mol N}_2}{28.02 \text{ g N}_2} \times \dfrac{2 \text{ mol NH}_3}{1 \text{ mol N}_2}$

$\times \dfrac{17.03 \text{ g NH}_3}{1 \text{ mol NH}_3} \times \dfrac{1 \text{ kg NH}_3}{1000 \text{ g NH}_3} = 38.3 \text{ kg NH}_3$

$5.22 \text{ kg H}_2 \times \dfrac{1000 \text{ g H}_2}{1 \text{ kg H}_2} \times \dfrac{1 \text{ mol H}_2}{2.02 \text{ g H}_2} \times \dfrac{2 \text{ mol NH}_3}{3 \text{ mol H}_2}$

$\times \dfrac{17.04 \text{ g NH}_3}{1 \text{ mol NH}_3} \times \dfrac{1 \text{ kg NH}_3}{1000 \text{ g NH}_3} = 29.4 \text{ g NH}_3$

Therefore, the H_2 is the limiting reagent, and the theoretical yield is 29.4 kg of NH_3.

8.6. $185 \text{ g Fe}_2\text{O}_3 \times \dfrac{1 \text{ mol Fe}_2\text{O}_3}{159.70 \text{ g Fe}_2\text{O}_3} \times \dfrac{2 \text{ mol Fe}}{1 \text{ mol Fe}_2\text{O}_3} \times \dfrac{55.85 \text{ g Fe}}{1 \text{ mol Fe}} = 129 \text{ g Fe}$

$95.3 \text{ g CO} \times \dfrac{1 \text{ mol CO}}{28.01 \text{ g CO}} \times \dfrac{2 \text{ mol Fe}}{3 \text{ mol CO}} \times \dfrac{55.85 \text{ g Fe}}{1 \text{ mol Fe}} = 127 \text{ g Fe}$

The limiting reagent is the CO, with a theoretical yield of 127 g Fe.

The percent yield is thus $\dfrac{87.4 \text{ g Fe}}{127 \text{ g Fe}} \times 100\% = 68.8\%$

8.7 $155 \text{ g NH}_3 \times \dfrac{1 \text{ mol NH}_3}{17.04 \text{ g NH}_3} \times \dfrac{-906 \text{ kJ}}{4 \text{ mol NH}_3} = -2.06 \times 10^3 \text{ kJ}$

Plus. $1.5 \times 10^3 \text{ kJ} \times \dfrac{1 \text{ mol C}_4\text{H}_{10}}{2658 \text{ kJ}} \times \dfrac{58.14 \text{ g C}_4\text{H}_{10}}{1 \text{ mol C}_4\text{H}_{10}} = 33. \text{ g C}_4\text{H}_{10}$

$1.5 \times 10^3 \text{ kJ} \times \dfrac{4 \text{ mol CO}_2}{2658 \text{ kJ}} \times \dfrac{44.01 \text{ g CO}_2}{1 \text{ mol CO}_2} = 99. \text{ g CO}_2$

Suggested Demonstrations

Come in to class with all the ingredients for making ham sandwiches. Put the balanced equation on the board and do some sample calculations involving limiting reactants, theoretical yield, and percent yield.

Ignite a balloon containing only H_2 gas, and then ignite a balloon containing H_2 and O_2 in their stoichiometric amounts. This is a much larger explosion and can explain the idea of stoichiometry. This experiment is dangerous, and extreme caution should be exercised.

Burn a piece of Mg ribbon in open air, place a burning piece of Mg ribbon in a flask, and then seal the flask. The Mg will stop burning before it is consumed. Explain why this occurs. This experiment is dangerous, and extreme caution should be exercised.

Guided Inquiry Ideas

Below are a few example questions that students answer in the guided inquiry activities provided in the Guided Activity Workbook.

Frequently, chemists will want to predict how much product can be formed from a particular amount of reactant. Can the coefficients in the balanced chemical reaction be used directly to convert from mass product to mass reactant? Explain.

If not mass, what can the coefficients in a balanced chemical equation convert between?

Consider the following reaction: $Ti(s) + 2\ Cl_2(g) \rightarrow TiCl_4(s)$
We begin with 1.8 mol Ti and 3.2 mol Cl_2.
    What is the theoretical yield of $TiCl_4$ in moles? Explain your answer.

    What is the limiting reactant?

    Is the substance you have the least amount of always the limiting reactant? If not, how do you find the limiting reactant?

Electrons in Atoms and the Periodic Table

9

Chapter Overview

This chapter gives us the background of atomic theory. All atomic theories attempt to explain experimental observations. The current quantum-mechanical model is much more complicated than the Bohr model. The periodic trends based on electronic structure are also discussed.

Lecture Outline

9.1 Blimps, Balloons, and Models for the Atom
    A. Some atoms have similar properties
    B. Atomic structure ideas have evolved over the last century
9.2 Light: Electromagnetic Radiation
    <u>Learning Objective:</u> Understand and explain the nature of electromagnetic radiation.
    A. Light is not matter; it is pure energy
    B. A photon is a particle of light
        1. Wavelength is distance between adjacent wave crests
        2. Frequency is the number of cycles that pass a point per unit time
9.3 The Electromagnetic Spectrum
    <u>Learning Objective:</u> Predict the relative wavelength, energy, and frequency of different types of light.
    A. Gamma rays
        1. High energy
        2. Low frequency
    B. X-rays
    C. Ultraviolet
    D. Visible
        1. Very narrow band
        2. Does not damage biological molecules
    E. Infrared
    F. Microwaves
    G. Radio waves
9.4 The Bohr Model: Atoms with Orbits
    <u>Learning Objective:</u> Understand and explain the key characteristics of the Bohr model of the atom.
    A. Electrons exist in quantized orbits
    B. Electrons get excited to a higher-energy level by absorbing energy
    C. Electrons emit light when relaxing from higher levels
    D. Absorbed and emitted light correspond to energy difference between states

9.5 The Quantum-Mechanical Model: Atoms with Orbitals
  Learning Objective: Understand and explain the key characteristics of the quantum-mechanical model of the atom.
  A. Orbitals instead of orbits
  B. Electrons do not move in predictable paths
  C. Probability maps
9.6 Quantum-Mechanical Orbits and Electron Configurations
  Learning Objective: Write electron configurations and orbital diagrams for atoms.
  A. Quantum numbers
    1. Principal shell
    2. Subshell
  B. Electronic configuration
    1. Gives orbital "address" of each electron in the atom
    2. Li: $1s^2 2s^1$
  C. How the electrons fill in the atom's orbitals
    1. Electron spin
    2. Pauli exclusion principle
    3. Hund's rule
9.7 Electron Configurations and the Periodic Table
  Learning Objective: Identify valence electrons and core electrons.
  Learning Objective: Write electron configurations for elements based on their positions in the periodic table.
  A. Valence electrons
  B. Core electrons
  C. Writing electronic configurations
    1. Noble gas core
    2. Orbital assignments of valence electrons
    3. Highest n value is equal to row value
    4. d electron group is equal to row number - 1
9.8 The Explanatory Power of the Quantum-Mechanical Model
  Learning Objective: Explain why the chemical properties of elements are largely determined by the number of valence electrons they contain.
  A. Valence electrons dictate chemical reactivity
  B. Elements in the same column have similar reactivity
9.9 Periodic Trends: Atomic Size, Ionization Energy, and Metallic Character
  Learning Objective: Identify and understand periodic trends in atomic size, ionization energy, and metallic character.
  A. Atomic size
    1. Decreases as you move to the right along a row
    2. Increases as you go down a column
  B. Ionization energy
    1. Increases as you move to the right along a row
    2. Decreases as you go down a column
  C. Metallic character
    1. Decreases as you move to the right
    2. Increases as you go down

Chemical Principle Teaching Ideas

Light

Emphasize for the students that all light behaves like the light we see with our eyes. Radio waves, infrared radiation, and gamma rays are just different "colors" of light. Our eyes are radiation detectors of a very narrow frequency range. Our bodies can be thought of as infrared detectors. The molecules in our skin absorb the infrared radiation and get excited, making us feel warm. However, radio waves pass through us with no reaction.

Bohr Model

As a first model of the nuclear atom, the Bohr model explained many observed properties and was a good starting point to understanding the nature of electronic structure. However, the electrons do not follow circular orbits. This is another opportunity to emphasize that chemistry as a science is still evolving and that many questions remain to be answered.

The Quantum-Mechanical Model

The idea of not knowing the exact locations of the electrons is sometimes difficult for students. Try blowing up a latex balloon with helium. Have them think about one particular helium atom in the balloon. We know it is inside the balloon, but we do not know exactly where inside. However, the balloon will slowly deflate, so sometimes the helium atom is outside the balloon. You are more than 90% sure the atom is inside the balloon, but there is a small chance it is out. This is the same idea as the probability map of an electron orbital.

The Periodic Table

Not only do species in the same column or group have similar properties, but elements in electron blocks have similar group properties. Giving examples of these trends will help students down the road to see the periodic table as the ultimate cheat sheet, showing us trends, properties, and how the world interacts at the atomic level.

Skill Builder Solutions

9.1. a. Blue (~390 nm), green, red (~680 nm)
  b. Red, green, blue. Frequency is inversely proportional to wavelength; therefore, the order is reversed.
  c. Red, green, blue. Energy per photon increases with increasing frequency.

9.2. a. Al has 13 electrons. Putting them in order gives $1s^22s^22p^63s^23p^1$ or [Ne] $3s^23p^1$
  b. Br has 35 electrons. Putting them in order gives $1s^22s^22p^63s^23p^64s^23d^{10}4p^5$ or [Ar] $4s^23d^{10}4p^5$.
  c. Sr has 38 electrons. Putting them in order gives $1s^22s^22p^63s^23p^64s^23d^{10}4p^65s^2$ or [Kr] $5s^2$.

Plus. a. A neutral Al atom has 13 electrons, so a +3 ion has 13 − 3 = 10 electrons, and the electron configuration is the same as Ne: $1s^22s^22p^6$.
  b. A neutral chlorine atom has 17 electrons, but adding one for the −1 charge gives 18, just like argon: $1s^22s^22p^63s^23p^6$.
  c. A neutral oxygen atom has 8 electrons. Adding 2 electrons to make it a −2 charge, the electron configuration is $1s^22s^22p^6$.

9.3. Argon has 18 electrons. Let us first draw the orbitals up to the 3p.

□	□	□ □ □	□	□ □ □
1s	2s	2p	3s	3p

We then distribute the 18 electrons in the orbitals, going from left to right remembering Hund's rule:

↑↓	↑↓	↑↓ ↑↓ ↑↓	↑↓	↑↓ ↑↓ ↑↓
1s	2s	2p	3s	3p

9.4. Chlorine has the electronic configuration $1s^2 2s^2 2p^6 3s^2 3p^5$. The valence electrons are those in the outermost principal shell, $3s^2 3p^5$. All other electrons are core electrons.

9.5. Tin has 50 electrons. We go back to the last noble gas, which is Kr, and start with that core. The next electron shell is the 5s, which is filled. Then the 4d orbital set completely fills. Finally, two 5p electrons are included in the configuration, giving $[Kr]5s^2 4d^{10} 5p^2$.

9.6. a. Lead and polonium are in the same row, and by following the periodic trend, lead has the larger atomic radius.
b. Rubidium and sodium are in the same group; therefore, the lower species has the larger radius, in this case rubidium.
c. Tin and bismuth are not in the same row or group, and their relative position in the periodic table cannot tell us which is larger in radius.
d. Fluorine and selenium are not in the same row or group, however, fluorine is to the right *and* is higher. Both the row and column trends indicate that fluorine is smaller, so Se is the larger atom.

9.7. a. Magnesium is higher on the periodic table and in the same group. By the periodic trend, it has a higher ionization energy.
b. Tellurium and indium are in the same row, so the right most species has the higher ionization energy, in this case tellurium.
c. By following the periodic trends alone, we cannot determine which has the higher ionization energy.
d. Fluorine and sulfur are not in the same row or group; however, the periodic trend indicates that as you move either up or right in the periodic table, the ionization energy increases. By both of these trends, fluorine has the higher ionization energy.

9.8. a. Indium is closer to the lower left corner of the periodic table, and the periodic trend indicates that it has the larger metallic character.
b. Gallium and tin are equally distant from the lower left corner of the periodic table; therefore, we cannot predict which has a more metallic character on the basis of position alone.
c. Bismuth is lower on the periodic table and in the same group as phosphorus. Therefore, it has a more metallic character.
d. Boron is to the left of nitrogen on the periodic table, and following the trend, it has a more metallic character.

Guided Inquiry Ideas

Below are a few example questions that students answer in the guided inquiry activities provided in the Guided Activity Workbook.

What is the value of n for the outermost principle shell of silicon?

Recall that the alkali metals (first column of the periodic table) all demonstrate similar chemical properties. How do their electron configurations compare?

The fluoride ion, F^-, has one more electron than a fluorine atom. Write the electron configurations for F and F^-.

Why is the electron configuration for sodium (Na, 11 electrons) $1s^2\ 2s^2\ 2p^6\ 3s^1$ instead of $1s^2\ 2s^2\ 2p^7$?

Explain why the fluoride ion is particularly stable and relatively inert.

Chemical Bonding

10

Chapter Overview

Understanding how atoms bond gives us the power of predicting chemical behavior. Drawing Lewis structures and predicting the resulting molecular shapes are discussed. The chemical interactions based on shape, including electronegativity and polarity, are also explained.

Lecture Outline

10.1 Bonding Models and AIDS Drugs
    A. Bonding theories attempt to explain actual structures
    B. Some theories work for some molecules but not for others
10.2 Representing Valence Electrons with Dots
    <u>Learning Objective:</u> Write Lewis structures for elements.
    A. Valence electrons are outermost shell electrons
    B. Lewis structure of an atom is simply chemical symbol + valence electrons around it
    C. Most atoms "want" eight electrons around them (octet rule)
    D. Exception: duet rule for hydrogen and helium
    E. A covalent bond is a shared pair of electrons to achieve an octet around each atom
10.3 Lewis Structures of Ionic Compounds: Electrons Transferred
    <u>Learning Objective:</u> Write Lewis structures for ionic compounds.
    <u>Learning Objective:</u> Use the Lewis model to predict the chemical formula of an ionic compound.
    A. Metal atom effectively gives valence electrons to the nonmetal
    B. Metal is positively charged, nonmetal is negatively charged
    C. Attraction of opposite charges constitutes the ionic bond
10.4 Covalent Lewis Structures: Electrons Shared
    <u>Learning Objective:</u> Write Lewis structures for covalent compounds.
    A. Two bonded nonmetals share electrons such that both get an octet (or duet)
    B. Both species get "credit" for all electrons in bond
    C. Two species can share two, four, or six electrons
        1. Two electrons shared is a single bond
        2. Four electrons shared is a double bond
        3. Six electrons shared is a triple bond
10.5 Writing Lewis Structures for Covalent Compounds
    <u>Learning Objective:</u> Write Lewis structures for covalent compounds.
    A. Write the correct skeletal structure for the molecule
    B. Calculate the total number of valence electrons
    C. Distribute electrons among atoms, giving each an octet (or duet)

D. If any atoms then lack an octet, form double or triple bonds as necessary
   E. Exceptions to the octet rule
      1. Duet rule for hydrogen
      2. Odd number of electrons
      3. Boron usually has only six electrons about it
      4. Some elements in the third row and beyond may have more than eight electrons around them
10.6 Resonance: Equivalent Lewis Structures for the Same Molecule
   Learning Objective: Write resonance structures.
   A. More than one possible Lewis structure
   B. True structure is the average of all Lewis structures
10.7 Predicting the Shapes of Molecules
   Learning Objective: Predict the shapes of molecules.
   A. VSEPR theory
      1. Electron groups repel each other
      2. Electron geometry
         a. Linear
         b. Trigonal planar
         c. Tetrahedral
   B. Predicting structures
      1. Draw the Lewis structure
      2. Count the number of electron groups
      3. Determine the number of bonding groups and lone pair groups
      4. Determine electron geometry and molecular geometry
10.8 Electronegativity and Polarity: Why Oil and Water Don't Mix
   Learning Objective: Determine whether a molecule is polar.
   A. Electronegativity
   B. Bond dipoles
   C. Dipole moment
   D. Polar and nonpolar bonds
   E. Polar and nonpolar molecules

Chemical Principle Teaching Ideas

Lewis Structure
   All covalent bonds involve the sharing of electrons. By understanding how the atoms bond to each other, the students can begin to understand why species react the way they do.

Molecular Shapes
   The most effective way for students to remember the various molecular shapes is to memorize one example of each. For example, remember that NH_3 is trigonal pyramidal and has three bonds and one lone pair. H_2O has two lone pairs and two bonds, and the geometry is bent.

Electronegativity

Some atoms hold on to electrons tighter than others. In some interactions, the bonding is therefore uneven. Atoms are involved in a sort of tug-of-war with the electrons. A purely covalent bond involves two equal-strength opponents, resulting in equal sharing. An ionic bond involves one species that is much stronger than the other, resulting in complete electron transfer with no sharing.

Skill Builder Solutions

10.1. Magnesium has just 2 valence electrons, so the Lewis structure is •Mg•

10.2. NaBr is an ionic compound, so Na donates the 1 valence shell electron it has to bromine, which then has an octet in its valence shell. Sodium has a +1 charge and Br has a −1 charge. The Lewis structure is thus Na$^+$ [:B̈r̈:]$^-$

10.3. Since Mg has a +2 charge and N has a −3 charge, the molecular formula is Mg_3N_2. The Lewis structure is Mg^{+2}[:N̈:]$^{-3}$ Mg^{+2}[:N̈:]$^{-3}$ Mg^{+2}

10.4. Carbon monoxide has a total of 4 + 6 = 10 valence electrons. The skeletal structure is C—O, and then we add electrons around the outer atoms, giving them octets. We can start with :C—O:, but carbon does not have an octet, so we must form a triple bond with the oxygen atom, giving :C≡O: or :C:::O:

10.5. There are a total of 12 valence electron in this species. Following the symmetry guidelines and placing 2 electrons in for each bond, we get

O
H:C:H

Now add the remaining electrons around the outer oxygen atom

:Ö:
H:C:H

To give the carbon atom an octet, we must move a lone pair from the oxygen

:Ö:↶
H:C:H

Which gives our final Lewis structure

:Ö
::
H:C:H

10.6. The species has 7 electrons coming from the Cl and 6 coming from the O atom. This makes a total of 13, but one more comes from the −1 charge of the ion, for a total of 14. The two species share one pair of electrons, to give each an octet. The Lewis structure is

[:C̈l̈:Ö:]$^-$

10.7. The base Lewis structure is $[:\ddot{O}-\ddot{N}-\ddot{O}:]^-$. The nitrogen has only 6 electrons around it, so it wishes to make a multiple bond with one of the oxygen atoms. It does not matter from which oxygen it comes, so there are two possible resonance structures:

$$[\ddot{O}=\ddot{N}-\ddot{O}:]^- \leftrightarrow [:\ddot{O}-\ddot{N}=\ddot{O}]^-$$

10.8. The central nitrogen has three groups of electrons around it, two of which are bonds and one is a lone pair. The electron geometry is then trigonal planar, and the molecular structure is bent.

10.9. The central sulfur atom has four groups of electrons around it: one a lone pair and three bonds. This gives tetrahedral electron structure and trigonal pyramidal molecular geometry.

10.10. a. Because two iodine atoms have the exact same electronegativity, neither is stronger than the other. Therefore, the bond is pure covalent.
b. Cesium is a group IA metal, and bromine is a group VIIA nonmetal. When they bond, there is a transfer of the electrons, giving an ionic bond.
c. Phosphorus and oxygen are both nonmetals and are significantly different in electronegativity. When they bond, the electrons are shared, but oxygen has a larger pull, so there is a small dipole moment. The bond is polar covalent.

10.11. CH_4 has a tetrahedral electron geometry and a tetrahedral molecular geometry. Since all of the bonds are of the same slight polarity in terms of electronegativity difference and the bonding is symmetric, the bond dipoles cancel each other out. Thus, the overall molecule is nonpolar.

Suggested Demonstrations

Blow up four equally sized balloons and tie the knots together. The resulting structure is tetrahedral in geometry, and you can explain how the balloons try to get as far apart as possible. Then pop one of the balloons to show how three orbitals (balloons) orient themselves. Then pop another balloon and explain the resulting structure change.

Have a few students (of various sizes) come to the front of the room and have them make various molecular geometries by holding their arms in various orientations. This is an effective method for showing bond dipoles, dipole moments, and polarity.

Guided Inquiry Ideas

Below are a few example questions that students answer in the guided inquiry activities provided in the Guided Activity Workbook.

Which atom do you think is central in carbon dioxide? Why?

Carbon dioxide has two groups of electrons surrounding the carbon atom. Why is the OCO bond angle in CO_2 180°?

The four groups of electrons in methane get as far from each other as possible. Is the angle between them 90°? If not, what is it?

Can a molecule with no polar bonds be polar?

Is a linear molecule the only one in which all the polar bonds cancel? What other geometric arrangement of polar bonds also leads to the cancellation of all the polar bonds?

Gases

11

Chapter Overview

This chapter introduces the relationships between the four measurable quantities of a sample of gas: pressure, volume, temperature, and the number of molecules. Their relationships are shown, applications to stoichiometry are illustrated, and calculations involving them are explained.

Lecture Outline

11.1 Extra-Long Straws
11.2 Kinetic Molecular Theory: A Model for Gases
    Learning Objective: Describe how kinetic molecular theory predicts the main properties of a gas.
    A. A gas is a collection of particles in constant straight-line motion
    B. Particles neither attract nor repel one another
    C. Between species is mostly empty space
    D. Average kinetic energy is proportional to absolute (Kelvin) temperature
11.3 Pressure: The Result of Constant Molecular Collisions
    Learning Objective: Identify and explain the relationship between pressure, force, and area.
    Learning Objective: Convert among pressure units.
    A. Pressure = force/area
    B. Units of pressure
        1. Atmosphere (atm)
        2. Pascal (Pa)
        3. Millimeters of mercury (mm Hg)
        4. Torr
        5. Pounds per square inch (psi)
11.4 Boyle's Law: Pressure and Volume
    Learning Objective: Restate and apply Boyle's law.
    A. $P \propto 1/V$
    B. Temperature and number of moles must be constant
11.5 Charles's Law: Volume and Temperature
    Learning Objective: Restate and apply Charles's law.
    A. $V \propto T$
    B. Pressure and number of moles of gas must be constant
11.6 The Combined Gas Law: Pressure, Volume, and Temperature
    Learning Objective: Restate and apply the combined gas law.
    A. $PV \propto T$
    B. Number of moles of gas must be constant
    C. If you know two of the quantities, you can calculate the third

11.7 Avogadro's Law: Volume and Moles
　　Learning Objective: Restate and apply Avogadro's law.
　　A. V α n
　　B. Pressure and temperature must be constant
11.8 The Ideal Gas Law: Pressure, Volume, Temperature, and Moles
　　Learning Objective: Restate and apply the ideal gas law.
　　A. PV = nRT
　　B. R = 0.0821 $\frac{\text{L atm}}{\text{mol K}}$
　　C. Can be used to calculate the molar mass of a gas, if the mass of a sample is known
11.9 Mixtures of Gases
　　Learning Objective: Restate and apply Dalton's law of partial pressures.
　　A. P_A = fractional component of A × P_{total}
　　B. $P_{total} = P_a + P_b + P_c + \ldots$
　　C. Deep-sea diving and partial pressure
　　D. Collecting gases over water
11.10 Gases in Chemical Reactions
　　Learning Objective: Apply the principles of stoichiometry to chemical reactions involving gases.
　　A. Use the ideal gas law to calculate any one of the four measurables knowing the other three
　　B. Standard temperature and pressure (STP) : 0 °C, 1 atm
　　C. Molar volume: volume of gas under STP = 22.4 L

Chemical Principle Teaching Ideas

Kinetic Molecular Theory
　　Although this is just a theory and is not 100% correct, it is a good estimation when talking about gases under low-pressure and high-temperature conditions. Explain to students that these conditions make molecules move very fast and have a lot of space in between species, which means that there are very few interactions.

Pressure
　　Talking about why balloons and tires stay inflated gives the students a real-world example with which to associate. Why do airplanes need to have pressurized cabins? Why do your ears pop when climbing up tall hills? Explaining these simple examples will go far in helping the students grasp the concept of pressure.

Simple Gas Laws
　　Giving the students real-world examples will help them remember the gas laws. Reminding them that car tires deflate in winter helps them remember Charles's law. Pumping up a basketball explains the relationship between number of moles and volume, Avogadro's law.

The Combined Gas Law
   Getting the students familiar with the gas laws is most easily done by simply giving them plenty of simple calculations to do.

The Ideal Gas Law
   Once you understand the ideal gas law, the individual gas laws need not be memorized, as you can use the ideal gas law to convert between any of the quantities. Emphasis should be placed on the fact that this is a very simplified version of gas behavior. This law is only accurate when the sample of gas is at low pressure and high temperature.

Mixtures of Gases
   Talking about the partial pressure of the individual air molecules in the classroom is an easy way for the students to see Dalton's law. If only 21% of the gas molecules are oxygen atoms, that means that 21% of the molecules striking the wall (pressure) are oxygen molecules, so the partial pressure of oxygen is 0.21 that of atmospheric pressure.

Gases in Chemical Reactions
   Like any concept involving calculations, the more problems done by the students, the more comfortable they will be with the concept. This is a good opportunity to show students how a calculation can be set up by looking solely at the units.

Skill Builder Solutions

11.1. $173 \text{ in Hg} \times \dfrac{25.4 \text{ mm Hg}}{1 \text{ in Hg}} \times \dfrac{14.7 \text{ psi}}{760 \text{ mm Hg}} = 85.0 \text{ psi}$

Plus. $23.8 \text{ in Hg} \times \dfrac{25.4 \text{ mm Hg}}{1 \text{ in Hg}} \times \dfrac{101325 \text{ Pa}}{760 \text{ mm Hg}} \times \dfrac{1 \text{ kPa}}{1000 \text{ Pa}} = 80.6 \text{ kPa}$

11.2. $P_1V_1 = P_2V_2$; $(1.0 \text{ atm})(16. \text{ mL}) = 7.5 \text{ mL} \times P_2 \therefore P_2 = \dfrac{1.0 \text{ atm} \times 16 \text{ mL}}{7.5 \text{ mL}} = 2.1 \text{ atm}.$

   If it increases 1 atm per 10 m, and the pressure has increased by 1.1 atm, the snorkeler is approximately 11 m below the surface.

11.3. $\dfrac{V_1}{T_1} = \dfrac{V_2}{T_2}$; $\dfrac{88.2 \text{ mL}}{(35 + 273) \text{ K}} = \dfrac{V_2}{(155 + 273) \text{ K}}$; $V_2 = \dfrac{88.2 \text{ mL} \times 428 \text{ K}}{308 \text{ K}} = 123 \text{ mL}$

11.4. $\dfrac{P_1V_1}{T_1} = \dfrac{P_2V_2}{T_2}$; $\dfrac{1.1 \text{ atm} \times 3.7 \text{ L}}{(30. + 273) \text{ K}} = \dfrac{4.7 \text{ atm} \times V_2}{(15 + 273) \text{ K}}$; $V_2 = \dfrac{1.1 \text{ atm} \times 3.7 \text{ L} \times 288 \text{ K}}{4.7 \text{ atm} \times 303 \text{ K}} = 0.82 \text{ L}$

11.5. $\dfrac{V_1}{n_1} = \dfrac{V_2}{n_2}$; $\dfrac{2.1 \text{ L}}{0.11 \text{ mol}} = \dfrac{V_2}{0.69 \text{ mol}} \therefore V_2 = \dfrac{2.1 \text{ L} \times 0.69 \text{ mol}}{0.11 \text{ mol}} = 13 \text{ L}$

11.6. $PV = nRT$; $P \times 8.5 \text{ L} = 0.55 \text{ mol} \times 0.0821 \dfrac{\text{L atm}}{\text{K mol}} \times 305 \text{ K}$

$$\therefore P = \dfrac{0.55 \text{ mol} \times 0.0821 \dfrac{\text{L atm}}{\text{K mol}} \times 305 \text{ K}}{8.5 \text{ L}} = 1.6 \text{ atm}$$

11.7. We must convert temperature to K and pressure to atm before we use the ideal gas law.
$T = 58 + 273 = 331 \text{ K}$

$P = 715 \text{ mm Hg} \times \dfrac{1 \text{ atm}}{760 \text{ mm Hg}} = .941 \text{ atm}$

$PV = nRT$; $V = \dfrac{nRT}{P} = \dfrac{0.556 \text{ mol} \times 0.0821 \dfrac{\text{L atm}}{\text{K mol}} \times 331 \text{ K}}{0.941 \text{ atm}} = 16.1 \text{ L}$

Plus. We need to calculate the number of moles of He, the temperature in K, and volume in L.

$n = 0.133 \text{ g He} \dfrac{1 \text{ mol He}}{4.00 \text{ g He}} = 0.0332 \text{ mol He}$

$T = 32 + 273 = 305 \text{ K}$

$V = 648 \text{ mL} \times \dfrac{1 \text{ L}}{1000 \text{ mL}} = 0.648 \text{ L}$

$P = \dfrac{0.0332 \text{ mol He} \times 0.0821 \dfrac{\text{L atm}}{\text{K mol He}} \times \dfrac{760 \text{ torr}}{1 \text{ atm}} \times 305 \text{ K}}{0.648 \text{ L}} = 977 \text{ mm Hg}$

11.8. To calculate the molar mass of a gas, we need to know the number of moles and the grams of the sample.
$T = 273 + 88 = 361 \text{ K}$

$\text{Mass} = 827 \text{ mg} \times \dfrac{1 \text{ g}}{1000 \text{ mg}} = 0.827 \text{ g}$

$P = 975 \text{ mm Hg} \times \dfrac{1 \text{ atm}}{760 \text{ mm Hg}} = 1.28 \text{ atm}$

$PV = nRT$; $n = \dfrac{1.28 \text{ atm} \times 0.270 \text{ L}}{0.0821 \dfrac{\text{L atm}}{\text{K mol}} \times 361 \text{ K}} = 0.0117 \text{ mol of gas}$

$\text{Molar mass} = \dfrac{\text{mass of sample}}{\text{moles of sample}} = \dfrac{0.827 \text{ g}}{0.0117 \text{ mol}} = 70.8 \text{ g/mol}$

11.9. The total pressure is 745 torr and is due to water and hydrogen only. $P_{total} = P_{water} + P_{H2}$; $P_{H2} = P_{total} - P_{water} = 745 \text{ torr} - 24 \text{ torr} = 721 \text{ torr}$.

11.10. P_{O_2} = Fractional composition of $O_2 \times P_{total}$;

$$P_{tot} = \frac{P_{O_2}}{\text{fractional composition}} = \frac{0.21 \text{ atm}}{\left(\dfrac{5.0}{100.}\right)} = 4.2 \text{ atm}$$

11.11. The first step is to calculate the number of moles of O_2 that were formed. Then, using the balanced chemical equation, we can calculate the grams of Ag_2O that were formed.

$$PV = nRT \text{ ; } P = 745 \text{ mm Hg} \times \frac{1 \text{ atm}}{760 \text{ mm Hg}} = 0.980 \text{ atm}$$

$$n = \frac{0.980 \text{ atm} \times 4.58 \text{ L}}{0.0821 \dfrac{\text{L atm}}{\text{K mol}} \times 308 \text{ K}} = 0.177 \text{ mol of } O_2$$

$$g \text{ Ag}_2O = 0.177 \text{ mol } O_2 \times \frac{2 \text{ mol Ag}_2O}{1 \text{ mol } O_2} \times \frac{231.74 \text{ g Ag}_2O}{1 \text{ mol Ag}_2O} = 82.3 \text{ g Ag}_2O$$

11.12. We must first calculate the number of moles of oxygen required and then use the ideal gas law to calculate the volume of oxygen.

$$10.5 \text{ g H}_2O \times \frac{1 \text{ mol H}_2O}{18.02 \text{ g H}_2O} \times \frac{1 \text{ mol } O_2}{2 \text{ mol H}_2O} = 0.291 \text{ mol } O_2$$

at STP, T = 273 K, P = 1 atm

$$V = \frac{nRT}{P} = \frac{0.292 \text{ mol} \times 0.0821 \dfrac{\text{L atm}}{\text{K mol}} \times 273 \text{ K}}{1 \text{ atm}} = 6.53 \text{ L } O_2 \text{ gas reacted}$$

Suggested Demonstrations

Collapsing Can, *Chemical Demonstrations* 2:6, Shakhashiri, B. Z. University of Wisconsin Press, 1985.

Boyle's Law, *Chemical Demonstrations* 2:14, Shakhashiri, B. Z. University of Wisconsin Press, 1985.

Charles's Law, *Chemical Demonstrations* 2:28, Shakhashiri, B. Z. University of Wisconsin Press, 1985.

Avogadro's Hypothesis, *Chemical Demonstrations* 2:44, Shakhashiri, B. Z. University of Wisconsin Press, 1985.

Guided Inquiry Ideas

Below are a few example questions that students answer in the guided inquiry activities provided in the Guided Activity Workbook.

Complete the following sentence: If the amount of gas and the temperature are constant, the _____ is always the same.

The proportionality constant, R, is called the **ideal gas constant** and is known to have a value 0.0821 L·atm/K·mol. Of the variables pressure, volume, number of moles, and temperature, how many would you have to know to calculate them all?

Which of the following is true? (Subscripts refer to experiment numbers.)

$$P_1 V_1 = P_2 V_2 \qquad\qquad \frac{P_1}{V_1} = \frac{P_2}{V_2}$$

Which of the following incorporates all the gas behavior observed so far?

$$V \propto nTP \qquad\qquad V \propto \frac{nT}{P} \qquad\qquad V \propto \frac{P}{nT}$$

Which unit is equal to "Newtons per square meter."

Liquids, Solids, and Intermolecular Forces

12

Chapter Overview

The molecular-level interactions of solids and liquids are discussed in this chapter. Phase changes are also discussed, and appropriate calculations are performed. Intermolecular forces are explained, and their bulk phase influence is put into context.

Lecture Outline

12.1 Spherical Water
    A. Intermolecular forces
    B. Thermal energy = energy of motion
12.2 Properties of Liquids and Solids
    <u>Learning Objective:</u> Describe the properties of solids and liquids and relate them to their constituent atoms and molecules.
    A. Properties of liquids
        1. High densities compared to gases
        2. Indefinite shape, fluid
        3. Definite volume
    B. Properties of solids
        1. High densities compared to gases
        2. Definite shape
        3. Definite volume
        4. May be crystalline or amorphous
12.3 Intermolecular Forces in Action: Surface Tension and Viscosity
    <u>Learning Objective:</u> Describe how surface tension and viscosity are manifestations of the intermolecular forces in liquids.
    A. Surface tension: why some things more dense than water can still float
    B. Viscosity: resistance of liquid to flow
12.4 Evaporation and Condensation
    <u>Learning Objective:</u> Describe and explain the processes of evaporation and condensation.
    <u>Learning Objective:</u> Use the heat of vaporization in calculations.
    A. Evaporation and vaporization – liquid to gas
    B. Condensation – gas to liquid
    C. Dynamic equilibrium
    D. Boiling
    E. Energy change = heat of vaporization

12.5 Melting, Freezing, and Sublimation
  Learning Objective: Describe the process of melting, freezing, and sublimation.
  Learning Objective: Use the heat of fusion in calculations.
  A. Melting – solid to liquid
  B. Freezing – liquid to solid
  C. Energy change = heat of fusion
  D. Sublimation – solid to gas
12.6 Types of Intermolecular Forces: Dispersion, Dipole–Dipole, Hydrogen Bonding, and Ion–Dipole
  Learning Objective: Compare and contrast four types of intermolecular forces: dispersion, dipole–dipole, hydrogen bonds, and ion–dipole.
  Learning Objective: Determine the types of intermolecular forces in compounds.
  Learning Objective: Use intermolecular forces to determine relative melting and/or boiling points.
  A. Dispersion force, also known as London force
    1. Weakest intermolecular force
    2. Instantaneous dipole
    3. Involved in every intermolecular interaction
  B. Dipole–dipole force
    1. Usually stronger than dispersion forces
    2. Strength is a function of dipole moment and structure
  C. Hydrogen bonding
    1. Strongest intermolecular forces
    2. Only involving F, O, or N bonded to a hydrogen atom
  D. Ion–dipole force
    1. Mixture of ionic compounds and polar compounds
    2. Important in aqueous solutions of ionic compounds
12.7 Types of Crystalline Solids: Molecular, Ionic, and Atomic
  Learning Objective: Identify types of crystalline solids.
  A. Molecular solids
    1. Composite units are molecules
    2. Examples, dry ice ($CO_{2(s)}$), ice ($H_2O_{(s)}$)
  B. Ionic solids
    1. Composite units are formula units
    2. Examples, NaCl, CaF_2
  C. Atomic solids
    1. Composite units are individual atoms
    2. Covalent atomic solids
    3. Nonbonding atomic solids
    4. Metallic atomic solids
12.8 Water: A Remarkable Molecule
  Learning Objective: Describe the properties that make water unique among molecules.
  A. Boiling point very high for its molar mass
  B. Solid less dense than liquid

Chemical Principle Teaching Ideas

Properties of Liquids
    Look at a sample of liquid that students see every day, such as a bottle of soda. Ask the students about the liquid and measurable properties. Can it be compressed? What shape does it take? This real-world example will help them understand the properties and will also help them remember the ideas presented.

Properties of Solids
    Bring out a large hunk of a common metal such as lead. Ask the students about the properties of the sample. Can it be compressed? Is the volume fixed?

Manifestations of Intermolecular Forces
    Again, real-world examples are a good tool for these concepts. Showing a sample of thick oil slowly pouring out of a container will show them viscosity. Pointing out that some animals depend on surface tension for their very existence (water bugs) will show them how important the concept is for biological species.

Evaporation and Condensation
    Talk about what affects the rate of evaporation. What happens when you heat the liquid? The molecules are moving faster, so more of them have enough energy to leave the liquid state. When the temperature reaches the boiling point, evaporation is taking place at a furious pace. To show them how much energy is actually involved in this phase change, have the students calculate how much energy would be required to boil 1 L of water.

Melting and Freezing
    Do a sample calculation of how much energy is required to melt 1 L of solid water. Compare this to the energy needed to boil 1 L of liquid water. You can use this as an opportunity to explain why the intermolecular forces in liquids and solids are close to one another, but to overcome that force to separate the molecules (liquid to gas) requires a lot more energy.

Types of Intermolecular Forces
    To help show the relative magnitude of each kind of intermolecular force, give the measured dipole moments of a variety of species. Students frequently forget that dispersion forces are always around but are usually weak. Explain that this is why even hydrogen, which has no dipole at all, will form a solid at very low temperatures.

Types of Crystalline Solids
    Giving several real-world examples of each kind of bonding type will help the students remember their respective properties. Make sure to point out the fundamental measurable differences between the types. For example, ionic solids have high melting points but do not conduct electricity, whereas atomic solids tend to conduct electricity and have variable boiling points.

Water
Remind the students that their very existence depends on this strange behavior of water. Explain what would happen to lakes in the cold winter if solid water (ice) sank instead of floated. If the ice sank to the bottom, the cold air would make more ice at the surface that would then sink. The lakes would freeze solid every year, wreaking havoc on the biological organisms in the water. Since ice floats, it forms an insulating barrier above a small amount of liquid water, allowing fish and other species to survive the winter season.

Skill Builder Solutions

12.1. ΔH_{vap} is a conversion factor between amount of a substance and energy.
Heat required = mass × ΔH_{vap} =

$$2.58 \text{ kg H}_2\text{O} \times \frac{1000 \text{ g H}_2\text{O}}{1 \text{ kg H}_2\text{O}} \times \frac{1 \text{ mol H}_2\text{O}}{18.02 \text{ g H}_2\text{O}} \times \frac{40.7 \text{ kJ}}{1 \text{ mol H}_2\text{O}} = 5.83 \times 10^3 \text{ kJ}$$

Plus. First, calculate the amount of heat released from the condensation. Then calculate the change in temperature of the aluminum metal.

$$q = \Delta H_{vap} \times m_{H2O} = 0.48 \text{ g H}_2\text{O} \times \frac{1 \text{ mol H}_2\text{O}}{18.02 \text{ g H}_2\text{O}} \times \frac{40.7 \text{ kJ}}{1 \text{ mol H}_2\text{O}} = 1.1 \text{ kJ}$$

$q = \Delta T \times$ specific heat $\times$ mass$_{Al}$;

$$\Delta T = \frac{q}{\text{specific heat}_{Al} \times \text{mass}_{Al}} = \frac{1100 \text{ J}}{0.903 \frac{\text{J}}{\text{g} \cdot {}^\circ\text{C}} \times 55 \text{ g}} = 22 \text{ }^\circ\text{C}$$

$T_f = T_i + \Delta T = 25 \text{ }^\circ\text{C} + 22 \text{ }^\circ\text{C} = 47 \text{ }^\circ\text{C}$.

12.2. $$15.5 \text{ g H}_2\text{O} \times \frac{1 \text{ mol H}_2\text{O}}{18.02 \text{ g H}_2\text{O}} = 0.861 \text{ mol H}_2\text{O}$$

$$q = \text{mol}_{H2O} \times \Delta H_{fusion} = 0.861 \text{ mol H}_2\text{O} \times \frac{6.02 \text{ kJ}}{1 \text{ mol H}_2\text{O}} = 5.18 \text{ kJ}$$

Plus. Heat from melting of ice = $\Delta H_{fusion} \times \text{mol}_{H2O}$

$$5.6 \text{ g H}_2\text{O} \times \frac{1 \text{ mol H}_2\text{O}}{18.02 \text{ g H}_2\text{O}} \times \frac{6.02 \text{ kJ}}{1 \text{ mol H}_2\text{O}} = 1.87 \text{ kJ of energy from warm water}$$

$$\Delta T = \frac{q}{\text{specific heat}_{H2O} \times \text{mass}_{H2O}} = \frac{-1870 \text{ J}}{4.184 \frac{\text{J}}{\text{g} \cdot {}^\circ\text{C}} \times 195 \text{ g}} = -2.3 \text{ }^\circ\text{C}$$

12.3. Methane has a molecular weight of 16.04, and C_2H_6 has a weight of 30.07 g/mol. Since C_2H_6 has the higher molar mass, the dispersion forces are greater, giving a higher boiling point.

12.4. a. CI_4 is tetrahedral in geometry, and the C and all the I have identical electronegativities. Thus, the individual bonds are nonpolar, and there are no intermolecular dipole–dipole interactions.
 b. CH_3Cl is tetrahedral in geometry, but the bond dipoles are not equivalent; therefore, dipole–dipole intermolecular forces do exist.
 c. HCl is a linear molecule and the two species have different electronegativities. Therefore, dipole–dipole forces attract the HCl species to one another.

12.5. Both species have strong dipole–dipole intermolecular forces; however, HF has hydrogen bonding, which is much stronger than dipole–dipole forces, and thus has a higher boiling point.

12.6. a. Ammonia (NH_3) is a polar covalent molecule; thus, it is a molecular solid.
 b. Calcium oxide (CaO) is an ionic solid formed between a metal and a nonmetal. Thus, it is an ionic solid.
 c. Krypton (being a monatomic species), when cooled to a solid phase, is an atomic solid.

Suggested Demonstrations

Measure the mass of a piece of ice and place it in a beaker on an activated hot plate. Have the students calculate how much energy will be required to change the ice into water gas.

Density and Miscibility of Liquids, *Chemical Demonstrations* 3:233, Shakhashiri, B. Z. University of Wisconsin Press, 1989.

Etching Glass with Hydrogen Fluoride, *Chemical Demonstrations* 3:80, Shakhashiri, B. Z. University of Wisconsin Press, 1989.

Evaporations as an Endothermic Process, *Chemical Demonstrations* 3:233, Shakhashiri, B. Z. University of Wisconsin Press, 1989.

Guided Inquiry Ideas

Below are a few example questions that students answer in the guided inquiry activities provided in the Guided Activity Workbook.

Define intermolecular force using a complete sentence.

Is the covalent bond between a hydrogen atom and an oxygen atom within a single water molecule held together by an intermolecular force? Why or why not?

What intermolecular forces do all atoms and molecules have?

What is the necessary criterion for a molecule to have a dipole–dipole intermolecular force?

Draw the Lewis structures for CO_2, CH_2O, CH_2Cl_2, and NH_3. Using VSEPR theory, determine whether each is polar or nonpolar.

Solutions

13

Chapter Overview

Chapter 13 discusses what a solution is and how we characterize its properties and concentrations. Background is given, and appropriate calculations are shown. Practical applications of solutions phase reactions are described.

Lecture Outline

13.1 Tragedy in Cameroon
13.2 Solutions: Homogeneous Mixtures
    <u>Learning Objective:</u> Define solution, solute, and solvent.
    A. Any homogeneous mixture is a solution
    B. Solute: major component
    C. Solute: minor component, can be more than one
13.3 Solutions of Solids Dissolved in Water: How to Make Rock Candy
    <u>Learning Objective:</u> Relate the solubility of solids in water to temperature.
    A. Solubility and saturation
        1. Saturated: solvent holding as much solute as it can
        2. Unsaturated: solvent can hold more solute
        3. Supersaturated: solvent holding more than the maximum amount of solute
    B. Electrolyte solutions: dissolved ionic solids
        1. Strong electrolyte solution
        2. Nonelectrolyte solution
    C. Solubility is temperature dependent
    D. Rock candy
13.4 Solutions of Gases in Water: How Soda Pop Gets Its Fizz
    <u>Learning Objective:</u> Relate the solubility of gases in liquids to temperature and pressure.
    A. Solubility is a function of temperature and pressure
    B. Henry's law
13.5 Specifying Solution Concentration: Mass Percent
    <u>Learning Objective:</u> Calculate mass percent.
    <u>Learning Objective:</u> Use mass percent in calculations.
    A. Mass percent = $\dfrac{\text{mass solute}}{\text{mass solution}} \times 100\%$
    B. Conversion factor between amount of solution and amount of a particular solute
13.6 Specifying Solution Concentration: Molarity
    <u>Learning Objective:</u> Calculate molarity.
    <u>Learning Objective:</u> Use molarity in calculations.
    <u>Learning Objective:</u> Calculate ion concentration.

  A. Molarity (M) = $\dfrac{\text{moles solute}}{\text{liters solution}}$
  B. Volume of solution, not just solvent
  C. Unit always mol/L
  D. Ion concentrations
13.7 Solution Dilution
  Learning Objective: Use the dilution equation in calculations.
  A. When adding water to a solution, the number of moles of solute remains unchanged
  B. $M_1V_1 = M_2V_2$
13.8 Solution Stoichiometry
  Learning Objective: Use volume and concentration to calculate the number of moles of reactants or products and then use stoichiometric coefficients to convert to other quantities in a reaction.
  A. Balanced chemical equations give molar ratios only
  B. Convert volume to moles using molarity, then use balanced chemical equation
13.9 Freezing Point Depression and Boiling-Point Elevation: Making Water Freeze Colder and Boil Hotter
  Learning Objective: Calculate molality.
  Learning Objective: Calculate freezing points and boiling points for solutions.
  A. Colligative property depends on the amount of solute, not identity
  B. Molality (*m*) = $\dfrac{\text{moles solute}}{\text{kg solvent}}$
  C. $\Delta T_f = m \times K_f$, where K_f is the freezing point depression constant
  D. $\Delta T_b = m \times K_b$, where K_b is the boiling point elevation constant
13.10 Osmosis: Why Drinking Saltwater Causes Dehydration
  Learning Objective: Summarize and explain the process of osmosis.
  A. Semipermeable membrane
  B. Osmotic pressure

Chemical Principle Teaching Ideas

Solutions
  Students usually associate solutions as liquid (usually water) with solid dissolved in it, but in fact any two species mixed can make a solution. Give an example and show each different combination to the students.

Solid and Water Solutions
  Use the idea of making Jell-O® by dissolving sugar in water to explain the temperature dependence of solubility. To speed up the dissolving process, you need to increase the temperature of the solution.

Gas and Water Solutions
  The solubility of gases is temperature dependent but not in the same way as solids and liquids. This fact sometimes confuses students, but again a simple, real-world example comes to the rescue. Why does warm soda taste flatter than cold soda? It has to do with the solubility of

the carbon dioxide in the soda solution and how much "fizz" goes out when you originally open the container.

Solution Concentration
There are many different ways to express the concentration of a solution, and each of them has its place in various kinds of applications. The most common place for a mistake by students is traditionally the denominator of the unit molality. It is a difficult concept for students to understand, so particular emphasis should be made here.

Solution Dilution
Because adding water to a solution does not change the moles of solute, we can use the fact that M_1V_1 gives an answer in the unit of moles of solute and is equal to M_2V_2.

Freezing Point Depression and Boiling Point Elevation
Why do we put antifreeze in our cars? Why do we put salt on icy roads? Why does soup boil hotter than pure water? All of these are practical, real-world examples of these concepts and are easy for students to understand.

Osmosis
To emphasize the importance of this idea, be sure to include a discussion of what would happen if you placed pure water in your bloodstream. The water would be soaked into the individual blood cells, and they would expand, burst, and die.

Skill Builder Solutions

13.1. $412.1 \text{ mL H}_2\text{O} \times \dfrac{1.00 \text{ g H}_2\text{O}}{1.0 \text{ mL H}_2\text{O}} = 412.1 \text{ g H}_2\text{O}$

11.3 g sucrose + 412.1 g H₂O = 423.4 g solution

$\dfrac{11.3 \text{ g sucrose}}{423.4 \text{ g solution}} \times 100\% = 2.67 \%$ by mass sucrose

13.2. $355 \text{ mL solution} \times \dfrac{1.04 \text{ g solution}}{1 \text{ mL solution}} \times \dfrac{11.5 \text{ g C}_{12}\text{H}_{22}\text{O}_{11}}{100 \text{ g solution}} = 42.5 \text{ g C}_{12}\text{H}_{22}\text{O}_{11}$

13.3. $55.8 \text{ g NaNO}_3 \times \dfrac{1 \text{ mol NaNO}_3}{845.00 \text{ g NaNO}_3} \times \dfrac{1}{2.50 \text{ L solution}} = 0.263 \dfrac{\text{mol NaNO}_3}{\text{L solution}} = 0.263 \text{ M}$

13.4. $55.8 \text{ g KCl} \times \dfrac{1 \text{ mol KCl}}{74.55 \text{ g KCl}} \times \dfrac{1 \text{ L solution}}{0.225 \text{ mol KCl}} = 3.33 \text{ L solution}$

13.5. $\dfrac{0.75 \text{ mol CaCl}_2}{1 \text{ L solution}} \times \dfrac{1 \text{ mol Ca}^{2+}}{1 \text{ mol CaCl}_2} = \dfrac{0.75 \text{ mol Ca}^{2+}}{1 \text{ L solution}} = 0.75 \text{ M Ca}^{2+}$

$\dfrac{0.75 \text{ mol CaCl}_2}{1 \text{ L solution}} \times \dfrac{2 \text{ mol Cl}^-}{1 \text{ mol CaCl}_2} = \dfrac{1.50 \text{ mol Cl}^-}{1 \text{ L solution}} = 1.5 \text{ M Cl}^-$

13.6. $M_1V_1 = M_2V_2$; $(1.2 \text{ M})(0.585 \text{ L}) = 6.0 \text{M } V_2$; $V_2 = \dfrac{1.2 \text{ M} \times 0.585 \text{ L}}{6.0 \text{ M}} = 0.12 \text{ L}$

13.7. $27.2 \text{ mL HNO}_3 \text{ soln} \times \dfrac{0.135 \text{ mol HNO}_3}{1000 \text{ mL HNO}_3 \text{ soln}} \times \dfrac{1 \text{ mol Na}_2\text{CO}_3}{2 \text{ mol HNO}_3}$

$\times \dfrac{1000 \text{ mL Na}_2\text{CO}_3 \text{ soln}}{0.112 \text{ mol Na}_2\text{CO}_3} = 16.4 \text{ mL Na}_2\text{CO}_3$

Plus.

$35.7 \text{ ml Na}_2\text{CO}_3 \text{ soln} \times \dfrac{0.108 \text{ mol Na}_2\text{CO}_3}{1000 \text{ ml Na}_2\text{CO}_3 \text{ soln}} \times \dfrac{2 \text{ mol HNO}_3}{1 \text{ mol Na}_2\text{CO}_3} \times \dfrac{1}{0.0250 \text{ L HNO}_3} = 0.308 \text{ M}$

13.8. $50.4 \text{ g C}_{12}\text{H}_{22}\text{O}_{11} \times \dfrac{1 \text{ mol C}_{12}\text{H}_{22}\text{O}_{11}}{342.34 \text{ g C}_{12}\text{H}_{22}\text{O}_{11}} \times \dfrac{1}{0.332 \text{ kg H}_2\text{O}} = 0.443 \, m$

13.9. $\Delta T_f = m \times K_f = \dfrac{2.6 \text{ mol sucrose}}{1 \text{ kg solvent}} \times \dfrac{1.86 \, °\text{C kg solvent}}{1 \text{ mol sucrose}} = 4.8 \, °\text{C}$

$T = 0.0 \, °\text{C} - \Delta T = 0.0 - 4.8 \, °\text{C} = -4.8 \, °\text{C}$

13.10. $\Delta T_b = m \times K_b = \dfrac{3.5 \text{ mol glucose}}{1 \text{ kg solvent}} \times \dfrac{0.512 \, °\text{C kg solvent}}{1 \text{ mol glucose}} = 1.8 \, °\text{C}$

$T = 100.0 \, °\text{C} + \Delta T = 100.0 \, °\text{C} + 1.8 \, °\text{C} = 101.8 \, °\text{C}$

Suggested Demonstrations

Bring examples of each combination of solids, liquids, and gases in solutions. Explain each example, how the solutions are of various concentrations, and how each behaves.

Ice cream is made by freezing cream using a salt–ice bath. Explain how it is made, and make some in the classroom.

Crystallization from Supersaturated Solutions of Sodium Acetate, *Chemical Demonstrations* 1:27, Shakhashiri, B. Z. University of Wisconsin Press, 1983.

Density and Miscibility of Liquids, *Chemical Demonstrations* 3:233, Shakhashiri, B. Z. University of Wisconsin Press, 1989.

Osmotic Pressure of a Sugar Solution, *Chemical Demonstrations* 3:283, Shakhashiri, B. Z. University of Wisconsin Press, 1989.

Guided Inquiry Ideas

Below are a few example questions that students answer in the guided inquiry activities provided in the Guided Activity Workbook.

Are solids (e.g., salt) more soluble in hot water or cold water?

Trout only live in very cold mountain lakes and streams. How much oxygen is in that water compared to water in the tropics?

Explain why bubbles form in a glass of cold tap water as it warms up to room temperature.

In a complete sentence or two, summarize the solubility trends for solids and gases.

Explain the effect that a solute has on the range of temperatures over which a substance is a liquid.

Acids and Bases

14

Chapter Overview

    What is an acid? Are all acids bad? What is the difference between an acid and a base? The definitions of acids and bases are given in this chapter, as well as how they interact with one another and other species. The biological implications and socioeconomic consequences of their reactivity are also discussed. In addition, calculations involving acid and base reactivity are performed and explained.

Lecture Outline

14.1 Sour Patch Kids and International Spy Movies
14.2 Acids: Properties and Examples
    <u>Learning Objective:</u> Identify common acids and describe their key characteristics.
    A. Sour taste
    B. Dissolve many metals
    C. Turns blue litmus paper red
    D. Common acids
        1. HCl is hydrochloric acid
        2. H_2SO_4 is sulfuric acid
        3. HNO_3 is nitric acid
14.3 Bases: Properties and Examples
    <u>Learning Objective:</u> Identify common bases and describe their key characteristics.
    A. Bitter taste
    B. Feel slippery
    C. Turns red litmus paper blue
    D. Common bases
        1. Ammonia
        2. Sodium hydroxide
        3. Sodium bicarbonate
14.4 Molecular Definitions of Acids and Bases
    <u>Learning Objective:</u> Identify Arrhenius acids and bases.
    <u>Learning Objective:</u> Identify Brønsted–Lowry acids and bases and their conjugates.
    A. Arrhenius definition
        1. Acids produce H^+ ions in aqueous solution
        2. Bases produce OH^- ions in aqueous solution
    B. H^+ is actually H_3O^+
    C. Brønsted–Lowry definition
        1. Acids are proton (H^+ ion) donors
        2. Bases are proton (H^+ ion) acceptors
    D. Conjugate acid–base pairs

14.5 Reactions of Acids and Bases
    <u>Learning Objective:</u> Write equations for neutralization reactions.
    <u>Learning Objective:</u> Write equations for the reactions of acids with metals and with metal oxides.
    A. Neutralization reactions
    B. Acid reactions
    C. Base reactions

14.6 Acid–Base Titrations: A Way to Quantify the Amount of Acid or Base in a Solution
    <u>Learning Objective:</u> Use acid-base titration to determine the concentration of an unknown solution.
    A. Add solution of known concentration to solution of unknown concentration
    B. Equivalence point
    C. Indicator

14.7 Strong and Weak Acids and Bases
    <u>Learning Objective:</u> Identify strong and weak acids and strong and weak bases.
    <u>Learning Objective:</u> Determine $[H_3O^+]$ in acid solutions.
    <u>Learning Objective:</u> Determine $[OH^-]$ in base solutions.
    A. Strong acids are strong electrolytes
    B. Weak acids are weak electrolytes
    C. Diprotic and triprotic acids
    D. Strong bases
    E. Weak bases

14.8 Water: Acid and Base in One
    <u>Learning Objective:</u> Calculate $[H_3O^+]$ or $[OH^-]$ from Kw.
    A. Amphoteric
    B. All aqueous solution contains both H_3O^+ and OH^-
    C. $K_w = [H_3O^+][OH^-]$
    D. Acid : $[H_3O^+] > 1.0 \times 10^{-7}\,M$
    E. Base : $[H_3O^+] < 1.0 \times 10^{-7}\,M$

14.9 The pH and pOH Scale: Ways to Express Acidity and Basicity
    <u>Learning Objective:</u> Calculate pH from $[H_3O^+]$.
    <u>Learning Objective:</u> Calculate $[H_3O^+]$ from pH.
    <u>Learning Objective:</u> Calculate $[OH^-]$ from pOH.
    <u>Learning Objective:</u> Compare and contrast the pOH scale and the pH scale.
    A. $pH = -\log[H_3O^+]$; $pOH = -\log[OH^-]$
    B. $pH + pOH = 14$
    C. In an acidic solution, pH < 7
    D. In a neutral solution, pH = 7
    E. In a basic solution, pH > 7

14.10 Buffers: Solutions That Resist pH Change
    <u>Learning Objective:</u> Describe how buffers resist pH change.
    A. Contain both a weak acid and its conjugate base
    B. Weak acid neutralizes added base
    C. Conjugate base neutralizes added acid

Chemical Principle Teaching Ideas

Properties of Acids and Bases
   Give the students an example of real-world acids and bases and ask them to describe their various properties.

Reactions of Acids and Bases
   Doing a few demonstrations in class helps, but discussing some acid-base reactions that they are familiar with, for example, drinking Mylanta ($Mg(OH)_2$) to react with the excess HCl in your upset stomach, is even more effective.

Strong and Weak Acids and Bases
   Give examples of the acids and bases (strong and weak) that students see and use every day. Explain how their action is important to biological organisms.

pH Scale
   There are examples of household chemicals at all pH levels. Bring in samples of each kind and measure their pH. Emphasize that pH is a logarithmic scale and that if we know the pH of a solution, we can easily calculate both $[H_3O^+]$ and $[OH^-]$.

Buffers
   Buffers can be made from weak acids and their conjugate bases or from a weak base and its conjugate acid. Every acid has a conjugate as does every base, not just weak ones. Given many examples, the students will recognize that they only differ in an H^+.

Skill Builder Solutions

14.1. a. H_2O loses an H^+, so it is the Brønsted–Lowry acid. The conjugate base for H_2O is OH^-.
       C_5H_5N gains the proton, so it is the Brønsted–Lowry base with conjugate acid CH_5NH^+.
   b. H_2O gains a H^+, so it is the Brønsted–Lowry base. The conjugate acid for H_2O is H_3O^+.
       HNO_3 loses the proton, so it is the Brønsted–Lowry acid with conjugate base NO_3^-.

14.2. H_3PO_4 is a triprotic acid, meaning that 1 mol of H_3PO_4 requires 3 mol of OH^- to completely react with it. Thus, the equation is
       $3\ NaOH(aq) + H_3PO_4(aq) \rightarrow 3\ H_2O(l) + Na_3PO_4(aq)$

14.3. a. The reaction between HCl and Sr metal will form hydrogen gas and a salt. The salt contains the anion from the acid and the metal ion. Thus, $Sr(s) + HCl(aq) \rightarrow SrCl_2(aq) + H_2(g)$ is the skeletal reaction, and balancing gives $Sr(s) + 2\ HCl(aq) \rightarrow SrCl_2(aq) + H_2(g)$.
   b. The reaction of HI and BaO forms water and a neutral salt. The salt contains the anion of the acid and the metal cation. The skeletal reaction is $BaO(s) + HI(aq) \rightarrow H_2O(l) + BaI_2(aq)$, and when balanced, it is $BaO(s) + 2\ HI(aq) \rightarrow H_2O(l) + BaI_2(aq)$.

14.4. The balanced chemical equation is $H_2SO_4(aq) + 2\ KOH(aq) \rightarrow 2\ H_2O(l) + K_2SO_4(aq)$

$22.87\ \text{mL KOH} \times \dfrac{0.158\ \text{mol KOH}}{1000\ \text{mL KOH}} \times \dfrac{1\ \text{mol H}_2\text{SO}_4}{2\ \text{mol KOH}} \times \dfrac{1}{0.020\ \text{L H}_2\text{SO}_4}$

$= 9.03 \times 10^{-2}\ M\ H_2SO_4$.

14.5. a. $HCHO_2$ is a weak acid, so it does not completely ionize in solution: $[H_3O^+] < 0.5$ M.
b. HI is a strong acid, so it completely ionizes: $[H_3O^+] = 1.25$ M.
c. HF is a weak acid, so it does not completely ionize in solution: $[H_3O^+] < 0.75$ M.

14.6. a. $Ba(OH)_2$ is a strong base, so it completely ionizes in solution. For every one $Ba(OH)_2$ that dissociates, 2 $OH^-(aq)$ are formed, so the $[OH^-] = 2 \times 0.055\ M = 0.11$ M.
b. C_5H_5N is a weak base, so it does not completely ionize in solution. Therefore, the $[OH^-] < 1.05$ M.
c. NaOH is a strong base, so it completely ionizes in solution giving $[OH^-] = 0.45$ M.

14.7. $K_w = [OH^-][H_3O^+] = 1.0 \times 10^{-14}$; $[H_3O^+] = \dfrac{K_w}{[OH^-]} = \dfrac{1.0 \times 10^{-14}}{[OH^-]}$

a. $[H_3O^+] = \dfrac{K_w}{[OH^-]} = \dfrac{1.0 \times 10^{-14}}{1.5 \times 10^{-2}} = 6.7 \times 10^{-13}$ M

Since $[H_3O^+] < 1.0 \times 10^{-7}$, the solution is basic.

b. $[H_3O^+] = \dfrac{K_w}{[OH^-]} = \dfrac{1.0 \times 10^{-14}}{1.0 \times 10^{-7}} = 1.0 \times 10^{-7}$ M

Since $[H_3O^+] = 1.0 \times 10^{-7}$, the solution is neutral.

c. $[H_3O^+] = \dfrac{K_w}{[OH^-]} = \dfrac{1.0 \times 10^{-14}}{8.2 \times 10^{-10}} = 1.2 \times 10^{-5}$ M

Since $[H_3O^+] > 1.0 \times 10^{-7}$, the solution is acidic.

14.8. a. $pH = -\log[H_3O^+] = -\log(9.5 \times 10^{-9}\ M) = 8.02$. With a pH > 7, the solution is basic.
b. $pH = -\log[H_3O^+] = -\log(6.1 \times 10^{-3}\ M) = 2.21$. With a pH < 7, the solution is acidic.

Plus. We must first calculate the $[H_3O^+]$:

$[H_3O^+] = \dfrac{K_w}{[OH^-]} = \dfrac{1.0 \times 10^{-14}}{1.3 \times 10^{-2}} = 7.7 \times 10^{-13}$ M

$pH = -\log[H_3O^+] = -\log(7.7 \times 10^{-13}\ M) = 12.11$; therefore, the solution is basic.

14.9. $pH = -\log[H_3O^+]$; $8.37 = -\log[H_3O^+]$; $\log^{-1}(-8.37) = \log^{-1}(\log[H_3O^+])$;
$\log^{-pH}(-8.37) = [H_3O^+]$; $[H_3O^+] = 4.3 \times 10^{-9}$ M.

Plus. $pH = -\log[H_3O^+]$; $3.66 = -\log[H_3O^+]$; $10^{-3.66} = 10^{\log [H_3O^+]}$; $[H_3O^+] = 2.2 \times 10^{-4}$ M;

$[OH^-] = \dfrac{K_w}{[H_3O^+]} = \dfrac{1.0 \times 10^{-14}}{2.188 \times 10^{-4}} = 4.6 \times 10^{-11}$ M

14.10. pOH = −log[OH⁻] ; 4.25 = −log[OH⁻] ; $\log^{-1}(-4.25) = \log^{-1}(\log[OH^-])$;
$\log^{-pOH}(-4.25) = [OH^-]$; $[OH^-] = 5.6 \times 10^{-5}$ M.

Plus. pOH = −log[OH⁻] ; 5.68 = −log[OH⁻] ; $10^{-5.68} = 10^{\log[OH^-]}$; $[OH^-] = 2.1 \times 10^{-6}$ M;

$$[OH^-] = \frac{K_w}{[H_3O^+]} = \frac{1.0 \times 10^{-14}}{2.1 \times 10^{-4}} = 4.8 \times 10^{-9} \text{ M}$$

Suggested Demonstrations

Make up a concentrated solution of baking soda (sodium bicarbonate) and pass it around the class. Let the students touch it to feel the slippery nature of bases.

Make 50 mL of a 6M NaOH solution and 50 mL of a 6 M HCl solution. Test each with a piece of litmus paper and talk about their reactivities. Mix the two solutions and test the resulting solution of saltwater. To emphasize the neutrality of the neutral solution, stick your fingers in the solution.

Colorful Acid–Base Indicators, *Chemical Demonstrations* 3:33, Shakhashiri, B. Z. University of Wisconsin Press, 1989.

Food Is Usually Acidic, Cleaners Are Usually Basic, *Chemical Demonstrations* 3:65, Shakhashiri, B. Z. University of Wisconsin Press, 1989.

Determination of Neutralizing Capacity of Antacids, *Chemical Demonstrations* 3:162, Shakhashiri, B. Z. University of Wisconsin Press, 1989.

Guided Inquiry Ideas

Below are a few example questions that students answer in the guided inquiry activities provided in the Guided Activity Workbook.

In a complete sentence or two, describe the relationship between the strength of an acid and the strength of its conjugate base.

In a complete sentence, state why it is commonly said that acids "dissolve" metals.

Explain how to find the H_3O^+ concentration of a solution from its pH. Make sure you address the sign.

What is the pH of a solution with $[H_3O^+] = 0.01$? $[H_3O^+] = 1.0 \times 10^{-7}$?

Explain how to find the pH of a solution from its H_3O^+ concentration. Make sure you address the sign.

Chapter 15: Chemical Equilibrium

Chapter Overview

Equilibrium is defined and shown to be all around us every day. The chemical definition, as well as "real-world" references, are presented. Appropriate calculations involved in equilibrium chemistry are shown and explained. In addition, solubility products and catalyst functions are introduced.

Lecture Outline

15.1 Life: Controlled Disequilibrium
15.2 The Rate of a Chemical Reaction
  Learning Objective: Identify and explain the relationship between concentration and temperature and the rate of a chemical reaction.
  A. Collision theory
  B. The effect of concentration on reaction rate
  C. The effect of temperature on reaction rate
15.3 The Idea of Dynamic Chemical Equilibrium
  Learning Objective: Define dynamic equilibrium.
  A. Reversible reactions
  B. Rate of forward reaction the same as rate of reverse reaction
15.4 The Equilibrium Constant: A Measure of How Far a Reaction Goes
  Learning Objective: Write equilibrium constant expressions for chemical reactions.
  A. Not the same amount of stuff on each side of equation
  B. Can predict equilibrium amounts using K_{eq}
  C. Products over reactants
  D. K_{eq} is temperature dependent
15.5 Heterogeneous Equilibria: The Equilibrium Expression for Reactions Involving a Solid or a Liquid
  Learning Objective: Write equilibrium expressions for chemical reactions involving a solid or a liquid.
  A. Pure solids and liquids are not included in K_{eq} expression
  B. Solid and liquid concentrations stay essentially constant during reaction
15.6 Calculating and Using Equilibrium Constants
  Learning Objective: Calculate equilibrium constants.
  Learning Objective: Use the equilibrium constant to find the concentration of a reactant or product at equilibrium.
  A. Calculating K_{eq} values
  B. Using K_{eq} values to predict equilibrium concentrations

15.7 Disturbing a Reaction at Equilibrium: Le Châtelier's Principle
    Learning Objective: Restate Le Châtelier's principle.
    A. When an equilibrium is disturbed, the system will attempt to get back to where it was
    B. No relationship to reaction rate

15.8 The Effect of Concentration Change on Equilibrium
    Learning Objective: Apply Le Châtelier's principle in the case of a change in concentration.
    A. Increasing the concentration of one or more reactants causes the reaction to shift right
    B. Increasing the concentration of one or more products causes the reaction to shift left

15.9 The Effect of Volume Change on Equilibrium
    Learning Objective: Apply Le Châtelier's principle in the case of a change in volume.
    A. Decreasing V causes the reaction to shift in the direction of fewer gas molecules
    B. Increasing V causes the reaction to shift in the direction of more gas molecules

15.10 The Effect of Temperature Change on Equilibrium
    Learning Objective: Apply Le Châtelier's principle in the case of a change in temperature.
    A. Exothermic: energy moves from system to surroundings
    B. Endothermic: energy moves from surroundings to system
    C. Increasing temperature shifts endothermic reactions right
    D. Increasing temperature shifts exothermic reactions left

15.11 The Solubility-Product Constant
    Learning Objective: Use Ksp to determine molar solubility.
    Learning Objective: Write an expression for the solubility-product constant.
    A. Used to calculate how much solid dissolves
    B. Temperature dependent
    C. K_{sp} related to molar solubility

15.12 The Path of a Reaction and the Effect of a Catalyst
    Learning Objective: Describe the relationship between activation energy and reaction rates and the role catalysts play in reactions.
    A. Activation energy
    B. Catalysts lower activation energy
    C. Enzymes: biological catalysts

Chemical Principle Teaching Ideas

Equilibrium
    All chemical reactions are really equilibria, but many of them lie far to one side of the reaction. Discuss chemical equilibria that the students would see in the real world so that they can internalize the idea.

Rates of Chemical Reactions
    Some reactions, such as the rusting of metal, proceed very slowly. Some reactions such as the combustion of hydrogen, go very quickly. Many students think that reactions either happen or do not happen. Emphasize that rate of reaction has no correlation to final equilibrium concentrations.

Dynamic Chemical Equilibrium

Draw a basketball court on the board and draw nine X's representing players on the bench. What happens at the beginning of the game? Five people get up from the bench (react) and get on to the court. This leaves four people on the bench. This represents the initial forward reaction. During the game (equilibrium) some players may sit down, and others will go in from the bench. What is important is that the number of players on the court should be constant. The players on the court might change, but throughout the entire game, there are five players on the court and four on the bench.

Le Chatelier's Principle

Most systems, when left to their own devices, reach a state of "happiness" in which each reactant has some fixed concentration. Left alone, they will stay "happy." When you disturb the system, however (for example, if you remove some part of one of the reactants), the system will become "unhappy" and will shift in the direction necessary to "undo" the change you made. It should be emphasized that you can never get back to the "old" concentrations before the change, but a new equilibrium is established.

Effect of Changes on Equilibrium

Some changes will not alter the equilibrium of a system. If the same number of moles of gas is present on the reactant and product sides, then a shift in the V (or P) will have no effect on the system.

Reaction Paths and Catalysts

The reaction pathway is sometimes referred to as the steps that a reaction goes through (intermediate products) to arrive at the end products. It should be made clear that the addition of a catalyst does not change the activation energy of one particular reaction pathway, but it facilitates the reaction taking a different path with a smaller activation energy, making the reaction "easier" for the system to initiate and sustain.

Skill Builder Solutions

15.1. Because the coefficient in front of the HF(g) is a 2, the [HF] in the numerator must be squared. The terms in the denominator are all single power as they both have a coefficient of 1.

$$K_{eq} = \frac{[HF]^2}{[H_2][F_2]}$$

15.2. Water is a pure liquid, so it is not included in the equilibrium expression. The Cl_2 has a coefficient of 2, so $[Cl_2]$ must be squared in the expression. Likewise, the [HCl] must be to the fourth power and the $[O_2]$ to the first, giving

$$K_{eq} = \frac{[Cl_2]^2}{[HCl]^4[O_2]}.$$

15.3. $K_{eq} = \dfrac{[CH_3OH]}{[H_2]^2[CO]} = \dfrac{0.151}{(0.146)^2(0.489)} = 14.5$

Plus.   The concentration of CO went down by 0.35 to 0.15, so [H$_2$] went down by 0.35 × 2 = 0.70 to the equilibrium concentration of 1.00 − 0.70 = 0.30. For every 1 mole of CO that reacts, 1 mole of CH$_3$OH is formed, so [CH$_3$OH] = 0.35. Plugging these numbers into the K_{eq} expression, we have

$K_{eq} = \dfrac{[CH_3OH]}{[H_2]^2[CO]} = \dfrac{0.35}{(0.30)^2(0.15)} = 26$

15.4. $K_{eq} = 0.011 = \dfrac{[I]^2}{[I_2]}$; $[I] = \sqrt{K_{eq}[I_2]} = \sqrt{0.011 \times .10} = 0.033$ M

15.5.   When you add Br$_2$ to the system, the reaction will try to reverse the change made, so it wants to reduce the concentration of Br$_2$. Thus, it will shift to the left. If you add BrNO, the reaction will shift right to decrease the amount of BrNO in the system, offsetting your perturbation.

Plus.   The reaction will try to get back to the concentration of [Br$_2$] that existed before the perturbations, so the reaction will shift to the right.

15.6.   The initial decrease in volume at constant temperature gives an increase in pressure (Boyle's law). Then in order to get the pressure back to a lower value, the system shifts right, where there are fewer moles of gas. Conversely, an increase in volume would cause the system to shift to the left.

15.7.   The reaction releases energy when proceeding to the right. To offset an increase in the temperature of the system, the reaction will shift in the direction that absorbs energy, which is to the left. Conversely, a decrease in the temperature of the reaction mixture would cause a shift to the right.

15.8.   a. AgI dissociates into 1 Ag$^+$ and 1 I$^-$, so $K_{sp} = [Ag^+][I^-]$
        b. Calcium hydroxide dissociates into 1 Ca^{2+} and 2 OH$^-$ species, so $K_{sp} = [Ca^{2+}][OH^-]^2$

15.9. $K_{sp} = 7.10 \times 10^{-5} = [Ca^{2+}][SO_4^{2-}] = S \times S = S^2$
        $S = \sqrt{K_{sp}} = \sqrt{7.10 \times 10^{-5}} = 8.43 \times 10^{-3}$ M

Suggested Demonstrations

Carefully fill a syringe with NO$_2$ gas from the reaction of HNO$_3$ and Cu. In front of the students, quickly adjust the volume of the syringe (with the syringe tightly capped). The color will change suddenly and then adjust. This comes from the equilibrium between NO$_2$ (brown) and N$_2$O$_4$ (colorless).

Disproportionation of Acidified Sodium Thiosulfate, *Chemical Demonstrations* 4:77, Shakhashiri, B. Z. University of Wisconsin Press, 1992.

Guided Inquiry Ideas

Below are a few example questions that students answer in the guided inquiry activities provided in the Guided Activity Workbook.

In a complete sentence, describe how the rate of a reaction depends on the concentrations of the reactants?

In a complete sentence, describe how the rate of a reaction depends on the temperature.

Which factor seems to make a bigger difference in the rate: concentration or temperature? Explain your answer.

Calcium sulfate, $CaSO_4$, is only very slightly soluble in water. Write the equilibrium constant expression for the dissolution of calcium sulfate.

Do you expect the numerical value for the K_{sp} for calcium sulfate to be a big number or a small number? Why?

Oxidation and Reduction 16

Chapter Overview

The definition of oxidation–reduction reactions is given, using several examples. The many ways to detect such a reaction are also explained. Everyday oxidation–reduction reactions are shown.

Lecture Outline

16.1 The End of the Internal Combustion Engine?
16.2 Oxidation and Reduction: Some Definitions
    <u>Learning Objective:</u> Define and identify oxidation and reduction.
    <u>Learning Objective:</u> Identify oxidizing agents and reducing agents.
    A. Oxidation is a loss of electrons
    B. Reduction is a gain of electrons
    C. Oxidizing agent is the substance being reduced
    D. Reducing agent is the substance being oxidized
16.3 Oxidation States: Electron Bookkeeping
    <u>Learning Objective:</u> Assign oxidation states.
    <u>Learning Objective:</u> Use oxidation states to identify oxidation and reduction.
    A. Atom in free element = 0
    B. Oxidation states of all atoms in species add up to charge of species
    C. Group I metals in compounds are +1
    D. Group II metals in compounds are +2
    E. Oxidation is an increase in oxidation state
    F. Reduction is a decrease in oxidation state
16.4 Balancing Redox Equations
    <u>Learning Objective:</u> Balance redox reactions.
    A. Assign oxidation states to all atoms
    B. Separate into half-reactions and balance
    C. Make the number of electrons gained equal to the number lost
    D. Add the half-reactions together and cancel electrons
16.5 The Activity Series: Predicting Spontaneous Redox Reactions
    <u>Learning Objective:</u> Predict spontaneous redox reactions.
    <u>Learning Objective:</u> Predict whether a metal will dissolve in acid.
    A. Some species are more apt to oxidize
    B. Use activity series to predict reaction spontaneity
    C. Use activity series to predict if a metal will dissolve in acid

16.6 Batteries: Using Chemistry to Generate Electricity
    Learning Objective: Describe how a voltaic cell functions.
    Learning Objective: Compare and contrast the various types of batteries.
    A. Electrochemical cell
    B. Two half-cells, wire, and a salt bridge
    C. Anode versus cathode
    D. Galvanic = voltaic cell
    E. Dry cells
    F. Lead–acid storage battery
    G. Fuel cells
16.7 Electrolysis: Using Electricity to Do Chemistry
    Learning Objective: Describe the process of electrolysis and how an electrolytic cell functions.
    A. Reverse a spontaneous reaction
    B. Get pure metal from metal oxides
16.8 Corrosion: Undesirable Redox Reactions
    Learning Objective: Describe the process of corrosion and the various methods used to prevent rust.

Chemical Principle Teaching Ideas

Oxidation and Reduction
    Many reactions that the students have already seen are redox equations, but the students do not necessarily see the change in oxidation state. The combustion of hydrogen and oxygen is a good example of a simple reaction involving oxidation and reduction. Emphasize that if you have oxidation occurring in a reaction, you must have a reduction occurring. You cannot have one without the other.

Oxidation States
    Make sure the students do not confuse oxidation state with ion charge, as they are not the same. Also mention that, in a redox reaction, you need at least one species to be reduced and at least one species to be oxidized. However, there is no rule that all of the species in a redox reaction gain or lose electrons.

The Activity Series
    The activity series ranks substances from most easily oxidized to least easily oxidized.

Batteries
    Explain how energy is released when a spontaneous redox reaction occurs and shows that the farther away the species are on the activity series, the more voltage you get.

Electrolysis
	What many students do not understand is the prevalence of electrolysis in their everyday lives. Coatings on jewelry, bridges, and metal dishes are all achieved by electrolysis.

Corrosion
	An example of corrosion with which the students would be familiar is the oxidation of Al in aluminum soda or beer cans. The acidity of the soda combined with the dissolved oxygen give canned drinks their distinctive metallic taste. Not all corrosion is bad. Some bridges are coated with a layer of metal that corrodes to form a protective layer covering the rest of the piece of metal.

Skill Builder Solutions

16.1. a. The charge of K on the reactant side is 0, and the Cl_2 has a charge of 0 as well. On the right-hand side, K is in a +1 state and Cl is in a −1 state. K lost one electron, so it was oxidized; Cl gained the electron, so it was reduced.
   b. Al goes from a 0 charge on the reactant side to a +3 state. Hence, by losing 3 electrons, it is oxidized. The Sn^{2+} ions gain electrons and are thus reduced.
   c. Carbon gains oxygen and is thus oxidized. Oxygen, the other reactant, must therefore be reduced.

16.2. a. K was oxidized, so it was the reducing agent. Conversely, Cl was reduced, so it was the oxidizing agent.
   b. Al was oxidized, so it was the reducing agent. Conversely, Sn^{+2} was being reduced, so it was the oxidizing agent.
   c. C was oxidized, so it was the reducing agent. Conversely, the O_2 was reduced, and it was the oxidizing agent.

16.3. a. 0; lone neutral atoms are always 0 by rule 1.
   b. +2; lone atomic ion is the charge of the ion by rule 2.
   c. Cl has an oxidation state of −1 by rule 5. Ca is +2 by rule 4.
   d. F has an oxidation state of −1 by rule 5. Since there are four fluorine atoms for every one carbon atom, the carbon must have an oxidation state of +4 by rule 3.
   e. Oxygen is in a −2 state by rule 5. There are two oxygen atoms for a total negative charge of −4. By rule 3, the oxidation states of all the atoms must add to the charge of the species, so the N must have a charge of +3.
   f. Oxygen is in a −2 state by rule 5, for a total negative charge of −6. The one sulfur must therefore be in a +6 state by rule 3.

16.4. $\overset{0}{Sn}(s) + 4\ \overset{+1\ +5\ -2}{HNO_3}(aq) \rightarrow \overset{+4\ -2}{SnO_2}(s) + 4\ \overset{+4\ -2}{NO_2}(g) + 2\ \overset{+1\ -2}{H_2O}(g)$

Therefore, the Sn is oxidized and the N is reduced.

16.5. Write the two skeletal half-reactions:
$H^+(aq) \rightarrow H_2(g)$
$Cr(s) \rightarrow Cr^{+3}(aq)$
Balance the hydrogen in the first half-reaction; then balance the charge in both:
$2 H^+(aq) + 2 e^- \rightarrow H_2(g)$
$Cr(s) \rightarrow Cr^{+3}(aq) + 3 e^-$
To make the number of electrons in each half-reaction the same, we must multiply the first half-reaction by 3 and the second by 2, giving
$6 H^+(aq) + 6 e^- \rightarrow 3 H_2(g)$
$2 Cr(s) \rightarrow 2 Cr^{+3}(aq) + 6 e^-$
Adding them together and canceling electrons, yield
$6 H^+(aq) + 2 Cr(s) \rightarrow 2 Cr^{+3}(aq) + 3 H_2(g)$

16.6. Two skeletal half-reactions:
$Cu(s) \rightarrow Cu^{+2}(aq)$
$NO_3^-(aq) \rightarrow NO_2(g)$
The oxygens are balanced in the second equation by adding 1 H_2O to the right:
$NO_3^-(aq) \rightarrow NO_2(g) + H_2O(l)$
To balance the hydrogen atoms, we put 2 H^+ on the left side:
$2 H^+(aq) + NO_3^-(aq) \rightarrow NO_2(g) + H_2O(l)$
Placing electrons in the appropriate places to balance charge, we get
$2 H^+(aq) + NO_3^-(aq) + 1 e^- \rightarrow NO_2(g) + H_2O(l)$
$Cu(s) \rightarrow Cu^{+2}(aq) + 2 e^-$
To have the same number of electrons in each half-reaction, we must multiply the first half-reaction by 2:
$4 H^+(aq) + 2 NO_3^-(aq) + 2 e^- \rightarrow 2 NO_2(g) + 2 H_2O(l)$
$Cu(s) \rightarrow Cu^{+2}(aq) + 2 e^-$
Combining the two reactions gives
$4 H^+(aq) + 2 NO_3^-(aq) + Cu(s) \rightarrow 2 NO_2(g) + 2 H_2O(l) + Cu^{+2}(aq)$

16.7. Two skeletal half-reactions:
$Sn(s) \rightarrow Sn^{+2}(aq)$
$MnO_4^-(aq) \rightarrow Mn^{+2}(aq)$
To balance the second equation with respect to O, add 4 H_2O on the right:
$MnO_4^-(aq) \rightarrow Mn^{+2}(aq) + 4 H_2O(l)$
To balance the same equation now with respect to H atoms, add 8 H^+ to the left:
$8 H^+(aq) + MnO_4^-(aq) \rightarrow Mn^{+2}(aq) + 4 H_2O(l)$
Add electrons to balance the charge in each half-reaction:
$8 H^+(aq) + MnO_4^-(aq) + 5 e^- \rightarrow Mn^{+2}(aq) + 4 H_2O(l) +$
$Sn(s) \rightarrow Sn^{+2}(aq) + 2 e^-(aq)$
The lowest multiple of 2 and 5 is 10, so we must multiply the first half-reaction by 2, and the second half-reaction by 5, yielding
$16 H^+(aq) + 2 MnO_4^-(aq) + 10 e^- \rightarrow 2 Mn^{+2}(aq) + 8 H_2O(l)$
$5 Sn(s) \rightarrow 5 Sn^{+2}(aq) + 10 e^-(aq)$
Adding the two equations yields
$5 Sn(s) + 2 MnO_4^-(aq) + 16 H^+(aq) \rightarrow 2 Mn^{+2}(aq) + 8 H_2O(l) + 5 Sn^{+2}(aq)$

16.8. Two skeletal half-reactions:
$ClO_2(aq) \rightarrow ClO_2^-(aq)$
$H_2O_2(aq) \rightarrow O_2(g)$
The first reaction is already balanced with respect to O and H.
The second equation is already balanced with respect to O. To balance the H atoms, add 2 H^+ on the right:
$H_2O_2(aq) \rightarrow O_2(g) + 2\ H^+(aq)$
Because the reaction is in a basic solution, now add enough OH^- to both sides to cancel the H^+ ions, in this case 2 OH^-, which form 2 $H_2O(l)$ on the right:
$H_2O_2(aq) + 2\ OH^-(aq) \rightarrow O_2(g) + $ **2 $H^+(aq)$ + 2 $OH^-(aq)$**
$H_2O_2(aq) + 2\ OH^-(aq) \rightarrow O_2(g) + $ **2 $H_2O(l)$**
Add electrons to balance the charge in each half-reaction:
$ClO_2(aq) + 1\ e^-(aq) \rightarrow ClO_2^-(aq)$
$H_2O_2(aq) + 2\ OH^-(aq) \rightarrow O_2(g) + 2\ H_2O(l) + 2\ e^-$
The lowest multiple of 1 and 2 is 2, so we must multiply the first half-reaction by 2, yielding
$H_2O_2(aq) + 2\ OH^-(aq) \rightarrow O_2(g) + 2\ H_2O(l) + 2\ e^-$
$2\ ClO_2(aq) + 2\ e^- \rightarrow 2\ ClO_2^-(aq)$
Adding the two equations yields
$H_2O_2(aq) + 2\ OH^-(aq) + 2\ ClO_2(aq) \rightarrow O_2(g) + 2\ H_2O(l) + 2\ ClO_2^-(aq)$

16.9. a. The oxidation of Zn is above the reaction of Ni (Ni $\rightarrow$ Ni^{+2}) in the activity series, so the reaction is spontaneous.
b. The oxidation of Zn is below the reaction of Ca (Ca $\rightarrow$ Ca^{+2}) in the activity series, so the reaction is not spontaneous.

16.10. Ag $\rightarrow$ Ag$^+$ + e$^-$ is below H$_2$ on the activity series, so Ag will not dissolve in HBr.

Suggested Demonstrations

Fill a test tube with a very dilute solution of Cu^{+2} ion, place a small amount of solid Zn inside, and shake. Explain why the blue color is leaving the solution as the electrons are transferred from the Zn to the Cu^{2+}.

Galvanizing: Zinc Plating, *Chemical Demonstrations* 4:244, Shakhashiri, B. Z. University of Wisconsin Press, 1992.

Electricity from a Fuel Cell, *Chemical Demonstrations* 4:123, Shakhashiri, B. Z. University of Wisconsin Press, 1992.

Forming a Copper Mirror, *Chemical Demonstrations* 4:224, Shakhashiri, B. Z. University of Wisconsin Press, 1992.

Guided Inquiry Ideas

Below are a few example questions that students answer in the guided inquiry activities provided in the Guided Activity Workbook.

Does oxidation always involve combination with oxygen? If not, give a counter example.

When a lithium atom changes from being in elemental Li(*s*) to being in LiCl, does it gain an electron from chlorine or does it loose an electron to chlorine? Explain.

Does electronegativity increase as you go up or down on the periodic table?

What is the most electronegative element in CO_2?

If CO_2 were ionic (it isn't, of course) what would the charge be on each oxygen?

Radioactivity and Nuclear Chemistry

17

Chapter Overview

Chapter 17 gives us an introduction to the history of radiation, from its discovery to how it has changed the history of humankind in the last 100 years. The main kinds of radiation are explained, and appropriate reactions are discussed. The current and future implications of fission and fusion are also discussed.

Lecture Outline

17.1 Diagnosing Appendicitis
17.2 The Discovery of Radioactivity
    <u>Learning Objective:</u> Explain how the experiments of Becquerel and Curie led to the discovery of radioactivity.
    A. Becquerel
    B. Curie
17.3 Types of Radioactivity: Alpha, Beta, and Gamma Decay
    <u>Learning Objective:</u> Write nuclear equations for alpha decay.
    <u>Learning Objective:</u> Write nuclear equations for beta decay.
    <u>Learning Objective:</u> Write nuclear equations for positron emission.
    A. Alpha: 2 protons + 2 neutrons
    B. Beta: high-energy electron
    C. Gamma: electromagnetic radiation with no mass
    D. Positron: mass of electron but +1 charge
17.4 Detecting Radioactivity
    <u>Learning Objective:</u> Describe and explain the methods used to detect radioactivity.
    A. Film badge dosimeter
    B. Geiger–Müller counter
    C. Scintillation counter
17.5 Natural Radioactivity and Half-Life
    <u>Learning Objective:</u> Use half-life to relate radioactive sample amounts to elapsed time.
    A. Half-life: time for half of sample to disintegrate
    B. Radioactive decay series
17.6 Radiocarbon Dating: Using Radioactivity to Measure the Age of Fossils and Other Artifacts
    <u>Learning Objective:</u> Use carbon-14 content to determine the age of fossils or artifacts.
    A. Living organisms have constant concentration of carbon-14
    B. When no longer "living," the amount of carbon-14 slowly decreases

17.7 The Discovery of Fission and the Atomic Bomb
  Learning Objective: Explain how the experiments of Fermi, Meitner, Strassmann, and Hahn led to the discovery of nuclear fission.
  A. Fermi, Meitner, Strassmann, and Hahn
  B. Fission: splitting of large atom
  C. Chain reaction
  D. Critical mass
  E. Manhattan Project
17.8 Nuclear Power: Using Fission to Generate Electricity
  Learning Objective: Explain how nuclear power plants generate electricity using fission.
  A. Reactor
  B. Control rods
  C. Steam generator
  D. Containment building
17.9 Nuclear Fusion: The Power of the Sun
  Learning Objective: Compare and contrast nuclear fission and nuclear fusion.
  A. Fusion: two light atoms combine to form a heavier atom
  B. Reaction on the sun
  C. Much more energetic than fission
17.10 The Effects of Radiation on Life
  Learning Objective: Describe how radiation exposure affects biological molecules and how exposure is measured.
  A. Acute radiation damage
  B. Increased cancer risk
  C. Genetic defects
  D. Measuring radiation exposure
17.11 Radioactivity in Medicine
  Learning Objective: Describe how radioactivity is used in the diagnosis and treatment of disease.
  A. Isotope scanning
  B. Radiotherapy

Chemical Principle Teaching Ideas

Forms of Radiation
	Students rarely realize that they are bombarded with radiation every day as they go about their normal routines. Giving them real-world examples of radiation will help them internalize the concept.

Detecting Radioactivity
	The methods of detecting radiation range from very expensive measurement tools to a piece of unexposed film. Many of the most sophisticated measurement tools are used to look at the radiation coming from outer space to help scientists determine how the universe was formed.

Half-Life and Radiocarbon Dating
	Only species that were once alive can be radiocarbon dated. Non living things such as rocks are not in equilibrium with the carbon dioxide in the atmosphere, so they never get

carbon-14 in them to decay and be measured. It should be emphasized that anything over 50,000 years old cannot be radiocarbon dated, but there are other isotopes that can be used to measure time besides carbon-14. Some of these isotopes are useful for artifacts that never lived.

Fission, the Atomic Bomb, and Nuclear Power
   This is a delicate subject, as many students will already have formed a previous opinion about nuclear power–usually negative. This view generally stems from slanted background information they may have gotten from the media. Use this as an opportunity to give the students both sides of this controversial issue.

Nuclear Fusion
   Fusion is much more powerful than fission and, at this point, is so energetic that we cannot harness the power as we can do with fission. Many projects are being carried out in an attempt to harness fusion energy for the traditional energy needs of the world population. If this can be achieved, power will become very cheap because the raw ingredients for fusion reactions are inexpensive and can be obtained from ocean water.

The Effects of Radiation on Life and Nuclear Medicine
   Most students consider radiation a "dirty" word, having only negative connotations. What they sometimes don't understand is that radiation has many applications that have a positive impact on our everyday lives.

Skill Builder Solutions

17.1. Let us first write what we know from the following question:
$^{216}_{84}Po \rightarrow \, ^{4}_{2}He + \, ?$
The sum of the atomic numbers on both sides of a nuclear equation must be equal, and the sum of the mass numbers must also be equal. So the other species on the product side must have an atomic number of 84 – 2 = 82 and a mass number of 216 – 4 = 212. Any species that has an atomic number of 82 is lead, giving
$^{216}_{84}Po \rightarrow \, ^{4}_{2}He + \, ^{212}_{82}Pb$.

17.2. What is given in the following question:
$^{228}_{89}Ac \rightarrow \, ^{0}_{-1}e + \, ?$
The sum of the atomic numbers on both sides of a nuclear equation must be equal, and the sum of the mass numbers must also be equal. So the product must have an atomic number of 89 – (–1) = 90, and a mass number of 228 – 0 = 228. Any species with 90 protons is a thorium, giving
$^{228}_{89}Ac \rightarrow \, ^{0}_{-1}e + \, ^{228}_{90}Th$.

Plus. Step 1: $^{235}_{92}U \rightarrow \, ^{4}_{2}He + \, ?$ The other product must have an atomic number of 92 – 2 = 90 and a mass number of 235 – 4 = 231, giving
$^{235}_{92}U \rightarrow \, ^{4}_{2}He + \, ^{231}_{90}Th$.

Step 2: $^{231}_{90}Th \rightarrow \, ^{0}_{-1}e + \, ?$ The other product must have an atomic number of 90 – (–1) = 91 and a mass number of 231, giving

$${}^{231}_{90}\text{Th} \rightarrow {}^{0}_{-1}\text{e} + {}^{231}_{91}\text{Pa}.$$

Step 3: ${}^{231}_{91}\text{Pa} \rightarrow {}^{4}_{2}\text{He} + ?$ The other product has a mass number of $231 - 4 = 227$ and an atomic number of $91 - 2 = 89$, yielding

$${}^{231}_{92}\text{Pa} \rightarrow {}^{4}_{2}\text{He} + {}^{227}_{89}\text{Ac}.$$

17.3. From the given information, we have
$${}^{22}_{11}\text{Na} \rightarrow {}^{0}_{+1}\text{e} + ?$$
The sum of the atomic numbers on both sides of a nuclear equation must be equal, and the sum of the mass numbers must also be equal. Therefore, the other product must have an atomic number of $11 - 1 = 10$, (Ne) and a mass number of $22 - 0 = 22$, giving
$${}^{22}_{11}\text{Na} \rightarrow {}^{0}_{+1}\text{e} + {}^{22}_{10}\text{Ne}.$$

17.4. After every half life, the amount of substance has been reduced by half. After 1600 years, $\dfrac{0.112 \text{ mol Ra}}{2} = 0.0560$ mol Ra remains. After another half-life (3200 years total), only $\dfrac{0.0560 \text{ mol Ra}}{2} = 0.0280$ mol Ra remains. After another half-life ($3200 + 1600 = 4800$ years), there remains $\dfrac{0.0280 \text{ mol Ra}}{2} = 0.0140$ mol Ra. After one more half-life ($4800 + 1600 = 6400$ years), there is only $\dfrac{0.0140 \text{ mol Ra}}{2} = 7.00 \times 10^{-3}$ mol Ra remaining.

17.5. The fact that none of the carbon-14 in the scroll has disintegrated implies that the object has not been "dead" for a significant period of time, meaning the scroll was recently made.

Guided Inquiry Ideas

Below are a few example questions that students answer in the guided inquiry activities provided in the Guided Activity Workbook.

What is the total charge before ${}^{238}_{92}\text{U}$ undergoes α decay?

What is the total charge after ${}^{238}_{92}\text{U}$ undergoes α decay?

In a complete sentence, write a rule regarding charge and mass during nuclear reactions.

If half of a sample of thorium-232 decays in 14 billion years, does it *all* decay in 28 years? Explain.

Organic Chemistry

18

Chapter Overview

Organic chemistry is a broad domain of chemistry, and it is very important to our world, especially to us humans but to other biological organisms as well. The simplest groups of organic molecules are defined in this chapter and named using a standard system.

Lecture Outline

18.1 What Do I Smell?
18.2 Vitalism: The Difference Between Organic and Inorganic
    Learning Objective: Explain how the experiments of Wöhler were fundamental to the development of organic chemistry.
    A. Vitalism
    B. Wöhler
18.3 Carbon: A Versatile Atom
    Learning Objective: Identify the unique properties that allow carbon to form such a large number of compounds.
    A. Many combinations of four bonds
    B. Chains, branch structures, and rings
18.4 Hydrocarbons: Compounds Containing Only Carbon and Hydrogen
    Learning Objective: Differentiate between alkanes, alkenes, and alkynes based on molecular formulas.
    A. Saturated–alkanes
    B. Unsaturated–alkenes, alkynes, and aromatics
18.5 Alkanes: Saturated Hydrocarbons
    Learning Objective: Write formulas for *n*-alkanes.
    A. Carbon containing only single bonds
    B. Condensed structural formula
    C. Normal alkanes
18.6 Isomers: Same Formula, Different Structure
    Learning Objective: Write structural formulas for hydrocarbon isomers.
18.7 Naming Alkanes
    Learning Objective: Name alkanes.
    A. Find base chain
    B. Add substituents, called alkyl groups, giving the lowest total of location numbers
    C. Use prefixes with alkyl group names when two or more of the same are present

18.8 Alkenes and Alkynes
    <u>Learning Objective:</u> Name alkenes and alkynes.
    A. Alkenes have at least one double bond
    B. Alkynes have at least one triple bond
    C. Unsaturated hydrocarbon has at least one multiple bond
    D. Naming
        1. Alkenes: Base chain must contain the double bond; base name ends with *-ene*
        2. Alkynes: Base chain must contain the triple bond; base name ends with *-yne*

18.9 Hydrocarbon Reactions
    <u>Learning Objective:</u> Compare and contrast combustion, substitution, and addition reactions.
    A. Combustion
    B. Substitution
    C. Addition

18.10 Aromatic Hydrocarbons
    <u>Learning Objective:</u> Name aromatic hydrocarbons.
    A. Contain benzene ring
    B. Monosubstituted
    C. Disubstituted
    D. Naming

18.11 Functional Groups
    <u>Learning Objective:</u> Identify common functional groups and families of organic compounds.
    A. Alcohols – ROH
    B. Ethers – ROR
    C. Aldehydes – RCHO
    D. Ketones – RCOR
    E. Carboxylic acids – RCOOH
    F. Esters – RCOOR
    G. Amines – R_3N

18.12 Alcohols
    <u>Learning Objective:</u> Identify alcohols and their properties.
    A. Naming alcohols
        1. Base chain must contain the –OH group
        2. Base name ends with *–ol*
    B. Properties of alcohols

18.13 Ethers
    <u>Learning Objective:</u> Identify common ethers.
    A. Naming of ethers
    B. Properties of ethers

18.14 Aldehydes and Ketones
    <u>Learning Objective:</u> Identify aldehydes and ketones and their properties.
    A. Naming
        1. Aldehyde names end with *–al*
        2. Ketone names end with *–one*
    B. Properties of aldehydes and ketones

18.15 Carboxylic Acids and Esters
　　Learning Objective: Identify carboxylic acids and esters and their properties.
　　A. Naming
　　　　1. Carboxylic acid names end with –*oic acid*
　　　　2. Esters end with –*ate*
　　B. Properties of carboxylic acids and esters
18.16 Amines
　　Learning Objective: Identify amines and their properties.
　　A. Naming
　　B. Properties of amines
18.17 Polymers
　　Learning Objective: Identify the unique properties of polymers.
　　A. Monomers
　　B. Addition polymer
　　C. Copolymer
　　D. Condensation polymer
　　E. Dimer

Chemical Principle Teaching Ideas

Organic Chemistry
　　Organic chemistry is the study of carbon-containing compounds. Almost all biological processes involve organic chemistry. Students can appreciate that their bodies are stockpiles of organic chemistry reactions. Carbon is the most versatile of elements, and you can emphasize its versatility by showing a few reactions to illustrate its unique properties.

Isomers
　　The easiest way to talk about isomers is to bring in a chemical model kit with some large, simple hydrocarbons in a straight-chain formation. Hold it up and have students name the species. Then start switching a central H with a methyl or an ethyl group removed from the end of the chain. Have students name the new species, and then do it again. This is good practice in naming organic molecules, and it will give them a visual definition of isomer.

Functional Groups
　　The easiest way to remember these functional groups is to memorize the simplest form of each different kind, using them as reference species for naming.

Polymers
　　All plastics are polymers, and most plastics will tell you what kind they are if you find the number inside of the triangle. For example, a plastic recycling code of 2 is for high-density polyethylene (HDPE). Explain the difference between the various kinds of plastics and what they actually are.

Skill Builder Solutions

18.1.　a. The number of hydrogen atoms is 2 times the number of carbons, so C_6H_{12} must be an alkene.
　　　b. The number of hydrogen atoms is 2 times the number of carbons – 2, so C_8H_{14} is an alkyne.
　　　c. The number of hydrogen atoms is 2 times the number of carbon atoms + 2, so C_5H_{12} is an alkane.

18.2. First, we draw five carbons connected by single bonds:

C—C—C—C—C

The next step is to draw hydrogen atoms around each carbon giving them all four bonds.

```
    H   H   H   H   H
    |   |   |   |   |
H—C—C—C—C—C—H
    |   |   |   |   |
    H   H   H   H   H
```

The condensed formula contains each C atom, followed by the number of hydrogen atoms bonded to it, $CH_3CH_2CH_2CH_2CH_3$.

18.3. First, draw the three possible carbon backbones:

```
C—C—C—C—C              C
                        |
                    C—C—C
C—C—C—C                 |
    |                   C
    C
```

Then add hydrogen atoms to each carbon giving each four bonds.

```
    H   H   H   H   H              H
    |   |   |   |   |              |
H—C—C—C—C—C—H              H—C—H
    |   |   |   |   |          H   |   H
    H   H   H   H   H          |   |   |
                           H—C—C—C—H
                               |   |   |
                               H   |   H
                                 H—C—H
    H   H   H   H                  |
    |   |   |   |                  H
H—C—C—C—C—H
    |   |   |   |
    H   H   |   H
          H—C—H
            |
            H
```

18.4. The longest chain is 3 carbons long, so the base name is propane. There is one substituent with one carbon (hence called methyl) not located on the longest chain. Normally, you would need to include the location of that methyl group, but since it can only be on the second carbon and still not be in the longest chain, the name is simply methylpropane.

18.5. The longest chain is 5 carbons long and will be numbered from the left. The base name for 5 carbon atoms is pentane. There are two substituents, one a methyl group on carbon 2 and one an ethyl group on carbon atom 3. They are placed in the molecule name in alphabetical order, so the name is 3-ethyl-2-methylpentane.

18.6. The longest chain is 6 atoms long, so the base name is hexane. There are two possible numbering methods, one from each side. From one side, the numbers of the substituents would be 2, 3, and 5. From the other direction, the numbers would be 2, 4, and 5. We chose the one that gives the smallest total, so we will count from right to left. There are three methyl groups (hence the prefix *tri-*), giving the name 2,3,5-trimethylhexane.

18.7. a. The longest carbon chain is 5 atoms long, so the base name is pentane. There is a triple bond between carbon atoms 2 and 3, so the base name changes to 2-pentyne. There are two methyl groups (hence the prefix *di-*), both bonded to carbon 4, so the name is 4,4-dimethyl-2-pentyne.
b. The longest carbon chain is 7 atoms long, so the base name is heptane. There is a double bond between carbon atom 1 and 2, so the base name changes to 1-heptene. There is one ethyl group bonded to atom 3 and two (*di-*) methyl groups bonded to carbons 4 and 6. Putting these in alphabetical order gives 3-ethyl-4,6-dimethyl-1-heptene.

18.8. There are two of the same species bonded to the benzene, so it does not matter which C (-Br) we start numbering with, as the numbering will be 1,3 either way. The name is therefore 1,3-dibromobenzene. By the other naming scheme, the molecule is *meta*-dibromobenzene or *m*-dibromobenzene because the Br atoms are separated by one C atom.

Suggested Demonstrations

Mix a small amount (~ 1 tsp) of sodium polyacrylate (used in some baby diapers) with ~ 50 mL of distilled water. The sodium polyacrylate will make a polymer chain with the water. The addition of solid sodium chloride will break down the polymer.

Combustion of Methane, *Chemical Demonstrations* 1:113, Shakhashiri, B. Z. University of Wisconsin Press, 1983.

Guided Inquiry Ideas

Below are a few example questions that students answer in the guided inquiry activities provided in the Guided Activity Workbook.

What do the structures of all ethers have in common?

What do the names of all ethers have in common?

What do the structures of all alcohols have in common?

What do the names of all alcohols have in common?

Draw 3-heptanol and methyl ethyl ether.

Biochemistry 19

Chapter Overview

Chapter 19 introduces us to the complex world of the chemistry of biological organisms. The complete study is well beyond the scope of this text, but the introduction will give an appreciation for the complexities of life on the molecular scale. The primary components of biological organisms and their molecular interactions are given.

Lecture Outline

19.1 The Human Genome Project
    A. Blueprint for life
    B. Biochemistry
19.2 The Cell and Its Main Chemical Components
    <u>Learning Objective:</u> Identify the key chemical components of the cell.
    A. Nucleus
    B. Cell membrane
    C. Cytoplasm
19.3 Carbohydrates: Sugar, Starch, and Fiber
    <u>Learning Objective:</u> Identify carbohydrates and compare and contrast monosaccharides, disaccharides, and polysaccharides.
    A. Monosaccharides
    B. Disaccharides
    C. Polysaccharides
        1. Starch
        2. Cellulose
        3. Glycogen
19.4 Lipids
    <u>Learning Objective:</u> Identify lipids.
    <u>Learning Objective:</u> Compare and contrast saturated and unsaturated triglycerides.
    A. Fatty acids
    B. Fats and oils
    C. Other lipids
        1. Phospholipids
        2. Glycolipids
        3. Steroids
19.5 Proteins
    <u>Learning Objective:</u> Identify proteins.
    <u>Learning Objective:</u> Describe how amino acids link together to form proteins.
    A. Polymers of amino acids
    B. Enzymes
    C. Peptides

19.6 Protein Structure
　　Learning Objective: Describe primary structure, secondary structure, tertiary structure, and quaternary structure in proteins.
　　A. Primary structure
　　B. Secondary structure
　　　　1. Pleated sheet
　　　　2. Helix
　　　　3. Random coil
　　C. Tertiary structure
　　　　1. Hydrogen bonds
　　　　2. Disulfide linkages
　　　　3. Hydrophilic interactions
　　　　4. Salt bridges
　　D. Quaternary structure
19.7 Nucleic Acids: Molecular Blueprints
　　Learning Objective: Describe the role that nucleic acids play in determining the order of amino acids in a protein.
　　A. DNA
　　B. RNA
　　C. Nucleotides
　　　　1. Codon
　　　　2. Gene
　　　　3. Chromosome
19.8 DNA Structure, DNA Replication, and Protein Synthesis
　　Learning Objective: Summarize the process of DNA replication and protein synthesis.
　　A. DNA structure
　　B. DNA replication
　　C. Protein synthesis

Chemical Principle Teaching Ideas

The Cell
　　Many students are familiar with the most basic ideas of cells from biology, but they fail to see that there is something going on in cells that is much smaller than what they learned in biology class, the chemistry happening on the molecular level.

Carbohydrates
　　Students are familiar with carbohydrates, having read about them on a food ingredient label. Show a few such labels and talk about the differences between simple carbohydrates and more complex ones.

Lipids
　　Different lipids do different things, but all lipids have chemical similarities. Some of them are very important for life (fatty acids), and some are illegally taken by athletes (steroids).

Proteins
Most students think of protein simply as a necessary component of their everyday diet. Make sure to emphasize that this is a very narrow view of a most complex group of molecules.

Nucleic Acids and DNA Replication
Understanding life means understanding DNA replication. DNA replication makes life possible as we know it; yet how it occurs on the molecular level has not been known for very long. By understanding the number of possibilities for combining the four bases, the student can begin to understand how complex life really is.

Skill Builder Solutions

19.1.   a. Carboxylic acids are not carbohydrates.
b. This species has an oxygen atom in the ring, with several OH groups also attached to the ring. Its formula, $C_6H_{12}O_6$, matches the generic form $(CH_2O)_m$ with m = 6. Therefore, it is a carbohydrate. There is only one ring, making it a monosaccharide.
c. Carboxylic acids are not carbohydrates.
d. There are two O-containing rings, both containing multiple OH groups. There are two rings, making this a disaccharide.

19.2.   a. There is no three-carbon glycerol backbone to this species, so it is not a triglyceride.
b. There is a three-carbon glycerol backbone with a long fatty acid tail, so it is a triglyceride. The carbon chains contain double bonds, so this is an unsaturated fat.
c. There is no three-carbon glycerol backbone to this species, so it is not a triglyceride.
d. There is a three-carbon glycerol backbone with a long fatty acid tail; hence, it is a triglyceride. The carbon chains contain only single bonds, so this is a saturated fat.

19.3.   Peptide bonds are formed when the carboxylic end of one amino acid reacts with the amine end of a second amino acid to form a dipeptide and water as shown here.

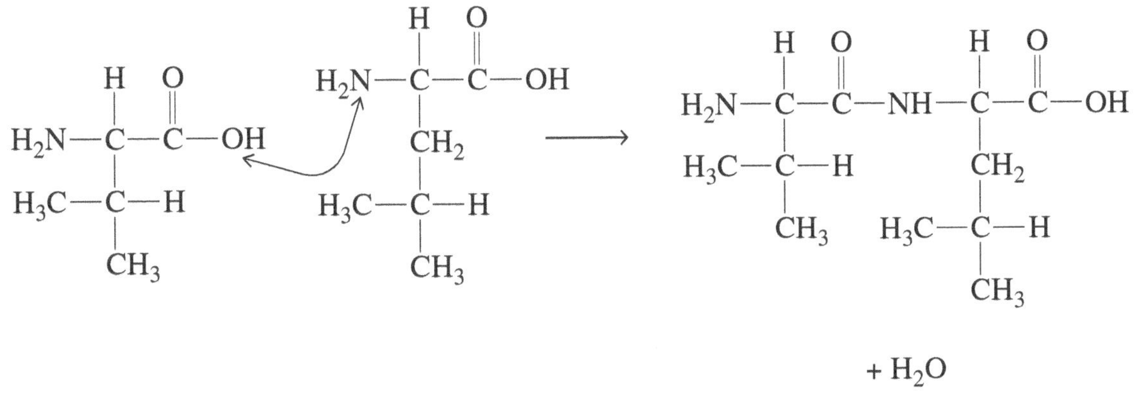

19.4. A pairs with T and C pairs with G, giving

Guided Inquiry Ideas

Below are a few example questions that students answer in the guided inquiry activities provided in the Guided Activity Workbook.

What holds monosaccharides together in a polysaccharide?

Is starch a carbohydrate? Why or why not?

Are most carbohydrates polar or non polar compounds?

Will most carbohydrates tend to be soluble in water? Why or why not?